Optogenetics

Hiromu Yawo • Hideki Kandori
Amane Koizumi
Editors

Optogenetics

Light-Sensing Proteins and Their Applications

Springer

Editors
Hiromu Yawo
Department of Developmental Biology and Neuroscience
Tohoku University Graduate School of Life Sciences
Sendai, Japan

Hideki Kandori
Department of Frontier Materials
Nagoya Institute of Technology
Nagoya, Japan

Amane Koizumi
National Institutes of Natural Sciences
Tokyo, Japan

ISBN 978-4-431-56197-2 ISBN 978-4-431-55516-2 (eBook)
DOI 10.1007/978-4-431-55516-2

Springer Tokyo Heidelberg New York Dordrecht London

Softcover reprint of the hardcover 1st edition 2015

Printed on acid-free paper

Springer Japan KK is part of Springer Science+Business Media (www.springer.com)

Preface

This book is a genuine miracle consisting of 28 chapters in five parts: "Light-Sensing Proteins", "Optogenetics in Biological Systems", "Optogenetics in Neuroscience", "Medical Optogenetics", and "Optogenetics and Opto-electronics". However, several recent topics were not included because the fields related to optogenetics have been expanding explosively. There is also much room for science and technology to employ various areas of optogenetics. In October 2014, while we were meeting at the International Conference on Retinal Proteins in Nagahama, Japan (ICRP2014), we received news that the 2014 Nobel Prize in Physics had been awarded to three scientists who invented blue LED. The "Era of Light" is just beginning in the twenty-first century. People have begun to learn how to manipulate light as their ancestors learned to handle fire over 500,000 years ago. Optogenetics will also mature with the progress of basic technologies and will soon make important contributions to human life and society. We will be very happy if this book contributes to that revolution.

While editing this book, we noticed that the same important keywords are frequently termed differently depending on the author (Table 1). For example, genetic products that manipulate cellular functions when stimulated by light, such as channelrhodopsin-2, are frequently termed "optogenetic tools". But the same term is also used collectively to include genetic products that report cellular functions by emitting light, such as GCaMP. This is because optogenetics is a young field of science. Here, we would like to propose a systematic nomenclature by analogy with that of electronic devices such as "optogenetic biosensor (OBS)", "optogenetic bioactuator (OBA)", and "optogenetic bioware (OBW)".

This present volume is based on a Japanese book published by NTS Inc. Reading through it, we had the strong impression that that book would become a milestone of optogenetics as well as a compass for any researchers, from beginners to experts, to explore this uncharted world. No previous book had been as comprehensive as that one, from the molecular basis of photoreceptor proteins to clinical and industrial applications. The book should be read by any scientist or student doing research in any way related to optogenetics all over the world. For that reason, we have re-edited the book in English with the collaboration of the original and new authors.

Sendai, Japan Hiromu Yawo
Nagoya, Japan Hideki Kandori
Tokyo, Japan Amane Koizumi
November 2014

Table 1 Nomenclature for optogenetics

Proposed nomenclature	Definition	Current terms	Chapters
Optogenetic biosensor (OBS)	A genetic product that reports cellular function by emitting light (Chameleon, GCaMP, etc.)	Fluorescent probe molecule	9
		Genetically encoded indicator	10
		Optical (molecular) probe	8, 12
		Optogenetic sensor	11
Optogenetic bioactuator (OBA)	A genetic product that manipulates cellular function when stimulated by light (Channelrhodopsins, NpHR, Arch, PAC, etc.)	Light-sensitive protein	19
		Optogenetic actuator	21
		Optogenetic tool	2, 3, 5, 6, 7, 22
		Photoactivatable protein	10, 18
		Photo-actuator molecule	9
Optogenetic bioware (OBW)	Collective term covering the above two items	Light-activated molecule	16
		Optogenetic molecular reagent	8
		Optogenetic protein	13
		Optogenetic tool	9, 10

Contents

Part I
Light-Sensing Proteins

Chapter 1
History and Perspectives of Light-Sensing Proteins

Hideki Kandori

Abstract Light-sensing proteins used in optogenetics are rhodopsins. The word 'rhodopsin' originates from the Greek words 'rhodo' and 'opsis', indicating rose and sight, respectively. Although the classical meaning of rhodopsin is the red-colored pigment in our eyes, the modern meaning of rhodopsin encompasses photoactive proteins containing a retinal chromophore in animals and microbes. Animal and microbial rhodopsins possess 11-*cis* and all-*trans* retinal, respectively, to capture light in seven-transmembrane α-helices; photoisomerizations into all-*trans* and 13-*cis* forms, respectively, initiate each function. We can find ion-transporting proteins in microbial rhodopsins, such as light-gated channels and light-driven pumps, which are the main tools in optogenetics. In this chapter, historical aspects and molecular properties of rhodopsins are introduced. The first part, *What is Rhodopsin?*, gives a general introduction to rhodopsin. The molecular mechanism of bacteriorhodopsin, a light-driven proton pump and the best studied microbial rhodopsin, is then described. In the section *Channelrhodopsin, the Light-Gated Ion Channel by Light*, molecular properties and several variants are introduced. History has proven that an understanding of the molecular mechanisms of microbial rhodopsins is a prerequisite for useful functional design of optogenetics tools in future.

Keywords Animal rhodopsin • Microbial rhodopsin • Retinal • Photoisomerizatoin • Pump • Channel • Photocycle • Proton transfer • Hydrogen bond • Structural change

1.1 What Is Rhodopsin?

The word 'rhodopsin' originates from the Greek words 'rhodo' and 'opsis', which indicate rose and sight, respectively. Thus, the classical meaning of rhodopsin is the red-colored pigment in the retinal rods of eyes. The chromophore molecule that absorbs light is retinal, which is the origin of the red color. Similar-colored

---

H. Kandori (✉)
Department of Frontier Materials and OptoBioTechnology Research Center, Nagoya Institute of Technology, Showa-ku, Nagoya 466-8555, Japan
e-mail: kandori@nitech.ac.jp

H. Yawo et al. (eds.), *Optogenetics*, DOI 10.1007/978-4-431-55516-2_1

retinal-binding proteins were found in microbes, expanding the definition of the word rhodopsin. The modern meaning of rhodopsin encompasses photoactive proteins containing a retinal chromophore in animals and microbes (Ernst et al. 2014). Rhodopsins are now found in all domains of life and are classified into two groups. While lower organisms utilize the microbial rhodopsin family for light–energy conversion and intracellular signaling, animals use the photosensory functions of a different family of rhodopsins (animal rhodopsin), a specialized subset of G protein-coupled receptors (GPCRs). Microbial and animal rhodopsins share a common architecture of seven-transmembrane α-helices with the N- and C-terminus facing outwards from and inside of the cell, respectively, but have almost no sequence homology and differ largely in their functions (Ernst et al. 2014). Retinal, the aldehyde of vitamin A, is derived from β-carotene and is bound to the protein in the shape of all-*trans* and 11-*cis* forms in microbial and animal rhodopsins, respectively (Fig. 1.1). Retinal is attached by a Schiff base linkage to the ε-amino group of a Lysine side chain in the middle of the 7th helix, and this retinal Schiff base (RSB) is protonated ($RSBH^+$) in most cases (Fig. 1.1).

Microbial rhodopsins were first found in the Archaea (*Halobacterium salinarum*) (Oesterhelt and Stoeckenius 1971) and were therefore initially termed archaeal rhodopsins. *H. salinarum* contains bacteriorhodopsin (BR) (Oesterhelt and Stoeckenius 1971) and halorhodopsin (HR) (Matsuno-Yagi and Mukohata 1977),

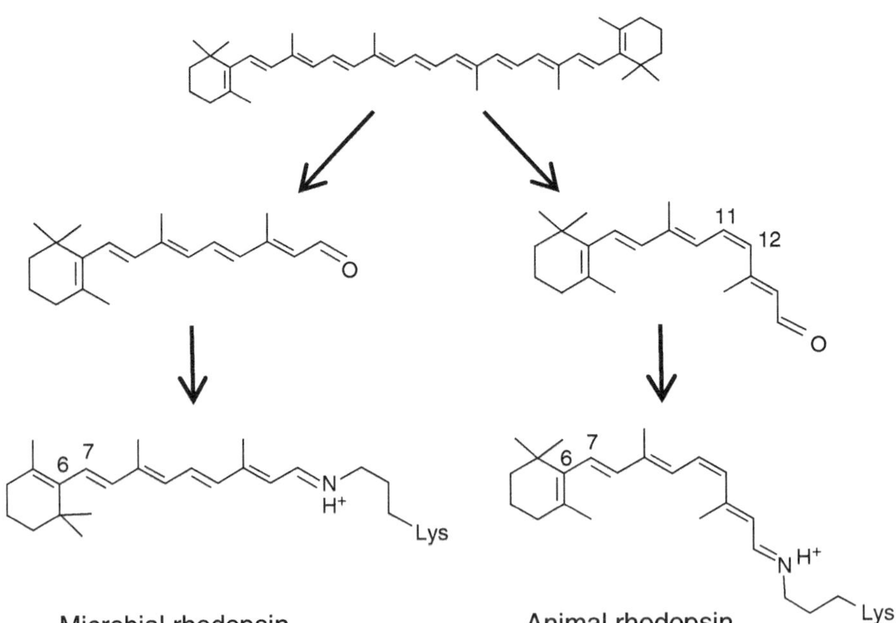

Fig. 1.1 Chromophore molecules of microbial (*left*) and animal (*right*) rhodopsins. β-carotene (*top*) is the source of the chromophore, and all-*trans* and 11-*cis* retinal is bound to protein to form microbial and animal rhodopsins, respectively

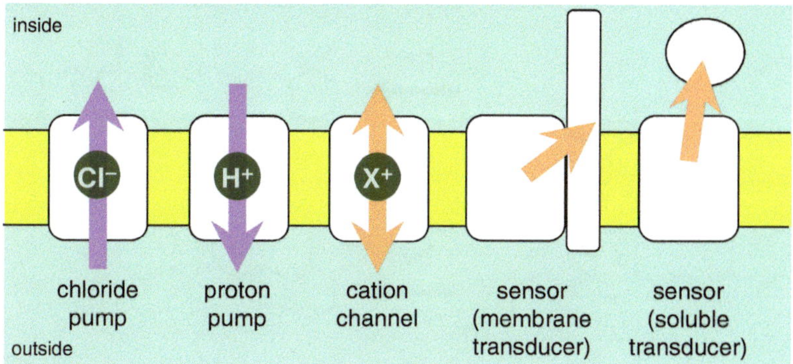

Fig. 1.2 Function of microbial rhodopsins. Microbial rhodopsins can act as pumps, channel, and light-sensors. *Arrows* indicate the direction of transport or flow of a signal

which act as a light-driven outward proton pump or an inward Cl^- ion pump, respectively (Fig. 1.2). BR and HR contribute to the formation of a membrane potential and thus function in light-energy conversion. The two other *H. salinarum* rhodopsins are sensory rhodopsin I and II (SRI and SRII) (Spudich and Bogomolni 1984; Jung et al. 2003). They act as positive and negative phototaxis sensors, respectively (Fig. 1.2). For the first 30 years since the early 1970s, microbial rhodopsins were epitomized by haloarchaeal proteins, the first-discovered and best-studied light-driven proton pump BR and its close relatives (HR, SRI, and SRII). However, over the past 15 years, thousands of related photoactive proteins with similar or different functions have been identified in *Archaea*, *Eubacteria*, and *Eukaryota*. Since the original discovery of BR in *H. salinarum*, similar rhodopsins have been found in *Eubacteria* and lower *Eukaryota*, and are now called microbial rhodopsins. Channelrhodopsins (ChRs), another group of microbial rhodopsins, were discovered in green algae where they function as light-gated cation channels within the algal eye to depolarize the plasma membrane upon light absorption (Fig. 1.2) (Nagel et al. 2002, 2003). Thus, ChRs also naturally function as signaling photoreceptors. The discovery of ChR led to the emergence of a new field, optogenetics (Miesenbock 2011), in which light-gated ion channels and light-driven ion pumps are used to depolarize and hyperpolarize selected cells of neuronal networks. This new method is strongly expected to aid in understand the circuitry of the brain (Deisseroth 2011; Diester et al. 2011).

It should be noted that microbial rhodopsins are used in optogenetics in which animal brain functions are studied by incorporating microbial rhodopsins, but not animal rhodopsins, into the animal brain. There are two reasons for this. One is the isomeric structure of the chromophore (Fig. 1.1). An 11-*cis* retinal, the chromophore molecule of animal rhodopsins, is not generally abundant in animal cells (Ernst et al. 2014). In contrast, endogenous all-*trans* retinal, the chromophore molecule of microbial rhodopsins, is sufficient for optogenetics in animal cells (Ernst et al. 2014). The second reason is 'bleaching'. Upon light absorption, animal

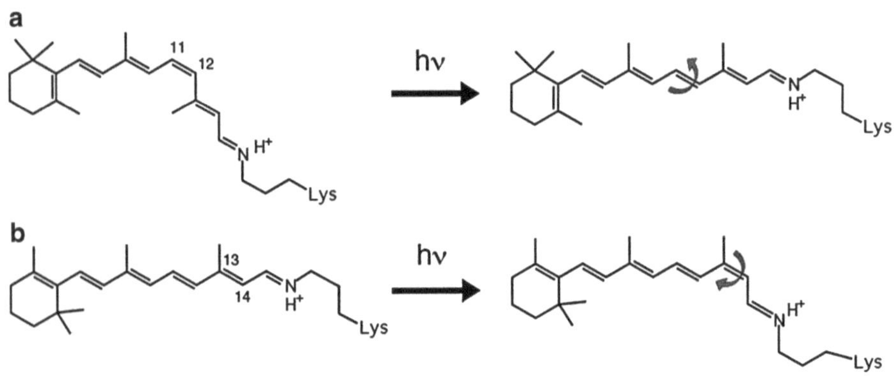

Fig. 1.3 Photoreactions in animal (**a**) and microbial (**b**) rhodopsins. Photoisomerizations from the 11-*cis* to the all-*trans* form and from the all-*trans* to the 13-*cis* form are the primary reactions in animal and microbial rhodopsins, respectively

and microbial rhodopsins exhibit retinal isomerization from the 11-*cis* to all-*trans*, and all-*trans* to 13-*cis* forms, respectively (Fig. 1.3) (Ernst et al. 2014). While the isomerization reaction initiates each function, the final part of the photoreaction differs between animal and microbial rhodopsins. The isomerized all-*trans* retinal does not return to the 11-*cis* form in animal rhodopsins, and is thus called 'photobleaching' (Ernst et al. 2014). This is not a problem in human visual cells because enzymatically isomerized 11-*cis* retinal is newly supplied, but this is not the case in other animal cells. In contrast, the 13-*cis* form is thermally isomerized into the all-*trans* form, and the spontaneous return is termed the 'photocycle' in microbial rhodopsins (Ernst et al. 2014). The existence of naturally abundant all-*trans* retinal and its photocycle feature have led microbial rhodopsins to become a tool in optogenetics. Previously, optical tools called 'optoXRs' were proposed, in which animal rhodopsin chimera containing the cytoplasmic loop of other GPCRs was used to respond to light (Airan et al. 2009). However, there have been few reports of 'optoXRs' in optogenetics since then, presumably because of the two reasons listed above.

Despite a variety of sequences and functions, the structural and mechanistic principles of microbial rhodopsin architecture have one common structure: a tight alpha-helical bundle of seven-transmembrane helices surrounding the retinal chromophore (Ernst et al. 2014; Klare et al. 2008; Zhang et al. 2011; Brown 2013). Figure 1.4 illustrates the overall structure of BR, highlighting the conserved aromatic amino acids with an important function. The retinal-binding pocket is the most conserved common element of the structure. Strongly conserved Trp86 and Trp182 constitute an important part of the chromophore-binding site by vertically sandwiching all-*trans*-retinal (Fig. 1.4). The presence of these bulky groups possibly determines the isomerization pathway from the all-*trans* to the 13-*cis* form. Moreover, the interaction of photoisomerized retinal with Trp182 may serve as the mechanical transducer for passing the energy stored in retinal deformation into the functionally important changes of the helical tilts necessary for function

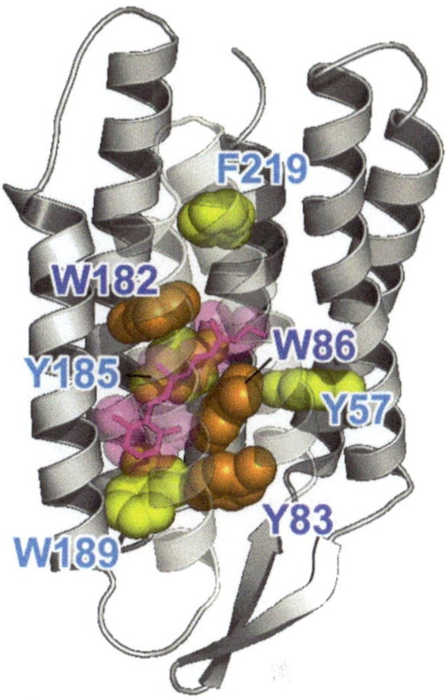

Fig. 1.4 Structure of bacteriorhodopsin (BR) with conserved aromatic residues (PDB: 1QM8). Tyr83, Trp86, and Trp182 are strongly conserved among microbial rhodopsins (*orange*). Aromatic residues are strongly conserved at the Tyr185, Trp189, and Phe219 positions (*yellow*). In BR, Trp86, Trp182, Tyr185, and Trp189 constitute the chromophore-binding pocket for all-*trans* retinal (*red*)

(Ernst et al. 2014). Another important position occupied by aromatic amino acids in the retinal-binding pocket is that of Tyr185 in BR (Fig. 1.4), which participates in hydrogen-bonding stabilization of the Schiff base counter-ion for many rhodopsins. This is replaced by a Phe in ChRs, suggesting that the lack of a hydrogen-bonding interaction at this position is important for channel function (Ernst et al. 2014).

In addition to the aromatic sidechain rings, electrostatic and hydrogen-bonding interactions in the proximal part of retinal are crucial in defining the functionality of microbial rhodopsins (Ernst et al. 2014). The sidechain of BR Lys216 (or its homologs in other microbial rhodopsins) forms a covalent bond with the retinal molecule through the Schiff base (Fig. 1.1). Since the Schiff base is usually protonated, Lys216 and the super-conserved Arg82 of helix C in BR provide two positive charges within the protein (Fig. 1.5), which requires two negative charges for electrostatic stabilization. This dictates the most common configuration of the Schiff base counter-ion, which includes two carboxylic acids (Asp85 and Asp212 in BR) that are perfectly conserved for proton-pumping microbial rhodopsins (Fig. 1.5) (Ernst et al. 2014).

Any deviation from this arrangement has strong functional consequences. For example, the negatively charged Asp at position 85 is replaced by Thr in the chloride-pumping HR, while the corresponding amino acid is Glu in ChR (Ernst et al. 2014). Tyr at position 185 of BR is conserved in HR, but is replaced by Phe in ChR. Importantly, all microbial rhodopsins contain protein-bound water molecules near the Schiff base (Fig. 1.5), presumably contributing to the stabilization of the protonated Schiff base in the hydrophobic protein interior. These water molecules play a crucial role in protein function, and have been extensively studied via X-ray

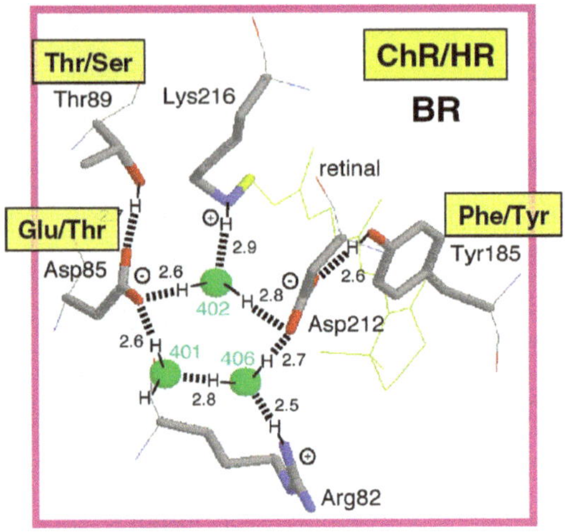

Fig. 1.5 Structure of the Schiff base region in bacteriorhodopsin (BR). This is the side view of the Protein Data Bank structure 1C3W, which has a resolution of 1.55 Å. The membrane normal is approximately in the vertical direction of this figure. Hydrogen atoms and hydrogen bonds (*dashed lines*) are assumed from the structure, while the numbers indicate hydrogen-bonding distances in Å. Arg, Asp, and Lys at positions 82, 212, and 216, respectively, are fully conserved in channelrhodopsin (ChR) and halorhodopsin (HR), while Asp85, Thr89, and Tyr185 in BR are replaced in ChR and HR, as shown in the figure

crystallography of photointermediates, Fourier transform infrared spectroscopy (FTIR) spectroscopy and computational methods (Ernst et al. 2014). The electrostatic quadrupole in the Schiff base region is characteristic of most microbial rhodopsins, and light-induced retinal isomerization causes hydrogen-bonding alteration of this region as well as steric effects, leading to various functions of microbial rhodopsins.

1.2 Bacteriorhodopsin, the First Ion Transport by Light

In 1971, the first microbial rhodopsin was discovered: BR from *H. salinarum* (Oesterhelt and Stoeckenius 1971). Electron microscopy showed that the structure of this membrane protein was composed of seven helices (Henderson and Unwin 1975); it was also the first membrane protein for which the amino acid sequence was determined (Khorana et al. 1979). As the best studied microbial rhodopsin, it serves as a paradigm of a light-driven retinal-binding ion pump and aids in studies of novel rhodopsins (Ernst et al. 2014). Archaerhodopsin-3 (Arch), a homologous proton-pump protein, is the best used in optogenetics as a neural silencer (Chow et al. 2010), sharing 58 % amino acid identity with BR (Ernst et al. 2014). The molecular mechanism of Arch is similar to that of BR.

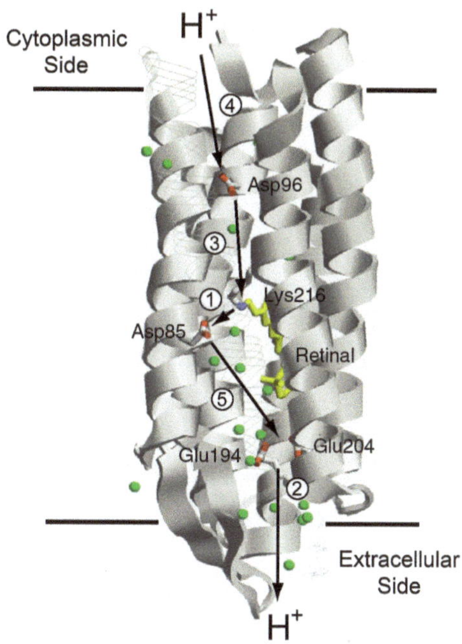

Fig. 1.6 Proton transport pathway in bacteriorhodopsin (BR). *Arrows* indicate each proton transfer, and the numbers indicate a temporal order: (*1*) Schiff base to Asp85, (*2*) proton release, (*3*) Asp96 to Schiff base, (*4*) uptake, and (*5*) Asp85 to the proton-release group

Figure 1.6 shows the proton pathway across the membrane from the cytoplasmic to the extracellular region in BR, together with protonatable groups and the order of respective proton transfers. A summary of the photocycle is shown in Fig. 1.7, which illustrates key intermediate states for most microbial rhodopsins. Although the photocycle of BR contains six intermediates, named alphabetically as J, K, L, M, N, and O states, only three states (K, M, and N) are shown in Figure 1.7 for clarity. After light absorption, photoisomerization occurs from the all-*trans*- to 13-*cis*-form in 10^{-13} s (Ernst et al. 2014; Kandori 2011). The ultrafast retinal isomerization yields the formation of red-shifted J and K intermediates in which J is the precursor of the K state. As the protein cavity, which accommodates retinal, cannot change its shape promptly, the K intermediate contains twisted 13-*cis* retinal and an altered hydrogen-bonding network in the Schiff base region, which yields higher free energy in K than in the original state (Kandori 2011). This leads to subsequent protein structural changes that accompany relaxation.

In the case of BR, relaxation of the K intermediate leads to the formation of the blue-shifted L intermediate. For proton pumping (and some of the photosensory) rhodopsins, the L intermediate serves as the precursor of the proton transfer reaction from the Schiff base to its primary carboxylic proton acceptor, by which the M intermediate is formed. This is a key step in proton transport. Since the M intermediate has a deprotonated 13-*cis* chromophore, it exhibits a characteristically strong blue-shifted absorption (λ_{max} at 360–410 nm) and is well isolated from other intermediates

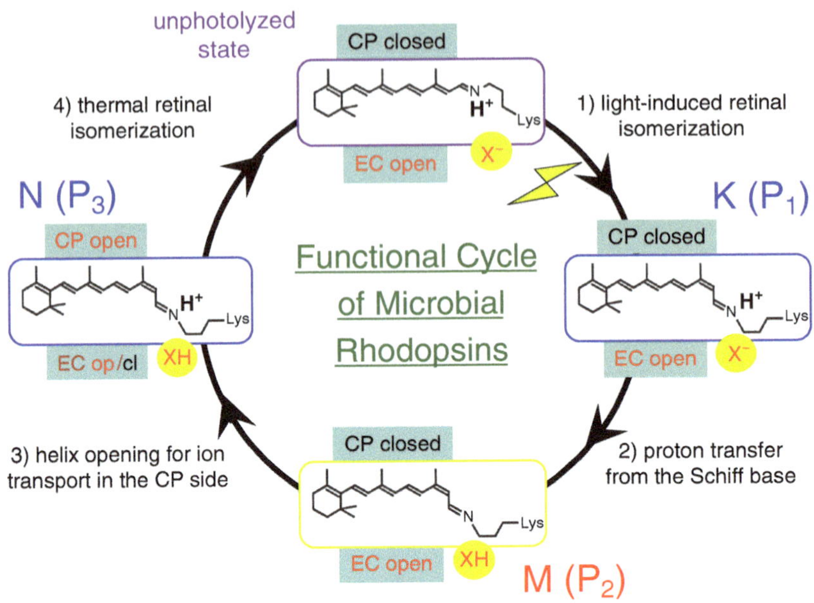

Fig. 1.7 Typical photocycle of microbial rhodopsins, showing the isomeric and protonation states of retinal. Names of the photocycle intermediates are for bacteriorhodopsin (BR), while those in parenthesis are for channelrhodopsin (ChR). X^- represents the Schiff base counter-ion, and Asp85 in BR also acts as the proton acceptor from the Schiff base. X^- is Glu123 in ChR2, whereas the proton acceptor of the Schiff base is not well established. In the chloride pump halorhodopsin (HR), X^- is a chloride ion, so that the M intermediate is not formed because the Schiff base is not deprotonated. Instead, the chloride ion is transported upwards (in this figure) in HR. CP and EC indicate cytoplasmic and extracellular domains, respectively. In the unphotolyzed state of microbial rhodopsins, the EC side is open through a hydrogen-bonding network, but the CP side is closed. While this is persistent in the K (P1) and M (P2) states, the CP side is open in the N (P3) state. When the EC side is closed (*black*), the CP side is open, as is the case for an ion pump, as occurs in the N intermediate of BR. When the EC side is open (*red*), the CP side is open, as is the case for an ion channel, as occurs for the P3 intermediate of ChR

(Ernst et al. 2014). In BR, the proton acceptor (X^- in Fig. 1.7) is Asp85, so the primary proton transfer takes place from the Schiff base to Asp85. In the case of chloride pump HR, the Schiff base does not deprotonate during the photocycle, because Asp85 in BR is replaced by Thr. In HR, X^- in Fig. 1.7 is a chloride ion, which is directly translocated upon decay of the L intermediate (Ernst et al. 2014).

If the Schiff base of M is reprotonated from Asp85 in BR, no proton transport occurs. In reality, the Schiff base is reprotonated from Asp96 in the cytoplasmic region (Fig. 1.6), by which the N intermediate is formed (Ernst et al. 2014). The molecular mechanism of unidirectional transport of protons in BR has attracted the attention of many researchers, and it is believed that the primary proton transfer from the Schiff base to Asp85, and the subsequent proton transfer from Asp96 to the Schiff base, determine the unidirectionality from the cytoplasmic to the extracellular region. The crystal structure of BR exhibits an asymmetric hydration pattern; while seven internal water molecules are found in the extracellular half, only two are observed in the cytoplasmic half (Fig. 1.6). Such asymmetry makes sense in view of the function of BR, as the water molecules build a hydrogen-bonding network on the extracellular side for fast proton release, while the cytoplasmic side is likely inaccessible in the dark and allows proton uptake only after the light-induced accessibility switch (Ernst et al. 2014). Such asymmetric access (EC open and CP closed) is the case not only for the unphotolyzed state, but also for the K and M intermediates (Fig. 1.7).

For proton conduction in the cytoplasmic region to be possible (CP open), an additional conformational change allowing the entrance of water into the vicinity of Asp96 should take place. Such a conformational alteration is realized mainly by changes in helical tilts (especially of the cytoplasmic half of helix F), and the N intermediate is often characterized by the largest changes in the protein backbone conformation, most notably, outward tilts of the cytoplasmic end of helix F. Such helical motions are functionally significant both for ion transport and for interactions with transducers of sensory rhodopsins. The photocycle usually ends with another red-shifted intermediate, known as the O intermediate, serving as a last step in resetting the original unphotolyzed conformation.

1.3 Channelrhodopsin, the Light-Gated Ion Channel by Light

The molecular mechanism of the light-driven proton pump in BR is well established, although the detailed directionality remains uncertain. In pumps, the transport pathways between the two sides of the membrane cannot be fully connected because the gradient formed by active transport will collapse. This view is clearly explained in the case of BR. This is an important aspect when distinguishing pumps from channels. The former needs energy input, which ensures the unidirectionality of transport across the membrane. The protein architecture of BR in Figs. 1.4 and 1.6 effectively explains the unidirectional transport in which the extracellular region contains many water molecules, while the cytoplasmic side has no pathway (EC open and CP closed in Fig. 1.7). Therefore, an alternative pump-specific conformation must take place, in which the protein first releases a proton on one side, and then takes up a proton from the other side (EC closed and CP open in Fig. 1.7). In contrast, a channel needs a fully connected ion pathway for passive transport of ions upon opening (EC and CP open). Given the structure of BR (Figs. 1.4 and 1.6), the channel function is most likely impossible for microbial rhodopsins.

In 2002–2003, three groups independently identified novel DNA sequences that encode microbial-type rhodopsins in *Chlamydomonas* (Nagel et al. 2002; Sineshchekov et al. 2002; Suzuki et al. 2003). Furthermore, a light-gated ion channel function was proved for two proteins, ChR1 and ChR2, in *Xenopus* oocytes using two electrode voltage clamp measurements (Nagel et al. 2002, 2003). ChR2 was also shown to be expressed and used to depolarize mammalian cells in response to light (Boyden et al. 2005; Ishizuka et al. 2006). Several groups then began to work with ChRs, primarily with a truncated version of ChR2 that expressed more effectively than the full-length protein and far more effectively than ChR1 (Ernst et al. 2014; Deisseroth 2011). This was the dawn of optogenetics. Surprisingly, the mammalian brain contains sufficient retinoid levels to allow wild-type ChR2 to function without the addition of exogenous retinoid cofactors.

ChR is a light-gated ion channel, implying that the channel is closed in the dark, and opens upon light absorption. Initially, researchers suspected that the channel pathway lay inside the seven-transmembrane helices as well as the proton pathway in BR (Fig. 1.6). Instead, it is more likely that ChR forms a dimer and that the intra-dimer cavity constitutes the ion conduction pathway. However, mutation studies of ChR suggested that the ion conduction pathway lay inside the seven-transmembrane helices (Ernst et al. 2014). In addition, X-ray crystallography of a chimeric protein of ChR1 and ChR2 showed the presence of the channel cavity at the extracellular domain, strongly suggesting that the ion conduction pathway lay inside the seven-transmembrane helices (Kato et al. 2012). It is now believed that a monomer is the functional unit of ChR. The crystal structure of ChR showed that the protein architecture is common to all microbial rhodopsins but that some structural modification must be linked to their unique function. For instance, in all microbial rhodopsins, the hydrogen-bonding acceptor of the Schiff base is a water molecule (Fig. 1.5), whereas crystal structure and FTIR spectroscopy revealed the direct interaction of the Schiff base with the counter-ion (Glu at the position of Asp85 in BR; Fig. 1.5) (Kato et al. 2012; Ito et al. 2014).

Like BR, the ChR photocycle has been studied via various methods, and it is now established that ChR has a photocycle similar to that of other microbial rhodopsins (Ernst et al. 2014). After the absorption of light, photoisomerization occurs from the all-*trans* to the 13-*cis* form very rapidly, and forms the red-shifted K-like intermediate (P1 in Fig. 1.7). As the protein cavity, which accommodates retinal, cannot respond promptly, the P1 intermediate contains twisted 13-*cis* retinal and an altered hydrogen-bonding network in the Schiff base region, which possesses higher free energy than in the original state. This yields subsequent protein structural changes that accompany relaxation as is also the case in BR. Then, the proton is transferred from the Schiff base, forming the M-like P2 intermediate. The P2 intermediate has a deprotonated 13-*cis* chromophore, the absorption of which is strongly blue-shifted (λ_{max} at 380 nm). The N-like P3 intermediate is formed by reprotonation of the Schiff base, and is believed to be the ion-conducting state. It is reasonable to assume that the N-like state exhibits the largest conformational changes in microbial rhodopsins. When the CP side is open, the EC side must also be open in ChR, so that a transient ion-conduction pathway is created. Therefore, in Fig. 1.7, pumps and channels show the N-like intermediate in CP-open/EC-closed and CP-open/EC-open conformations, respectively.

The channel property of ChR is important for optogenetic applications. In particular, absorption color, ion selectivity, and open/close dynamics should be taken into account. Thus far, various mutants have been shown to improve the properties of ChR. How then were these variants designed before structural determination? Knowledge of BR has contributed significantly to the design of mutants. For instance, two important Asp in BR, Asp85 and Asp96, correspond to Glu123 and His134 in ChR2, respectively, to which mutations were attempted. In the case of E123T/A, channel closure is faster than that of the wild type, allowing for a ten times higher repetition rate (200 Hz) than in the wild type in optogenetics (Fig. 1.8) (Gunaydin et al. 2010). The E123T/A mutant also shows red-shifted absorption, caused by neutralization of the Schiff base counter-ion (Glu123; X^- in Fig. 1.7) (Ernst et al. 2014). In the case of H134R, channel closure is slower than that of the wild type, improving conductivity and changing ion selectivity (Nagel et al. 2005). Asp96 and His134 in BR and ChR2, respectively, are located at important positions of the cytoplasmic domain (Figs. 1.6 and 1.8), presumably influencing ion transport. In BR, Thr90, Asp115, and Met118 (Cys128, Asp156, and Thr159 in ChR2, respectively) are located near the retinal chromophore (Fig. 1.8). D156A/C128S exhibits an extremely slow photocycle, leading to a permanently open channel that can be photoconverted to the original state by orange light (Berndt et al. 2011). Since ion conductance was increased for T159C, it was used under weak light conditions (Berndt et al. 2009). These examples show how specific variants of ChR2 have been applied in optogenetics.

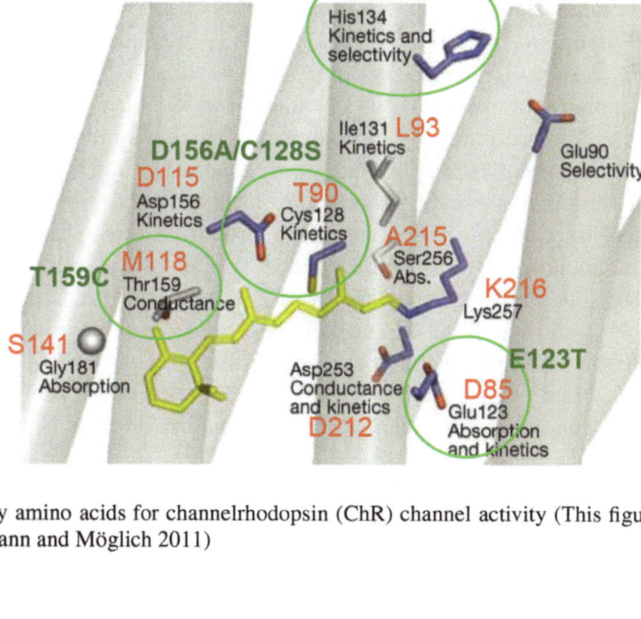

Fig. 1.8 Key amino acids for channelrhodopsin (ChR) channel activity (This figure is modified from Hegemann and Möglich 2011)

1.4 Perspectives: Towards a Novel Design

A decade has passed since the emergence of optogenetics. Despite the creation of useful ChR2 variants, there are still many requirements for channel function. Limited conductance of ChR needs to be improved. However, unlike other channel proteins, ions are conducted inside seven-transmembrane helices in ChR. The creation of wider pores through molecular engineering may be a challenge. Considering color absorption, ChR2 absorbs blue light maximally (470–480 nm), which is harmful to organisms and also penetrates them less. Therefore, red-absorbing proteins, particularly from algae, are being actively searched (Klapoetke et al. 2014). Strict control of ion selectivity, such as K^+ and Na^+, is highly sought after but not easy to achieve. A mechanism to conduct Ca^{2+} is eagerly anticipated (Kleinlogel et al. 2011). And although open/close dynamics are reasonably well controlled, a greater variety is desired.

Thus far, ChR2 and its variants have been mostly used for neural excitation by depolarizing cells. For neural silencing, ion pumps such as HR (Zhang et al. 2007) and Arch (Chow et al. 2010) have been used, in which chloride ions and protons, respectively, are transported. Compared with ChR, molecular engineering of neural silencers has been less examined, but is needed for the future development of optogenetics. In this sense, the recent discovery of sodium-pumping rhodopsin is noted. As the retinal chromophore is normally positively charged in rhodopsins, it is believed that cation-pumping rhodopsins, except for the proton pump, are impossible because of electrostatic repulsion. However, a sodium pump in the ocean has already been naturally created (Inoue et al. 2013), which will be used as a novel neural silencer.

For the functional design of new optogenetic tools, the mutation of microbial rhodopsins is important. In particular, functional conversion by mutation is useful, not only for basic research but also for optogenetic applications. The first functional conversion of a microbial rhodopsin was achieved in 1995 for a light-driven proton pump BR into a chloride pump by replacement of a single amino acid (Asp85 to Thr) (Sasaki et al. 1995). However, since then, successful functional conversions have been very limited, suggesting that nature has optimized each microbial rhodopsin for each function (Muroda et al. 2012). Examples of successful functional conversions include that of BR into a light sensor by the replacement of three amino acids (Sudo and Spudich 2006), and that of ChR2 into a chloride channel by replacement of a single amino acid (Glu90 to a positively charged residue; Fig. 1.8) (Wietek et al. 2014). The functional conversion of ChR2 into a chloride channel has been reported on the basis of structure-based engineering (Berndt et al. 2014). Thus, understanding the molecular mechanism of microbial rhodopsins is a prerequisite for useful functional design (Ernst et al. 2014).

Finally, a note on optogenetic tools with non ion-transporting proteins. To research the brain's function, rapid temporal response is necessary, and ion-transporting rhodopsins, such as light-gated ion channels and light-driven ion pumps, have been exclusively used (Ernst et al. 2014). On the other hand, photoactivation of an intracellular signaling cascade is useful, which is also recognized as one aspect of optogenetics. To achieve this aim, plant photoreceptors such as flavin-binding protein (Wu et al. 2009) and phytochrome (Levskaya et al. 2009) were used, by regulating intracellular GTPase activities of Rac and Rho, respectively. Another interesting and demanding target is hetero-trimeric G-protein, which is activated by GPCR, because various signaling processes in mammalian cells are regulated by GPCR. Animal rhodopsin activates a G-protein in human visual cells but optogenetic tools are needed to activate various G-proteins in other cells by light. A bi-stable pigment of animal rhodopsin, such as melanopsin, is a candidate application tool (Ye et al. 2011), and microbial rhodopsin can be used as a template for light activation of G-protein (Sasaki et al. 2014).

References

Airan RD, Thompson KR, Fenno LE et al (2009) Temporally precise in vivo control of intracellular signaling. Nature 458:1025–1029

Berndt A, Yizhar O, Gunaydin LA et al (2009) Bistable neural state switches. Nat Neurosci 12:229–234

Berndt A, Schoenenberger P, Mattis J et al (2011) High-efficiency channelrhodopsins for fast neuronal stimulation at low light levels. Proc Natl Acad Sci U S A 108:7595–7600

Berndt A, Lee SY, Ramakrishnan C et al (2014) Structure-guided transformation of channelrhodopsin into a light-activated chloride channel. Science 344:420–424

Boyden ES, Zhang F, Bamberg E et al (2005) Millisecond-timescale, genetically targeted optical control of neural activity. Nat Neurosci 8:1263–1268

Brown LS (2013) Eubacterial rhodopsins – unique photosensors and diverse ion pumps. Biochim Biophys Acta 1837:553–561

Chow BY, Han X, Dobry AS et al (2010) High-performance genetically targetable optical neural silencing by light-driven proton pumps. Nature 463:98–102

Deisseroth K (2011) Optogenetics. Nat Methods 8:26–29

Diester I, Kaufman MT, Mogri M et al (2011) An optogenetic toolbox designed for primates. Nat Neurosci 14:387–397

Ernst OP, Lodowski DT, Elstner M et al (2014) Microbial and animal rhodopsins: structures, functions, and molecular mechanisms. Chem Rev 114:126–163

Gunaydin LA, Yizhar O, Berndt A et al (2010) Ultrafast optogenetic control. Nat Neurosci 13:387–392

Hegemann P, Möglich A (2011) Channelrhodopsin engineering and exploration of new optogenetic tools. Nat Methods 8:39–42

Henderson R, Unwin PN (1975) Three-dimensional model of purple membrane obtained by electron microscopy. Nature 257:28–32

Inoue K, Ono H, Abe-Yoshizumi R et al (2013) A light-driven sodium ion pump in marine bacteria. Nat Commun 4:1678

Ishizuka T, Kakuda M, Araki R et al (2006) Kinetic evaluation of photosensitivity in genetically engineered neurons expressing green algae light-gated channels. Neurosci Res 54:85–94

Ito S, Kato HE, Taniguchi R et al (2014) Water-containing hydrogen-bonding network in the active center of channelrhodopsin. J Am Chem Soc 136:3475–3482

Jung KH, Trivedi VD, Spudich JL (2003) Demonstration of a sensory rhodopsin in eubacteria. Mol Microbiol 47:1513–1522

Kandori H (2011) Protein-controlled ultrafast photoisomerization in rhodopsin and bacteriorhodopsin. In: Ramamurthy V, Inoue Y (eds) Supramolecular photochemistry: controlling photochemical processes. Wiley, Hoboken, pp 571–596
Kato HE, Zhang F, Yizhar O et al (2012) Crystal structure of the channelrhodopsin light-gated cation channel. Nature 482:369–374
Khorana HG, Gerber GE, Herlihy WC et al (1979) Amino acid sequence of bacteriorhodopsin. Proc Natl Acad Sci U S A 76:5046–5050
Klapoetke NC, Murata Y, Kim SS et al (2014) Indepeendent optical excitation of distinct neural populations. Nat Methods 11:338–346
Klare JP, Chizhov I, Engelhard M (2008) Microbial rhodopsins: scaffolds for ion pumps, channels, and sensors. Results Probl Cell Differ 45:73–122
Kleinlogel S, Feldbauer K, Dempski RE et al (2011) Ultra light-sensitive and fast neuronal activation with the Ca^{2+}-permiable channelrhodopsin CatCh. Nat Neurosci 14:513–518
Levskaya A, Weiner OD, Lim WA et al (2009) Spatiotemporal control of cell signalling using a light-switcheable protein interaction. Nature 461:997–1001
Matsuno-Yagi A, Mukohata Y (1977) Two possible roles of bacteriorhodopsin; a comparative study of strains of *Halobacterium halobium* differing in pigmentation. Biochem Biophys Res Commun 78:237–243
Miesenbock G (2011) Optogenetic control of cells and circuits. Annu Rev Cell Dev Biol 27:731–758
Muroda K, Nakashima K, Shibata M et al (2012) Protein-bound water as the determinant of asymmetric functional conversion between light-driven proton and chloride pumps. Biochemistry 51:4677–4684
Nagel G, Ollig D, Fuhrmann M et al (2002) Channelrhodopsin-1: a light-gated proton channel in green algae. Science 296:2395–2398
Nagel G, Szellas T, Huhn W et al (2003) Channelrhodopsin-2, a directly light-gated cation-selective membrane channel. Proc Natl Acad Sci U S A 100:13940–13945
Nagel G, Brauner M, Liewald JF et al (2005) Light activation of channelrhodopsin-2 in excitable cells of *Caenorhabditis elegans* triggers rapid behavioral responses. Curr Biol 15:2279–2284
Oesterhelt D, Stoeckenius W (1971) Rhodopsin-like protein from the purple membrane of *Halobacterium halobium*. Nat New Biol 233:149–152
Sasaki J, Brown LS, Chon YS et al (1995) Conversion of bacteriorhodopsin into a chloride ion pump. Science 269:73–75
Sasaki K, Yamashita T, Yoshida K et al (2014) Chimeric proton-pumping rhodopsins containing the cytoplasmic loop of bovine rhodopsin. PLoS One 9:e91323
Sineshchekov OA, Jung KH, Spudich JL (2002) Two rhodopsins mediate phototaxis to low- and high-intensity light in *Chlamydomonas reinhardtii*. Proc Natl Acad Sci U S A 99:8689–8694
Spudich JL, Bogomolni RA (1984) Mechanism of colour discrimination by a bacterial sensory rhodopsin. Nature 312:509–513
Sudo Y, Spudich JL (2006) Three strategically placed hydrogen-bonding residues convert a proton pump into a sensory receptor. Proc Natl Acad Sci U S A 103:16129–12134
Suzuki T, Yamasaki K, Fujita S et al (2003) Archaeal-type rhodopsins in *Chlamydomonas*: model structure and intracellular localization. Biochem Biophys Res Commun 301:711–717
Wietek J, Wiegert JS, Adeeishvili N et al (2014) Conversion of channelrhodopsin into a light-gated chloride channel. Science 344:409–412
Wu YI, Frey D, Lungu OI et al (2009) A genetically encoded photoactivatable Rac controls the motility of living cells. Nature 461:104–108
Ye H, Daoud-El Baba M, Peng RW et al (2011) A synthetic optogenetic transcription device enhances blood-glucose homeostasis in mice. Science 332:1565–1568
Zhang F, Wang LP, Brauner M et al (2007) Multimodal fast optical interrogation of neural circuitry. Nature 446:633–639
Zhang F, Vierock J, Yizhar O et al (2011) The microbial opsin family of optogenetic tools. Cell 147:1446–1457

Chapter 2
Biology of Light-Sensing Proteins in Plants and Microorganisms

Mineo Iseki and Tetsuo Takahashi

Abstract A wide variety of light-sensing proteins that are found in plants and microorganisms and that provide natural resources for engineering optogenetic tools are briefly reviewed. We include microbial rhodopsins, which absorb blue/green light; phytochromes, which absorb red/far-red light; UV-A/blue-absorbing flavoproteins (cryptochromes, LOV-domain proteins, BLUF-domain proteins); and the recently discovered UV-B sensor UVR8. Among them, the significance of channelrhodopsins and photoactivated adenylyl cyclases is emphasized.

Keywords Rhodopsin • Phytochrome • Cryptochrome • LOV domain • BLUF domain • UVR8 • Channelrhodopsin • Photoactivated adenylyl cyclase

2.1 Introduction

In recent years, techniques for controlling biological functions with light via so-called optogenetics have grown in popularity, particularly in the field of neuroscience. A major factor influencing this trend is presumably attributable to the wide variety of light-sensing proteins in plants and microorganisms that have been identified in the last two decades. In this chapter, we give a general overview of the light-sensing proteins in plants and microorganisms, emphasizing the brief histories of their discoveries. Among them channelrhodopsins and photoactivated adenylyl cyclases are described in independent sections. The former, of course is the most practical and widely used optogenetic tool, and the latter has the potential to control various biological functions other than neuronal activities, including gene expression, metabolism, and development.

This chapter is dedicated to the late professor Masakatsu Watanabe.

M. Iseki (✉) • T. Takahashi
Faculty of Pharmaceutical Sciences, Toho University, Chiba, Japan
e-mail: mineo.iseki@phar.toho-u.ac.jp

H. Yawo et al. (eds.), *Optogenetics*, DOI 10.1007/978-4-431-55516-2_2

2.2 A Variety of Light-Sensing Proteins in Plants and Microorganisms

2.2.1 *Microbial (Type-I) Rhodopsins*

An extremely halophilic Archaea, *Halobacterium* acquires energy from light using light-dependent ion pumps, including bacteriorhodopsin (Oesterhelt and Stoeckenius 1971; Grote 2013) and halorhodopsin (Matsuno-Yagi and Mukohata 1977; Ihara et al. 1999). Halobacterial cells also show photoaccumulation to an orange light spot and are repelled by UV-A/blue light, as a consequence of at least two color-discriminating light-sensing systems (Hildebrand and Dencher 1975). The photobehavior is mediated by two bacteriorhodopsin-like proteins, sensory rhodopsin I and II (also called phoborhodopsin) (Spudich and Spudich 1982; Bogomolni and Spudich 1982; Takahashi et al. 1985; Tomioka et al. 1986). These microbial rhodopsins, similar to animal (also called type-II) rhodopsins, comprise seven-transmembrane helices where a retinal molecule is held as a chromophore. The primary photoreception process of the animal rhodopsins is photoisomerization of retinal from 11-*cis* to all-*trans*, whereas that of the microbial rhodopsins is from all-*trans* to 13-*cis* (for comprehensive review of both microbial and animal rhodopsins, see Ernst et al. 2014). Microbial rhodopsins are also found in other phyla such as eubacteria, fungi, and algae. For example, proteorhodopsins (Béjà et al. 2000) and xanthorhodopsins (Balashov et al. 2005) in eubacteria act as ion pumps like bacteriorhodopsin, whereas *Anabaena* sensory rhodopsin (Jung et al. 2003) is considered to be associated with light signal transduction in the cyanobacterium. The microbial rhodopsin found in the unicellular green alga *Chlamydomonas* functions as a light-gated cation channel and thus is well-known as a 'channelrhodopsin'. From an evolutionary point of view, it is noteworthy that animal rhodopsins have not been found in any phyla other than metazoa. The two protein families, microbial and animal rhodopsins, might have arisen via convergent evolution (Spudich et al. 2000; Mackin et al. 2014).

2.2.2 *Phytochromes*

Seed germination in higher plants is affected by various environmental factors such as temperature, water content, and light. For example, imbibed lettuce seeds germinate upon red light irradiation at an appropriate temperature, but this response is canceled by subsequent irradiation with far-red light. This effect is reversible, and the wavelength of light received last determines whether or not the seeds germinate (Borthwick et al. 1952). This phenomenon can be explained by assuming a photochromic pigment that interconverts between a red-absorbing form and a far-red absorbing form upon red/far-red irradiation. Such a pigment, phytochrome, was first demonstrated spectroscopically (Butler et al. 1959), followed by purification of the chromoprotein in the 1970s (Gardner et al. 1971; Rice et al. 1973; Vierstra and Quail 1982) and by complementary DNA (cDNA) cloning in the 1980s (Hershey et al. 1984).

The primary structure of phytochrome exhibits several functional domains: a GAF domain that binds a bilin chromophore, a PHY domain that is required for conformational stability of the far-red absorbing form, a PAS domain that mediates dimerization, and a histidine kinase-related domain (Fig. 2.1). Phytochrome regulates gene expression by interacting with several partner proteins, including phytochrome-interacting factor 3 (PIF3), a transcription factor (Ni et al. 1998), and phytochrome kinase substrate 1 (PKS1), which accepts phosphate from phytochrome (Fankhauser et al. 1999).

Phytochrome was long considered to be a plant-specific protein until a similar gene was found in the genome sequence of the cyanobacterium *Synechocystis*

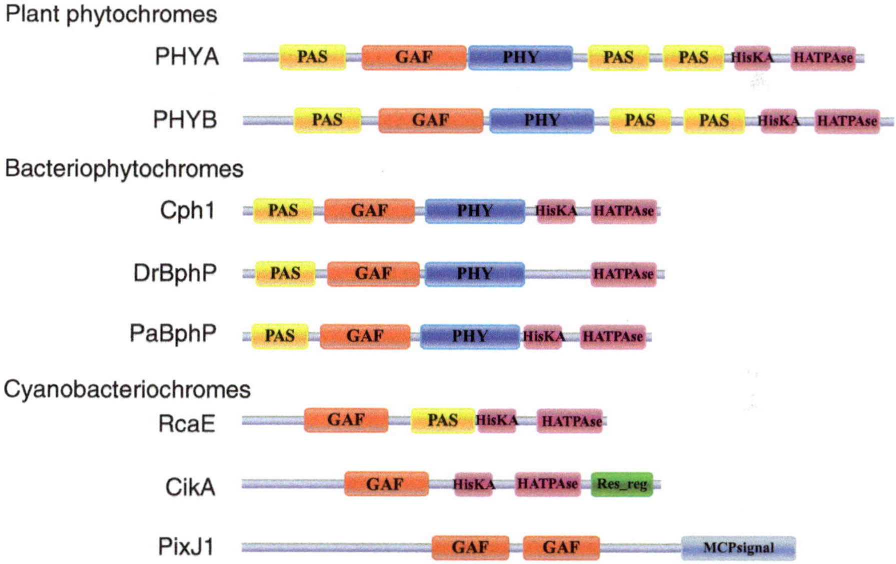

Fig. 2.1 *Domain architecture of plant phytochromes and related protein families in bacteria.* PHYA and PHYB, phytochrome A and B from *Arabidopsiss thaliana*; Cph1, a phytochrome-like protein from the cyanobacterium *Synechocyctis* sp. PCC6803 that binds phycocyanobilin as a chromophore and shows red/far-red photoconversion; DrBphP and PaBphP, phytochrome-like protein from the bacteria *Deinococcus radiodurans* and *Pseudomonas aeruginosa*, respectively (both bind biliverdin); RcaE, a putative photosensor for chromatic adaptation in the cyanobacterium *Fremyella diplosiphon*; CikA, a cyanobacteriochrome involved in circadian rhythms in *Synechococcus elongatus* PCC7942 that shows violet/yellow photoconversion; PixJ1, a cyanobacteriochrome involved in phototaxis in *Synechocytis* sp. PCC6803, which shows blue/green photoconversion. A domain search was conducted on Pfam (http://pfam.xfam.org/): PAS, Per-Arnt-Sim domain; GAF, cGMP-specific phosphodiesterases, adenylyl cyclases and FhlA domain; PHY, phytochrome region; HisK, His Kinase A (phospho-acceptor) domain; HATPAse, Histidine kinase-, DNA gyrase B-, and HSP90-like ATPase; Res_reg, Response regulator receiver domain; MCPsignal, Methyl-accepting chemotaxis protein (MCP) signaling domain

(Hughes et al. 1997). To date, various photochromic pigments have been identified in cyanobacteria, and are called cyanobacteriochromes (Ikeuchi and Ishizuka 2008). Phytochrome-like pigments have also been found in non-photosynthetic bacteria (Davis et al. 1999), providing a good source for the engineering of infrared fluorescent probes (Shu et al. 2009).

2.2.3 *Cryptochromes*

Although blue light has long been known to affect a wide variety of biological processes in plants and microorganisms, the molecular identity of the blue light photoreceptor has long remained unidentified. Ahmad and Cashmore (1993) isolated the causal gene in the *hy4* mutation of *Arabidopsis*, which shows a long-hypocotyl phenotype under blue light, and found that the gene product was similar to bacterial DNA photolyases that repair ultraviolet (UV) light-induced lesions in DNA using light energy at a longer wavelength (Fig. 2.2). The gene product, later named cryptochrome, binds FAD and possibly a pterin as chromophores, like *Escherichia coli* photolyase. Although the mechanism of light signal transduction by cryptochrome is not well understood, cryptochrome is phosphorylated upon photoexcitation and is assumed to regulate gene expression by interacting with other proteins such as the cryptochrome-interacting basic-helix–loop–helix 1 (CIB1) and the suppressor of phytochrome A 1/constitutively photomorphogenic 1 (SPA1/COP1) (Liu et al. 2008, 2011). They are widely distributed in animals and humans and act as a component of the circadian clock (van der Horst et al. 1999).

2.2.4 *LOV-Domain Proteins*

Phototropism, considered the most typical blue-light response and extensively studied by many investigators, is not mediated by the first identified blue-light receptor cryptochrome. In 1997, Huala et al. isolated the causal gene for the non-phototropic mutant of *Arabidopsis* and found that the gene encodes a protein that has two flavin-binding domains in the N-terminal region and a serine/threonine kinase domain in the C-terminal region (Fig. 2.3). The gene product, later named phototropin 1 (phot1), is autophosphorylated upon blue light irradiation, activating the subsequent signal transduction pathway (Christie et al. 1998). A similar protein, named phototropin 2 (phot2), has been identified in *Arabidopsis*. Both phototropins have been shown to be involved in not only phototropism but also chloroplast location (Kagawa et al. 2001) and stomata opening (Kinoshita et al. 2001). Given that the flavin-binding domain in phototropin is similar to that of a redox sensor in bacteria and that of a voltage-sensitive potassium-channel subunit, it was named the light, oxygen, or voltage (LOV) domain.

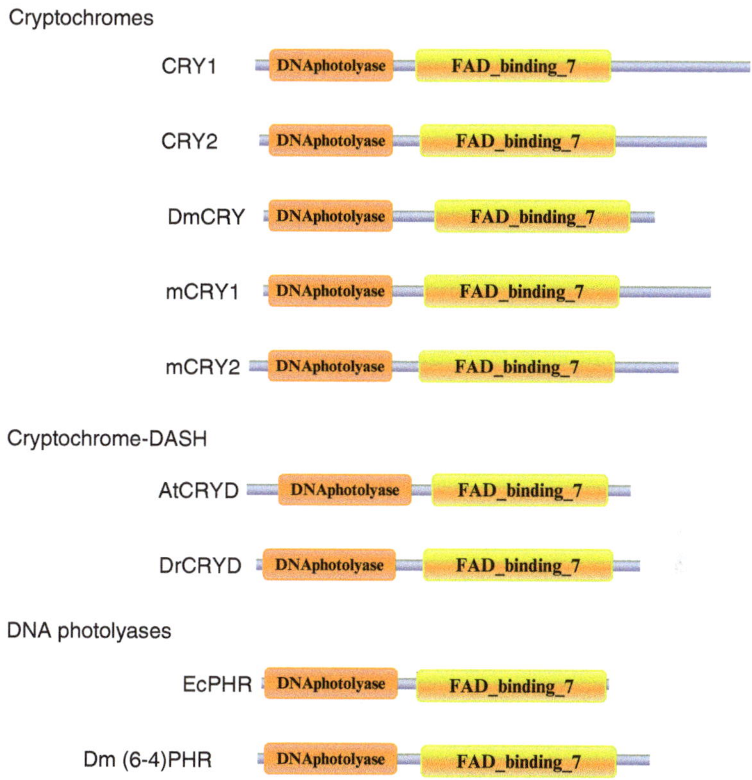

Fig. 2.2 *Domain architecture of cryptochromes and related protein families.* CRY1 and CRY2, cryptochrome 1 and 2 from *Arabidopsis thaliana*; DmCRY, cryptochrome in *Drosophila melanogaster*; mCRY1 and mCRY2, mouse cryptochrome 1 and 2; AtCRYD, cryptochrome-DASH from *A. thaliana*; DrCRYD, cryptochrome-DASH from *Danio rerio*; EcPHR, DNA photolyase from *E. coli*; Dm(6–4)PHR, (6–4)-photolyase from *D. melanogaster*. DNA_photolyase, DNA photolylase domain that may bind a second chromophore, such as methenyltetrahydrofolate (MTHF); FAD_binding_7, FAD-binding domain of DNA photolyase

Light-sensing proteins harboring LOV domains are widely distributed in plants and microorganisms. In higher plants, the ZTL/FKF1/LKP2 family of proteins, which function as E3 ubiquitin ligases, are involved in photoperiodism and circadian rhythms (Demarsy and Fankhauser 2009). Aureochrome, which was first discovered in the yellow-green alga *Vaucheria*, has a LOV domain in its C-terminal region, a bZIP domain in its N-terminal region, and acts as a light-activated transcription factor (Takahashi et al. 2007). White collar 1 (WC-1) and Vivid (VVD) are light-sensing LOV domain proteins in the fungus *Neurospora* and both are involved in circadian rhythms (He et al. 2002; Heintzen et al. 2001). LOV-histidine kinases identified in bacteria transduce light signals as sensors of the bacterial two-component system, inducing various biological responses (Swartz et al. 2007). Recently, a light-activated adenylyl cyclase containing a LOV domain in a cyanobacterium has been reported (Raffelberg et al. 2013).

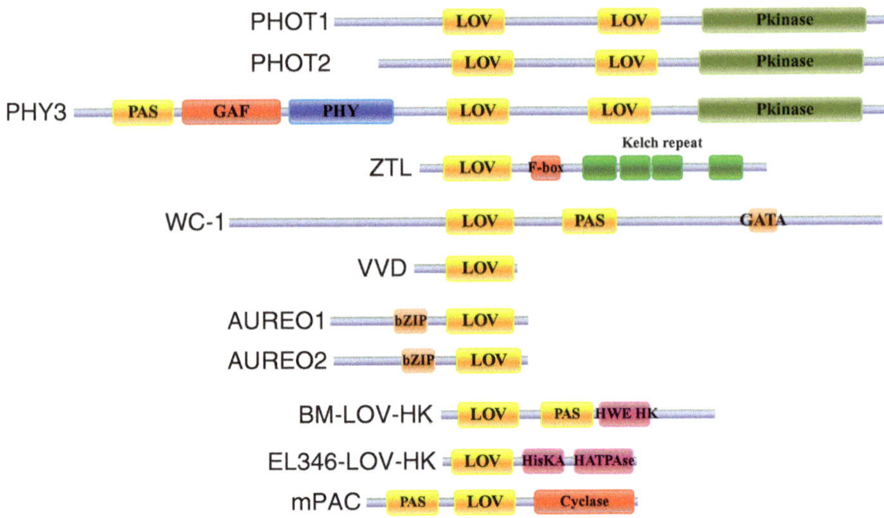

Fig. 2.3 *Domain architecture of LOV-domain proteins*. PHOT1 and PHOT2, phototropin 1 and 2 from *Arabidopsis thaliana*; PHY3, phytochrome 3 from the fern *Adiantum capillus-veneris*; ZTL, zeitlupe from *A. thaliana*; WC-1, white collar-1 from the fungus *Neurospora crassa*; VVD, vivid from *N. crassa*; AUREO1 and AUREO2, aureochrome 1 and 2 from the yellow-green alga *Vaucheria frigida*; BM-LOV-HK, LOV-histidine kinase from the bacterium *Brucella melitensis*; EL346-LOV-HK, LOV-histidine kinase from the bacterium *Erythrobacter litoralis*; mPAC, a LOV domain containing adenylyl cyclase from the cyanobacterium *Microcoleus chthonoplastes* PCC 7420. LOV, LOV (designated as PAS_9 in Pfam) domain that binds a chromophore FMN or FAD; Pkinase, protein kinase domain; F-box, F-box domain; GATA, GATA zinc finger domain; bZIP, basic leucine zipper domain; HisK and HATPAse, histidine kinase domains; cyclase, adenylyl and guanylyl cyclase catalytic domain (designated as Guanylate_cyc in Pfam)

2.2.5 BLUF-Domain Proteins

The third family of blue light-sensing flavoproteins was discovered via investigation of the photomovement responses in the flagellate *Euglena* (Iseki et al. 2002) and of the mechanism of gene expression in the photosynthetic bacterium *Rhodobacter* (Gomelsky and Kaplan 1998; Masuda and Bauer 2002). The light-sensing proteins associated with these phenomena have a flavin-binding domain that usually binds FAD and were accordingly named sensors of Blue Light Using FAD (BLUF) domain (Gomelsky and Klug 2002) (Fig. 2.4). The BLUF domain proteins are found mainly in bacteria, though they are not ubiquitous, whereas few BLUF-domain proteins are found in eukaryotes. Many of the BLUF-domain proteins in bacteria are as small as approximately 15 kDa and are considered to interact with other proteins to transduce light signals. BlrP1, the BLUF-domain protein in *Klebsiella*, acts as a light-dependent phosphodiesterase that catalyzes the destruction of cyclic di-guanosine 5′-monophosphate (c-di-GMP), a ubiquitous bacterial second messenger (Barends et al. 2009).

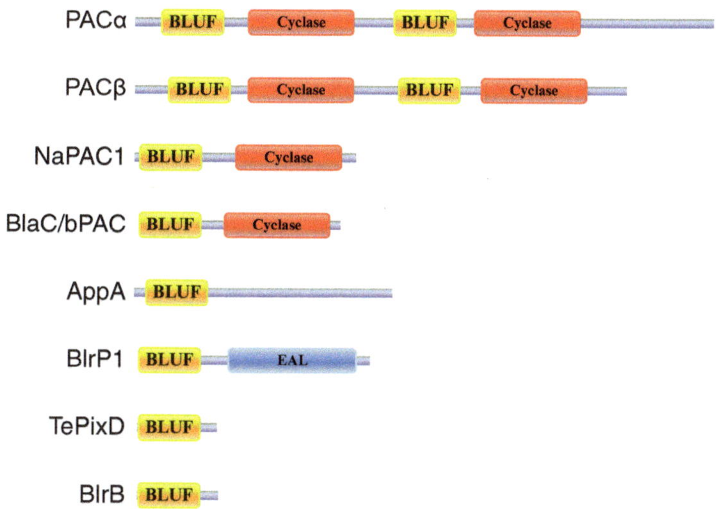

Fig. 2.4 *Domain architecture of BLUF-domain proteins*. PACα and PACβ, subunits of photoactivated adenylyl cyclase from the unicellular flagellate *Euglena gracilis*; NaPAC1, a photoactivated adenylyl cyclase from the amoeba *Naegleria australiensis*; BlaC/bPAC, a photoactivated adenylyl cyclase from the bacterium *Beggiatoa* sp.; AppA, a regulator for photosynthesis gene expression in the bacterium *Rhodobactor sphaeroides*; BlrP1, a light-dependent c-di-GMP phosphodiesterase in the bacterium *Klebsiella pneumoniae*; TePixD, a short BLUF protein from the thermophilic cyanobacterium *Thermosynechococcus elongatus*; BlrB, a short BLUF protein from *R. sphaeroides*. BLUF, BLUF domain; cyclase, adenylyl and guanylyl cyclase catalytic domain; EAL, EAL domain

2.2.6 UVR8

Plants produce various UV-absorbing substances, such as anthocyanins and flavonoids, to protect themselves from strong UV radiation. Given that biosynthesis of these substances is induced by UV-B (280–315 nm) light, it has been postulated that a specific UV-B sensor plays a role in such UV-B–induced responses (Yatsuhashi et al. 1982). In 2002, Kliebenstein et al. isolated the UV-B–sensitive *Arabidopsis* mutant, *UV resistance locus 8* (*uvr8*) and found that the gene product UVR8 was similar to the human regulator of chromatin condensation 1 (RCC1) and involved in UV-B signal transduction. Rizzini et al. (2011) showed that UVR8 interacted with COP1 in vitro upon UV-B irradiation, a discovery that constituted strong evidence that UVR8 acts as a UV-B sensor. UVR8 is a unique light-sensing protein, in the sense that it needs no prosthetic molecule but uses intrinsic tryptophan residues as chromophores. This feature of UVR8 seems suitable for the optogenetic use of the protein. Interaction between UVR8 and COP1 expressed in mammalian cells was recently reported (Crefcoeur et al. 2013; Müller et al. 2013).

2.3 Channelrhodopsins

2.3.1 Discovery of Channelrhodopsin

The unicellular green alga *Chlamydomonas reinhardtii* exhibits conspicuous phototaxis to blue/green light. The phototaxis sensor of *Chlamydomonas* had not been identified and was thought to be related to blue light photoreceptors in plants. However, an article that suggested the involvement of rhodopsin in *Chlamydomonas* phototaxis marked a turning point (Foster et al. 1984). Following the article, several research groups reported that all-*trans* retinal was effective in restoring the phototaxis of a retinal-deficient mutant, indicating that the sensor was type I rhodopsin (Hegemann et al. 1991; Lawson et al. 1991; Takahashi et al. 1991; Derguini et al. 1991). However, the molecular identity of the phototaxis sensor remained vague until some 10 years later when the Kazusa DNA institute in Japan released an Expressed Sequence Tag (EST) database of *Chlamydomonas* (Asamizu et al. 1999; Fuhrmann et al. 2001; Sakamoto et al. 1998). The EST data released in the spring of 2000 enabled three research groups to clone and sequence the cDNA independently and register them in GenBank under different names. Sineshchekov et al. (2002) used RNA interference (RNAi) experiments to demonstrate that the rhodopsins from *Chlamydomonas* act as phototaxis sensors and subsequently named them *Chlamydomonas* sensory opsin-A, –B (CSOA, CSOB). Nagel et al. (2002) showed that the rhodopsins from *Chlamydomonas* act as a light-gated ion channel and named the protein channelopsin-1 and -2 (chop-1, chop-2). Suzuki et al. (2003) showed localization of the rhodopsins from *Chlamydomonas* near the eyespot and named them archeal-type *Chlamydomonas* opsin-1, and -2 (Acop-1, Acop-2). Currently, the most widely acknowledged names are 'channelopsin' for the apoprotein and 'channelrhodopsin' (ChR1, ChR2) for the holoprotein.

2.3.2 Diversity of Channelrhodopsins

Channelrhodopsins are considered to be commonly distributed in photomotile green algae and to date have been identified in several green algae other than *C. reinhardtii*. Zhang et al. (2008) isolated a channelrhodopsin variant (VcChR1) from *Volvox carteri*, a spherical colony of flagellated cells similar to *Chlamydomonas*, and showed that it acts as a light-gated ion channel with spectral sensitivity red-shifted by approximately 70 nm from that of ChR2. Hou et al. (2012) identified channelrhodopsin variants by homology cloning in three species of *Chlamydomonas*: *C. augustae*, *C. raudensis*, and *C. yellowstonensis*, which were originally isolated from habitats permanently covered with snow or ice. The channelrhodopsin variants from *C. augustae* (CaChR1) and *C. yellowstonensis* (CyChR1) were successfully expressed in HEK293 cells, where they acted as light-gated ion channels like ChR1 and ChR2, though the features, including current kinetics, light intensity dependence, and spectral sensitivity, differ from species to species. The phylogenetically most distant channelrhodopsin variant was obtained from *Mesostigma viride*, a Prasinophyceae alga, distinct from Chlorophyceae algae such as *Chlamydomonas* and *Volvox* (Govorunova et al. 2011). The *Mesostigma* channelrhodopsin (MChR1) showed maximal spectral sensitivity at 531 nm, a wavelength longer than that of VcChR1.

2.4 Photoactivated Adenylyl Cyclases

2.4.1 Discovery of Photoactivated Adenylyl Cyclase

Euglena, a photosynthetic unicellular flagellate, responds to light stimuli to locate itself in an appropriate light environment. Given that the action spectra for the photoaccumulation and photoavoidance of *Euglena* showed peaks in the UV-A/blue region of the spectrum, a flavin-based photoreceptor was postulated as a strong candidate for the sensor of this organism (Diehn 1969; Matsunaga et al. 1998). Iseki et al. (2002) isolated the light-sensing organelle of *Euglena*, from which they purified a 400-kDa flavoprotein that binds FAD as a chromophore. The 400-kDa flavoprotein showed significant adenylyl cyclase activity that was drastically elevated upon blue light irradiation. It was accordingly named photoactivated adenylyl cyclase (PAC). PAC is a heterotetramer of alpha- and beta-subunits that contain two BLUF domains each followed by an adenylyl cyclase catalytic domain. RNAi experiments have revealed that PAC acts as the sensor for photoavoidance of *Euglena* (Iseki et al. 2002).

Because PAC is activated directly by light without any intervening signal transducers such as G-proteins, PAC was expected to serve as a tool to control various biological processes mediated by cyclic adenosine monophosphate (cAMP) using light. Such an optogenetic approach has been reported in *Aplysia* neurons (Nagahama et al. 2007), *Xenopus* oocytes (Schröder-Lang et al. 2007) and embryo (Hong et al. 2011), cultured mammalian cells (Schröder-Lang et al. 2007), *Drosophila* brains (Schröder-Lang et al. 2007; Bucher and Buchner 2009), *Caenorhabditis* neurons (Weissenberger et al. 2011), and mammalian neurons (Nicol et al. 2011).

2.4.2 Diversity of Photoactivated Adenylyl Cyclases

A PAC-like gene was recently found in the genome sequence of the sulfur bacterium *Beggiatoa* (Losi and Gärtner 2008), though PAC had been considered to be a light-sensing protein specific in euglenids (Koumura et al. 2004). The *Beggiatoa* PAC gene has a BLUF domain and a cyclase domain showing high similarity to the C-terminal half of α- and β-subunits of *Euglena* PAC. The *Beggiatoa* PAC has been heterologously expressed in *E. coli* (Ryu et al. 2010), *Xenopus* oocytes, and mammalian neurons (Stierl et al. 2011), where it acted as a light-dependent adenylyl cyclase. Moreover, Ryu et al. (2010) succeeded in converting the *Beggiatoa* PAC into photoactivated guanylyl cyclase by substituting three amino acid residues. Other PAC-like genes were found in the recently released genome sequence of the free-living amoeba *Naegleria gruberi* (Fritz-Laylin et al. 2010). Yasukawa et al. (2013) isolated similar genes from the potentially pathogenic *Naegleria* species, *N. australiensis,* and showed that their products act as photoactivated adenylyl cyclases when expressed in *E. coli.*

2.5 Concluding Remarks

Plants and microorganisms have developed a wide variety of photosensors to acclimate themselves to given light environments. Given that the spectral sensitivity and the effector functions of the photosensors are diverse, we can choose any of them for engineering new optogenetic tools. In fact, they are widely used by neuroscientists today. However, we should remember that most of the light-sensing proteins are the product of pure biological research motivated by intellectual curiosity. We hope that the story of photoreceptor research in plants and microorganisms, and the development of optogenetics, will go down in history as a good example of the value of basic science.

References

Ahmad M, Cashmore AR (1993) *HY4* gene of *A. thaliana* encodes a protein with characteristics of a blue-light photoreceptor. Nature 366:162–166

Asamizu E, Nakamura Y, Sato S et al (1999) A large scale structural analysis of cDNAs in a unicellular green alga, *Chlamydomonas reinhardtii*. I. Generation of 3433 non-redundant expressed sequence tags. DNA Res 6:369–373

Balashov SP, Imasheva ES, Boichenko VA et al (2005) Xanthorhodopsin: a proton pump with a light-harvesting carotenoid antenna. Science 309:2061–2064

Barends TR, Hartmann E, Griese JJ et al (2009) Structure and mechanism of a bacterial light-regulated cyclic nucleotide phosphodiesterase. Nature 459:1015–1018

Béjà O, Aravind L, Koonin EV et al (2000) Bacterial rhodopsin: evidence for a new type of phototrophy in the sea. Science 289:1902–1906

Bogomolni RA, Spudich JL (1982) Identification of a third rhodopsin-like pigment in phototactic *Halobacterium halobium*. Proc Natl Acad Sci U S A 79:6250–6254

Borthwick HA, Hendricks SB, Parker MW et al (1952) A reversible photoreaction controlling seed germination. Proc Natl Acad Sci U S A 38:662–666

Bucher D, Buchner E (2009) Stimulating PACalpha increases miniature excitatory junction potential frequency at the *Drosophila* neuromuscular junction. J Neurogenet 23:220–224

Butler WL, Norris KH, Siegelman HW et al (1959) Detection, assay, and preliminary purification of the pigment controlling photoresponsive development of plants. Proc Natl Acad Sci U S A 45:1703–1708

Christie JM, Reymond P, Powell GK et al (1998) *Arabidopsis* NPH1: a flavoprotein with the properties of a photoreceptor for phototropism. Science 282:1698–1701

Crefcoeur RP, Yin R, Ulm R et al (2013) Ultraviolet-B-mediated induction of protein-protein interactions in mammalian cells. Nat Commun 4:1779

Davis SJ, Vener AV, Vierstra RD (1999) Bacteriophytochromes: phytochrome-like photoreceptors from nonphotosynthetic eubacteria. Science 286:2517–2520

Demarsy E, Fankhauser C (2009) Higher plants use LOV to perceive blue light. Curr Opin Plant Biol 12:69–74

Derguini F, Mazur P, Nakanishi K et al (1991) All-*trans*-retinal is the chromophore bound to the photoreceptor of the alga *Chlamydomonas reinhardtii*. Photochem Photobiol 54:1017–1021

Diehn B (1969) Action spectra of the phototactic responses in *Euglena*. Biochim Biophys Acta 177:136–143

Ernst OP, Lodowski DT, Elstner M et al (2014) Microbial and animal rhodopsins: structures, functions, and molecular mechanisms. Chem Rev 114:126–163

Fankhauser C, Yeh KC, Lagarias JC et al (1999) PKS1, a substrate phosphorylated by phytochrome that modulates light signaling in *Arabidopsis*. Science 284:1539–1541

Foster KW, Saranak J, Patel N et al (1984) A rhodopsin is the functional photoreceptor for phototaxis in the unicellular eukaryote *Chlamydomonas*. Nature 311:756–759

Fritz-Laylin LK, Prochnik SE, Ginger ML, Dacks JB, Carpenter ML, Field MC, Kuo A, Paredez A, Chapman J, Pham J, Shu S, Neupane R, Cipriano M, Mancuso J, Tu H, Salamov A, Lindquist E, Shapiro H, Lucas S, Grigoriev IV, Cande WZ, Fulton C, Rokhsar DS, Dawson SC (2010) The genome of Naegleria gruberi illuminates early eukaryotic versatility. Cell 140:631–642

Fuhrmann M, Stahlberg A, Govorunova E et al (2001) The abundant retinal protein of the *Chlamydomonas* eye is not the photoreceptor for phototaxis and photophobic responses. J Cell Sci 114:3857–3863

Gardner G, Pike CS, Rice HV et al (1971) "Disaggregation" of phytochrome in vitro-a consequence of proteolysis. Plant Physiol 48:686–693

Gomelsky M, Kaplan S (1998) AppA, a redox regulator of photosystem formation in *Rhodobacter sphaeroides* 2.4.1, is a flavoprotein. Identification of a novel FAD binding domain. J Biol Chem 273:35319–35325

Gomelsky M, Klug G (2002) BLUF: a novel FAD-binding domain involved in sensory transduction in microorganisms. Trends Biochem Sci 27:497–500

Govorunova EG, Spudich EN, Lane CE et al (2011) New channelrhodopsin with a red-shifted spectrum and rapid kinetics from *Mesostigma viride*. MBio 2:e00115-11

Grote M (2013) Purple matter, membranes and 'molecular pumps' in rhodopsin research (1960s–1980s). J Hist Biol 46:331–368

He Q, Cheng P, Yang Y et al (2002) White collar-1, a DNA binding transcription factor and a light sensor. Science 297:840–843

Hegemann P, Gärtner W, Uhl R (1991) All-*trans* retinal constitutes the functional chromophore in *Chlamydomonas* rhodopsin. Biophys J 60:1477–1489

Heintzen C, Loros JJ, Dunlap JC (2001) The PAS protein VIVID defines a clock-associated feedback loop that represses light input, modulates gating, and regulates clock resetting. Cell 104:453–464

Hershey HP, Colbert JT, Lissemore JL et al (1984) Molecular cloning of cDNA for *Avena* phytochrome. Proc Natl Acad Sci U S A 81:2332–2336

Hildebrand E, Dencher N (1975) Two photosystems controlling behavioural responses of *Halobacterium halobium*. Nature 257:46–48

Hong KP, Spitzer NC, Nicol X (2011) Improved molecular toolkit for cAMP studies in live cells. BMC Res Notes 4:241

Hou SY, Govorunova EG, Ntefidou M et al (2012) Diversity of *Chlamydomonas*. Photochem Photobiol 88:119–128

Huala E, Oeller PW, Liscum E et al (1997) *Arabidopsis* NPH1: a protein kinase with a putative redox-sensing domain. Science 278:2120–2123

Hughes J, Lamparter T, Mittmann F et al (1997) A prokaryotic phytochrome. Nature 386:663
Ihara K, Umemura T, Katagiri I et al (1999) Evolution of the archaeal rhodopsins: evolution rate changes by gene duplication and functional differentiation. J Mol Biol 285:163–174
Ikeuchi M, Ishizuka T (2008) Cyanobacteriochromes: a new superfamily of tetrapyrrole-binding photoreceptors in cyanobacteria. Photochem Photobiol Sci 7:1159–1167
Iseki M, Matsunaga S, Murakami A et al (2002) A blue-light-activated adenylyl cyclase mediates photoavoidance in *Euglena gracilis*. Nature 415:1047–1051
Jung KH, Trivedi VD, Spudich JL (2003) Demonstration of a sensory rhodopsin in eubacteria. Mol Microbiol 47:1513–1522
Kagawa T, Sakai T, Suetsugu N et al (2001) *Arabidopsis* NPL1: a phototropin homolog controlling the chloroplast high-light avoidance response. Science 291:2138–2141
Kinoshita T, Doi M, Suetsugu N et al (2001) Phot1 and phot2 mediate blue light regulation of stomatal opening. Nature 414:656–660
Kliebenstein DJ, Lim JE, Landry LG et al (2002) *Arabidopsis* UVR8 regulates ultraviolet-B signal transduction and tolerance and contains sequence similarity to human regulator of chromatin condensation 1. Plant Physiol 130:234–243
Koumura Y, Suzuki T, Yoshikawa S, Watanabe M, Iseki M (2004) The origin of photoactivated adenylyl cyclase (PAC), the Euglena blue-light receptor: phylogenetic analysis of orthologues of PAC subunits from several euglenoids and trypanosome-type adenylyl cyclases from Euglena gracilis. Photochem Photobiol Sci 3:580–586
Lawson MA, Zacks DN, Derguini F et al (1991) Retinal analog restoration of photophobic responses in a blind *Chlamydomonas reinhardtii* mutant. Evidence for an archaebacterial like chromophore in a eukaryotic rhodopsin. Biophys J 60:1490–1498
Liu B, Zuo Z, Liu H et al (2011) *Arabidopsis* cryptochrome 1 interacts with SPA1 to suppress COP1 activity in response to blue light. Genes Dev 25:1029–1034
Liu H, Yu X, Li K et al (2008) Photoexcited CRY2 interacts with CIB1 to regulate transcription and floral initiation in *Arabidopsis*. Science 322:1535–1539
Losi A, Gärtner W (2008) Bacterial bilin- and flavin-binding photoreceptors. Photochem Photobiol Sci 7:1168–1178
Mackin KA, Roy RA, Theobald DL (2014) An empirical test of convergent evolution in rhodopsins. Mol Biol Evol 31:85–95
Masuda S, Bauer CE (2002) AppA is a blue light photoreceptor that antirepresses photosynthesis gene expression in *Rhodobacter sphaeroides*. Cell 110:613–623
Matsunaga S, Hori T, Takahashi T et al (1998) Discovery of signaling effect of UV-B/C light in the extended UV-A/blue-type action spectra for stepdown and step-up photophobic responses in the unicellular flagellate alga *Euglena gracilis*. Protoplasma 201:45–52
Matsuno-Yagi A, Mukohata Y (1977) Two possible roles of bacteriorhodopsin; a comparative study of strains of *Halobacterium halobium* differing in pigmentation. Biochem Biophys Res Commun 78:237–243
Müller K, Engesser R, Schulz S et al (2013) Multi-chromatic control of mammalian gene expression and signaling. Nucleic Acids Res 41:e124
Nagahama T, Suzuki T, Yoshikawa S et al (2007) Functional transplant of photoactivated adenylyl cyclase (PAC) into *Aplysia* sensory neurons. Neurosci Res 59:81–88
Nagel G, Ollig D, Fuhrmann M et al (2002) Channelrhodopsin-1: a light-gated proton channel in green algae. Science 296:2395–2398
Ni M, Tepperman JM, Quail PH (1998) PIF3, a phytochrome-interacting factor necessary for normal photoinduced signal transduction, is a novel basic helix-loop-helix protein. Cell 95:657–667
Nicol X, Hong KP, Spitzer NC (2011) Spatial and temporal second messenger codes for growth cone turning. Proc Natl Acad Sci U S A 108:13776–13781
Oesterhelt D, Stoeckenius W (1971) Rhodopsin-like protein from the purple membrane of *Halobacterium halobium*. Nat New Biol 233:149–152
Raffelberg S, Wang L, Gao S et al (2013) A LOV-domain-mediated blue-light-activated adenylate (adenylyl) cyclase from the cyanobacterium *Microcoleus chthonoplastes* PCC 7420. Biochem J 55:359–365

Rice HV, Briggs WR, Jackson-White CJ (1973) Purification of oat and rye phytochrome. Plant Physiol 51:917–926
Rizzini L, Favory JJ, Cloix C et al (2011) Perception of UV-B by the *Arabidopsis* UVR8 protein. Science 332:103–106
Ryu MH, Moskvin OV, Siltberg-Liberles J et al (2010) Natural and engineered photoactivated nucleotidyl cyclases for optogenetic applications. J Biol Chem 285:41501–41508
Sakamoto M, Wada A, Akai A et al (1998) Evidence for the archaebacterial-type conformation about the bond between the beta-ionone ring and the polyene chain of the chromophore retinal in chlamyrhodopsin. FEBS Lett 434:335–338
Schröder-Lang S, Schwärzel M, Seifert R et al (2007) Fast manipulation of cellular cAMP level by light in vivo. Nat Methods 4:39–42
Shu X, Royant A, Lin MZ et al (2009) Mammalian expression of infrared fluorescent proteins engineered from a bacterial phytochrome. Science 324:804–807
Sineshchekov OA, Jung KH, Spudich JL (2002) Two rhodopsins mediate phototaxis to low- and high-intensity light in *Chlamydomonas reinhardtii*. Proc Natl Acad Sci U S A 99:8689–8694
Spudich EN, Spudich JL (1982) Control of transmembrane ion fluxes to select halorhodopsin-deficient and other energy-transduction mutants of *Halobacterium halobium*. Proc Natl Acad Sci U S A 79:4308–4312
Spudich JL, Yang CS, Jung KH et al (2000) Retinylidene proteins: structures and functions from archaea to humans. Annu Rev Cell Dev Biol 16:365–392
Stierl M, Stumpf P, Udwari D et al (2011) Light modulation of cellular cAMP by a small bacterial photoactivated adenylyl cyclase, bPAC, of the soil bacterium *Beggiatoa*. J Biol Chem 286:1181–1188
Suzuki T, Yamasaki K, Fujita S et al (2003) Archaeal-type rhodopsins in *Chlamydomonas*: model structure and intracellular localization. Biochem Biophys Res Commun 301:711–717
Swartz TE, Tseng TS, Frederickson MA et al (2007) Blue-light-activated histidine kinases: two-component sensors in bacteria. Science 317:1090–1093
Takahashi F, Yamagata D, Ishikawa M et al (2007) AUREOCHROME, a photoreceptor required for photomorphogenesis in stramenopiles. Proc Natl Acad Sci U S A 104:19625–19630
Takahashi T, Tomioka H, Kamo N et al (1985) A photosystem other than PS370 also mediates the negative phototaxis of *Halobacterium halobium*. FEMS Microbiol Lett 28:161–164
Takahashi T, Yoshihara K, Watanabe M et al (1991) Photoisomerization of retinal at 13-ene is important for phototaxis of *Chlamydomonas reinhardtii*: simultaneous measurements of phototactic and photophobic responses. Biochem Biophys Res Commun 178:1273–1279
Tomioka H, Takahashi T, Kamo N et al (1986) Flash spectrophotometric identification of a fourth rhodopsin-like pigment in *Halobacterium halobium*. Biochem Biophys Res Commun 139:389–395
van der Horst GT, Muijtjens M, Kobayashi K et al (1999) Mammalian Cry1 and Cry2 are essential for maintenance of circadian rhythms. Nature 398:627–630
Vierstra RD, Quail PH (1982) Native phytochrome: inhibition of proteolysis yields a homogeneous monomer of 124 kilodaltons from *Avena*. Proc Natl Acad Sci USA 79:5272–5276
Weissenberger S, Schultheis C, Liewald JF et al (2011) PACα–an optogenetic tool for in vivo manipulation of cellular cAMP levels, neurotransmitter release, and behavior in *Caenorhabditis elegans*. J Neurochem 116:616–625
Yasukawa H, Sato A, Kita A, Kodaira K, Iseki M, Takahashi T, Shibusawa M, Watanabe M, Yagita K (2013) Identification of photoactivated adenylyl cyclases in Naegleria australiensis and BLUF-containing protein in Naegleria fowleri. J Gen Appl Microbiol 59:361–369
Yatsuhashi H, Hashimoto T, Shimizu S (1982) Ultraviolet action spectrum for anthocyanin formation in broom sorghum first internodes. Plant Physiol 70:735–741
Zhang F, Prigge M, Beyrière F et al (2008) Red-shifted optogenetic excitation: a tool for fast neural control derived from *Volvox carteri*. Nat Neurosci 11:631–633

Chapter 3
Structure-Functional Analysis of Channelrhodopsins

Hideaki E. Kato, Ryuichiro Ishitani, and Osamu Nureki

Abstract Channelrhodopsin (ChR) was the first light-gated cation channel to be discovered from green algae. Since the inward flow of cations triggers neuron firing, neurons expressing ChRs can be optically controlled, even within freely moving mammals. Although ChR has been broadly applied to neuroscience research, little is known about its molecular mechanisms. In this chapter, we first describe the simple background of rhodopsin family proteins including ChR, and how optogenetics technology has been established since the discovery of ChR in 2002. We later introduce recent findings about the structure-functional relationship of ChR by especially focusing on a paper about the crystal structure of chimeric ChR (C1C2). After we explain the molecular architecture, the initial photoreactions, the ion-conducting pathway, and the putative channel gates of C1C2, we use three recent studies as examples to further explore the possibility of the structure-based engineering of ChR variants with properties that are more ideal for use as optogenetics tools.

Keywords Membrane protein • Light-gated ion channel • Channelrhodopsin • Optogenetics • X-ray crystallography • Structure-based engineering • Electrophysiology

H.E. Kato (✉)
Department of Molecular and Cellular Physiology, Stanford University, Stanford, CA 94305, USA
e-mail: hekato@stanford.edu

R. Ishitani • O. Nureki
Department of Biophysics and Biochemistry, Graduate School of Science, The University of Tokyo, Bunkyo-ku, Tokyo 113-0032, Japan
e-mail: nureki@bs.s.u-tokyo.ac.jp

H. Yawo et al. (eds.), *Optogenetics*, DOI 10.1007/978-4-431-55516-2_3

3.1 Classification of Rhodopsin

When you hear the word 'rhodopsin', the first picture you have in mind would be the visual pigment rhodopsin that exists in our eyes. Visual rhodopsin, which is expressed in the rod cells in the retina, consists of seven-transmembrane (7-TM) helix proteins called opsin and a covalently bound small molecule called retinal. When the retinal is illuminated by light, this small molecule is isomerized from *11-cis* to *all-trans*. The protein moiety changes its structure and activates the heterotrimeric G proteins; the activated G proteins transduce signals to various intracellular effectors. Although the most famous 'rhodopsin' would be the visual pigment rhodopsin, rhodopsin proteins form a large superfamily and have been isolated not only from animal eyes, but also from various lower eukaryotes, bacteria, and even archaea living under extreme environmental conditions.

Dr. Oesterhelt and Dr. Stoeckenius discovered the first rhodopsin from prokaryotes in 1971 (Oesterhelt and Stoeckenius 1971), 100 years after the discovery of the visual pigment rhodopsin by Dr. Boll in 1876 (Stabell and Stabell 2009). Because of its similarity to visual pigments, the rhodopsin isolated from haloarchaea *Halobacterium halobium* was named bacteriorhodopsin (BR). Two years later, Dr. Oesterhelt and Dr. Stoeckenius found that BR works as a light-driven proton pump, and we found that the rhodopsin family proteins had much more diverse functions than we had expected.

On the basis of primary sequences, the rhodopsin family proteins are now classified into two groups: microbial (type I) and animal (type II). Type I rhodopsins are found in archaea, eubacteria, fungi, and algae, whereas type II rhodopsins are expressed in animals, including human beings (Fig. 3.1). Type I and type II rhodopsins both have 7-TM domains, a highly conserved lysine residue on TM7, and the retinal chromophore covalently bound to the lysine via a Schiff base. However, the primary and tertiary structures of type II rhodopsins are more similar to other receptor proteins (e.g., beta2 adrenergic receptor and adenosine A2A receptor) than to type I (Dixon et al. 1986; Rasmussen et al. 2007), and type I and type II rhodopsins use different retinal isomers. Type I rhodopsins have more divergent functions than type II, and the photoreactions between type I and type II are significantly different. Thus, whether these two protein families diverge from a common ancestral protein or whether they have converged from independent origins is still controversial (Spudich et al. 2000). (Recently, a new rhodopsin called middle rhodopsin [MR] was discovered by Dr. Sudo's group. While this rhodopsin is a type I, it can bind *11-cis* retinal, the isomer used by type II rhodopsins (Sudo et al. 2011). Thus, it is expected that MR would provide us insight into the evolutionary relationship between type I and type II rhodopsins.)

In this chapter, I focus on type I rhodopsin proteins that are widely used as optogenetics tools. As described above, type I rhodopsins have divergent functions compared with type II. It was 6 years after the discovery of BR that Dr. Mukohata's group discovered the light-driven Cl^- pump 'halorhodopsin' (HR) (Matsuno-Yagi and Mukohata 1977), and Dr. Spudich's group identified the light-sensitive signal

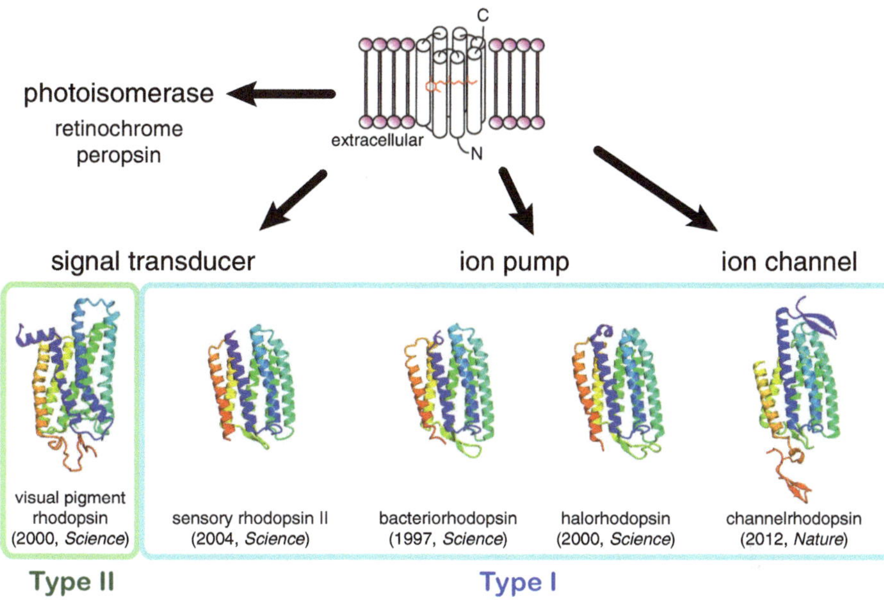

Fig. 3.1 Classification of rhodopsin family proteins

transducer 'sensory rhodopsin I' (SRI) in 1982 (Bogomolni and Spudich 1982). It was 20 years later that the light-gated cation channel 'channelrhodopsin' (ChR) was isolated by Dr. Hegemann's group (Nagel et al. 2002) (Fig. 3.1).

3.2 Discovery of ChR and the Development of Optogenetics

In the 1990s, Dr. Hegemann's and Dr. Keszthelyi's groups found that the fast electrical current was involved in the rhodopsin-mediated phototactic responses in tiny green algae (Deininger et al. 1995; Sineshchekov et al. 1990). Since the currents occurred only within 30 μs after light illumination, it was suggested that the photoreceptor and the channel formed a protein complex or were the same protein. In 2002–03, Dr. Hegemann et al., Dr. Spudich et al., and Dr. Takahashi et al. isolated new microbial rhodopsins from the algae (Nagel et al. 2002; Sineshchekov et al. 2002; Suzuki et al. 2003). They discovered that two archaeal-type new rhodopsins were expressed in an organelle called eyespot, and Dr. Hegemann's group identified them as light-gated cation channels by electrophysiological experiments (Nagel et al. 2002). Using both oocytes of *Xenopus laevis* and mammalian cells, his group demonstrated that these two proteins permeate various monovalent and divalent cations, including H^+, Na^+, K^+, and Ca^{2+} (Berthold et al. 2008; Nagel et al. 2002, 2003). Since these proteins were the first channel-type rhodopsin family proteins, they named these proteins channelrhodopsins (ChR1 and ChR2). (When ChR1 was first characterized in 2002, it was thought that ChR1 could only permeate H^+. However, Dr. Hegemann's group re-analyzed the functional properties of ChR1 and reported in 2008 that, like ChR2, ChR1 permeated Na^+, K^+, and Ca^{2+} (Berthold et al. 2008).)

"… The use of ChR1 as a tool for measuring and/or manipulating electrical and proton gradients across cell membranes, simply by illumination." As Dr. Hegemann said in his 2002 paper (Nagel et al. 2002), he seemed to have an idea that this protein could be useful to manipulate the membrane potential by light. Indeed, taking his suggestion into account, several groups began to work with ChRs, and German, American, and Japanese groups successfully activated neurons in a light-dependent manner during 2005–2006 (Bi et al. 2006; Boyden et al. 2005; Ishizuka et al. 2006; Li et al. 2005; Nagel et al. 2005). Only 1 year later, using ChR2 and fiber optic technologies, Dr. Deisseroth's group finally succeeded in controlling sleep–wake behavior in freely moving mice (Adamantidis et al. 2007). Since these studies, several neuroscientists have begun to use ChRs. The number of publications about optogenetics has rapidly increased, and the optogenetics technique was nominated as 'method of the year' in 2010 (Deisseroth 2011). ChRs have been established as a new powerful tool for neuroscience, and have now been used to control neuronal activity in a wide range of animals, with resulting insights into fundamental aspects of circuit function as well as dysfunction and treatment in pathological states (Chaudhury et al. 2012; Ciocchi et al. 2010; Ji et al. 2012; Kravitz et al. 2010).

However, despite the rapid progress of optogenetics, little was known about the molecular mechanism of ChR itself. Although the previous electrophysiological and spectroscopic analyses provided us with insight into its ion selectivity, kinetics, and cyclic photoreaction (photocycle), many important questions were unanswered: Where is the ion-conducting pathway located in the protein moiety? How do light illumination and retinal isomerization cause the structural change of ChR and allow the cation to pass through its conducting pathway? The rough helical arrangement was reported in the recently published electron microscopic (EM) structure of ChR at 6 Å resolution (Muller et al. 2011); a lack of understanding of accurate amino acid positioning and insights into channel function remained.

In parallel with this research, some neuroscientists and biophysicists have tried to engineer ChR variants with improved properties. Even with limited structural models, ChR variants have been engineered with faster or slower kinetics (Berndt et al. 2009, 2011; Gunaydin et al. 2010; Kleinlogel et al. 2011; Yizhar et al. 2011), shifted absorption spectra (Gunaydin et al. 2010; Wen et al. 2010; Yizhar et al. 2011), reduced desensitization (Lin et al. 2009; Nagel et al. 2005; Wang et al. 2009; Wen et al. 2010), and increased expression and photocurrent amplitude (Berndt et al. 2011; Kleinlogel et al. 2011; Lin et al. 2009; Nagel et al. 2005; Yizhar et al. 2011). These advances still represent only the tip of the iceberg, if detailed structural knowledge could be obtained to rationally design the mutations to alter the channel properties and functions. Thus, a high-resolution structure would be of enormous value, not only to enhance our understanding of the mechanism of this first discovered light-gated ion channel, but also to guide the way to the design of ChR variants with novel functions related to absorption spectrum, selectivity, and kinetics; therefore, we tried to solve the crystal structure of ChR.

3.3 Overall Structure of C1C2

Generally speaking, membrane proteins, especially eukaryotic membrane proteins, are notoriously difficult to express, purify, and crystallize. They are often flexible and unstable and easily lose their functions when removed from the lipid bilayer. Thus, to find a promising candidate for structural studies, we tried to express various ChRs from their natural sources and their chimeric ChRs in every expression system, including *Escherichia coli*, yeast, mammalian, and insect cells. Although the expression levels and/or the stabilities of the most screened proteins were poor, we found that one chimeric ChR (TMs 1–5 are derived from *Chlamydomonas reinhardtii* ChR1 and TMs 6–7 are from *C. reinhardtii* ChR2) showed excellent expression level and stability. ChRs from *C. reinhardtii* consist of ~700 amino acids, and it was well known that the C-terminal ~400 amino acids were disordered and the photocurrent functionality were all contained within the N-terminal ~300 amino acids. Thus, we truncated the non-structured C-terminal ~400 amino acid residues, and expressed, purified, crystallized, and solved the structure of this chimera, named C1C2 (residues 1–356). (Since C1C2 retains the channel activity, and the sequence similarity between ChR1 and ChR2 are very high, it is assumed that most structural features described below are common to other ChRs, including ChR2, the most widely used optogenetics tool.)

C1C2 is composed of an N-terminal extracellular domain (49–83), the 7-TM domains connected by three cytoplasmic and extracellular loops (84–317), and the C-terminal intracellular domain (318–342) (Fig. 3.2) (Kato et al. 2012). Residues 24–48, 110–117, and 343–356 are disordered and not visible in the crystal structure. (Residues 1–23 are also not visible since this region works as signal sequence and is cleaved after the translation.) Notably, as previously predicted from EM (Muller et al. 2011), ChR is tightly associated with a dimer via interfacial interactions between the N-terminal domain, ECL1, TMs 3 and 4 of each molecule. Since three cysteine residues in the N-terminal extracellular domain form three disulfide bonds between protomers, it is assumed that C1C2 forms dimers not only in the crystal but also in the physiological condition. This result is a little bit surprising because most microbial rhodopsins form homo-trimer or hetero-tetramer with its signal transducers (Gordeliy et al. 2002; Pebay-Peyroula et al. 1997). Although it has recently been reported that some proteorhodopsins (PRs) form homo-pentamer or homo-hexamer (Ran et al. 2013), ChR is the first example of a microbial rhodopsin forming a dimer conformation.

3.4 Structural Comparison Between C1C2 and BR

To further understand the C1C2 structure, we compared our C1C2 with the most well studied microbial rhodopsin, BR. Although the primary sequence identity between C1C2 and BR is as low as 15 %, the overall structure of C1C2 is well superimposed on that of BR (Fig. 3.3). TMs 3–6 are very similar, and the position

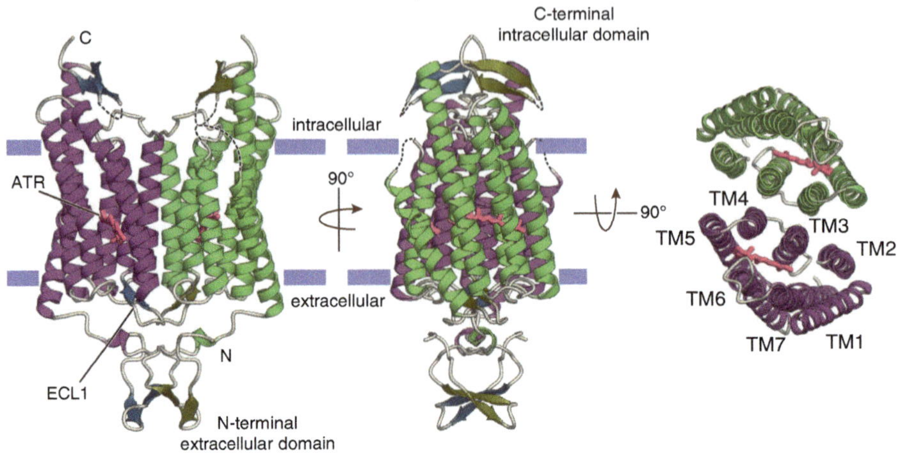

Fig. 3.2 Overall structure of C1C2. The disordered regions are represented by the *dashed black lines*. The approximate positioning of the lipid bilayers are indicated by *thick blue lines*

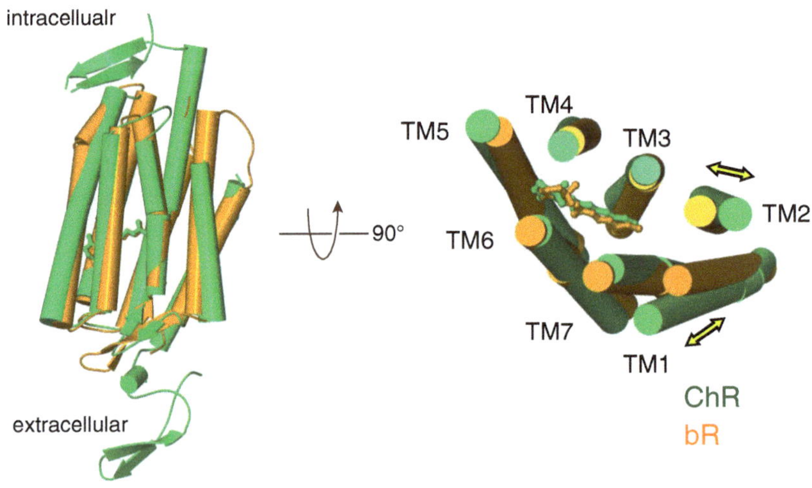

Fig. 3.3 The structural comparison between C1C2 and bacteriorhodopsin (BR)

of the retinal is well conserved, whereas there are two distinct features between C1C2 and BR. First, C1C2 has additional N-terminal and C-terminal domains. Although it is assumed that the N-terminal domain is contributed to the dimer formation of C1C2, the physiological role of the intracellular C-terminal domain is still elusive. One possibility is that the C-terminal domain and truncated ~400 amino acid residues cooperatively help ChRs to localize within the eyespot in *C. reinhardtii*. Indeed, the recent study showed that there are several conserved motifs in the C-terminal region, and the authors proposed in the study that the unknown proteins would interact with such a conserved motif and allow ChR to localize in the eyespot (Kianianmomeni et al. 2009). Second, the extracellular side of TMs 1 and 2 are tilted compared with those of BR. Since the sequence similarities of TM1 and 2 between C1C2 and BR are very low, and these helices have important roles in channel function as described later, TM1 and 2 are thought to be key helices in ChRs.

3.5 The Initial Photocycle Reaction in ChRs

Unlike type II rhodopsins, when type I rhodopsins absorb light energy, they undergo cyclic reactions comprising a series of intermediate states. This cyclic photoreaction is called the photocycle. ChR is known to have the complex photocycle; there is still no consensus on details, but previous studies have revealed most events that occur in its initial stage. In all microbial rhodopsins, *all-trans* retinal (ATR) is covalently bound to the lysine residue on TM7, forming the Schiff base. In the dark/ground state, the nitrogen atom of the Schiff base is protonated, and when the retinal absorbs light, ATR is isomerized from *all-trans* to *13-cis* (P^{500} state), and the proton is transferred from the Schiff base to a negatively charged carboxylate located in the extracellular region (P^{390} state). These are the initial reactions of the photocycle in most microbial rhodopsins, including ChR (Fig. 3.4).

In BR, Asp85 and Asp212 are located near the Schiff base and Asp85 works as a primary proton acceptor from the Schiff base in the M intermediate (corresponding to P^{390} in ChR). This can be explained with two reasons. First, Asp85 is nearer to the Schiff base than Asp212. Second, Asp212 retains a low pK_a and cannot change its protonation state because it is hydrogen bonded to Tyr57 and Tyr185, which do not move during the photocycle (Fig. 3.5a). However, in C1C2, the distances from the protonated Schiff base to Glu162 (Glu123 in ChR2 and Asp85 in BR) and Asp292 (Asp253 in ChR2 and Asp212 in BR) are 3.4 and 3.0 Å, respectively, and Tyr57 and

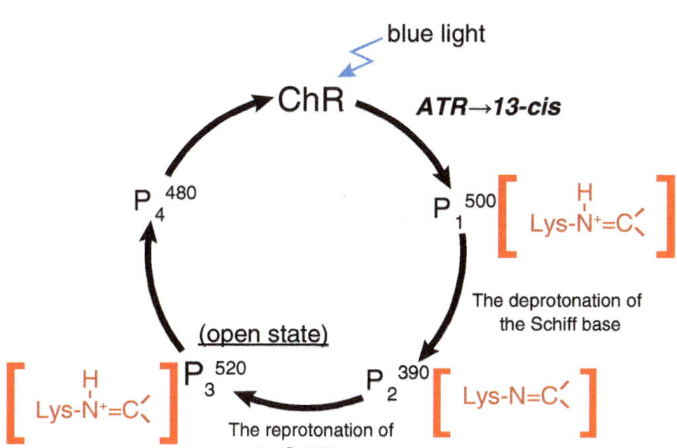

Fig. 3.4 Simplified photocycle of channelrhodopsin (ChR)

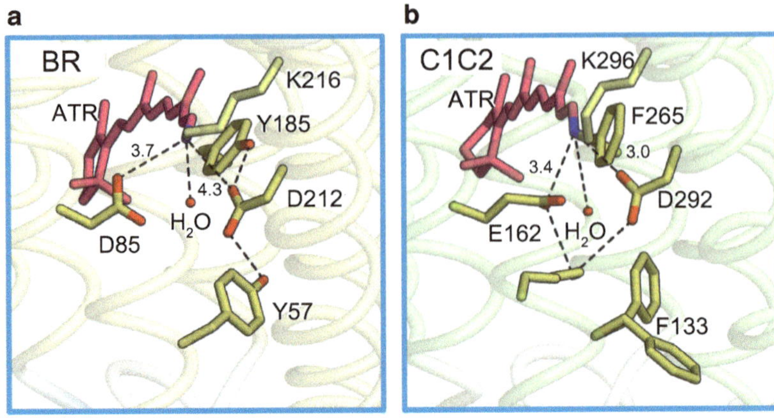

Fig. 3.5 The Schiff base regions in C1C2 and bacteriorhodopsin (BR). The structures of the Schiff base and its counter ions in (**a**) C1C2 and (**b**) BR. *Dashed black lines* indicate the hydrogen bonds, and the *numbers* are the distances between two atoms

Tyr185 in BR are substituted with Phe133 and Phe265 (Val94 and Phe226 in ChR2) (Fig. 3.5b). Thus, it was assumed that Asp292 could move relatively freely, change its pK_a, and receive and release the proton in the photocycle. Then, we prepared and analyzed the properties of E162A and D292A mutants. We found that, while the E162A mutant showed moderately decreased photocurrents, the D292A mutant almost completely lost channel activity. Moreover, the substitution of Asp292 with alanine decreased the time required to open the channel. These results are consistent with the crystal structure, suggesting that Asp292, rather than Glu162, works as the major proton acceptor from the protonated Schiff base in P^{390} state in C1C2. To analyze the functional importance of the flexibility of Asp292, we next prepared F133Y and F265Y mutants and measured their photocurrents. We found that only the F265Y mutant showed significantly decreased current (Kato et al., unpublished), and this is also consistent with the structure in which only the side chain of Phe265 points toward Asp292. The side chain of Phe133 is flexible and has two alternative conformations in the structure.

3.6 Cation-Conducting Pathway

To understand the molecular mechanism of ChR, it is essential to identify the cation-conducting pathway. While 10 years passed after the discovery of ChR in 2002, the position of the ion-conducting pathway remained controversial. Some studies suggested that ChR oligomerizes into a dimer or trimer, and cations pass through the interface of the oligomer (Muller et al. 2011) like potassium, sodium, and calcium channels (Doyle et al. 1998; Hou et al. 2012; Payandeh et al. 2011); others suggested that the ion-conducting pathway is placed within a monomer, like some chloride transporters and proton channels (Dutzler et al. 2002; Takeshita et al. 2014). Even

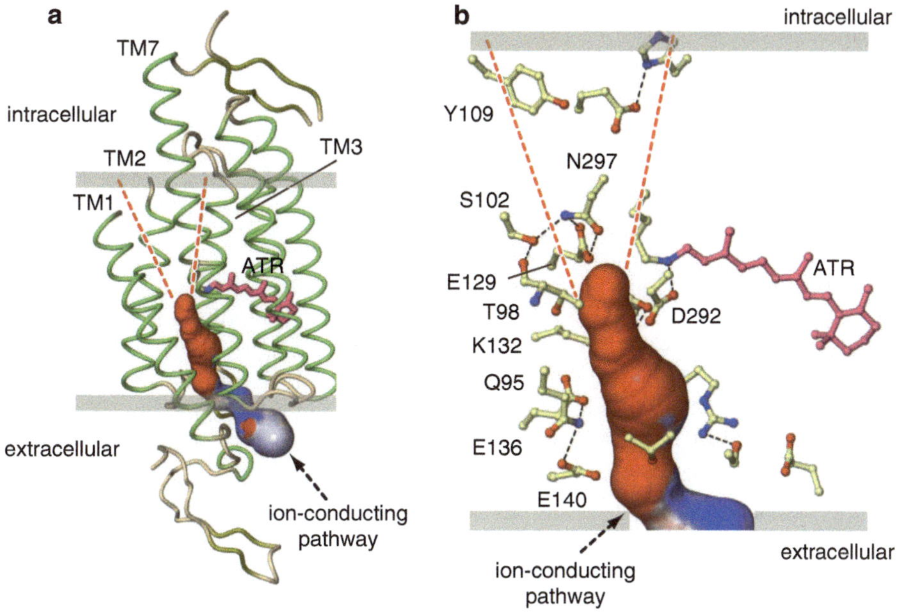

Fig. 3.6 The putative ion-conducting pathway. *Dashed red lines* are putative intracellular vestibules of the pathway. (**a**) Overall view and (**b**) magnified view of the pathway

among the researchers supporting the latter hypothesis, there was no consensus about where cations are permeated within a monomer (Bamann et al. 2010; Muller et al. 2011). When we observed the cross-section of ChR from the extracellular side, we found that although there is no space to allow the cations to pass through the dimer interface, there is a pore between TMs 1, 2, 3, and 7 within the monomer. Since the pore is formed by the tilts of TMs 1 and 2 mentioned previously, this pore is unique to ChR. Moreover, the calculated electrostatic surface potential reveals that this pore is strongly electronegative, so we assume that this pore would work as the cation-conducting pathway in ChR (Fig. 3.6a). In this putative cation-conducting pathway, there are 12 polar residues (Fig. 3.6b). Thus, to verify this hypothesis, we substituted 4 of 12 pore-lining residues Gln95, Lys132, Glu136, and Glu140 (Gln56, Lys94, Glu97, Glu101 in ChR2) with alanine, expressed them in HEK293 cells, and recorded photocurrents in response to blue light pulses. We found that only the mutant of the positively charged amino acid, K132A, had similar current amplitude relative to wild type, while the Q95A, E136A, and E140A mutants exhibited reduced currents. Moreover, three of the four mutants (Q95A, K132A, and E136A) showed significantly altered ion selectivities. These results strongly support our hypothesis that this pore actually works as the cation-conducting pathway. After we published the crystal structure of C1C2, several biophysical, computational, and electrophysiological studies have been reported; as far as we know, all these studies support this ion-conducting pathway (Gaiko and Dempski 2013; Krause et al. 2013; Richards and Dempski 2012; Sattig et al. 2013; Tanimoto et al. 2012; Watanabe et al. 2013).

3.7 The Channel Gates in the Conducting Pathway

The crystal structure revealed that an extracellular vestibule expands to a diameter of about 8 Å and suggested that this passage would allow water molecules to move from the extracellular side to the middle of the pathway. However, in contrast to the extracellular side, the intracellular side of the pathway is occluded at two constriction sites, revealing a closed state consistent with the experimental conditions, in which C1C2 was crystallized and harvested in the dark.

The outer constriction is formed by Tyr109, Glu122, and His173 (Tyr60, Glu82, and His134 in ChR2), and the inner constriction near the middle of the pathway is formed by Thr98, Ser102, Glu129, and Asn297 (Ala59, Ser63, Glu90, and Asn258 in ChR2) (Fig. 3.7). In the outer constriction, the hydrogen bond between Glu122 and His173 links TM2 to TM3, and Tyr109 and Glu122 occlude the channel space surrounded by TMs 1, 2, 3, and 7 (Fig. 3.7a). In the inner constriction, Ser102, with its β-OH group hydrogen bonded to the main-chain carbonyl oxygen of Thr98, fixes the position of Asn297, which in turn anchors Glu129 with bipartite hydrogen bonds. This hydrogen bond tethers TMs 2 and 7 together and occludes the pore (Fig. 3.7b). Here, we focus on the inner constriction. Using the electrophysiological analysis, we found that, although E129A showed a similar current to that of wild type, the E129Q mutant showed a decreased photocurrent. Since the calculated pK_a of Glu129 suggests that this residue is protonated in the closed/dark state, we assume that Glu129 is deprotonated once during the photocycle and the change of the protonation state has an important role in the channel activity. The spectroscopic study reported by Dr. Gerwert's group, showing that Glu90 in ChR2 (Glu129 in C1C2) is protonated in the closed/dark state and deprotonated early in

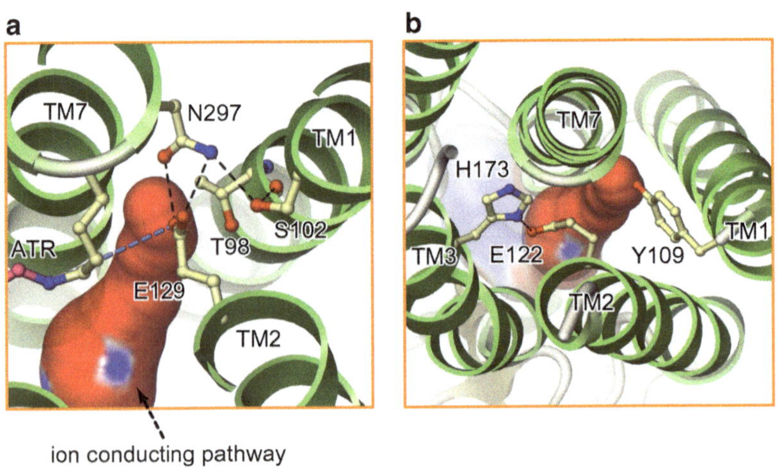

Fig. 3.7 Two constriction sites located in the putative ion-conducting pathway. (**a**) The first constriction mainly formed by conserved Glu129. (**b**) The second constriction mainly formed by Tyr109 and Glu122. *Dashed black lines* indicate the hydrogen bonds

the photocycle, supports our idea (Eisenhauer et al. 2012). His group also performed molecular dynamics (MD) simulation using ChR2 homology model made from the crystal structure of BR and showed that when Glu90 is deprotonated, the diameter of our pathway would be wider (Eisenhauer et al. 2012). Because there are some differences between their ChR homology model and the crystal structure, we recently performed MD simulation in similar conditions using the actual crystal structure of C1C2. We verified that the deprotonated Glu129 lost its interaction with Asn297, the side chain of Glu129 changes its orientation, the diameter of our pathway is wider, and the water molecules penetrate deeper into the pore. These results suggest that Glu129 works as a putative channel gate in ChR and that it is regulated by its protonation change (Kato et al., unpublished). However, a new, recently published, spectroscopic paper proposes that Glu90 in ChR2 is deprotonated not at the early stage of the photocycle but after the channel pore is closed (Lorenz-Fonfria et al. 2013), so further studies are needed to completely understand how the inner constriction is opened during the photocycle.

3.8 Towards the Engineering of New Optogenetics Tools

While ChR has been used as a powerful tool in neuroscience, there is still much room for improvement. Many strategies have been applied to create ChR variants with improved properties for optogenetics, and these approaches have generated a number of ChR variants with useful properties. The structural determination of C1C2 has now opened the way to rational engineering of ChR variants with novel functionalities. An excellent example is the structure-based engineering of anion-selective ChRs. In April 2014, Dr. Hegemann's group and Dr. Deisseroth's group independently reported that they successfully engineered the first anion-selective ChR variants (Berndt et al. 2014; Wietek et al. 2014). Dr. Hegemann's group engineered chloride-conducting ChRs (ChloCs) only by introducing single E-to-R mutation to Glu90 in ChR2. Since Glu90 (Glu129 in C1C2) forms inner constriction with Thr98, Ser102, and Asn253, this study suggests that Glu90 works as the ion-selective filter as well as the channel gate. In contrast, Dr. Deisseroth's group engineered chloride-conducting C1C2 'iC1C2' by altering the overall electrostatic environment along an ion-conducting pathway. He introduced mutations to 6 of the 12 pore-lining hydrophilic residues mentioned previously, and substituted Val156 (Gln117 in ChR2), His173, and Val281 with positively charged residues. It should be noted that the two groups accomplished the remarkable functional conversion from cation to anion-selective channel in two very different ways. This suggests that other functional properties of ChR, as in this case of ion selectivity, are less strictly defined than we expected, and could be drastically changed in variable ways.

In addition to ion selectivity, one of the most promising candidate properties of ChR for structure-based engineering is the absorption spectrum. Color variants, which enable us to separately activate two neurons expressing different ChRs, would be of enormous value to neuroscientists. To date, most ChRs with a blue- or red-shifted absorption spectrum have been isolated from natural sources. The most red-shifted ChR ChrimsonR (λ_{max}=590 nm), and most blue-shifted ChR PsChR (λ_{max}=437 nm) were isolated from *Chlamydomonaus noctigama* and *Platymonas subsordiformis*, respectively (Govorunova et al. 2013; Klapoetke et al. 2014). Therefore, some researchers would assume that, in order to develop the color variants, the above strategy is faster and more effective than the rational design of mutants based on the structural information. However, recently, the structure-based variant design has enabled the engineering of a series of CRBPII (cellular retinol-binding protein II) mutants with absorption spectra in the range of 425–644 nm (Wang et al. 2012). This outstanding progress might not have been achieved without structural information. Moreover, we recently developed a rational design of blue-shifted color variants of type I rhodopsins and successfully obtained a new ChR variant with a blue-shifted absorption spectrum (λ_{max}=~450 nm), close to the most blue-shifted microbial protein, PsChR. Notably, the design is universally applicable to other type I rhodopsins, and we finally engineered the variant of archaerhodopsin-3 (Arch) with a ~100 nm blue-shifted absorption spectrum (Kato et al., unpublished). Because Arch is one of the most widely used optogenetics tools to silence neurons, it is expected that the combination of Arch and blue-shifted Arch will enable us to silence different types of neurons individually or simultaneously.

3.9 Concluding Remarks

In this chapter, we described the simple background of rhodopsin family proteins, the structure–function relationship of C1C2, and the structure-based variant design of ChR. The present crystal structure of C1C2 in the closed/dark state provided important insight into the molecular basis for the function of ChRs, and paved the way for the precise and principled design of ChR variants with novel properties. However, to understand the photocycle in more detail, further structural studies, including determination of crystal structures in the open state, as well as further electrophysiological, spectroscopic, and computational studies are clearly needed. The integration of these techniques will enable us to fully understand the molecular basis of the ChR function and to design more ideal research tools in optogenetics.

References

Adamantidis AR, Zhang F, Aravanis AM et al (2007) Neural substrates of awakening probed with optogenetic control of hypocretin neurons. Nature 450:420–424

Bamann C, Gueta R, Kleinlogel S et al (2010) Structural guidance of the photocycle of channelrhodopsin-2 by an interhelical hydrogen bond. Biochemistry (Mosc) 49:267–278

Berndt A, Yizhar O, Gunaydin LA et al (2009) Bi-stable neural state switches. Nat Neurosci 12:229–234

Berndt A, Schoenenberger P, Mattis J et al (2011) High-efficiency channelrhodopsins for fast neuronal stimulation at low light levels. Proc Natl Acad Sci U S A 108:7595–7600

Berndt A, Lee SY, Ramakrishnan C et al (2014) Structure-guided transformation of channelrhodopsin into a light-activated chloride channel. Science 344:420–424
Berthold P, Tsunoda SP, Ernst OP et al (2008) Channelrhodopsin-1 initiates phototaxis and photophobic responses in chlamydomonas by immediate light-induced depolarization. Plant Cell 20:1665–1677
Bi A, Cui J, Ma YP et al (2006) Ectopic expression of a microbial-type rhodopsin restores visual responses in mice with photoreceptor degeneration. Neuron 50:23–33
Bogomolni RA, Spudich JL (1982) Identification of a third rhodopsin-like pigment in phototactic Halobacterium halobium. Proc Natl Acad Sci U S A 79:6250–6254
Boyden ES, Zhang F, Bamberg E et al (2005) Millisecond-timescale, genetically targeted optical control of neural activity. Nat Neurosci 8:1263–1268
Chaudhury D, Walsh JJ, Friedman AK et al (2012) Rapid regulation of depression-related behaviours by control of midbrain dopamine neurons. Nature 493:532–6
Ciocchi S, Herry C, Grenier F et al (2010) Encoding of conditioned fear in central amygdala inhibitory circuits. Nature 468:277–282
Deininger W, Kroger P, Hegemann U et al (1995) Chlamyrhodopsin represents a new type of sensory photoreceptor. EMBO J 14:5849–5858
Deisseroth K (2011) Optogenetics. Nat Methods 8:26–29
Dixon RA, Kobilka BK, Strader DJ et al (1986) Cloning of the gene and cDNA for mammalian beta-adrenergic receptor and homology with rhodopsin. Nature 321:75–79
Doyle DA, Morais Cabral J, Pfuetzner RA et al (1998) The structure of the potassium channel: molecular basis of K+ conduction and selectivity. Science 280:69–77
Dutzler R, Campbell EB, Cadene M et al (2002) X-ray structure of a ClC chloride channel at 3.0 A reveals the molecular basis of anion selectivity. Nature 415:287–294
Eisenhauer K, Kuhne J, Ritter E et al (2012) In channelrhodopsin-2 Glu-90 is crucial for ion selectivity and is deprotonated during the photocycle. J Biol Chem 287:6904–6911
Gaiko O, Dempski RE (2013) Transmembrane domain three contributes to the ion conductance pathway of channelrhodopsin-2. Biophys J 104:1230–1237
Gordeliy VI, Labahn J, Moukhametzianov R et al (2002) Molecular basis of transmembrane signalling by sensory rhodopsin II-transducer complex. Nature 419:484–487
Govorunova EG, Sineshchekov OA, Li H et al (2013) Characterization of a highly efficient blue-shifted channelrhodopsin from the marine alga Platymonas subcordiformis. J Biol Chem 288:29911–29922
Gunaydin LA, Yizhar O, Berndt A et al (2010) Ultrafast optogenetic control. Nat Neurosci 13:387–392
Hou X, Pedi L, Diver MM et al (2012) Crystal structure of the calcium release-activated calcium channel Orai. Science 338:1308–1313
Ishizuka T, Kakuda M, Araki R et al (2006) Kinetic evaluation of photosensitivity in genetically engineered neurons expressing green algae light-gated channels. Neurosci Res 54:85–94
Ji ZG, Ishizuka T, Yawo H (2012) Channelrhodopsins-their potential in gene therapy for neurological disorders. Neurosci Res 196:29–47
Kato HE, Zhang F, Yizhar O et al (2012) Crystal structure of the channelrhodopsin light-gated cation channel. Nature 482:369–374
Kianianmomeni A, Stehfest K, Nematollahi G et al (2009) Channelrhodopsins of Volvox carteri are photochromic proteins that are specifically expressed in somatic cells under control of light, temperature, and the sex inducer. Plant Physiol 151:347–366
Klapoetke NC, Murata Y, Kim SS et al (2014) Independent optical excitation of distinct neural populations. Nat Methods 11:338–346
Kleinlogel S, Feldbauer K, Dempski RE et al (2011) Ultra light-sensitive and fast neuronal activation with the Ca(2)+-permeable channelrhodopsin CatCh. Nat Neurosci 14:513–518
Krause N, Engelhard C, Heberle J et al (2013) Structural differences between the closed and open states of channelrhodopsin-2 as observed by EPR spectroscopy. FEBS Lett 587:3309–3313
Kravitz AV, Freeze BS, Parker PR et al (2010) Regulation of parkinsonian motor behaviours by optogenetic control of basal ganglia circuitry. Nature 466:622–626

Li X, Gutierrez DV, Hanson MG et al (2005) Fast noninvasive activation and inhibition of neural and network activity by vertebrate rhodopsin and green algae channelrhodopsin. Proc Natl Acad Sci U S A 102:17816–17821

Lin JY, Lin MZ, Steinbach P et al (2009) Characterization of engineered channelrhodopsin variants with improved properties and kinetics. Biophys J 96:1803–1814

Lorenz-Fonfria VA, Resler T, Krause N et al (2013) Transient protonation changes in channelrhodopsin-2 and their relevance to channel gating. Proc Natl Acad Sci U S A 110:E1273–E1281

Matsuno-Yagi A, Mukohata Y (1977) Two possible roles of bacteriorhodopsin; a comparative study of strains of Halobacterium halobium differing in pigmentation. Biochem Biophys Res Commun 78:237–243

Muller M, Bamann C, Bamberg E et al (2011) Projection structure of channelrhodopsin-2 at 6 A resolution by electron crystallography. J Mol Biol 414:86–95

Nagel G, Ollig D, Fuhrmann M et al (2002) Channelrhodopsin-1: a light-gated proton channel in green algae. Science 296:2395–2398

Nagel G, Szellas T, Huhn W et al (2003) Channelrhodopsin-2, a directly light-gated cation-selective membrane channel. Proc Natl Acad Sci U S A 100:13940–13945

Nagel G, Brauner M, Liewald JF et al (2005) Light activation of channelrhodopsin-2 in excitable cells of *Caenorhabditis elegans* triggers rapid behavioral responses. Curr Biol 15:2279–2284

Oesterhelt D, Stoeckenius W (1971) Rhodopsin-like protein from the purple membrane of Halobacterium halobium. Nat New Biol 233:149–152

Payandeh J, Scheuer T, Zheng N et al (2011) The crystal structure of a voltage-gated sodium channel. Nature 475:353–358

Pebay-Peyroula E, Rummel G, Rosenbusch JP et al (1997) X-ray structure of bacteriorhodopsin at 2.5 angstroms from microcrystals grown in lipidic cubic phases. Science 277:1676–1681

Ran T, Ozorowski G, Gao Y et al (2013) Cross-protomer interaction with the photoactive site in oligomeric proteorhodopsin complexes. Acta Crystallogr D Biol Crystallogr 69:1965–1980

Rasmussen SG, Choi HJ, Rosenbaum DM et al (2007) Crystal structure of the human beta2 adrenergic G-protein-coupled receptor. Nature 450:383–387

Richards R, Dempski RE (2012) Re-introduction of transmembrane serine residues reduce the minimum pore diameter of channelrhodopsin-2. PLoS One 7:e50018

Sattig T, Rickert C, Bamberg E et al (2013) Light-induced movement of the transmembrane helix B in channelrhodopsin-2. Angewandte Chemie 52:9705–9708

Sineshchekov OA, Litvin FF, Keszthelyi L (1990) Two components of photoreceptor potential in phototaxis of the flagellated green alga Haematococcus pluvialis. Biophys J 57:33–39

Sineshchekov OA, Jung KH, Spudich JL (2002) Two rhodopsins mediate phototaxis to low- and high-intensity light in Chlamydomonas reinhardtii. Proc Natl Acad Sci U S A 99:8689–8694

Spudich JL, Yang CS, Jung KH et al (2000) Retinylidene proteins: structures and functions from archaea to humans. Annu Rev Cell Dev Biol 16:365–392

Stabell B, Stabell U (2009) Duplicity theory of vision. Cambridge University Press

Sudo Y, Ihara K, Kobayashi S et al (2011) A microbial rhodopsin with a unique retinal composition shows both sensory rhodopsin II and bacteriorhodopsin-like properties. J Biol Chem 286:5967–5976

Suzuki T, Yamasaki K, Fujita S et al (2003) Archaeal-type rhodopsins in Chlamydomonas: model structure and intracellular localization. Biochem Biophys Res Commun 301:711–717

Takeshita K, Sakata S, Yamashita E et al (2014) X-ray crystal structure of voltage-gated proton channel. Nat Struct Mol Biol 21:352–357

Tanimoto S, Sugiyama Y, Takahashi T et al (2012) Involvement of glutamate 97 in ion influx through photo-activated channelrhodopsin-2. Neurosci Res 75:13–22

Wang H, Sugiyama Y, Hikima T et al (2009) Molecular determinants differentiating photocurrent properties of two channelrhodopsins from chlamydomonas. J Biol Chem 284:5685–5696

Wang W, Nossoni Z, Berbasova T et al (2012) Tuning the electronic absorption of protein-embedded all-*trans*-retinal. Science 338:1340–1343

Watanabe HC, Welke K, Sindhikara DJ et al (2013) Towards an understanding of channelrhodopsin function: simulations lead to novel insights of the channel mechanism. J Mol Biol 425:1795–1814
Wen L, Wang H, Tanimoto S et al (2010) Opto-current-clamp actuation of cortical neurons using a strategically designed channelrhodopsin. PLoS One 5:e12893
Wietek J, Wiegert JS, Adeishvili N et al (2014) Conversion of channelrhodopsin into a light-gated chloride channel. Science 344:409–412
Yizhar O, Fenno LE, Prigge M et al (2011) Neocortical excitation/inhibition balance in information processing and social dysfunction. Nature 477:171–178

Chapter 4
Photochemistry of Halorhodopsin

Takashi Kikukawa, Naoki Kamo, and Makoto Demura

Abstract Halorhodopsin is a light-driven inward Cl^- pump found in the membrane of a halophilic archaeon called *Halobacterium salinarum*. While the physiological role of halorhodopsin has not been fully resolved, its functional mechanism has been studied as a model system for anion transport. Halorhodopsin has become widely used in optogenetics due to its light-induced neural-silencing ability. Here, we summarize the functional analyses of halorhodopsin since its discovery. Like other microbial rhodopsins, halorhodopsin contains all-*trans* retinal bound to a specific lysine residue through a protonated Schiff base. Proton-pumping rhodopsins utilize Asp residues as the counter-ions for the protonated Schiff bases. In halorhodopsin, this Asp residue is replaced by Thr, and Cl^- becomes the counter-ion. Photoexcited halorhodopsin undergoes a photocycle including several intermediates where sequential Cl^- movements occur. During the formation of the N-intermediate, Cl^- moves from its original position to the cytoplasmic channel. During the subsequent N decay, the Cl^- is released to the cytoplasmic medium. During the Cl^- release, the dissociation constant of Cl^- increases significantly compared with that at the dark state. Next, another Cl^- is captured from the extracellular medium to complete the net Cl^- translocation. This recapture process is not well defined.

Keywords Halorhodopsin • Bacteriorhodopsin • Photocycle • Ion pump • Chloride pump • Flash photolysis • Microbial rhodopsin

4.1 Discovery of Halorhodopsin

D. Oesterhelt and W. Stoeckenius discovered a purple-colored protein in the membrane of a highly halophilic archaeon called *Halobacterium salinarum* (formerly *halobium*) and revealed that it functions as an ion pump to move protons from the inside to the outside of the cell (Oesterhelt and Stoeckenius 1971, 1973). This purple protein was named bacteriorhodopsin (BR) through an analogy with the visual

T. Kikukawa • N. Kamo • M. Demura (✉)
Faculty of Advanced Life Science, Hokkaido University, Sapporo 060-0810, Japan
e-mail: kikukawa@sci.hokudai.ac.jp; nkamo@pharm.hokudai.ac.jp; demura@sci.hokudai.ac.jp

H. Yawo et al. (eds.), *Optogenetics*, DOI 10.1007/978-4-431-55516-2_4

pigment rhodopsin because it contains retinal bound to a specific lysine residue through a Schiff base linkage. In the cell membrane, BR aggregates form crystalline patches known as the purple membrane; these regions contain only BR as a protein. Therefore, BR is a proton pump that exists without contacting energy-transducing proteins. This discovery inspired Racker and Stoeckenius to utilize BR to test P. Michell's chemiosmotic hypothesis for energy coupling (Racker and Stoeckenius 1974), consequently performing a key experiment. The BR was easily isolated from the cell membrane and was extremely stable. In addition to these experimental advantages, the tertiary structure was highly similar to that of the G-protein coupled receptor. Subsequently, extensive studies have been conducted using BR as a model system for active transport and other biophysical aspects of membrane proteins.

H. salinarum is colored by the presence of BR and bacterioruberin (Brub). After plating, colonies with altered colors often occur through the spontaneous variability in the synthesis of these pigments. This high frequency of variability (~10^{-4}) originates from the transposable elements (Pfeifer et al. 1981). In 1977, Matsuno-Yagi and Mukohata isolated a red strain (later named R_1mR) from an *H. salinarum*-type strain R_1 and observed an increase in the pH of the cell suspension under illumination (Matsuno-Yagi and Mukohata 1977). Due to the secondary coupled processes, the original R_1 strain exhibits complicated pH changes while including the pH-decreasing phase, reflecting the BR activity. The pH increase with R_1mR disappeared after 5 min of incubation at 75 °C, but persisted after treatment with NH_2OH. However, BR is heat resistant but is bleached by NH_2OH treatment, eliminating its proton-pumping activity. R_1mR cells grown in the presence of nicotine, which is a retinal synthesis inhibitor, lost the ability for light-induced pH changes; however, this ability was restored after adding all-*trans* retinal. The action spectrum for the pH increase revealed a maximum at wavelengths longer than the absorption spectrum of BR. Therefore, the R_1 strain contains another retinal protein. Later, Mukohata et al. named this novel retinal protein halorhodopsin (HR) (Matsuno-Yagi and Mukohata 1980; Mukohata and Kaji 1981). As described below, HR pumps halogen ions. Therefore, 'halorhodopsin' correctly describes its function. However, 'halo' was probably named after '*Halobacterium*' because the anion-pumping function had not yet been revealed.

The next research focus involved the function of this new rhodopsin and why the pH increased in the extracellular (EC) medium with R_1mR. Lindley and MacDonald first reported results that addressed these problems (Lindley and MacDonald 1979). These researchers showed the following using R_1mR membrane vesicles: (1) illumination generated an interior negative membrane potential, (2) this potential was not eliminated by the uncouplers, (3) the coexistence of valinomycin and potassium ions eliminated the potential and consequently depressed the pH increase. These results indicated that the light-induced pH increase was passively driven by the membrane potential generated by the transport of a specific ion. This first transported ion was studied; sodium ion extrusion was detected, suggesting that HR was an outwardly directed sodium ion pump. This finding seemed physiologically relevant: *H. salinarum* maintains the cellular sodium concentration at a level lower than growth medium containing 4~4.3 M NaCl. However, previous research suggested that the sodium extrusion mainly occurred via an Na^+/H^+ antiporter (Luisi et al. 1980). The ions transported by HR create the membrane potential, which inhibits further ion transport. Therefore, HR must combine with another system to eliminate the membrane potential for effective ion extrusion.

By utilizing the high degree of spontaneous variation in *H. salinarum*, several mutant strains were isolated and used for various studies (Ogurusu et al. 1981; Weber and Bogomolni 1981; Tsuda et al. 1982; Kamo et al. 1983). By using strain L33, which contains a large amount of HR, Schobert and Lanyi proved that HR acts as inwardly directed chloride ion pump (Schobert and Lanyi 1982). Their observations could be summarized as follows: the absence of chloride ions inhibited the light responses, including the interior negative membrane potential and the passive proton uptake. The amino acid sequence of HR from *H. salinarum* (HsHR) was determined by Oesterhelt's group (Steiner and Oesterhelt 1983; Blanck and Oesterhelt 1987). As described below, the two Asp residues in BR, which are the key residues for proton transport, are replaced by Thr and Ala in HR, respectively.

In addition to the original HsHR, the HR from haloalkaliphilic archaeon *Natronomonas* (formerly *Natronobacterium*) *pharaonis* (NpHR) has also been utilized for studies on HR. For NpHR, a large amount of this protein was functionally expressed in the cell membrane of *Escherichia coli* (Hohenfeld et al. 1999; Sato et al. 2002). The functional expression of both HsHR and NpHR in halobacterial cell membranes was also reported (Otomo and Muramatsu 1995; Váró et al. 1995a). An NpHR-overproducing *N. pharaonis* mutant strain was also isolated (Ihara et al. 2008). HR studies are currently conducted using these expression systems and mutant strains.

The physiological significance of HR remains under debate. HR was proposed to be a driver for adenosine triphosphate (ATP) synthesis. However, this function is questionable because HR cannot continuously generate the proton-motive force needed to drive the H^+-ATPase. By coexisting with BR, HR may contribute to ATP synthesis. To understand the physiological role of HR, the destination of the Cl^- after transport should be clarified. Haloalkaliphile *N. pharaonis* contains only NpHR as an ion-pumping rhodopsin. The passive proton uptake induced by NpHR may help maintain neutral pH of the cell interior despite the alkaline environment.

4.2 Photochemical Reactions of Bacteriorhodopsin

Studies of microbial rhodopsins involving HR have been conducted based on the findings with BR (Haupts et al. 1999; Balashov 2000; Heberle 2000; Lanyi 2004, 2006). Therefore, the proton-pumping mechanism of BR will be overviewed to clarify the halide-pumping mechanism of HR. After illumination, the microbial rhodopsins undergo cyclic reactions called photocycles, which are the thermal relaxation processes of the photoexcited states that occur through various photochemical

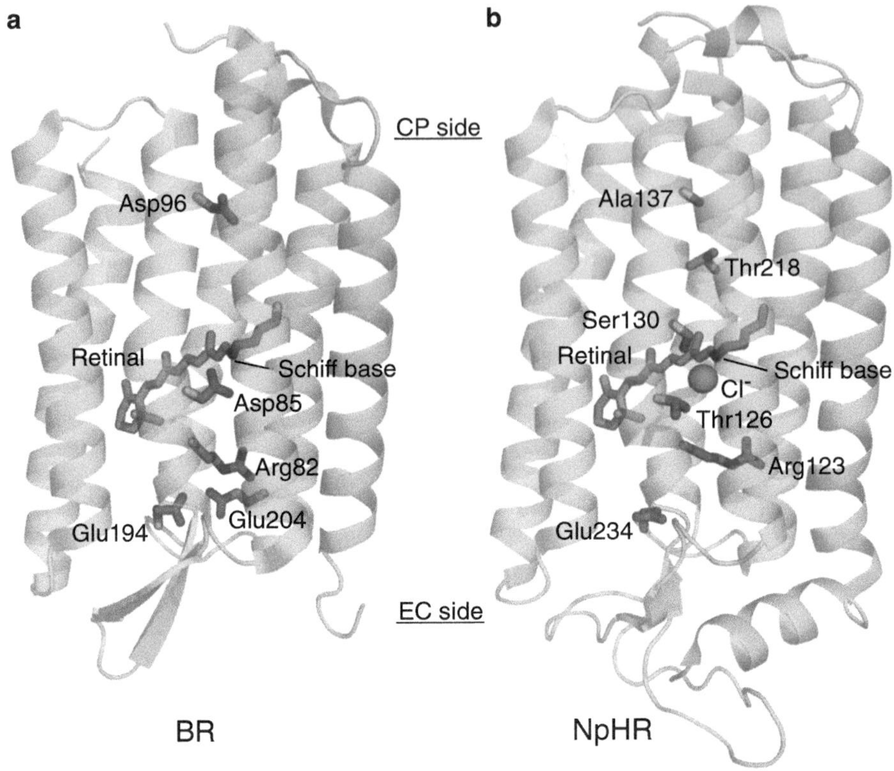

Fig. 4.1 Tertiary structures of bacteriorhodopsin (BR) and the halorhodopsin from *Natronomonas pharaonis* (NpHR). The cytoplasmic (CP) sides are at the top and the extracellular (EC) sides are at the bottom. The Protein Data Bank codes are 1C3W for BR and 3A7K for NpHR

intermediates. During these photocycles, BR pumps one proton outward, while HR pumps one chloride ion inward. Figure 4.1 illustrates their tertiary structures, showing the residues that are important for ion pumping, and Fig. 4.2 shows the photocycles. Under continuous illumination, these photocycles are triggered repeatedly.

In all microbial rhodopsins, retinal aldehydes are bound through Schiff base linkages with the lysine residues in the seventh transmembrane helix. The Schiff bases are ordinary protonated due to their high pK_a values (~11). For BR, the isomeric composition of retinal depends on the light/dark conditions. This behavior is called light-dark adaptation. The ratio of all-*trans* to 13-*cis* is approximately 2:1 in the dark, while, during illumination, the all-*trans* isomer is dominant. Only the BR with all-*trans* retinal undergoes the proton-pumping photocycle, which is triggered by the photoisomerization of retinal from all-*trans* to the 13-*cis* configuration. The photocycle includes the K, L, M, N, and O intermediates. These intermediates have absorption spectra that reflect their differences in retinal configurations, protein conformations, and protonation states of some residues. The photoisomerization completes at the K intermediate, but this reaction is so fast that the protein moiety

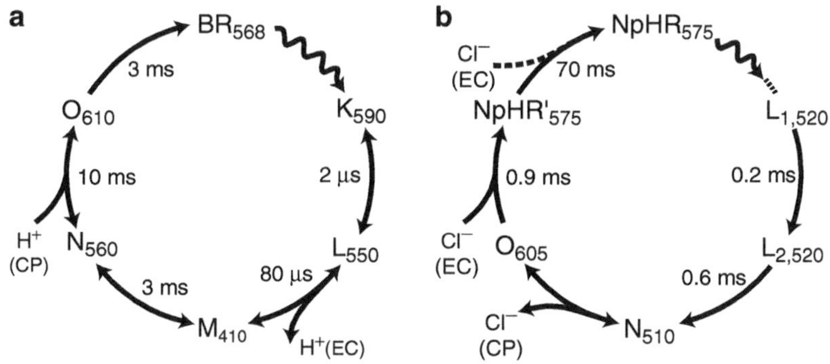

Fig. 4.2 Photocycles of bacteriorhodopsin (BR) and halorhodopsin from *Natronomonas pharaonis* (NpHR). The λ_{max} values of the dark state and the photo-intermediates are denoted by the three-digit numbers in their subscripts. The lifetimes of the intermediates are also shown. For NpHR, our model (Scheme 4.3) is shown with the lifetimes determined for NpHR in the native membrane and in the presence of 1 M NaCl (Kikukawa et al. 2012). The lifetime of N is not denoted because N forms a quasi-equilibrium with O and they decay together. For the Cl^- recapture from the extracellular (EC) medium, two possible sequences are illustrated. For details, see the text

does not undergo a significant conformational change. Afterwards, the conformational change induces the change in pK_a for the Schiff base and its counter-ion, which is Asp85. Subsequently, the deprotonation of the Schiff base occurs during the L–M transition. The proton is transferred to an 'acceptor' Asp85, and the subsequent proton release toward the EC medium occurs from the proton-releasing complex (PRC) consisting of Arg82, Glu194, Glu204, and water molecules. M has at least two sub-states, which are denoted M_1 and M_2. During the M_1–M_2 transition, the Schiff base nitrogen changes its orientation from the EC to the cytoplasmic (CP) side. During the subsequent M–N transition, the deprotonated Schiff base accepts a proton from the 'donor' Asp96, which is located at the CP half channel. During this reaction, the pK_a of Asp85 increases to avoid deprotonation and the consequent proton transfer to the Schiff base. Next, the deprotonated Asp96 captures proton from the CP medium during the N–O transition, which is accompanied the reisomerization of retinal to the all-*trans* configuration. During the recovery from O to the original unphotolyzed state, the final proton transfer occurs from the protonated Asp85 to the deprotonated PRC. Through this series of reactions, one proton is transferred from the CP to the EC side. The molecular aspects of the unphotolyzed BR and the intermediates are summarized in Table 4.1 for comparison with HR.

4.3 Molecular Mechanism of Halorhodopsin

As mentioned above, two HRs, specifically HsHR and NpHR, have been the most commonly used for HR studies. Therefore, the experimental results for the two HRs are described.

Table 4.1 Key features of the unphotolyzed state and photo-intermediates of bacteriorhodopsin (BR)

	BR	L	M	N	O
Retinal configuration	all-*trans*	13-*cis*	13-*cis*	13-*cis*	all-*trans*
Schiff base	+	+	0	+	+
Acceptor (Asp85)	–	–	0	0	0
Donor (Asp96)	0	0	0	–	0

'+', '–', and '0' denote the electric charges of three parts reflecting their protonation states, where '0' is an electrically neutral state

4.3.1 Cl^- Binding to the Dark State

The tertiary structure of NpHR is shown in Fig. 4.1b with the important residues. The significant difference between BR and NpHR is the replacement of Asp85 by Thr126; in HsHR, Asp85 is replaced with Thr111. As described above, the Asp85 in BR is the counter-ion for the protonated Schiff base and acts as the proton acceptor for the Schiff base during the first half of the photocycle. However, HR binds Cl^- near the Thr residue that acts as the counter-ion for the protonated Schiff base (Kolbe et al. 2000; Kouyama et al. 2010). During the photocycle, HR should release this Cl^- to the CP medium. After the Cl^- binds to the dark states, HsHR undergoes a rather small red shift in the visible absorption spectrum (λ_{max}, 570→580 nm) (Ogurusu et al. 1982; Lanyi 1986; Váró et al. 1995c; Yamashita et al. 2011), while NpHR undergoes a large blue shift (λ_{max}, 600→580 nm) (Váró et al. 1995a; Scharf and Engelhard 1994; Sato et al. 2005). The dissociation constant of Cl^- estimated from these spectral shifts ranges from 1 to ~20 mM. In physiological halophilic environments, the binding of Cl^- to the dark state is almost fully saturated. The mutation experiments revealed that, in addition to the Thr residue, Ser130 and Arg123 in NpHR (Ser115 and Arg108 in HsHR) are important for the binding of Cl^- to the dark state (Rüdiger et al. 1995; Rüdiger and Oesterhelt 1997; Sato et al. 2003a, b, 2005).

4.3.2 Photocycle

Although the photocycle of HsHR remained unclear and contradictory, Lanyi's group conducted careful experiments and assembled the following scheme (Váró et al. 1995c):

$$HsHR_{586} \Rightarrow K_{598} \leftrightarrow L_{1,529} \leftrightarrow L_{2,529} \leftrightarrow N_{590} \rightarrow HsHR \qquad \text{(Scheme 4.1)}$$

where the subscripted three-digit numbers are the λ_{max} of the intermediates and are the readouts from their figures. The double-headed arrows indicate the existence of backward reactions. HsHR shows a light-dark adaptation and can accommodate

13-*cis* retinal (Kamo et al. 1985). The HsHR with 13-*cis* retinal and the HsHR without Cl^- binding undergo similar photocycles with intermediates only at long wavelengths (Váró et al. 1995c).

NpHR does not exhibit a light-dark adaptation: the all-*trans* retinal dominates and remains unchanged by the light/dark conditions (Váró et al. 1995a). For the NpHR photocycle, three groups including Lanyi (Váró et al. 1995a, b; Ludmann et al. 2000), Engelhard (Chizhov and Engelhard 2001), and ourselves (Sato et al. 2005; Hasegawa et al. 2007) proposed schemes based primarily on the flash-induced changes in absorbance. These schemes are similar, but the details differ. Lanyi's group proposed the following scheme:

$$\mathrm{NpHR}_{580} \Rightarrow \mathrm{K}_{585} \leftrightarrow \mathrm{L}_{1,530} \leftrightarrow \mathrm{L}_{2,530} \leftrightarrow \mathrm{N}_{580} \leftrightarrow \mathrm{O}_{615} \leftrightarrow \mathrm{NpHR}'_{580} \rightarrow \mathrm{NpHR} \qquad \text{(Scheme 4.2)}$$

Our scheme is as follows:

$$\mathrm{NpHR}_{575} \Rightarrow (\mathrm{K}) \rightarrow \mathrm{L}_{1,520} \rightarrow \mathrm{L}_{2,520} \rightarrow \mathrm{N}_{510} \leftrightarrow \mathrm{O}_{605} \rightarrow \mathrm{NpHR}'_{575} \rightarrow \mathrm{NpHR} \qquad \text{(Scheme 4.3)}$$

where (K) means that K should appear but is not detected by our apparatus due to its rapid decay. The data analyses by Engelhard and ourselves are based on a sequential irreversible model that assumes the intermediates appear sequentially and that their transitions are irreversible (Chizhov et al. 1996). Here, the 'intermediate' is a kinetically distinguishable intermediate, which might include physically distinguishable intermediates if a quasi-equilibrium exists between them. The amount of intermediates, as well as their decay constants and absorption spectra have been determined from the global fitting results obtained using the measured wavelengths. According to the λ_{max} values of the absorption spectra, the names of the intermediates were determined based on those of the intermediates in BR. In our scheme, the N–O transition is described using a double-headed arrow, indicating that N and O appear almost simultaneously in a quasi-equilibrium state.

As seen in Schemes 4.1, 4.2, and 4.3, the HR photocycles do not include short wavelength intermediates, such as M, observed with BR. As described above, Asp85 of BR is replaced by Thr. HR lacks the proton-accepting residue and undergoes the photocycle without the deprotonation of the Schiff base (however, our unpublished results reveal that the protonated Schiff base decreases the light-induced pK_a values, similar to BR in certain cases). Moreover, HR does not require a proton donor for the deprotonated Schiff base. The Asp96 in BR is replaced by Ala in HR (Ala137 in NpHR, Ala122 in HsHR).

4.3.3 Cl^- Release and Recapture

Váró et al. showed that, for NpHR, the transition rates for $\mathrm{N} \leftarrow \mathrm{O}$ and $\mathrm{O} \rightarrow \mathrm{NpHR}'$ increase with Cl^- concentration (Váró et al. 1995b). Based on these observations, during the formation of O, Cl^- is released to the CP medium, while during the formation of NpHR′, another Cl^- is captured from the EC medium. These steps are credible based on the absorption spectra of O and NpHR′. O has a red-shifted absorption spectrum, which is consistent with the spectral redshift for the unphotolyzed NpHR after the dissociation of Cl^-. The spectrum for NpHR′ closely resembles that of the original unphotolyzed NpHR, implying that at NpHR′, a new Cl^- is already bound near the Schiff base. The release of Cl^- during the N–O transition of NpHR was also reported based on other experiments. The transient grating experiment detected a Cl^- diffusion signal simultaneously with the appearance of O (Inoue et al. 2009). Furthermore, Stopped-flow (Sato et al. 2002) and Fourier transform infrared (FTIR) spectroscopy experiments (Guijarro et al. 2006) suggested that O and the anion-free unphotolyzed state were structurally similar. The measured electrogenicity of the light-activated NpHR revealed large electrogenicity during the N–O transition (Ludmann et al. 2000). We recently analyzed the Cl^- release using *N. pharaonis* strain KM-1, which is an NpHR-overproducing mutant (Kikukawa et al. 2012). In its cell membrane, the NpHR forms a 3:3 complex with the carotenoid Brub (Kouyama et al. 2010). The carotenoids change their absorption spectra based on the surrounding electrostatic environment. Indeed, the Brub associated with NpHR induces spectral change when Cl^- binds to the unphotolyzed NpHR. An analysis of the flash-induced absorbance changes revealed that the spectral change of Brub, which reflect the release of Cl^- from NpHR, occurs during O formation. As described above, the subsequent O–NpHR′ transition was attributed to the Cl^- recapture process (Váró et al. 1995b). However, the details are not fully resolved, as described below. In Fig. 4.2b, Scheme 4.3 is illustrated with the timing for Cl^- release and recapture. For the Cl^- recapture, two possible timings are shown.

For HsHR, the transition rates of the intermediates do not depend on the Cl^- concentration (Váró et al. 1995c). Muneyuki et al. observed transient electrogenic production during the decay of N in HsHR (Muneyuki et al. 1999). As shown in Scheme 4.1, the HsHR photocycle lacks intermediates corresponding to O and NpHR′ in the NpHR photocycle. For HsHR, the decay of these two intermediates might be so fast that their formation could not be detected. HsHR might undergo both Cl^- release and uptake during the decay of N due to its kinetic differences from NpHR.

4.3.4 Cl^- Dependent N-O Equilibrium in NpHR Photocycle

During the photocycle of NpHR, N and O appear almost simultaneously to form a quasi-equilibrium state. Analyses by Engelhard's group and ourselves determined the absorption spectrum of this equilibrium state (Chizhov and Engelhard 2001; Sato et al. 2005; Hasegawa et al. 2007; Shibasaki et al. 2013). As the Cl^- concentration increases, the equilibrium shifts toward N, and the amount of O in the equilibrium becomes small. By decomposing the absorption spectra, the molar fractions of N and O could be determined. As mentioned above, Cl^- release occurs during the N–O transition. Therefore, the dissociation constant for Cl^- release ($K_{d,NO}$) could be determined from the Cl^- dependence of the molar fractions. $K_{d,NO}$ is approximately 1 M in the solubilized state in n-dodecyl-β-D-maltoside (DDM) (Hasegawa et al. 2007; Shibasaki et al. 2013) and approximately 4 M in the native *N. pharaonis* membrane (Kikukawa et al. 2012). The dissociation constant during the dark state ($K_{d,dark}$) is approximately 1–5 mM in the DDM-solubilized state (Scharf and Engelhard 1994; Chizhov and Engelhard 2001; Sato et al. 2005), and approximately 20 mM in the native membrane (Kikukawa et al. 2012). To release Cl^- on the CP side, the photo-excited NpHR causes at least a 200-fold increase in the dissociation constant. The crystal structures revealed that, during the dark state, the CP channel is narrow and hydrophobic. The translocation of Cl^- through this channel seems to require a conformational change and resultant hydration of the channel. Based on time-resolved FTIR measurements, Hackmann et al. detected changes in amide I alterations during O formation, suggesting a conformational change (Hackmann et al. 2001). From the hydrostatic pressure dependence of $K_{d,NO}$, we detected a volume increase (ΔV) of approximately 24 mL/mol during the N–O transition (Shibasaki et al. 2013). This change might reflect the water entry at O, as observed during the M–N transition in BR (Váró and Lanyi 1995; Sass et al. 2000; Schobert et al. 2003). Thr218 is one of a few hydrophilic residues in the CP channel and might assist during Cl^- translocation (Rüdiger and Oesterhelt 1997). We generated various mutants of Thr218 and examined the relationships between $K_{d,NO}$, ΔV and the pumping activity (Shibasaki et al. 2013). $K_{d,NO}$ and ΔV were changed significantly by the mutation. However, both values were the largest for the wild type. For some mutants, large ΔV values were found, while the $K_{d,NO}$ values were small. Therefore, the volume increase, which was probably due to the water entry, is not sufficient for generating a large $K_{d,NO}$ value. To increase this value, a water molecule might have to interact with the OH group at position 218. The Cl^- pumping activities of the mutants indicated that a large $K_{d,NO}$ is necessary to avoid decreasing the Cl^- pumping activity, even in the presence of an interior negative membrane potential. For the HsHR photocycle, O is not detected as described above. This difference might be caused by the substantial shift of the N–O equilibrium toward N due to the low $K_{d,NO}$ value. The residue corresponding to Thr218 is Thr203 in HsHR. Therefore, this Thr might not be the only factor that determines $K_{d,NO}$. Indeed, we found that disrupting the NpHR trimer structure significantly decreases $K_{d,NO}$, implying that this oligomeric structure contributes to the proper conformation change during the photolyzed state (Tsukamoto et al. 2012). For effective ion pumping, a low $K_{d,NO}$ is not preferable, especially in the presence of an interior negative membrane potential. HsHR might accommodate a mechanism that overcomes the issues caused by a low $K_{d,NO}$.

4.3.5 Cl^- Translocation Across the Schiff Base Region

In the unphotolyzed state, Cl^- is located near the Schiff base in the EC channel. Therefore, Cl^- is translocated from EC to CP across the Schiff base. The X-ray structure of L_1 in HsHR revealed that Cl^- remained close to its original position (Gmelin et al. 2007). A low-temperature FTIR spectrum of HsHR showed that

Cl^- remained in the EC channel close to the Schiff base at both L_1 and L_2 (Chon et al. 1999). For both HsHR and NpHR, Muneyuki et al. detected electrogenicities, probably due to the Cl^- movements during the formation and decay of N (Muneyuki et al. 1999). Similar results were reported by Ludmann et al. for NpHR (Ludmann et al. 2000). They observed large electrogenicities during the L_2–N and N–O transitions. As described above, the N–O transition of NpHR is the Cl^- release to the CP medium. Therefore, the electrogenicity production during the decay of N is conceivable. Another electrogenicity observed during the formation of N was attributed to the Cl^- translocation from EC to CP across the Schiff base. The same conclusion was obtained from studies using an NpHR mutant to bind with the halobacterial transducer II from *N. pharaonis* (NpHtrII) (Hasegawa et al. 2007). Sensory rhodopsins I and II show proton-pumping activities similar to those of BR. However, these activities disappear upon the complex formations with their respective halobacterial transducers, HtrI and HtrII (Bogomolni et al. 1994; Sudo et al. 2001). The complex formation inhibits proton translocation in the CP channels; therefore, protons should only circulate through the EC channel. Similarly, NpHR lost its Cl^- pumping activity through the formation of a complex with *Np*HtrII (Hasegawa et al. 2007). During the photocycle of the complex, O appeared after L_2 without N formation. During the formation and decay of O, Cl^- was expected to circulate through the EC channel, resulting in the loss of Cl^- pumping activity. Afterwards, no N formed, suggesting that Cl^- translocated from EC to CP during N formation.

What is the mechanism driving the movement of Cl^- from EC to CP? In the unphotolyzed HR, Cl^- is surrounded by several water molecules (Kolbe et al. 2000; Kouyama et al. 2010). Low-temperature FTIR experiments by Shibata et al. revealed that the hydrogen bonds between the Schiff base and water molecules strengthen during the formation of L_1 (Shibata et al. 2005). Because L_1 contains 13-*cis* retinal, the change in the hydrogen bond strength implies that the water molecule was displaced from its original position. Therefore, the resultant hydrophobic environment around Cl^- should induce the translocation of Cl^- toward CP. At L_2 in NpHR, Shibata et al. observed the deprotonation of Glu234, which is near the EC surface (Shibata et al. 2006). The resultant negative charge at this position should prevent the release of Cl^- to the EC medium. Guijarro et al. reported that the deprotonated Glu234 reprotonates during Cl^- capture (Guijarro et al. 2006). For BR, the M_1–M_2 (or M_2') transition accompanies the change in orientation for the Schiff base nitrogen from the EC to CP side, as described above. For HR, a similar change in the orientation of the Schiff base NH^+ might occur during the L_1–L_2 transition (Oesterhelt 1995). This change might induce the translocation of Cl^- across the Schiff base.

The contribution of water molecules to the translocation of Cl^- was also obtained from electrophysiological experiments using *Xenopus laevis* oocytes (Seki et al. 2007). When NpHR is expressed in the oocyte membrane, the ion-pumping activity can be measured as the photo-induced outward current. The photo-induced current showed Michaelis-Menten–like saturation kinetics against the concentrations of Cl^- and other transportable anions. The Michaelis constants (K_m), which reflect the binding affinities of the anions, occurred in the following order: $Br^- < I^- < Cl^- < SCN^- < NO_3^-$. This order seems to correlate with the size of the hydrated anions. However, the maximum currents (I_{max}) occurred in the following order: $Cl^- > Br^- > I^- > NO_3^- > SCN^-$. This trend seems related to the size of the dehydrated anions. Thus, the anions appear to bind in the hydrated states but to be transported in the dehydrated states.

4.3.6 Cl^- Recapture

As described above, the recapture of Cl^- should occur during the O–NpHR′ transition for NpHR and the N to original state transition for HsHR. However, the measured NpHR-induced electrogenicity did not indicate any Cl^- movement during the O–NpHR′ and subsequent NpHR′–NpHR transitions (Ludmann et al. 2000; Muneyuki et al. 1999). This behavior might be caused by the large dielectric constant of the EC channel, which consists of many hydrophilic residues and water molecules. Therefore, the Cl^- recapture process has not been fully elucidated.

Mevorat-Kaplan et al. suggested that, in the dark state, a Cl^- binding site is located near the EC surface with a dissociation constant of ~ 100 mM (Mevorat-Kaplan et al. 2006). A recently resolved X-ray crystal structure also revealed a weak Cl^- binding site on the EC surface (Kanada et al. 2011). By assuming that this Cl^- binding site contributes, the Cl^- recapturing process could be explained using the following two-step Cl^- movements. During the O–NpHR′ transition, the Cl^- first moves from the binding site near the EC surface to the primary binding site near the Schiff base. This movement should recover the absorption spectrum, leaving it close to the original dark state. Second, the vacant site near the EC surface might capture the Cl^- from the EC medium during the NpHR′–NpHR transition. Additional evidence was obtained using the O–NpHR′ transition rate, $k_{O \to NpHR'}$ (Shibasaki et al. 2013). If the Cl^- comes from the EC medium, this transition rate should be proportional to the concentrations of O ([O]) and Cl^- in the medium ($[Cl^-]$):

$$k_{O \to NpHR'} = k_0 \times [O] \times [Cl^-] \tag{4.1}$$

where k_0 is a constant. As described above, N and O form a quasi-equilibrium that depends on the Cl^- concentration. By using the dissociation constant $K_{d,NO}$, [O] is described by $K_{d,NO} / (K_{d,NO} + [Cl^-])$, and Eq. 4.1 is rewritten as follows:

$$k_{O \to NpHR'} = k_0 \times \frac{K_{d,NO}}{K_{d,NO} + [Cl^-]} \times [Cl^-] \tag{4.2}$$

$k_{O \to NpHR'}$ should show Michaelis-Menten–like kinetics against the Cl^- concentration. However, $k_{O \to NpHR'}$ did not follow this equation (Shibasaki et al. 2013). At lower Cl^- concentrations ($[Cl^-] < K_{d,NO}$), $k_{O \to NpHR'}$ followed the equation, but $k_{O \to NpHR'}$ began to decrease when increasing the Cl^- concentration further. This convex change in $k_{O \to NpHR'}$ could be explained by assuming that the Cl^- does not come directly from the EC medium; instead, it might come from the Cl^- binding site near the EC surface. Under this assumption, the [Cl⁻] in Eq. 4.1 is replaced by $\left[\mathrm{Cl}^-\right]/\left(K_{\mathrm{d,EC}}+\left[\mathrm{Cl}^-\right]\right)$, where $K_{\mathrm{d,EC}}$ is the dissociation constant of the site at the EC surface. Therefore, Eq. 4.2 could be modified as follows:

$$k_{\mathrm{O\to NpHR'}} = k_0 \times \frac{K_{\mathrm{d,NO}}}{K_{\mathrm{d,NO}}+\left[\mathrm{Cl}^-\right]} \times \frac{\left[\mathrm{Cl}^-\right]}{K_{\mathrm{d,EC}}+\left[\mathrm{Cl}^-\right]} \tag{4.3}$$

The third term in Eq. 4.3 exhibits Michaelis-Menten behavior, while the second term decreases above Cl^- concentrations near $K_{\mathrm{d,NO}}$. When $K_{\mathrm{d,EC}} < K_{\mathrm{d,NO}}$, $k_{\mathrm{O\to NpHR'}}$ shows a convex curve as obtained by the measurements (Shibasaki et al. 2013). This Cl^- recapture process is illustrated in Fig. 4.2b using a broken line, meaning that the Cl^- was uptaken by the site near the EC surface. At lower [Cl^-], the Cl^- binding site on the EC surface is most likely empty in the dark state. Through photoexcitation, this NpHR becomes O, which does not bind Cl^- at the site on the EC surface. This O has a chance to bind a Cl^-, although the probability of this depends on $K_{\mathrm{d,EC}}$ and [Cl^-]. After this Cl^- binding occurs, O transforms into NpHR′ through the movement of Cl^- to the primary binding site. During the subsequent NpHR′–NpHR transition, most of the NpHR′ becomes NpHR without Cl^- binding to the second site.

4.4 Conclusion and Perspectives

The molecular aspects of the unphotolyzed state and intermediates of NpHR are summarized in Table 4.2. As described above, HR has at least two Cl^- binding sites: the original site near the Schiff base, which is the 'primary binding site', and the transient binding site in the CP channel, which is the 'CP binding site'. Their charge states are shown in Table 4.2 as the counterparts of the acceptor (Asp85) and donor (Asp96) residues of BR in Table 4.1. These tables reveal the similarities between the photocycles of BR and HR. The presence of negative charge (Cl^-) in the CP binding site at N is similar to the case of N of BR: the deprotonated Asp96 (donor) is located in CP channel. The similarities in key features of the photocycles imply that BR and HR share a common transport mechanism. Reports indicate that BR became a

Table 4.2 Key features of the unphotolyzed state and photo-intermediates of halorhodopsin from *Natronomonas pharaonis* (NpHR)

	NpHR	L_1	L_2	N	O	NpHR′
Retinal configuration	all-*trans*	13-*cis*	13-*cis*	13-*cis*	all-*trans*	all-*trans*
Schiff base	+	+	+	+	+	+
Primary binding site	–	–	–	0	0	–
CP binding site	0	0	0	–	0	0

'+', '–', and '0' denote the electrical charges as shown in Table 4.1. Here, '–' indicates the binding of Cl^-.

chloride pump after Asp85 was replaced with Thr or Ser (Sasaki et al. 1995). However, HR was not converted to BR, even after mutations at ten positions (Muroda et al. 2012). Therefore, additional studies are needed to understand the mechanism that differentiates the function of this protein.

For NpHR, NpHR′ appears at the end of the photocycle. Currently, experimental data are rare for this intermediate. The similarities of the absorption spectra suggest that NpHR′ might be a precursor for the original dark state. However, the NpHR′–NpHR transition rate is slowest and therefore rate-limiting process of the photocycle. This slow transition might reflect the Cl^- capture by the Cl^- binding site near the EC surface, as described above. The details of NpHR′ and its functional role must still be clarified. The acceleration of NpHR′ decay might contribute to optogenetics by enabling NpHR to induce a rapid change in the membrane potential.

References

Balashov SP (2000) Protonation reactions and their coupling in bacteriorhodopsin. Biochim Biophys Acta 1460(1):75–94

Blanck A, Oesterhelt D (1987) The halo-opsin gene. II. Sequence, primary structure of halorhodopsin and comparison with bacteriorhodopsin. EMBO J 6(1):265–273

Bogomolni RA, Stoeckenius W, Szundi I et al (1994) Removal of transducer HtrI allows electrogenic proton translocation by sensory rhodopsin I. Proc Natl Acad Sci USA 91(21):10188–10192

Chizhov I, Engelhard M (2001) Temperature and halide dependence of the photocycle of halorhodopsin from *Natronobacterium pharaonis*. Biophys J 81(3):1600–1612

Chizhov I, Chernavskii DS, Engelhard M et al (1996) Spectrally silent transitions in the bacteriorhodopsin photocycle. Biophys J 71(5):2329–2345

Chon Y-S, Kandori H, Sasaki J et al (1999) Existence of two L photointermediates of halorhodopsin from *Halobacterium salinarum*, differing in their protein and water FTIR bands. Biochemistry 38(29):9449–9455

Gmelin W, Zeth K, Efremov R et al (2007) The crystal structure of the L1 intermediate of halorhodopsin at 1.9 Å resolution. Photochem Photobiol 83(2):369–377

Guijarro J, Engelhard M, Siebert F (2006) Anion uptake in halorhodopsin from *Natronomonas pharaonis* studied by FTIR spectroscopy: consequences for the anion transport mechanism. Biochemistry 45(38):11578–11588

Hackmann C, Guijarro J, Chizhov I et al (2001) Static and time-resolved step-scan Fourier transform infrared investigations of the photoreaction of halorhodopsin from *Natronobacterium pharaonis*: consequences for models of the anion translocation mechanism. Biophys J 81(1):394–406

Hasegawa C, Kikukawa T, Miyauchi S et al (2007) Interaction of the halobacterial transducer to a halorhodopsin mutant engineered so as to bind the transducer: Cl^- circulation within the extracellular channel. Photochem Photobiol 83(2):293–302

Haupts U, Tittor J, Oesterhelt D (1999) Closing in on bacteriorhodopsin: progress in understanding the molecule. Annu Rev Biophys Biomol Struct 28:367–399

Heberle J (2000) Proton transfer reactions across bacteriorhodopsin and along the membrane. Biochim Biophys Acta 1458(1):135–147

Hohenfeld IP, Wegener AA, Engelhard M (1999) Purification of histidine tagged bacteriorhodopsin, *pharaonis* halorhodopsin and *pharaonis* sensory rhodopsin II functionally expressed in *Escherichia coli*. FEBS Lett 442(2–3):198–202

Ihara K, Narusawa A, Maruyama K et al (2008) A halorhodopsin-overproducing mutant isolated from an extremely haloalkaliphilic archaeon *Natronomonas pharaonis*. FEBS Lett 582(19):2931–2936

Inoue K, Kubo M, Demura M et al (2009) Reaction dynamics of halorhodopsin studied by time-resolved diffusion. Biophys J 96(9):3724–3734
Kamo N, Takeuchi M, Hazemoto N et al (1983) Light-induced ΔpH of envelope vesicles containing halorhodopsin measured by use of a spin probe. Arch Biochem Biophys 221(2):514–525
Kamo N, Hazemoto N, Kobatake Y et al (1985) Light and dark adaptation of halorhodopsin. Arch Biochem Biophys 238(1):90–96
Kanada S, Takeguchi Y, Murakami M et al (2011) Crystal structures of an O-like blue form and an anion-free yellow form of *pharaonis* halorhodopsin. J Mol Biol 413(1):162–176
Kikukawa T, Kokubo A, Tsukamoto T et al (2012) When does halorhodopsin capture and release Cl^- during photocycle? Examination by spectrum changes of an intrinsic voltage-sensitive dye, bacterioruberin. In 15th international conference on retinal proteins, Ascona, 30 Sept–5 Oct 2012
Kolbe M, Besir H, Essen LO et al (2000) Structure of the light-driven chloride pump halorhodopsin at 1.8 Å resolution. Science 288(5470):1390–1396
Kouyama T, Kanada S, Takeguchi Y et al (2010) Crystal structure of the light-driven chloride pump halorhodopsin from *Natronomonas pharaonis*. J Mol Biol 396(3):564–579
Lanyi JK (1986) Photochromism of halorhodopsin. *cis/trans* isomerization of the retinal around the 13–14 double bond. J Biol Chem 261(30):14025–14030
Lanyi JK (2004) Bacteriorhodopsin. Annu Rev Physiol 66:665–688
Lanyi JK (2006) Proton transfers in the bacteriorhodopsin photocycle. Biochim Biophys Acta 1757(8):1012–1018
Lindley EV, MacDonald RE (1979) A second mechanism for sodium extrusion in *Halobacterium halobium*: a light-driven sodium pump. Biochem Biophys Res Commun 88(2):491–499
Ludmann K, Ibron G, Lanyi JK et al (2000) Charge motions during the photocycle of *pharaonis* halorhodopsin. Biophys J 78(2):959–966
Luisi BF, Lanyi JK, Weber HJ (1980) Na^+ transport *via* Na^+/H^+ antiport in *Halobacterium halobium* envelope vesicles. FEBS Lett 117(1):354–358
Matsuno-Yagi A, Mukohata Y (1977) Two possible roles of bacteriorhodopsin; a comparative study of strains of *Halobacterium halobium* differing in pigmentation. Biochem Biophys Res Commun 78(1):237–243
Matsuno-Yagi A, Mukohata Y (1980) ATP synthesis linked to light-dependent proton uptake in a red mutant strain of *Halobacterium* lacking bacteriorhodopsin. Arch Biochem Biophys 199(1):297–303
Mevorat-Kaplan K, Brumfeld V, Engelhard M et al (2006) The protonated Schiff base of halorhodopsin from *Natronobacterium pharaonis* is hydrolyzed at elevated temperatures. Photochem Photobiol 82(6):1414–1421
Mukohata Y, Kaji Y (1981) Light-induced membrane-potential increase, ATP synthesis, and proton uptake in *Halobacterium halobium* R_1mR catalyzed by halorhodopsin: Effects of *N, N′*-dicyclohexylcarbodiimide, triphenyltin chloride, and 3, 5-di-*tert*-butyl-4-hydroxybenzylidene-malononitrile (SF6847). Arch Biochem Biophys 206(1):72–76
Muneyuki E, Shibazaki C, Ohtani H et al (1999) Time-resolved measurements of photovoltage generation by bacteriorhodopsin and halorhodopsin adsorbed on a thin polymer film. J Biochem 125(2):270–276
Muroda K, Nakashima K, Shibata M et al (2012) Protein-bound water as the determinant of asymmetric functional conversion between light-driven proton and chloride pumps. Biochemistry 51:4677–4684
Oesterhelt D (1995) Structure and function of halorhodopsin. Isr J Chem 35(3–4):475–494
Oesterhelt D, Stoeckenius W (1971) Rhodopsin-like protein from the purple membrane of *Halobacterium halobium*. Nature 233(39):149–152
Oesterhelt D, Stoeckenius W (1973) Functions of a new photoreceptor membrane. Proc Natl Acad Sci USA 70(10):2853–2857
Ogurusu T, Maeda A, Sasaki N et al (1981) Light-induced reaction of halorhodopsin prepared under low salt conditions. J Biochem 90(5):1267–1274

Ogurusu T, Maeda A, Sasaki N et al (1982) Effects of chloride on the absorption spectrum and photoreactions of halorhodopsin. Biochim Biophys Acta 682(3):446–451

Otomo J, Muramatsu T (1995) Over-expression of a new photo-active halorhodopsin in *Halobacterium salinarium*. Biochim Biophys Acta 1240(2):248–256

Pfeifer F, Weidinger G, Goebel W (1981) Genetic variability in *Halobacterium halobium*. J Bacteriol 145(1):375–381

Racker E, Stoeckenius W (1974) Reconstitution of purple membrane vesicles catalyzing light-driven proton uptake and adenosine triphosphate formation. J Biol Chem 249(2):662–663

Rüdiger M, Oesterhelt D (1997) Specific arginine and threonine residues control anion binding and transport in the light-driven chloride pump halorhodopsin. EMBO J 16(13):3813–3821

Rüdiger M, Haupts U, Gerwert K et al (1995) Chemical reconstitution of a chloride pump inactivated by a single point mutation. EMBO J 14(8):1599–1606

Sasaki J, Brown LS, Chon YS et al (1995) Conversion of bacteriorhodopsin into a chloride ion pump. Science 269(5220):73–75

Sass HJ, Buldt G, Gessenich R et al (2000) Structural alterations for proton translocation in the M state of wild-type bacteriorhodopsin. Nature 406(6796):649–653

Sato M, Kanamori T, Kamo N et al (2002) Stopped-flow analysis on anion binding to blue-form halorhodopsin from *Natronobacterium pharaonis*: comparison with the anion-uptake process during the photocycle. Biochemistry 41(7):2452–2458

Sato M, Kikukawa T, Araiso T et al (2003a) Roles of Ser130 and Thr126 in chloride binding and photocycle of *pharaonis* halorhodopsin. J Biochem 134(1):151–158

Sato M, Kikukawa T, Araiso T et al (2003b) Ser-130 of *Natronobacterium pharaonis* halorhodopsin is important for the chloride binding. Biophys Chem 104(1):209–216

Sato M, Kubo M, Aizawa T et al (2005) Role of putative anion-binding sites in cytoplasmic and extracellular channels of *Natronomonas pharaonis* halorhodopsin. Biochemistry 44(12): 4775–4784

Scharf B, Engelhard M (1994) Blue halorhodopsin from *Natronobacterium pharaonis*: wavelength regulation by anions. Biochemistry 33(21):6387–6393

Schobert B, Lanyi JK (1982) Halorhodopsin is a light-driven chloride pump. J Biol Chem 257(17):10306–10313

Schobert B, Brown LS, Lanyi JK (2003) Crystallographic structures of the M and N intermediates of bacteriorhodopsin: assembly of a hydrogen-bonded chain of water molecules between Asp-96 and the retinal Schiff base. J Mol Biol 330(3):553–570

Seki A, Miyauchi S, Hayashi S et al (2007) Heterologous expression of *pharaonis* halorhodopsin in *Xenopus laevis* oocytes and electrophysiological characterization of its light-driven Cl^- pump activity. Biophys J 92(7):2559–2569

Shibasaki K, Shigemura H, Kikukawa T et al (2013) Role of Thr218 in the light-driven anion pump halorhodopsin from *Natronomonas pharaonis*. Biochemistry 52(51):9257–9268

Shibata M, Muneda N, Sasaki T et al (2005) Hydrogen-bonding alterations of the protonated Schiff base and water molecule in the chloride pump of *Natronobacterium pharaonis*. Biochemistry 44(37):12279–12286

Shibata M, Saito Y, Demura M et al (2006) Deprotonation of Glu234 during the photocycle of *Natronomonas pharaonis* halorhodopsin. Chem Phys Lett 432(4–6):545–547

Steiner M, Oesterhelt D (1983) Isolation and properties of the native chromoprotein halorhodopsin. EMBO J 2(8):1379–1385

Sudo Y, Iwamoto M, Shimono K et al (2001) Photo-induced proton transport of *pharaonis* phoborhodopsin (sensory rhodopsin II) is ceased by association with the transducer. Biophys J 80(2):916–922

Tsuda M, Hazemoto N, Kondo M et al (1982) Two photocycles in *Halobacterium halobium* that lacks bacteriorhodopsin. Biochem Biophys Res Commun 108(3):970–976

Tsukamoto T, Sasaki T, Fujimoto KJ et al (2012) Homotrimer formation and dissociation of *pharaonis* halorhodopsin in detergent system. Biophys J 102(12):2906–2915

Váró G, Lanyi JK (1995) Effects of hydrostatic pressure on the kinetics reveal a volume increase during the bacteriorhodopsin photocycle. Biochemistry 34(38):12161–12169

Váró G, Brown LS, Sasaki J et al (1995a) Light-driven chloride ion transport by halorhodopsin from *Natronobacterium pharaonis*. I. The photochemical cycle. Biochemistry 34(44):14490–14499

Váró G, Needleman R, Lanyi JK (1995b) Light-driven chloride ion transport by halorhodopsin from *Natronobacterium pharaonis*. II. Chloride release and uptake, protein conformation change, and thermodynamics. Biochemistry 34(44):14500–14507

Váró G, Zimányi L, Fan X et al (1995c) Photocycle of halorhodopsin from *Halobacterium salinarium*. Biophys J 68(5):2062–2072

Weber HJ, Bogomolni RA (1981) P_{588}, a second retinal-containing pigment in *Halobacterium halobium*. Photochem Photobiol 33(4):601–608

Yamashita Y, Kikukawa T, Tsukamoto T et al (2011) Expression of *salinarum* halorhodopsin in *Escherichia coli* cells: solubilization in the presence of retinal yields the natural state. Biochim Biophys Acta 1808(12):2905–2912

Chapter 5
Molecular Mechanisms for Ion Transportation of Microbial Rhodopsins Studied by Light-Induced Difference FTIR Spectroscopy

Yuji Furutani

Abstract Microbial rhodopsins, many variants of which were developed via genetic engineering, have been widely utilized as optogenetic tools. Understanding the molecular mechanisms is important for designing such tools more efficiently. The dynamics of these proteins upon photoactivation can be studied by light-induced difference Fourier transform infrared (FTIR) spectroscopy. As the structural information involves hydrogen, which is not readily accessible via X-ray crystallography, light-induced difference FTIR spectroscopy is a powerful tool to study the molecular mechanisms of these light-receptive proteins. Low-temperature and time-resolved FTIR spectroscopy on two major microbial rhodopsins—bacteriorhodopsin (BR) and halorhodopsin (HR)—are summarized in this review. The low-temperature method stabilizes the intermediate states by decreasing temperature, whereas the time-resolved method allows direct observation at physiological temperature by using a measurement time short enough for their detection. By measuring the difference spectra in X–H and X–D stretching regions, changes in the hydrogen-bonding networks, including water molecules, were elucidated. It was demonstrated that water molecules played an important role in proton and ion translocation in BR and HR. Therefore, for develop optogenetic tools with favorable molecular properties, it may be important to design protein structures in the presence of internal water molecules.

Keywords Vibrational spectroscopy • Bacteriorhodopsin • Halorhodopsin • Hydrogen bonds • Water molecules • Light energy

Y. Furutani (✉)
Department of Life and Coordination-Complex Molecular Science, Institute for Molecular Science, Myodaiji, Okazaki 444-8585, Japan

Department of Structural Molecular Science, The Graduate University for Advanced Studies (SOKENDAI), Myodaiji, Okazaki 444-8585, Japan

PRESTO, Japan Science and Technology Agency (JST),
4-1-8 Honcho, Kawaguchi, Saitama 332-0012, Japan
e-mail: furutani@ims.ac.jp

H. Yawo et al. (eds.), *Optogenetics*, DOI 10.1007/978-4-431-55516-2_5

5.1 Introduction

Optogenetic tools (Deisseroth 2011; Boyden 2011; Fenno et al. 2011; Bernstein and Boyden 2011) developed from microbial rhodopsins, such as channelrhodopsin (ChR) (activation of nerve cells), halorhodopsin (HR), bacteriorhodopsin (BR), and archerhodopsin (silencing of nerve cells), have played important roles in the initial phases of this field of research. ChR is widely used in optogenetics, due to its ability to activate neurons of interest in vitro and in vivo by focusing laser or light-emitting diode (LED) light, which increases spatiotemporal resolution relative to the conventional electrode method. Various derivatives of ChR have been developed by introducing site-specific mutations and making chimera of ChRs from different clades in the phylogenetic tree. The molecular mechanisms employed in the light-gated channel and light-driven ion pumps of microbial rhodopsins are not well understood, despite the availability of the atomic structural data. Many physicochemical researchers are therefore investigating the molecular mechanisms of rhodopsin molecules, primarily for their potential to innovate and improve the function of optogenetic tools. This review article summarizes how the molecular mechanisms of microbial rhodopsins have been elucidated, particularly focusing on studies using infrared spectroscopic techniques.

5.2 The Difference Infrared Spectroscopy for Analyzing Light-Induced Structural Changes in Microbial Rhodopsins

The three-dimensional atomic structural information of proteins can be obtained by X-ray crystallography, nuclear magnetic resonance (NMR) spectroscopy, and electron microscopy. The spatial resolution achieved by these methods is 0.1–10 nm, depending on the experimental techniques and conditions. The structural models are visualized by a molecular modeling software such as PyMOL used in this review. Microbial rhodopsin is a transmembrane protein with seven α-helices spanning the membrane. The crystallization of membrane proteins is generally considered more difficult than soluble proteins. Many X-ray crystal structures of microbial rhodopsins have been elucidated in the past quarter of a century. Through the development of crystallization techniques using microbial rhodopsins, the lipidic cubic phase method was developed (Rummel et al. 1998) and has become a standard method for the crystallization of membrane proteins (e.g., G-protein-coupled receptors [GPCRs]). In 2012, the X-ray crystal structure of ChR was resolved, with important implications for research in the optogenetic field (Kato et al. 2012).

Although various intermediate states of microbial rhodopsins have been analyzed via X-ray crystallography, especially for BR (Lanyi 2004; Pebay-Peyroula et al. 2000), it has been challenging to elucidate the active state that induces the large protein conformational changes coupled with the protein function, such as channel gating or ion transportation. For example, visual rhodopsin in crystal lattice forms a yellow species like metarhodopsin II, wherein the large conformational change of helix six was not observed in the crystal (Salom et al. 2006). The formation of the crystal lattice may in fact impair conformational changes of a protein that take place in physiological conditions (Efremov et al. 2006). Time-resolved X-ray measurements triggered by a pulsed laser light have been successfully applied in the picosecond time range (Srajer et al. 1996) to capture the ultrafast phenomenon of photoreceptive proteins, such as a *cis-trans* photo-isomerization of chromophore (Schotte et al. 2012). Although structural changes important for expression of protein functions take place in the micro-to-millisecond time range, they may be hampered by crystal packing (Efremov et al. 2006). Spectroscopic and infrared spectroscopic methods with sufficient time resolution have been used to detect protein conformational changes. Light-induced difference infrared spectroscopy, which calculates a difference spectrum between infrared spectra obtained before and after the photoreaction, has been employed to analyze structural changes of BR (Maeda 2001; Kandori 2000; Heberle 2000; Gerwert 1999; Siebert 1995; Rothschild 1992) and other microbial rhodopsins (Furutani and Kandori 2014; Furutani et al. 2008, 2005). The molecular mechanisms of microbial rhodopsins have been investigated by combining spectroscopic data with the static atomic structures of the proteins produced predominantly by X-ray crystallography.

A photon of infrared light has an equivalent energy to that of a molecular vibration. Therefore, infrared absorption spectra give us information about chemical bonds, such as C–C single bonds and C=C double bonds in a retinal molecule. Rhodopsin has a retinal molecule as the chromophore, which primarily comprises an ethylenic chain with alternate C–C and C=C bonds. Photoisomerization of the retinal chromophore changes the frequencies of C–C and C=C stretching vibrations, as observed in light-induced difference infrared spectra (Fig. 5.1; the lower side). Protein is a polymer composed of 20 types of amino acids connected by peptide bonds. The primary structure is the sequence of the amino acids, which provides secondary and tertiary structures through the protein-folding reactions. α helices and β sheets are typical secondary protein structures that are formed by hydrogen bonds between C=O and N–H groups on a peptide backbone. Amide I and II bands are observed in an infrared spectrum of protein (Fig. 5.1; the upper side). The amide I band in particular is generally utilized for analysis of secondary protein structures (Barth and Zscherp 2002). Amino acid side chains also show vibrational bands in their typical frequency regions (Barth 2000). Among them, the C=O stretching bond of a protonated carboxylate group of aspartic and glutamic acids residues appearing in the frequency region of 1,780–1,700 cm^{-1} has been utilized as an indicator of the protonation states in microbial rhodopsins.

Water strongly absorbs infrared light, disturbing the measurement of infrared absorption spectra of protein in the solution. Therefore, the thickness of a sample cell for infrared spectroscopy is limited to several micrometers; alternatively, a protein film formed on an infrared window is used for the measurement.

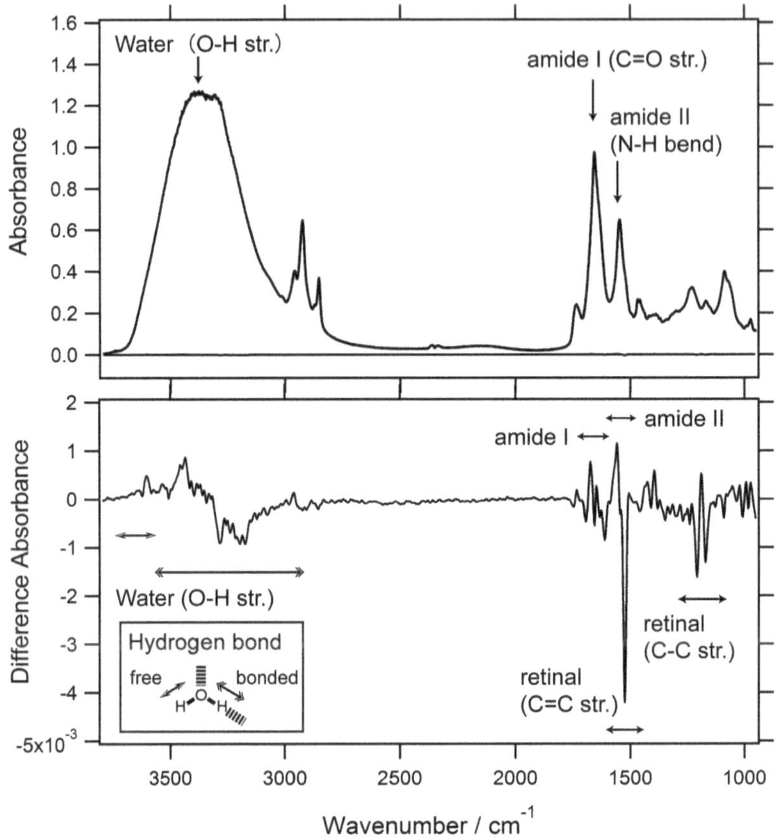

Fig. 5.1 An infrared absorption spectrum of *Natronomonas pharaonis* halorhodopsin (*Np*HR) in a hydrated film sample (the *upper side*) and its light-induced difference spectrum (the *lower side*). The difference spectrum is also shown in the upper side, yet it appears to be a zero line due to minimal spectral change caused by the photoreaction. The spectral change is attributed to the protein moiety (amide I and II), the retinal chromophore (C–C and C=C str.), and the internal water molecules (O–H str.) in *Np*HR

A dried protein film is favorable for measurement of infrared absorption spectra; however, protein molecules under such conditions fail to exhibit the proper structural changes observed in hydrated conditions (Nishimura et al. 1995). Considering this dilemma, the amount of water in protein samples should be carefully regulated. By appropriately reducing water content in the sample, O–H and O–D stretching vibrations of water molecules inside proteins can be observed (Furutani et al. 2005; Maeda 2001; Kandori 2000). By doing so, we can understand changes in the hydrogen bonds of water molecules in protein, which is considered to be a remarkable advantage of difference infrared spectroscopy (Fig. 5.1; the lower side).

5.3 Light-Driven Proton-Pumping Rhodopsin, Bacteriorhodopsin

BR was discovered in 1971 (Oesterhelt and Stoeckenius 1971) and has since been investigated using various experimental and theoretical methods. BR binds an all-*trans* retinal as the choromophore (λ_{max}=570 nm) and its photoisomerization to the 13-*cis* form occurs within 0.5 ps. Subsequently, several intermediate states of BR, called K, L, M, N, and O, are transiently formed (Lanyi 1993). During the transition from the N to the O state, the 13-*cis* chromophore thermally isomerizes to the all-*trans* form, and the O state returns to the original state. BR pumps a proton in each photocyclic reaction, which takes approximately 10 ms. The proton transfer reactions of BR have been extensively investigated by light-induced difference infrared spectroscopy, which revealed that the first proton transfer step involved deprotonation of the protonated Schiff base and protonation of the counter ion Asp85 upon the formation of the M intermediate (Braiman et al. 1991). After the protein's conformational change, the Schiff base can accept a proton from Asp96, which locates in the extracellular side and donates a proton on the formation of the N intermediate (Pfefferle et al. 1991). Moreover, time-resolved and low-temperature Fourier transform infrared (FTIR) spectroscopy revealed that water clusters play an important role in the proton-transfer processes (Garczarek and Gerwert 2006; Tanimoto et al. 2003).

The proton-transfer process from the Schiff base to Asp85 is a key step for the proton-pumping function of BR, and has been explained by several hypotheses describing changes in the hydrogen-bonding network, including a bridging water molecule. The existence of a water molecule between the Schiff base and Asp85 was suggested via low-temperature FTIR spectroscopy (Maeda et al. 1994) and later confirmed using high-resolution X-ray crystallography (Luecke et al. 1999) (Fig. 5.2). The hydrogen-bond strength of water molecules in each intermediate state has been extensively studied via low-temperature FTIR spectroscopy (Tanimoto et al. 2003; Maeda 2001; Kandori 2000). FTIR revealed that the bridging water has an extremely strong hydrogen bond (O–D stretching vibration at 2,171 cm^{-1}) in the ground state, which changes its strength on formation of the K intermediate (Shibata et al. 2003; Kandori and Shichida 2000) (Fig. 5.3; left). The water is sandwiched between a positive charge on the Schiff base and a negative charge on Asp85. This configuration of the hydrogen-bonding network probably strengthens hydrogen bonds compared with those found in a typical tetrahedral structure in ice according to the previous quantum mechanics/molecular mechanics (QM/MM) result (Hayashi et al. 2004). The QM/MM result also indicated that the light energy was predominantly captured in the retinal distortion, with weakening of the hydrogen bond an additional form of the energy capturing. Interestingly, the hydrogen bond returns to its original strength in the L and M states (Tanimoto et al. 2003) (Fig. 5.3; left). In the M state, Asp85 is protonated; therefore, a possible hydrogen-bonding partner for the water molecule is Asp212, which retains a negative charge in the M state. Based on our interpretation of these experimental results, the hydrogen-bonding switch reaction of the water molecules appears to be a key step for proton transfer reactions from the Schiff base to Asp85 (Fig. 5.3; right).

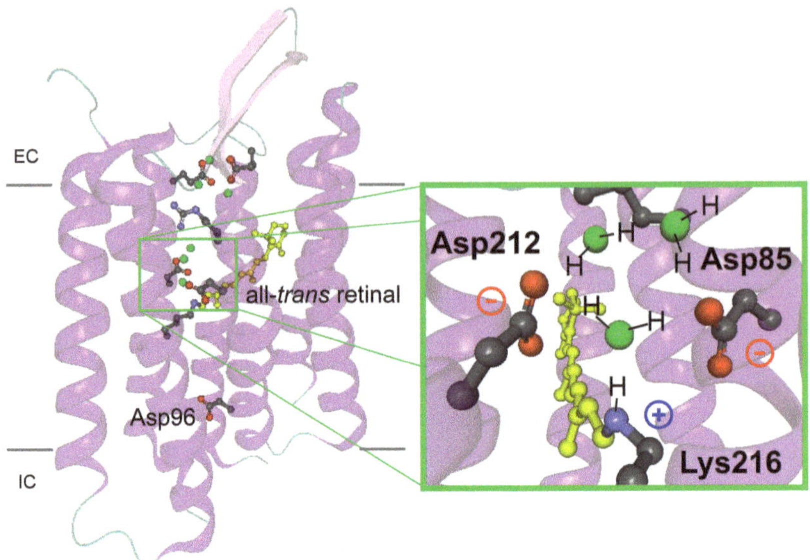

Fig. 5.2 The X-ray crystal structure of bacteriorhodopsin (BR) and the expanded view around the Schiff base region. The extracellular (EC) region is above and the intracellular (IC) region is below. The amino acid residues important for proton-pumping function are depicted

Since the turn of the century, many microbial rhodopsins have been discovered in a diverse variety of organisms; for example, proteorhodopsin, identified in microbacteria living in the ocean, was found to function as a light-driven proton pump and contributes to biomass generation by capturing solar energy (Beja et al. 2000, 2001). Moreover, sodium- (Inoue et al. 2013) and chloride-pumping (Yoshizawa et al. 2014) rhodopsins have also recently been found in the ocean. The molecular mechanisms of these pumping rhodopsins have not as yet been very well studied in comparison with BR. However, comparative studies of these microbial rhodopsins will provide us with a better understanding of their molecular mechanisms for pumping protons, and chloride and sodium ions. In the case of proton pumping, a water molecule with a strong hydrogen bond may be a crucial component (Furutani et al. 2005). From the extensive low-temperature FTIR studies available on many microbial rhodopsins, such water molecules were found in microbial rhodopsins that possess proton-pumping functions.

5.4 Light-Driven Chloride-Pumping Rhodopsin, Halorhodopsin

HR pumps extracellular chloride ions into the cell through a photocyclic reaction similar to that in BR (Varo 2000). Incorporation of chloride ions causes a negative membrane potential and suppresses depolarization of the cell. Therefore, HR,

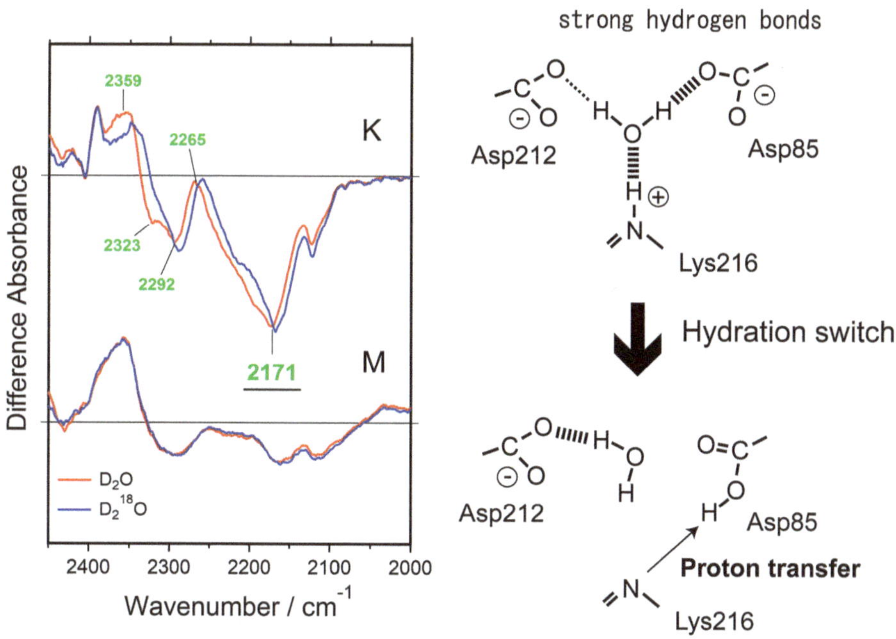

Fig. 5.3 The light-induced difference spectra of bacteriorhodopsin (BR) in the X–D stretching vibration in the 2,450–2,000 cm^{-1} region (*left*) and the proposed model of the initial proton transfer reaction in BR (hydration switch model) (*right*). The *green*-tagged bands were assigned to O–D stretching vibrations of internal water molecules in BR. The lowest band at 2,171 cm^{-1} was also assigned to the N–D stretching vibration of the Schiff base. The hydrogen bonds of the water molecules are weakened at the K intermediate and restore their strength at the M intermediate. On the basis of the experimental result, the model for the initial proton-transfer reaction is depicted

especially from *Natronomonas pharaonis* (*Np*HR), has been utilized as an optogenetic tool to silence action potentials in neurons (Zhang et al. 2007).

HR has an all-*trans* retinal as the chromophore, with an absorption maximum of approximately 580 nm. Photoisomerization of the retinal triggers the photocyclic reaction, where several intermediate states (K, L_1, L_2, N, and O) are formed (Varo 2000). Contrary to in BR, the protonated Schiff base retains its proton during the photocyclic reaction. BR has two negatively charged residues around the Schiff base (Asp85 and Asp212), whereas HR has one aspartate (Asp252 in *Np*HR) and a neutral threonine residue (Thr126 in *Np*HR) at the equivalent position of Asp85 in BR, which facilitates chloride ion binding. Mutation of Asp85 to Thr in BR is known to convert the proton pump to a chloride pump (Sasaki et al. 1995), making Asp85 in BR a crucial residue for determining types of ion pumps. In contrast, the inverse conversion of chloride-pumping function of HR to that of a proton pump by mutagenesis has not yet been successful, although HR can pump protons in the presence of azide molecules (Lakatos et al. 2002).

A chloride ion is a monoatomic ion, with no infrared absorption. Therefore, infrared spectroscopy cannot directly detect a chloride ion in HR. This is in clear contrast to the case of protons in BR. A proton exists as an oxonium ion in an aqueous solution, or binds to amino acid residues (e.g., Asp, Glu, and His) in protein. Transient protonation and deprotonation of the amino acid residues and water molecules enables us to develop a deep insight into the molecular mechanism behind the proton-pumping function of BR. Although a chloride ion cannot form a chemical bond in HR, it influences the hydrogen-bonding network and ionic interactions with amino acid residues and water molecules. Such slight effects caused by movement of a chloride ion in HR may be detected by infrared spectroscopy. In addition to chloride ions, HR can pump bromide, iodide, and nitrate ions. Comparison of infrared spectroscopic data obtained with these transportable ions was used to elucidate specific ion–protein interactions of water molecules and amino acid residues in HR (Shibata et al. 2005).

The molecular mechanism of HR has been investigated by various physicochemical methods and the X-ray crystal structures in the ground (Kouyama et al. 2010; Kolbe et al. 2000) (Fig. 5.4) and intermediate states (Nakanishi et al. 2013; Gmelin et al. 2007) solved with atomic resolution. The most important process of the HR chloride–pump reaction is the movement of a chloride ion bound near the Schiff base. The first X-ray crystal structure of the HR L_1 intermediate detected a chloride ion in the same position (Gmelin et al. 2007) and indicated that the chloride ion moves in the subsequent intermediate states (L_2 or N state). Previous low-temperature infrared spectroscopy studies also supported the observation and suggested that changes in hydrogen-bonding and the movement of water molecules around the chloride ion in K and L_1 states reduced polarity and hydrophilicity of the environment; thus, promoting movement of the chloride ion into the later intermediate states (Shibata et al. 2005).

In 2012, time-resolved FTIR spectroscopy on *Np*HR, in the entire mid-infrared region, revealed water molecule dynamics of water molecules inside the protein in accordance with protein conformational changes in L_1, L_2, N, and O intermediate states (Furutani et al. 2012). O–H stretching vibrations of water molecules were detected in a >3,600 cm^{-1} region, where the O–H groups were considered not to form hydrogen bonds. The difference in absorption spectra of the O–H stretching vibrations changed little in L_1 and L_2 intermediates but changed more significantly in the N and O states (Furutani et al. 2012) (Fig. 5.5). Based on the X-ray crystal structure, there are three water molecules near the chloride ion in the Schiff base region (Kouyama et al. 2010). The water molecules must change the hydrogen-bonding network when the chloride ion moves inside the cell after formation of the L_1 state. The O–H stretching vibration of water may be an intrinsic probe for chloride ion movement; therefore, the large spectral change of O–H stretching vibrations suggested that the chloride ion moves on formation of the N state. In the O state, the absence of a chloride ion may increase the number of water O–H groups free from a hydrogen bond, as shown by the large positive band in the difference spectra. In the future, when the signal-to-noise ratio in the lower frequency region (3,500–3,000 cm^{-1}) of difference spectra is improved, O–H stretching vibrations of water molecules and Thr residues, which directly interact with the chloride ion, will be more easily detected; thus, increasing our understanding of the molecular mechanism of chloride pumps.

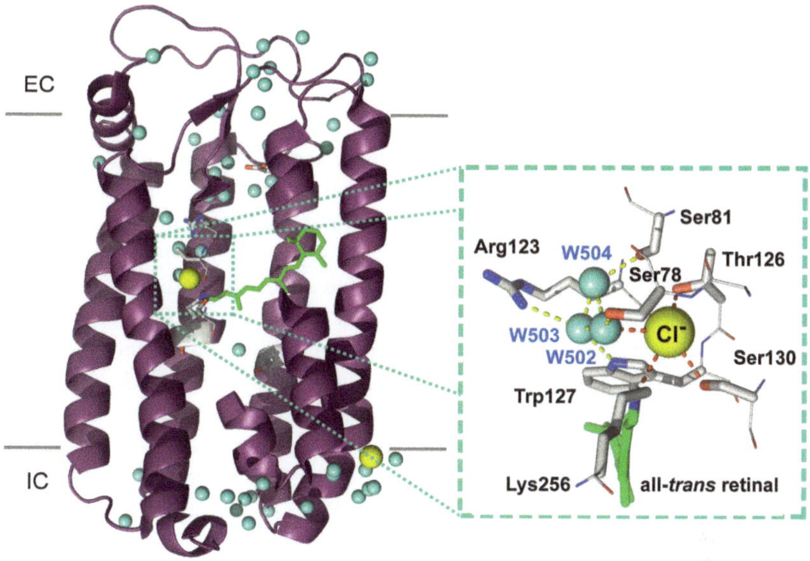

Fig. 5.4 X-ray crystal structure of *Natronomonas pharaonis* halorhodopsin (*Np*HR) and the expanded view around the Schiff base region. The extracellular (EC) side is shown at the *top* and the intracellular (IC) side at the *bottom*. The amino acid residues and water molecules surrounding the initial chloride-binding site are depicted

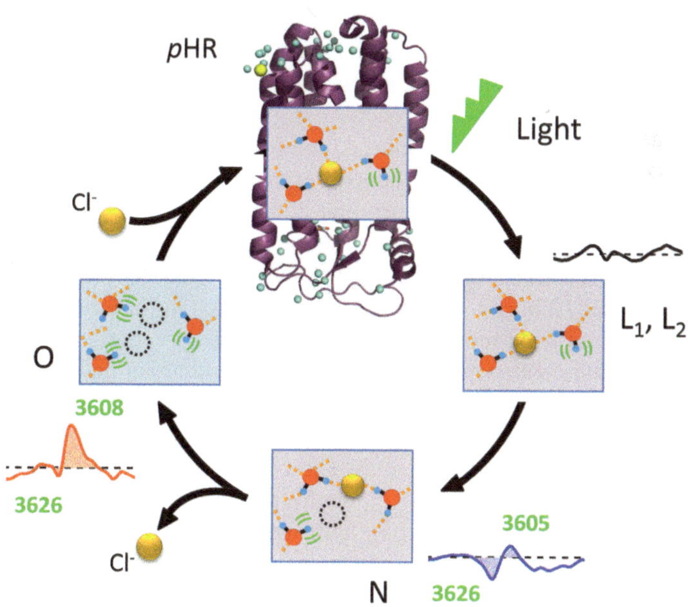

Fig. 5.5 Structural changes of water molecules during the photocyclic reaction of *Natronomonas pharaonis* halorhodopsin (*Np*HR) revealed by time-resolved Fourier transform infrared (FTIR) spectroscopy. Spectral changes of O–H stretching vibrations of water molecules without a hydrogen bond are shown for each intermediate state. It was demonstrated that translocation and release of a chloride ion induced larger spectral changes of water in the N and O intermediate states of *Np*HR

5.5 Light-Gated Ion Channel, Channelrhodopsin

ChR-1 and -2 were originally discovered in the green alga, *Chlamydomonas reinhardtii* (Nagel et al. 2003, 2002). Both function as light-gated cation channels, yet with different ion selectivity. Due to the enhanced electrophysiological properties of ChR-2 (ChR2) in various kinds of cell, it has been widely used as an optogenetic tool for activating neuronal cells by illumination with blue light. ChR2 binds an all-*trans* retinal as the chromophore (λ_{max} = ~470 nm) and with a similar photocyclic reaction to BR and HR. The intermediate states are called P_1 (500 nm), P_2 (390 nm), P_3 (520 nm), and P_4 (480 nm) states (Ritter et al. 2008; Bamann et al. 2008). Among them, the P_3 state is considered an active state where cations pass through ChR2. P_4 has a long lifetime (τ = ~20 s) and is considered to be a desensitized state as determined by electrophysiological measurement. In 2012, the X-ray crystal structure of a chimera of channelrhodpsin-1 and -2 (called C1C2) which was composed of the 1st–5th helices from ChR1 and the 6th–7th helices from ChR2, was revealed (Kato et al. 2012). Prior to the crystallographic study, light-induced difference spectroscopy was applied to ChR2, and elucidated structural changes of Glu90 and conformational changes of protein at 80 and 293 or 298 K (Radu et al. 2009; Ritter et al. 2008). Time-resolved FTIR measurement, with ~10 ms time resolution, was initially applied to ChR2, revealing the kinetics of its conformational changes and the deprotonation of Glu90 in P_3 and P_4 states (Eisenhauer et al. 2012). In early studies, low-temperature and time-resolved methods, with rapid scan mode, were applied to analyze structural changes in the intermediate and photostationary states. Due to the slow recovery kinetics of ChR2 caused by the P_4 intermediate, it was difficult to apply time-resolved FTIR spectroscopy with step-scan mode (reaching higher time resolutions: up to ~ μs), which required more than several hundred repetitive measurements. A pioneering work was reported in 2013 (Lorenz-Fonfria et al. 2013) that revealed conformational changes of the protein with the time constants and transient protonation changes of Glu90 and Asp156. Moreover, the time-resolved measurement exposed the proton transfer reaction from the Schiff base to Glu253, located at the equivalent position of Asp212 in BR. The protonation changes during the photocycle of ChR2 clearly differed from those in BR and HR. Therefore, there are likely to be key events that induce conformational changes of the channel and open the gate. Recently, a low-temperature FTIR study on C1C2 in the X–D stretching region characterized the water-containing hydrogen-bonding network around the Schiff base region (Ito et al. 2014). The study showed that the Schiff base interacts directly with Glu162 (Glu123 in ChR2), forming a strong hydrogen bond with it, while a water molecule formed a strong hydrogen bond with Asp292 (Asp253 in ChR2 and Asp212 in BR). The unique hydrogen-bonding network around the Schiff base region in ChR may be correlated with the unique protein structural changes on photoreaction. Interestingly, the distinctive structural changes of Glu129 (Glu90 in ChR2) in chimeric ChRs compared with that of ChR2 was recently revealed by a light-induced difference FTIR study (Inaguma et al. 2015).

5.6 Conclusion

Light-induced difference FTIR spectroscopy is a powerful tool to study structural changes of microbial rhodopsins. X–H and X–D stretching vibrations provide particularly fruitful information on dynamics of hydrogen-bonding networks inside proteins involved in the ion-pumping reactions, which is usually difficult to be accessed by other physicochemical methods. It was suggested that water molecules probably play an important role in transporting protons and ions inside proteins. Atomic structures determined by structural biological methods such as X-ray crystallography provide the basis of understanding of the molecular mechanisms of proteins; however, it is important to understand what happens to the structural dynamics of the protein when the protein is functioning. Understanding the molecular mechanisms of microbial rhodopsins would be helpful to design optogenetic tools with favorable molecular properties.

Acknowledgments The author thanks many collaborators, especially Professor Hideki Kandori, for studies in this review article and the research grants from KAKENHI (22770159, 22018030, 21023014, 21026016) and JST PRESTO 'Chemical conversion of light energy'. The authors would like to thank Enago (www.enago.jp) for the English language review.

References

Bamann C, Kirsch T, Nagel G et al (2008) Spectral characteristics of the photocycle of channelrhodopsin-2 and its implication for channel function. J Mol Biol 375(3):686–694. doi:10.1016/j.jmb.2007.10.072

Barth A (2000) The infrared absorption of amino acid side chains. Prog Biophys Mol Biol 74(3–5):141–173. doi:10.1016/S0079-6107(00), 00021-3 [pii]

Barth A, Zscherp C (2002) What vibrations tell us about proteins. Q Rev Biophys 35(4):369–430. doi:10.1017/S0033583502003815

Beja O, Aravind L, Koonin EV et al (2000) Bacterial rhodopsin: evidence for a new type of phototrophy in the sea. Science 289(5486):1902–1906

Beja O, Spudich EN, Spudich JL et al (2001) Proteorhodopsin phototrophy in the ocean. Nature 411(6839):786–789. doi:10.1038/35081051

Bernstein JG, Boyden ES (2011) Optogenetic tools for analyzing the neural circuits of behavior. Trends Cogn Sci 15(12):592–600. doi:10.1016/j.tics.2011.10.003, 00216-6 [pii]

Boyden ES (2011) A history of optogenetics: the development of tools for controlling brain circuits with light. Biol Rep 3:11. doi:10.3410/B3-11, 11 [pii]

Braiman MS, Bousche O, Rothschild KJ (1991) Protein dynamics in the bacteriorhodopsin photocycle: submillisecond Fourier transform infrared spectra of the L, M, and N photointermediates. Proc Natl Acad Sci U S A 88(6):2388–2392

Deisseroth K (2011) Optogenetics. Nat Methods 8(1):26–29. doi:10.1038/nmeth.f.324, nmeth.f.324 [pii]

Efremov R, Gordeliy VI, Heberle J et al (2006) Time-resolved microspectroscopy on a single crystal of bacteriorhodopsin reveals lattice-induced differences in the photocycle kinetics. Biophys J 91(4):1441–1451. doi:10.1529/biophysj.106.083345

Eisenhauer K, Kuhne J, Ritter E et al (2012) In channelrhodopsin-2 Glu-90 is crucial for ion selectivity and is deprotonated during the photocycle. J Biol Chem 287(9):6904–6911. doi:10.1074/jbc.M111.327700
Fenno L, Yizhar O, Deisseroth K (2011) The development and application of optogenetics. Annu Rev Neurosci 34:389–412. doi:10.1146/annurev-neuro-061010-113817
Furutani Y, Kandori H (2014) Hydrogen-bonding changes of internal water molecules upon the actions of microbial rhodopsins studied by FTIR spectroscopy. Biochim Biophys Acta 1837(5):598–605. doi:10.1016/j.bbabio.2013.09.004
Furutani Y, Shibata M, Kandori H (2005) Strongly hydrogen-bonded water molecules in the Schiff base region of rhodopsins. Photochem Photobiol Sci 4(9):661–666. doi:10.1039/b416698a
Furutani Y, Sudo Y, Kandori H (2008) FTIR studies of protein-protein interaction changes between pharaonis phoborhodopsin and its cognate transducer protein. Curr Top Biochem Res 10(2):63–77
Furutani Y, Fujiwara K, Kimura T et al (2012) Dynamics of dangling bonds of water molecules in pharaonis halorhodopsin during chloride ion transportation. J Phys Chem Lett 3(20):2964–2969. doi:10.1021/Jz301287n
Garczarek F, Gerwert K (2006) Functional waters in intraprotein proton transfer monitored by FTIR difference spectroscopy. Nature 439(7072):109–112. doi:10.1038/nature04231, nature04231 [pii]
Gerwert K (1999) Molecular reaction mechanisms of proteins monitored by time-resolved FTIR-spectroscopy. Biol Chem 380(7–8):931–935. doi:10.1515/Bc.1999.115
Gmelin W, Zeth K, Efremov R et al (2007) The crystal structure of the L1 intermediate of halorhodopsin at 1.9 angstroms resolution. Photochem Photobiol 83(2):369–377. doi:10.1562/2006-06-23-RA-947
Hayashi S, Tajkhorshid E, Kandori H et al (2004) Role of hydrogen-bond network in energy storage of bacteriorhodopsin's light-driven proton pump revealed by ab initio normal-mode analysis. J Am Chem Soc 126(34):10516–10517. doi:10.1021/Ja047506s
Heberle J (2000) Proton transfer reactions across bacteriorhodopsin and along the membrane. Biochim Biophys Acta 1458(1):135–147, 00064-5 [pii]
Inaguma A, Tsukamoto H, Kato HE et al (2015) Chimeras of channelrhodopsin-1 and -2 from Chlamydomonas reinhardtii exhibit distinctive light-induced structural changes from channelrhodopsin-2. J Biol Chem 290(18):11623–11634. doi:10.1074/jbc.M115.642256
Inoue K, Ono H, Abe-Yoshizumi R et al (2013) A light-driven sodium ion pump in marine bacteria. Nat Commun 4:1678. doi:10.1038/ncomms2689
Ito S, Kato HE, Taniguchi R et al (2014) Water-containing hydrogen-bonding network in the active center of channelrhodopsin. J Am Chem Soc 136(9):3475–3482. doi:10.1021/ja410836g
Kandori H (2000) Role of internal water molecules in bacteriorhodopsin. Biochim Biophys Acta 1460(1):177–191, 00138-9 [pii]
Kandori H, Shichida Y (2000) Direct observation of the bridged water stretching vibrations inside a protein. J Am Chem Soc 122(47):11745–11746. doi:10.1021/Ja0032069
Kato HE, Zhang F, Yizhar O et al (2012) Crystal structure of the channelrhodopsin light-gated cation channel. Nature 482(7385):369–374. doi:10.1038/nature10870
Kolbe M, Besir H, Essen LO et al (2000) Structure of the light-driven chloride pump halorhodopsin at 1.8 A resolution. Science 288(5470):1390–1396, 8554 [pii]
Kouyama T, Kanada S, Takeguchi Y et al (2010) Crystal structure of the light-driven chloride pump halorhodopsin from Natronomonas pharaonis. J Mol Biol 396(3):564–579, doi:S0022-2836(09)01455-7 [pii]
Lakatos M, Groma GI, Ganea C et al (2002) Characterization of the azide-dependent bacteriorhodopsin-like photocycle of salinarum halorhodopsin. Biophys J 82(4):1687–1695. doi:10.1016/S0006-3495(02)75521-5
Lanyi JK (1993) Proton translocation mechanism and energetics in the light-driven pump bacteriorhodopsin. Biochim Biophys Acta 1183(2):241–261
Lanyi JK (2004) X-ray diffraction of bacteriorhodopsin photocycle intermediates. Mol Membr Biol 21(3):143–150. doi:10.1080/09687680410001666345

Lorenz-Fonfria VA, Resler T, Krause N et al (2013) Transient protonation changes in channelrhodopsin-2 and their relevance to channel gating. Proc Natl Acad Sci U S A 110(14):E1273–E1281. doi:10.1073/pnas.1219502110

Luecke H, Schobert B, Richter HT et al (1999) Structure of bacteriorhodopsin at 1.55 A resolution. J Mol Biol 291(4):899–911. doi:10.1006/jmbi.1999.3027, S0022-2836(99)93027-9 [pii]

Maeda A (2001) Internal water molecules as mobile polar groups for light-induced proton translocation in bacteriorhodopsin and rhodopsin as studied by difference FTIR spectroscopy. Biochem Biokhimiia 66(11):1256–1268

Maeda A, Sasaki J, Yamazaki Y et al (1994) Interaction of aspartate-85 with a water molecule and the protonated Schiff base in the L intermediate of bacteriorhodopsin: a Fourier-transform infrared spectroscopic study. Biochemistry 33(7):1713–1717

Nagel G, Ollig D, Fuhrmann M et al (2002) Channelrhodopsin-1: a light-gated proton channel in green algae. Science 296(5577):2395–2398. doi:10.1126/science.1072068

Nagel G, Szellas T, Huhn W et al (2003) Channelrhodopsin-2, a directly light-gated cation-selective membrane channel. Proc Natl Acad Sci U S A 100(24):13940–13945. doi:10.1073/pnas.1936192100

Nakanishi T, Kanada S, Murakami M et al (2013) Large deformation of helix F during the photoreaction cycle of Pharaonis halorhodopsin in complex with azide. Biophys J 104(2):377–385. doi:10.1016/j.bpj.2012.12.018

Nishimura S, Sasaki J, Kandori H et al (1995) Structural changes in the lumirhodopsin-to-metarhodopsin I conversion of air-dried bovine rhodopsin. Biochemistry 34(51):16758–16763

Oesterhelt D, Stoeckenius W (1971) Rhodopsin-like protein from the purple membrane of Halobacterium halobium. Nat New Biol 233(39):149–152

Pebay-Peyroula E, Neutze R, Landau EM (2000) Lipidic cubic phase crystallization of bacteriorhodopsin and cryotrapping of intermediates: towards resolving a revolving photocycle. Biochim Biophys Acta 1460(1):119–132

Pfefferle JM, Maeda A, Sasaki J et al (1991) Fourier transform infrared study of the N intermediate of bacteriorhodopsin. Biochemistry 30(26):6548–6556

Radu I, Bamann C, Nack M et al (2009) Conformational changes of channelrhodopsin-2. J Am Chem Soc 131(21):7313–7319. doi:10.1021/ja8084274

Ritter E, Stehfest K, Berndt A et al (2008) Monitoring light-induced structural changes of Channelrhodopsin-2 by UV-visible and Fourier transform infrared spectroscopy. J Biol Chem 283(50):35033–35041, doi:M806353200 [pii]

Rothschild KJ (1992) FTIR difference spectroscopy of bacteriorhodopsin: toward a molecular model. J Bioenerg Biomembr 24(2):147–167

Rummel G, Hardmeyer A, Widmer C et al (1998) Lipidic cubic phases: new matrices for the three-dimensional crystallization of membrane proteins. J Struct Biol 121(2):82–91. doi:10.1006/jsbi.1997.3952

Salom D, Lodowski DT, Stenkamp RE et al (2006) Crystal structure of a photoactivated deprotonated intermediate of rhodopsin. Proc Natl Acad Sci U S A 103(44):16123–16128. doi:10.1073/pnas.0608022103

Sasaki J, Brown LS, Chon YS et al (1995) Conversion of bacteriorhodopsin into a chloride ion pump. Science 269(5220):73–75

Schotte F, Cho HS, Kaila VR et al (2012) Watching a signaling protein function in real time via 100-ps time-resolved Laue crystallography. Proc Natl Acad Sci U S A 109(47):19256–19261. doi:10.1073/pnas.1210938109

Shibata M, Tanimoto T, Kandori H (2003) Water molecules in the schiff base region of bacteriorhodopsin. J Am Chem Soc 125(44):13312–13313. doi:10.1021/ja037343s

Shibata M, Muneda N, Sasaki T et al (2005) Hydrogen-bonding alterations of the protonated Schiff base and water molecule in the chloride pump of Natronobacterium pharaonis. Biochemistry 44(37):12279–12286. doi:10.1021/bi050726d

Siebert F (1995) Infrared spectroscopy applied to biochemical and biological problems. Method Enzymol 246:501–526

Srajer V, Teng T, Ursby T et al (1996) Photolysis of the carbon monoxide complex of myoglobin: nanosecond time-resolved crystallography. Science 274(5293):1726–1729

Tanimoto T, Furutani Y, Kandori H (2003) Structural changes of water in the Schiff base region of bacteriorhodopsin: proposal of a hydration switch model. Biochemistry 42(8):2300–2306. doi:10.1021/bi026990d

Varo G (2000) Analogies between halorhodopsin and bacteriorhodopsin. Biochim Biophys Acta 1460(1):220–229, doi:S0005-2728(00)00141-9 [pii]

Yoshizawa S, Kumagai Y, Kim H et al (2014) Functional characterization of flavobacteria rhodopsins reveals a unique class of light-driven chloride pump in bacteria. Proc Natl Acad Sci U S A 111(18):6732–6737. doi:10.1073/pnas.1403051111

Zhang F, Wang LP, Brauner M et al (2007) Multimodal fast optical interrogation of neural circuitry. Nature 446(7136):633–639. doi:10.1038/nature05744

Chapter 6
Optogenetic Potentials of Diverse Animal Opsins

Akihisa Terakita, Takashi Nagata, Tomohiro Sugihara, and Mitsumasa Koyanagi

Abstract Animal opsin-based pigments are light-activated G-protein-coupled receptors (GPCRs), which drive signal transduction cascades via G proteins. Thousands of animal opsins have been identified, and molecular phylogenetic and biochemical analyses have revealed that opsin-based pigments have basically diversified in selective activation of G proteins (Gs, Gq, Gi, Go, and transducin). Here, we discuss the optogenetic potentials of diverse animal opsins, particularly Gq-coupled spider opsin, Gs-coupled jellyfish opsin, and Gi/Go-coupled mosquito opsin 3 (Opn3). After absorbing light, these purified opsin-based pigments do not release the chromophore retinal, indicating the bleach-resistant nature of their photoproducts. In addition, unlike vertebrate visual opsin-based pigments that have been conventionally used for optogenetic applications, the stable photoproducts of spider opsin- and mosquito Opn3-based pigments revert to their original dark states upon subsequent light absorption, which indicates their photoregeneration ability. Mammalian cultured cells that express spider opsin exhibit light-induced increases in Ca^{2+} levels, and jellyfish opsin- and mosquito Opn3-expressing cells exhibit light-dependent increases and decreases in cyclic adenosine monophosphate (cAMP) levels, respectively. These findings indicate that these pigments control different second messengers, Ca^{2+} and cAMP, in mammalian cultured cells, suggesting that these bleach-resistant opsins have an optogenetic potential.

Keywords Animal rhodopsin • Opsin • Bistable pigment • Photoregeneration • G protein-coupled receptor • Phototransduction • Second messenger • Cultured cells • Optogenetic potential

---

A. Terakita (✉) • T. Nagata • T. Sugihara • M. Koyanagi
Department of Biology and Geosciences, Graduate School of Science,
Osaka City University, 3-3-138 Sugimoto, Sumiyoshi-ku, Osaka 558-8585, Japan
e-mail: terakita@sci.osaka-cu.ac.jp

H. Yawo et al. (eds.), *Optogenetics*, DOI 10.1007/978-4-431-55516-2_6

6.1 Introduction

It is widely accepted that microbial rhodopsins are powerful tools for use in optogenetics: light-manipulating neural cells in which microbial rhodopsin genes are introduced and expressed (Hegemann and Nagel 2013; Zalocusky and Deisseroth 2013; Bernstein and Boyden 2011; Boyden 2011; Deisseroth 2011; Fenno et al. 2011). Microbial rhodopsins function as ion channels and ion pumps upon light irradiation and thereby control the membrane potentials of targeted neural cells in a light-dependent manner. On the other hand, animal opsin-based pigments (rhodopsins) are light-activated G-protein-coupled receptors (GPCRs), which drive signal transduction cascades via G proteins (Koyanagi and Terakita 2014; Terakita 2005; Terakita et al. 2012). To date, thousands of animal opsins have been identified, and molecular phylogenetic, biochemical, and spectroscopic analyses of them have revealed that the opsin-based pigments have diversified in wavelength sensitivity (ultraviolet [UV] to red) and G-protein selectivity (Gs, Gq, Gi, Go, and transducin [Gt]) (Koyanagi and Terakita 2014; Terakita 2005; Terakita et al. 2012).

Chemical signals, such as hormones and neurotransmitters, generally activate G-protein-mediated cascades and modulate various aspects of cellular physiology. Therefore, application of diverse animal opsins to optogenetics could provide an opportunity to control various cellular functions using light. In this review, we discuss the diversity of animal opsin-based pigments and their optogenetic potentials, particularly focusing on 'bleach-resistant' pigments, each of which is coupled to different G-protein-mediated signal transduction cascades.

6.2 Diversity of Animal Opsins

Animal visual pigments and related photosensitive pigments consist of a protein moiety opsin and a chromophore retinal (Fig. 6.1). Therefore, in this chapter, they are referred to as opsin-based pigments. To date, four different chromophore retinals (retinal, 3,4-dehydroretinal, 3-hydroxyretinal, and 4-hydroxyretinal) have been identified. Binding of each chromophore to an opsin basically forms a pigment that has a specific spectral sensitivity. On the other hand, thousands of opsins have been identified in a wide variety of animals thus far and this variety of opsins is primarily responsible for the diversity of opsin-based 'pigments'.

Opsins have seven transmembrane helices as GPCRs, and a highly conserved lysine residue in the seventh helix binds to a retinal as a chromophore to form a pigment. Most opsins bind to the chromophore 11-*cis* retinal, which is isomerized to an all-*trans* form upon light irradiation. This isomerization triggers a structural change in opsins near the chromophore and subsequently in cytoplasmic regions, which interact with a G protein. Therefore, the structural diversities of opsins near a chromophore and in the cytoplasmic region primarily contribute to the varying spectral sensitivities and selective activation of different G-protein-mediated signal transduction cascades, respectively.

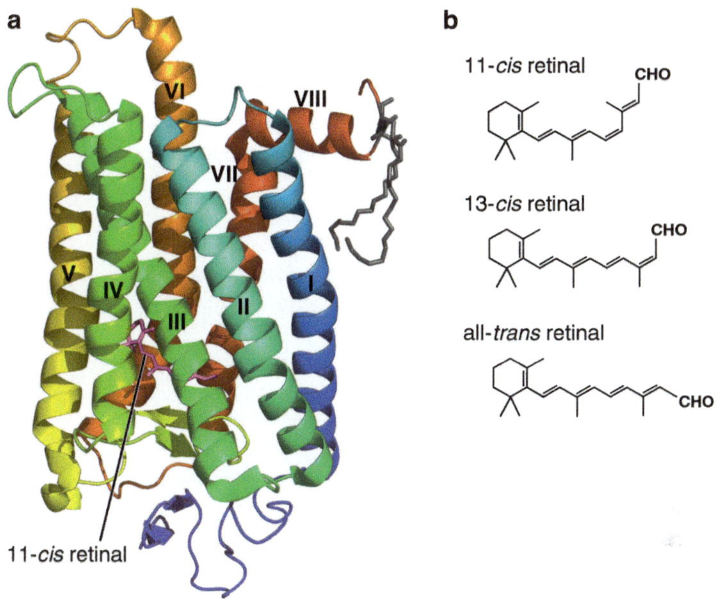

Fig. 6.1 (**a**) Crystal structure of bovine rhodopsin (Protein Data Bank ID: 1GZM). Its seven transmembrane alpha helices are numbered. Lys296 in helix VII covalently binds to 11-*cis* retinal via a Schiff base linkage. (**b**) Chemical structures of the 11-*cis*, 13-*cis* and all-*trans* forms of retinal

A molecular phylogenetic tree for opsins reveals that opsins can be divided into eight groups (Fig. 6.2, Terakita et al. 2012). For example, humans have nine opsin genes, each of which belongs to any one of six of these eight groups. The amino acid sequence identities among opsins that belong to different groups are 20–25 %. Interestingly, this classification also approximately corresponds to molecular functions. Members of six groups, including Gt-coupled opsins (vertebrate visual and non-visual opsins), Gq-coupled opsins (invertebrate Gq-coupled visual opsins and opsin 4 [Opn4] or melanopsin), invertebrate Go-coupled opsins, Gi/Go-coupled opsins (opsin 3 [Opn3] or encephalopsin and teleost multiple tissue [TMT] opsin), Gi-coupled opsin 5 (Opn5) or neuropsin, and cnidarian Gs-coupled opsins have been shown to function as light-sensing GPCRs. Members of the other two groups, peropsin and retinochrome/retinal G protein-coupled receptor (RGR), are considered to be retinal photoisomerases that generate 11-*cis* retinal.

6.3 Bleach-Resistant Pigments

Vertebrate visual opsins that have been conventionally used as optogenetic tools belong to the Gt-coupled opsin group (Fig. 6.2, Airan et al. 2009; Masseck et al. 2014). Bovine rhodopsin, which is one of the best studied members in this group,

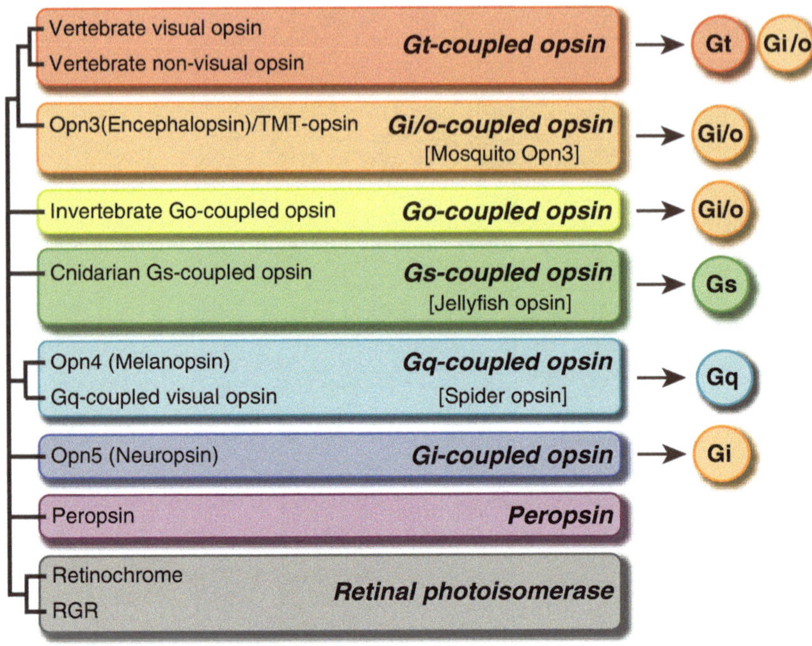

Fig. 6.2 Eight opsin groups and the particular G-protein subtypes to which they are predominantly coupled

releases a retinal chromophore and bleaches (becomes colorless) after light activation (Figs. 6.3 and 6.4). In contrast, most opsins, including spider opsin, jellyfish opsin, and mosquito Opn3, form bleach-resistant pigments that do not release a chromophore after light absorption. Moreover, most of the photoproducts of bleach-resistant pigments revert to their original dark states upon subsequent light absorption (Tsukamoto and Terakita 2010). These opsin-based pigments that have two stable states, a dark state and a photoproduct that has photoregeneration ability, are called bistable pigments.

Opsins require a chromophore retinal in order to have a photoreceptive function (i.e., capture light information and subsequently activate a G protein). As noted above, most opsin-based pigments bind to an 11-*cis* retinal as the chromophore. Tissues located outside photoreceptor organs, such as eyes and pineal organs, contain significantly less 11-*cis* retinal because these tissues do not express the enzymes that generate 11-*cis* retinoids, such as photoisomerases (RGR/retinochrome) and RPE65, which are specifically expressed in photoreceptor organs (Båvik et al. 1992; Jiang et al. 1993; Strauss 2005). The photoregeneration property of bistable pigments (i.e., reuse of a retinal chromophore) may allow an opsin to function in tissues located outside these photoreceptor organs. That is, the bistable nature of these particular pigments may be advantageous for their use as optogenetic tools.

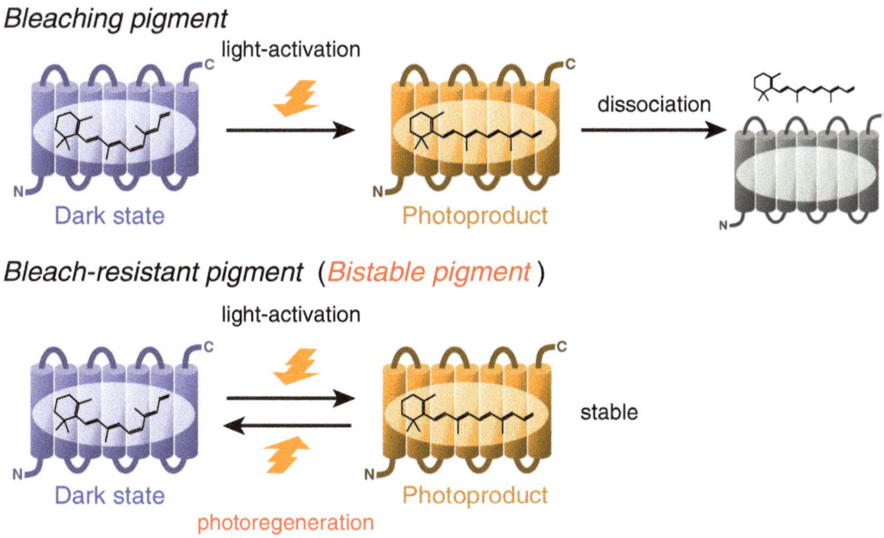

Fig. 6.3 Photoreactions of bleaching pigments and bleach-resistant pigments. The photoproducts of bleaching pigments are thermally unstable and release their chromophores. Bleach-resistant pigments photoconvert to stable photoproducts. The photoproducts of bistable pigments revert to their original dark state upon light absorption

Fig. 6.4 Absorption spectra of (**a**) bovine rhodopsin, (**b**) spider opsin, (**c**) jellyfish opsin, and (**d**) mosquito Opn3 in the dark (*black curves*) and after light irradiation (*broken curves*)

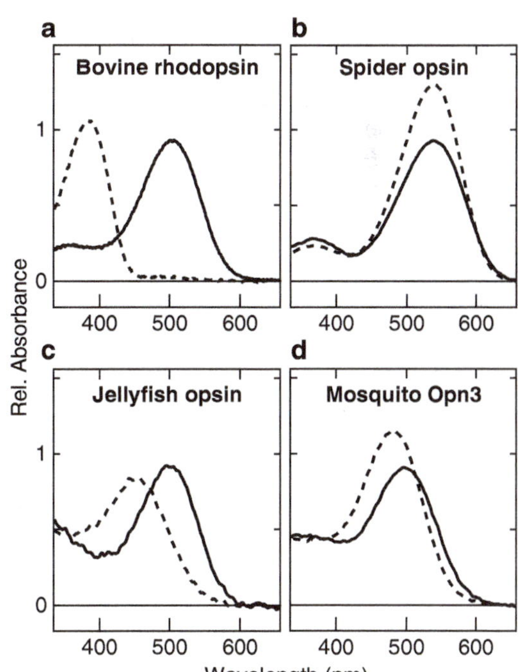

6.4 Molecular Characteristics of Gq, Gs, and Gi/Go-Coupled Bleach-Resistant Pigments

We have successfully obtained various functional opsins by expressing them in mammalian cultured cells and have characterized their opsin-based pigments spectroscopically and biochemically. Here, we review the molecular properties of three typical opsin-based pigments: spider opsin, jellyfish opsin, and mosquito Opn3-based pigments. Each of these has unique molecular properties in addition to its bistable nature.

6.4.1 Spider Opsin: Gq-Coupled Opsin

Spider opsin was isolated from the jumping spider *Hasarius adansoni* (Koyanagi et al. 2008a; Nagata et al. 2012) and belongs to the Gq-coupled opsin group, which comprises Gq-coupled visual opsins and melanopsins (Fig. 6.2) (Koyanagi and Terakita 2008). Gq-coupled visual opsins, including spider opsin, serve as visual pigments in invertebrates, including arthropods and molluscs. In the visual cells of invertebrates, Gq-coupled opsin-based pigments light dependently activate a Gq-type G protein (Terakita et al. 1993; Koyanagi and Terakita 2008), which results in the depolarization of these cells (Yau and Hardie 2009). Melanopsin also activates a Gq-mediated cascade in amphioxus photoreceptor cells (Koyanagi et al. 2005; Terakita et al. 2008). However, in vertebrates, including mammals, it remains unclear what transduction cascades melanopsins drive in vivo, although vertebrate melanopsin activates Gq *in vitro*.

We successfully obtained several Gq-coupled opsin-based pigments, which had different wavelength sensitivities, by expressing these opsins in cultured cells. These included UV- and blue-sensitive honeybee opsin-based pigments (Terakita et al. 2008), blue- and violet-sensitive butterfly opsin-based pigments (Wakakuwa et al. 2010), green-sensitive spider opsin-based pigment (Nagata et al. 2012), and blue-sensitive amphioxus melanopsin (Koyanagi et al. 2005).

In our expression system that used mammalian cultured cells, spider opsin could be expressed in the largest amount among the Gq-coupled opsins and thus, it could be characterized spectroscopically with high resolution. Spider opsin bound to 11-*cis* retinal as a chromophore and formed a bistable pigment that had an absorption maximum at 535 nm both while in the dark state and after light irradiation (Fig. 6.4b). As described below, spider opsin-based pigment initiates a Gq-signaling cascade in a light-dependent manner and causes increases in intracellular Ca^{2+} levels in mammalian cultured cells (Fig. 6.5).

6.4.2 *Jellyfish Opsin: Gs-Coupled Opsin*

Jellyfish opsin was identified from the lens eyes of a box jellyfish (Koyanagi et al. 2008b) and belongs to the Gs-coupled opsin group (Fig. 6.2), which also includes the opsins of cnidarians such as hydra and sea anemone. Jellyfish opsin-based pigment has an absorption maximum at approximately 500 nm in the dark state; upon irradiation, it converts to its photoproduct that has an absorption maximum in the visible region (Fig. 6.4c). In our experimental conditions for a purified sample, which included a detergent, the photoproduct was thermally stable but did not revert to the original dark state upon subsequent light irradiation. This suggested that jellyfish opsin was a bleach-resistant pigment but did not undergo a photoregeneration reaction in the detergent solution. It would be of interest to investigate whether jellyfish opsin-based pigment functions as a bistable pigment in photoreceptor cells.

Jellyfish opsin is co-localized with a large amount of Gs protein and adenylyl cyclase in jellyfish visual cells (Koyanagi et al. 2008b). Jellyfish opsin light dependently initiates a Gs and adenylyl cyclase signaling cascade in mammalian cultured cells (Koyanagi et al. 2008b) (Fig. 6.5). In addition, upon light irradiation, increased cyclic adenosine monophosphate (cAMP) levels were observed in jellyfish eyes, suggesting the activation of a Gs-adenylyl cyclase cascade upon light absorption by jellyfish opsin. These findings showed that jellyfish opsin functioned as a Gs-coupled visual opsin-based pigment in vivo. Recently, Balies et al. reported that jellyfish opsin had optogenetic potential as a Gs-coupled bleach-resistant pigment (Bailes et al. 2012).

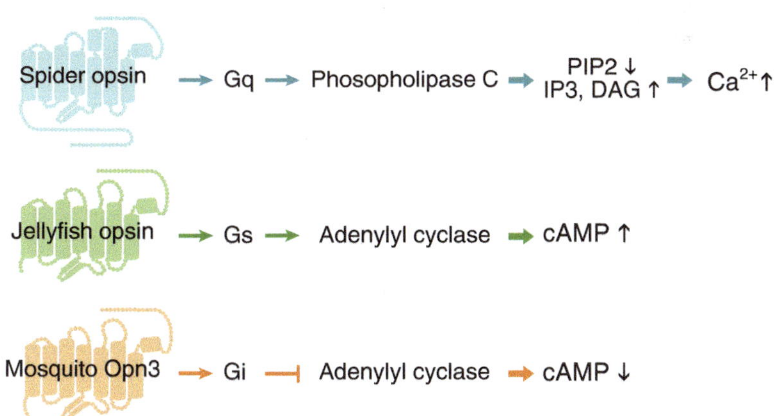

Fig. 6.5 G-protein signaling cascades initiated by each opsin-based pigment

6.4.3 Mosquito Opn3: Gi/Go-Coupled Opsin

Mosquito Opn3 is an invertebrate homolog of vertebrate Opn3 (Fig. 6.2) that was first identified in the mouse brain (encephalon) and therefore called encephalopsin (Blackshaw and Snyder 1999). Invertebrate Opn3 has also been found in insects, including mosquitoes, bees, moths, beetles (Hill et al. 2002; Nene et al. 2007; Velarde et al. 2005), an annelid worm (Arendt et al. 2004), and sea urchins (Ooka et al. 2010; Sodergren et al. 2006). Opn3 messenger RNA (mRNA) is expressed in the brains of annelid worms and honeybees (Arendt et al. 2004; Velarde et al. 2005). Mammals have only one Opn3, whereas most non-mammalian vertebrates have homologs of Opn3, TMT opsin(s), which was first identified from multiple tissues including the brain in teleosts (Moutsaki et al. 2003), in addition to Opn3. Although Opn3 homologs are found in a wide variety of animals, their physiological roles remain uncertain.

We recently succeeded in expressing mosquito Opn3 in mammalian cultured cells and determining its molecular properties (Koyanagi et al. 2013). Mosquito Opn3 binds to 11-*cis* retinal as a chromophore and forms a pigment with an absorption maximum at approximately 500 nm (Fig. 6.4d). Upon irradiation, this pigment converts to its photoproduct, the absorption maximum of which is blue-shifted in the visible region. This photoproduct is thermally stable and reverts to its original dark state upon irradiation, indicating that mosquito Opn3 can form a bistable pigment. Mosquito Opn3-based pigment light dependently activates Gi and Go in vitro and can initiate a Gi-signaling cascade in mammalian cultured cells (Fig. 6.5) (Koyanagi et al. 2013).

A potential advantage of mosquito Opn3 as an optogenetic tool is that it can bind to 13-*cis* retinal (Fig. 6.1). As described above, opsins, including the non-visual opsins that are expressed in extraocular tissues, generally require 11-*cis* retinal as the chromophore to function as light-sensing proteins. However, tissues located outside photoreceptor organs (eyes and pineal organs) contain significantly less 11-*cis* retinal because these tissues do not express the enzymes that generate 11-*cis* retinoids, such as photoisomerases (RGR/retinochrome) and RPE65, which are specifically expressed in photoreceptor organs.

Interestingly, mosquito Opn3 can bind to 13-*cis* retinal, which is ubiquitously present in the body, because the 13-*cis* form is generated in thermal equilibrium with the all-*trans* form, the predominant isomer. The 13-*cis* retinal-bearing mosquito Opn3 converts to its photoproduct upon light irradiation and this photoproduct, in turn, reverts to the 11-*cis* retinal-bearing pigment upon subsequent light irradiation. This 13-*cis* retinal-bearing pigment activates Gi and Go in a light-dependent manner similar to the 11-*cis* retinal-bearing pigment. These observations prompted us to speculate that mosquito Opn3 can form a pigment and retain its photosensitivity in extraocular tissues, including the brain, which do not contain sufficient 11-*cis* retinal.

6.5 Optogenetic Control of G-Protein Signaling by Animal Opsins

As described above, spider opsin, jellyfish opsin, and mosquito Opn3 activate Gq, Gs, and Gi/Go, respectively. In general, activating Gq, Gs, or Gi/Go results in changes in second messenger levels in cells, such as Ca^{2+} and cAMP levels (Fig. 6.5). To investigate the optogenetic potential of each opsin, we analyzed light-induced changes in second messenger levels in cultured cells that expressed spider opsin, jellyfish opsin, or mosquito Opn3.

Light-induced Ca^{2+} responses by HEK293 cells that expressed Gq-coupled spider opsin were investigated using fura-2, a ratiometric fluorescent Ca^{2+} indicator (Fig. 6.6a). The ratio of fluorescence with excitations at 340 and 380 nm increased

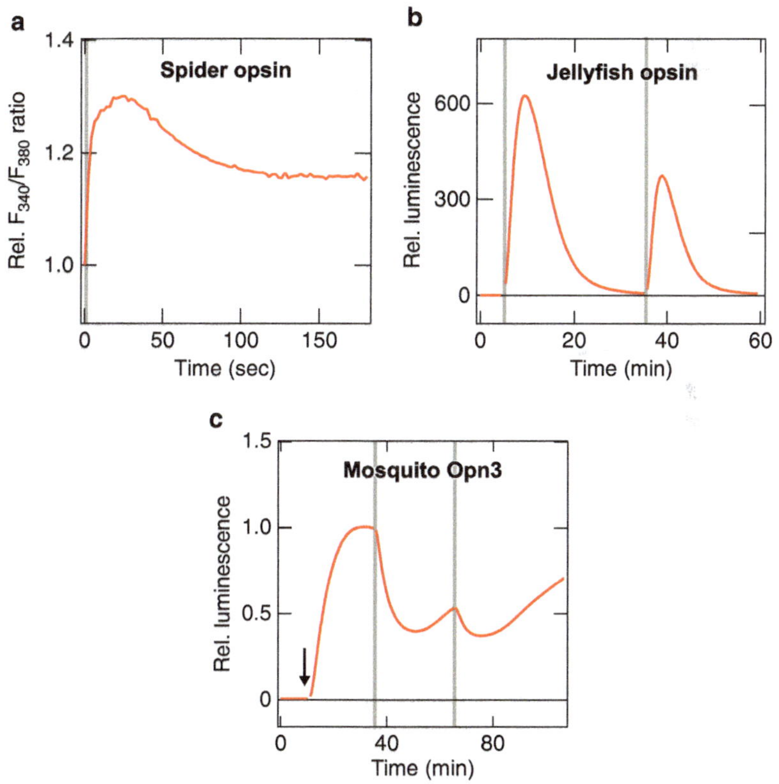

Fig. 6.6 Light regulation of G-protein signaling in mammalian cultured cells. (**a**) Light-induced increases in intracellular Ca^{2+} levels in HEK293 cells that express spider opsin. Ca^{2+} levels were probed with fura-2, a ratiometric fluorescent Ca^{2+} indicator. (**b, c**) Light-induced cyclic adenosine monophosphate (cAMP) increases and decreases probed with GloSensor, a cAMP-dependent luminescent protein, in HEK293 cells expressing jellyfish opsin and mosquito Opn3, respectively. The *arrow* indicates treatment with forskolin, an activator of adenylyl cyclase. The *grey bars* indicate visible light irradiation

immediately after light irradiation, which indicated a light-induced increase in intracellular Ca^{2+} levels. This clearly showed that spider opsin could initiate a Gq signaling cascade in cultured mammalian cells.

cAMP levels were monitored with GloSensor, a bioluminescent protein that emits luminescence when it binds to cAMP, for Gs-coupled jellyfish opsin- and Gi/Go-coupled mosquito Opn3-expressing cells (Fig. 6.6b and c). HEK293 cells that expressed jellyfish opsin exhibited increases in luminescence intensity upon light irradiation, which indicated that light absorption by a jellyfish opsin-based pigment resulted in increased intracellular cAMP levels via activating Gs, as reported previously (Bailes et al. 2012; Koyanagi and Terakita 2014). In contrast, mosquito Opn3-expressing cells exhibited light-dependent decreases in luminescence intensity (i.e., cAMP levels), which had been previously increased by forskolin treatment (Fig. 6.6c). This demonstrated light-dependent downregulation of cAMP via Gi activation by a mosquito Opn3-based pigment in cultured mammalian cells (Koyanagi et al. 2013).

6.6 Conclusions

Gq-coupled visual opsin-based pigments, having a bistable nature, vary in their wavelength sensitivities. In addition, bistable visual and non-visual opsin-based pigments are distributed among eight classes and this classification approximately corresponds to different molecular functions (e.g., selective activation of G proteins). Chimera mutants for these diverse opsins would allow us to generate opsins that activate particular G proteins on the basis of particular colors of light stimuli. These bistable opsins, including mutant opsins, could be powerful tools for optogenetic applications.

Acknowledgments This work was supported in part by grants-in-aid for Scientific Research from the Japanese Ministry of Education, Science, Sports and Culture (to A. T. and M. K.).

References

Airan RD, Thompson KR, Fenno LE et al (2009) Temporally precise in vivo control of intracellular signalling. Nature 458(7241):1025–1029

Arendt D, Tessmar-Raible K, Snyman H et al (2004) Ciliary photoreceptors with a vertebrate-type opsin in an invertebrate brain. Science 306(5697):869–871

Bailes HJ, Zhuang LY, Lucas RJ (2012) Reproducible and sustained regulation of Galphas signalling using a metazoan opsin as an optogenetic tool. PLoS One 7(1):e30774

Båvik CO, Busch C, Eriksson U (1992) Characterization of a plasma retinol-binding protein membrane receptor expressed in the retinal pigment epithelium. J Biol Chem 267(32):23035–23042

Bernstein JG, Boyden ES (2011) Optogenetic tools for analyzing the neural circuits of behavior. Trends Cogn Sci 15(12):592–600

Blackshaw S, Snyder SH (1999) Encephalopsin: a novel mammalian extraretinal opsin discretely localized in the brain. J Neurosci 19(10):3681–3690
Boyden ES (2011) A history of optogenetics: the development of tools for controlling brain circuits with light. F1000 Biol Rep 3:11
Deisseroth K (2011) Optogenetics. Nat Methods 8(1):26–29
Fenno L, Yizhar O, Deisseroth K (2011) The development and application of optogenetics. Annu Rev Neurosci 34:389–412
Hegemann P, Nagel G (2013) From channelrhodopsins to optogenetics. EMBO Mol Med 5(2):173–176
Hill CA, Fox AN, Pitts RJ et al (2002) G protein-coupled receptors in Anopheles gambiae. Science 298(5591):176–178
Jiang M, Pandey S, Fong HK (1993) An opsin homologue in the retina and pigment epithelium. Invest Ophthalmol Vis Sci 34(13):3669–3678
Koyanagi M, Terakita A (2008) Gq-coupled rhodopsin subfamily composed of invertebrate visual pigment and melanopsin. Photochem Photobiol 84(4):1024–1030
Koyanagi M, Terakita A (2014) Diversity of animal opsin-based pigments and their optogenetic potential. Biochim Biophys Acta 1837(5):710–716
Koyanagi M, Kubokawa K, Tsukamoto H et al (2005) Cephalochordate melanopsin: evolutionary linkage between invertebrate visual cells and vertebrate photosensitive retinal ganglion cells. Curr Biol 15(11):1065–1069
Koyanagi M, Nagata T, Katoh K et al (2008a) Molecular evolution of arthropod color vision deduced from multiple opsin genes of jumping spiders. J Mol Evol 66(2):130–137
Koyanagi M, Takano K, Tsukamoto H et al (2008b) Jellyfish vision starts with cAMP signaling mediated by opsin-G(s) cascade. Proc Natl Acad Sci U S A 105(40):15576–15580
Koyanagi M, Takada E, Nagata T et al (2013) Homologs of vertebrate Opn3 potentially serve as a light sensor in nonphotoreceptive tissue. Proc Natl Acad Sci U S A 110(13):4998–5003
Masseck OA, Spoida K, Dalkara D et al (2014) Vertebrate cone opsins enable sustained and highly sensitive rapid control of Gi/o signaling in anxiety circuitry. Neuron 81(6):1263–1273
Moutsaki P, Whitmore D, Bellingham J et al (2003) Teleost multiple tissue (tmt) opsin: a candidate photopigment regulating the peripheral clocks of zebrafish? Brain Res Mol Brain Res 112(1–2):135–145
Nagata T, Koyanagi M, Tsukamoto H et al (2012) Depth perception from image defocus in a jumping spider. Science 335(6067):469–471
Nene V, Wortman JR, Lawson D et al (2007) Genome sequence of Aedes aegypti, a major arbovirus vector. Science 316(5832):1718–1723
Ooka S, Katow T, Yaguchi S et al (2010) Spatiotemporal expression pattern of an encephalopsin orthologue of the sea urchin Hemicentrotus pulcherrimus during early development, and its potential role in larval vertical migration. Dev Growth Differ 52(2):195–207
Sodergren E, Weinstock GM, Davidson EH et al (2006) The genome of the sea urchin Strongylocentrotus purpuratus. Science 314(5801):941–952
Strauss O (2005) The retinal pigment epithelium in visual function. Physiol Rev 85(3):845–881
Terakita A (2005) The opsins. Genome Biol 6(3):213
Terakita A, Hariyama T, Tsukahara Y et al (1993) Interaction of GTP-binding protein Gq with photoactivated rhodopsin in the photoreceptor membranes of crayfish. FEBS Lett 330(2):197–200
Terakita A, Tsukamoto H, Koyanagi M et al (2008) Expression and comparative characterization of Gq-coupled invertebrate visual pigments and melanopsin. J Neurochem 105(3):883–890
Terakita A, Kawano-Yamashita E, Koyanagi M (2012) Evolution and diversity of opsins. Wiley Interdiscip Rev Membr Transp Signal 1(1):104–111
Tsukamoto H, Terakita A (2010) Diversity and functional properties of bistable pigments. Photochem Photobiol Sci 9(11):1435–1443
Velarde RA, Sauer CD, Walden KK et al (2005) Pteropsin: a vertebrate-like non-visual opsin expressed in the honey bee brain. Insect Biochem Mol Biol 35(12):1367–1377

Wakakuwa M, Terakita A, Koyanagi M et al (2010) Evolution and mechanism of spectral tuning of blue-absorbing visual pigments in butterflies. PLoS One 5(11):e15015
Yau KW, Hardie RC (2009) Phototransduction motifs and variations. Cell 139(2):246–264
Zalocusky K, Deisseroth K (2013) Optogenetics in the behaving rat: integration of diverse new technologies in a vital animal model. Optogenetics 1:1–17

Chapter 7
Color Tuning in Retinylidene Proteins

Kota Katayama, Sivakumar Sekharan, and Yuki Sudo

Abstract Retinylidene proteins (also called rhodopsins) are membrane-embedded photoreceptors that contain a vitamin A aldehyde linked to a lysine residue by a Schiff base as their light-sensing chromophore. The chromophore is surrounded by seven-transmembrane α-helices and absorbs light at different wavelengths due to differences in the electronic energy gap between its ground and excited states. The variation in the wavelength of maximal absorption (λ_{max}: 360–620 nm) of rhodopsins arises due to interaction between the apoprotein (opsin) and the retinyl chromophore, the 'opsin shift'. This chapter reviews the color tuning mechanisms in type-1 microbial and type-2 animal rhodopsins as revealed mainly by our experimental and theoretical studies.

Keywords Retinal • Color tuning • Rhodopsin • π-conjugation • Color variant • Visible light • Water molecule • Vitamin-A

7.1 Introduction

Light is one of the most important energy sources and environmental signals for organisms. To absorb the visible light, the photoactive proteins have cognate chromophore molecules within the protein moiety (Briggs and Spudich 2005). The light-absorbing photoactive proteins exist in various organisms, and they show a variety of colors originating from an energy gap between the electronic ground and excited states, which leads to changes in the color (absorption maximum:λ_{max}) (Fig. 7.1). It is important to know the molecular mechanism of these color changes,

K. Katayama
Department of Frontier Materials, Nagoya Institute of Technology, Nagoya 466-8555, Japan

S. Sekharan
Department of Chemistry, Yale University, New Haven, CT 06520-8107, USA

Y. Sudo, PhD (✉)
Division of Pharmaceutical Sciences, Graduate School of Medicine, Dentistry and Pharmaceutical Sciences, Okayama University,
1-1-1 Tsushima-naka, Kita-ku, Okayama 700-8530, Japan
e-mail: sudo@pharm.okayama-u.ac.jp

H. Yawo et al. (eds.), *Optogenetics*, DOI 10.1007/978-4-431-55516-2_7

not only to understand the biological functions, but also to utilize it as a tool in fields such as optogenetics (Williams and Deisseroth 2013).

Seven-transmembrane (TM) proteins with retinal (vitamin A aldehyde) as a chromophore are called 'retinylidene proteins' (or rhodopsin) and have been divided into two types (type-1 for microbial and type-2 for animal) (Spudich et al. 2000). The retinylidene proteins are covalently bound to an 11-*cis* for type-2 animal rhodopsins or to an all-*trans* retinyl chromophore for type-1 microbial rhodopsins at a conserved lysine residue on the 7th(G)-helix via a protonated Schiff base (PSB) linkage (a carbon-nitrogen double bond) (Fig. 7.1) (Spudich et al. 2000). Photo-absorption of the retinal molecule triggers isomerization at a specific double bond, resulting in biological functions such as vision, ion transportation, and morphogenesis (Spudich et al. 2000; Irieda et al. 2012). One of the most notable properties of retinylidene proteins is the large variation in their absorption spectrum depending on their interaction between the opsin (apoprotein) and the retinyl chromophore, which is called the opsin shift (Fig. 7.2a) (Spudich et al. 2000). Experimental, empirical, and theoretical studies of rhodopsin have suggested several mechanisms for spectral tuning (Fig. 7.2b). They are roughly categorized into three factors: (1) electrostatic interaction between the retinyl chromophore and its counter-ion; (2) an alteration in the polarity or polarizability of the environment of the chromophore-binding site, caused by the arrangement of polar and/or non-polar residues in the protein moiety; and (3) twisting of the 6-S bond (6,7-torsion angle) connecting the polyene chain to the β-ionone ring as well as distortion of the chromophore due to steric interactions with residues lining the binding pocket. With the combina-

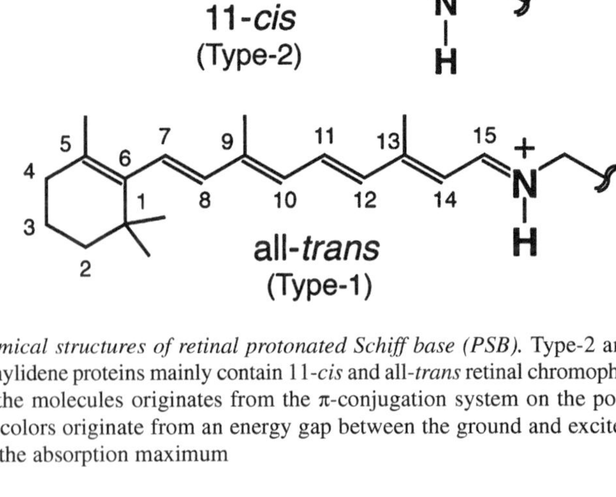

Fig. 7.1 *Chemical structures of retinal protonated Schiff base (PSB).* Type-2 animal and type-1 microbial retinylidene proteins mainly contain 11-*cis* and all-*trans* retinal chromophore, respectively. The color of the molecules originates from the π-conjugation system on the polyene chain. The characteristic colors originate from an energy gap between the ground and excited states, leading to changes in the absorption maximum

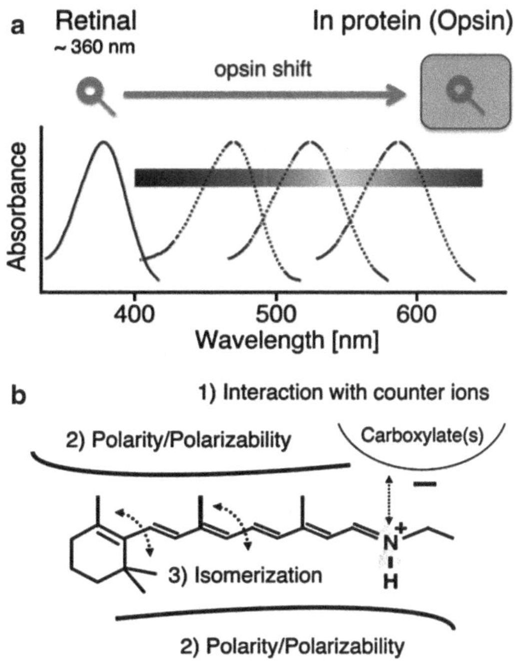

Fig. 7.2 *Spectral tuning in protein (opsin) and molecular origin of the opsin shift.* (**a**) The absorption spectra from retinal chromophore colored yellow in organic solution were greatly affected by a protein environment. This is called 'Opsin shift'. (**b**) Color tuning mechanisms in retinylidene proteins. There are some essential factors for the opsin shift, as shown in brackets *1*), *2*), and *3*)

tions of these effects, both type-1 and type-2 retinylidene proteins show various absorption maxima in visible light region (360–620 nm).

7.2 Animal (Type-2) Rhodopsins

Visual pigments found in the rod outer segments of vertebrates and invertebrates are called rhodopsins, and those that are found in the cone outer segments are called cone pigments (Birge 1990). Rhodopsin mediates the transformation of light into dim-light vision, and cone pigments mediate the transformation of light into trichromatic color vision (blue [~420 nm], green [~530 nm], and red [~560 nm) in humans and/or mammals (Wald 1935; Nathans et al. 1986; Oprian et al. 1991; Merbs and Nathans 1992; Shichida and Imai 1998).

Some freshwater amphibians, fishes, and crustaceans have been found to contain 3,4-hydroretinal or A_2-retinal (Wald 1936, 1937a, b, 1939; Vogt 1983), while insects have been found to possess 3-hydroxyretinal or A_3-retinal (Suzuki et al. 1985). As a result, visual pigments containing A_1-retinal are called rhodopsins, those containing A_2-retinal are porphyropsins, and those containing A_3-retinal are xanthopsins (Matsui et al. 1988; Nakanishi 1991), Firefly squid (Watasenia scintillans), which exhibit intense bioluminescence from large photophores, was the only animal found to be utilizing three visual pigments with three different chromophores for color vision, namely, A_1-, A_2-, and 4-hydroxyretinal or A_4-retinal (Seidou et al. 1990;

Kito et al. 1992; Michinomae et al. 1994; Provencio et al. 1998). In addition, several thousand opsins have been identified and their function characterized (Fig. 7.3), including the non-visual ocular photoreceptor melanopsin, and they all absorb blue light to mediate physiological and/or behavior responses, including the entrainment of circadian rhythms (Provencio et al. 2000; Yokoyama 2008).

One of the fundamental characteristics used to distinguish different visual pigments from each other is their λ_{max}. On the basis of the wavelength of their peak light sensitivity, visual pigments are classified into five subgroups: rhodopsin pigments RH1 (~500 nm) and RH2 (~535 nm), short-wavelength-sensitive pigments SWS2 (~450 nm), SWS1 (~360 nm), and middle/long-wavelength-sensitive pigments (M/LWS, 530–600 nm) (Fig. 7.3). While visible-absorbing pigments contain 11-*cis*-retinal PSB (PSB11), ultraviolet (UV)-absorbing pigments uniquely contain an 11-*cis*-retinal unprotonated Schiff base (USB11) (Hubbard and George 1958).

In the vertebrate rod and cone pigments, a negatively charged Glu residue at site 113, Glu113, serves as the counter-ion to the PSB11 (Fig. 7.4) (Hara et al. 1967; Hara-Nishimura et al. 1993; Okada et al. 2004; Shimamura et al. 2008), whereas in the invertebrate pigments, a neutral Tyr111 residue was found to occupy that position (Fig. 7.4). Furthermore, only one glutamate (Glu180) residue was located in the interior of the invertebrate (squid) pigment; it is situated ~5 Å away from PSB11 and corresponds to the position of Glu181 in the vertebrate (bovine) pigment. All other glutamates in the invertebrate pigment are located at the surface and more than 17 Å away from the PSB11 (Murakami and Kouyama 2008; Sekharan et al. 2010a). Computational investigation of the X-ray structure and spectral tuning mechanism using hybrid quantum mechanics and molecular mechanics (QM/MM) methods identified Glu180 as the counter-ion in the invertebrate (squid, butterfly, firefly squid) pigments (Jongejan et al. 2005; Murakami and Kouyama 2008; Sekharan et al. 2011).

Internal water molecules have been recently shown to mediate activation signals from the retinal binding pocket to the cytoplasmic surface of rhodopsin and rhodopsin-like G-protein-coupled receptors (GPCRs) (Yan et al. 2003; Xu et al. 2005; Garczarek and Gerwert 2006; Pardo et al. 2007). In bovine rhodopsin, the two water molecules near the Schiff base (SB) region are suggested to participate in the counter-ion switch mechanism from Glu113 to Glu181 during the formation of metarhodopsin-I, which is an early photointermediate and a precursor to metarhodopsin-II that binds and activates the G-protein, transducin (Pal et al. 2013). Among the type-1 microbial rhodopsins, bacteriorhodopsin (BR) and halorhodopsin pump ions across the cell membrane and contain three water molecules near the SB region that are shown to play a critical role in the chromophore counter-ion interactions (Garczarek and Gerwert 2006; Pal et al. 2013). Xanthorhodopsin, a proton pump with a light-harvesting carotenoid antenna, contains a bridged-water molecule between the SB and the counter-ions (Balashov et al. 2005; Luecke et al. 2008).

By using the highly accurate Fourier transform infrared (FTIR) spectra, protein-bound water molecules in the red and green cone pigments have been identified (Katayama et al. 2012). The water signals were found to differ not only from those of rhodopsin but also between the red and green cone pigments (Fig. 7.5). The

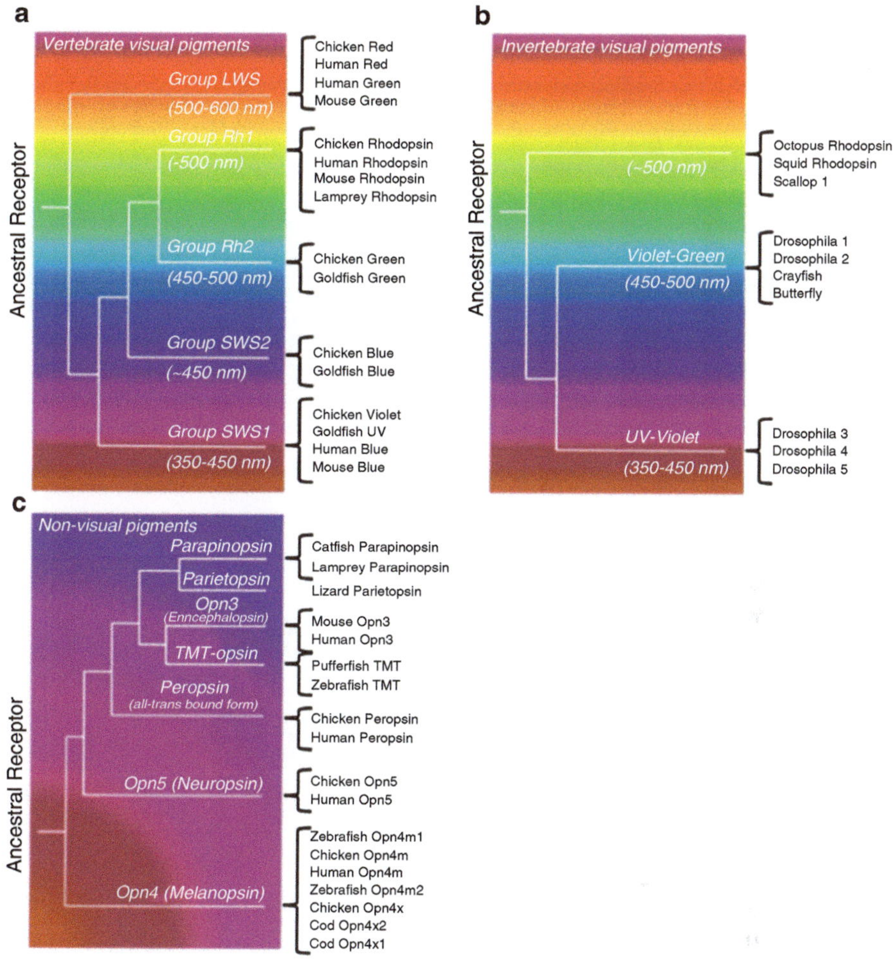

Fig. 7.3 *Phylogenetic tree showing the relationship of the various classes of vertebrate* (**a**)*, invertebrate* (**b**)*, and non-visual* (**c**) *opsins.* The tree was generated by the neighbor-joining method (Saitou and Nei 1987). It indicates that vertebrate opsin families are classified into two subfamilies: Opn1 for cone opsins and Opn2 for rod opsin, rhodopsin. Bovine rhodopsin was the first opsin to be cloned and sequenced (Nathans and Hogness 1983) and is the only vertebrate opsin for which the crystal structure has been determined (Palczewski et al. 2000). The cone opsins can be further classified into four subgroups, which correspond well with their absorption spectra: group long-wave sensitive (LWS), with a λ_{max} of 500–600 nm; group Rh2 or middle-wave sensitive (MWS), with a λ_{max} of 450–500 nm; group ultraviolet-violet sensitive (SWS1), with a λ_{max} of 350–450 nm; and group short-wave sensitive (SWS2), with a λ_{max} within 450 nm. The invertebrate opsin family is also divided into two subfamilies: cone opsins and rod opsin. Squid rhodopsin was the second opsin to have its crystal structure determined (Shimamura et al. 2008; Murakami and Kouyama 2008). The cone opsins of invertebrate can be only subdivided into two subgroups: violet-green sensitive with a λ_{max} of 450–500 nm, which includes drosophila1, 2, crayfish, and butterfly and ultraviolet-violet sensitive, with a λ_{max} of 350–450 nm, which includes drosophila3, 4, and 5. The non-visual opsin family, which also responds to light is divided into seven subfamilies: parapinopsin, first isolated from teleost parapineal; parietopsin; encephalopsin (Opn3) expressed in the retina and in all tissues like brain, testis, heart, liver, and kidney; teleost multiple tissue (TMT), which exhibit circadian oscillations in clock gene expression; peropsin; neuropsin (Opn5) identified in 2003 and expressed in mouse testis, brain, and eye and in human retina and brain; and melanopsin (Opn4). These non-visual opsin families almost contain their λ_{max} in ultraviolet-violet regions

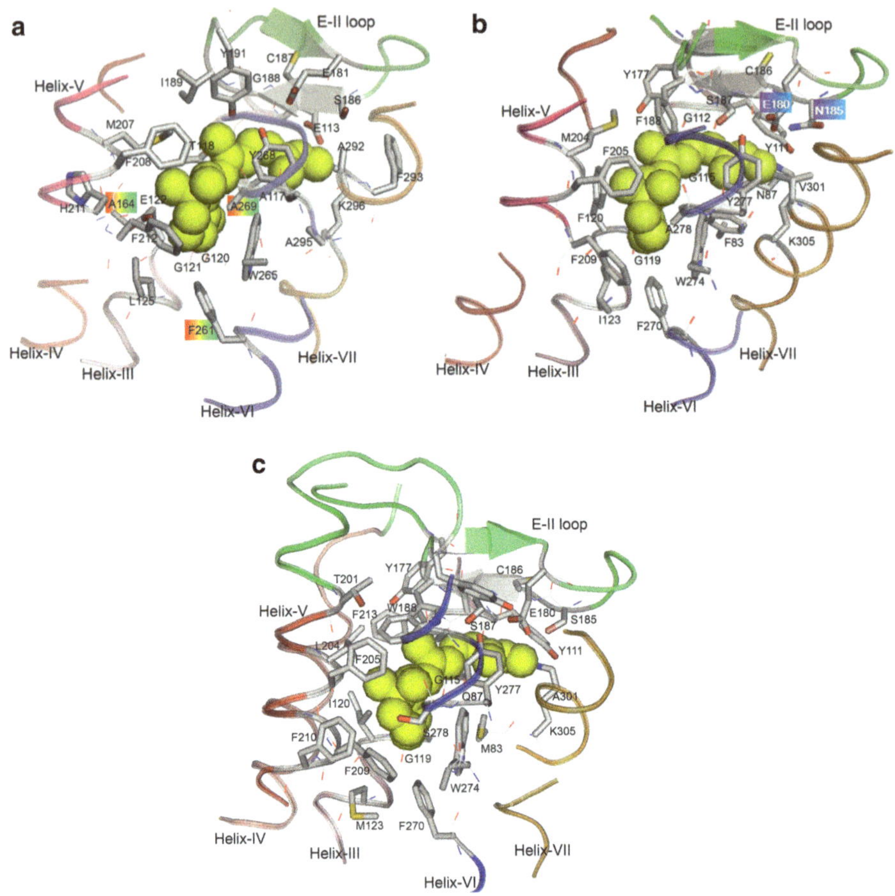

Fig. 7.4 *X-ray crystallographic structure of the chromophore-binding site of* (**a**) *bovine rhodopsin (Protein Data Bank [PDB] code: 1U19)* (Okada et al. 2002*),* (**b**) *squid rhodopsin (PDB code: 2Z73)* (Murakami and Kouyama 2008*),* (**c**) *and homology model of mouse melanopsin (PDB code: 5SJV)* (Sekharan et al. 2012a) *which is viewed from the helix VI side.* The upper and lower regions correspond to the extracellular and cytoplasmic sides, respectively. The retinal chromophore, which is bound to Lys296 or Lys305 or Lys305, is shown by yellow space-filling model, respectively. Side chains of the 27 amino acids within 5 Å from the retinal chromophore are shown by stick drawings, through some residues behind the retinal are hidden. Ribbon drawings illustrate the secondary structures around the retinal. Each root-mean-square deviation (RMSD) in which the RMS distance between corresponding residues for both bovine and squid rhodopsin for Helix-III, IV, V, VI, VII, and E-II loop are shown: 0.739, 1.634, 0.486, 0.188, 0.805, and 0.651 Å, respectively

protein-bound water molecules were found to be distributed randomly and the averaged frequencies of water confirmed the participation of water molecules in the wavelength regulation of visual pigments. In the absence of X-ray structures of non-visual opsins, the homology model of mouse melanopsin was built based on the QM/MM structure of squid rhodopsin (λ_{max} = 490 nm) (Sekharan et al. 2010a)

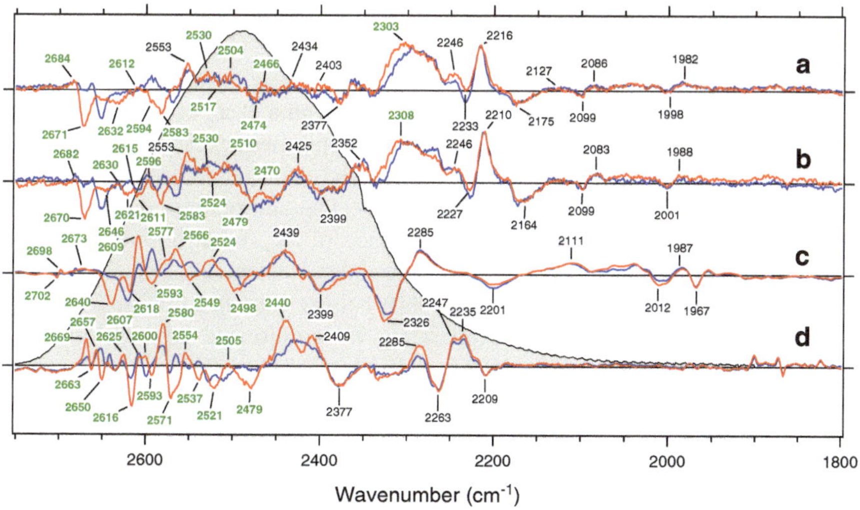

Fig. 7.5 *Water molecules in type-2 retinylidene proteins.* Light-minus-dark difference Fourier transform infrared (FTIR) spectra of monkey red (**a**), monkey green (**b**), bovine Rh (**c**) (Katayama et al. 2012), and squid Rh (**d**) (Ota et al. 2006), with 11-*cis*-retinyl, in the 2,750–1,800 cm^{-1} region measured at 77 K. *Red and blue lines* represent the spectra in D_2O and $D_2{}^{18}O$, respectively, and *green labeled* frequencies correspond to those identified as water-stretching vibrations. The *gray curve* in the 2,700–2,000 cm^{-1} region represents O–D stretching vibrations of D_2O at room temperature. One division of the *y*-axis corresponds to 0.00014 absorbance unit

because the photochemical properties of melanopsin are very similar to those of invertebrate photopigments (Koyanagi and Terakita 2008; Terakita et al. 2008). QM/MM studies revealed the presence of internal water molecules in mouse melanopsin, and the calculated maximal absorbencies at 447 and 484 nm are attributed to the differences in the presence and absence of water-mediated hydrogen-bonding interactions near the SB region (Sekharan et al. 2012a).

Color tuning of the visual pigments is defined as the difference in response between the excited and the ground states of the retinyl chromophore to the external perturbation from the protein environment. The spectroscopic oriented configuration interaction method with + Q Davidson correction (SORCI + Q) (Neese 2003) and 6-31G(d) basis set (Cornell et al. 1995) in conjunction with AMBER force field parameters was used to calculate the λ_{max} of PSB11 in the gas phase (QM-none) and in the protein environments (QM/MM). In the absence of any environmental perturbations (in *vacuo*), the *planar cis-* and *trans*-retinyl isomers absorb at around 610 nm (Andersen et al. 2005; Nielsen et al. 2006; Sekharan et al. 2006). Interaction of the chromophore with its counter-ion (Glu113 in vertebrates, Glu180 in invertebrates) induces a strong blue shift of ~120 nm and moves the calculated λ_{max} very close to the experimental value of ~500 nm for bovine and squid rhodopsins (Sekharan et al. 2007a, b). The shift is conceivable as the excited state charge density is shifted against the charge of the counter-ion (Sekharan and Buss 2008; Sekharan 2009) leading to change in the dipole moment $\Delta\mu_{(S1-S0)}$ calculated to be

~12.0 D, in excellent agreement with the experimental measurements (Sekharan and Morokuma 2011a, b).

In the case of M/LWS cone pigments, the molecular basis of variations in the λ_{max} of has been widely studied in the past. Yokoyama et al. compared the partial amino acid sequences of red and green color pigments from five mammalian orders, and showed that green pigments evolved from its red-sensitive ancestral pigments (Yokoyama and Radlwimmer 1998). The authors proposed the famous 'five-sites' rule for understanding the variance in λ_{max} of different M/LWS pigments. According to this rule, amino acid changes at sites 164, 181, 261, 269, and 292 would account for the 50 nm blue shift between the red (560 nm) and green (~510–530 nm) visual pigments. Apparently, the amino acid substitutions refer to replacement of SHYTA with AYFAS (S164A, H181Y, Y261F, T269A, A292S) composition. A simpler way to look at the rule is that, substitution of serine with alanine (S→A), histidine with tyrosine (H→Y), tyrosine with phenylalanine (Y→F), threonine with alanine (T→A), and alanine with serine (A→S) would lead to loss or gain of a hydroxyl group by the visual pigment, which in turn may cause the change in the observed λ_{max}.

One of the limitations of this rule is that it is pigment specific and thus can only be applied to the M/LWS pigments. Because all of the visual pigments contain the same PSB11 chromophore, any rule that helps us to understand the mechanism of color vision should be applicable to all visual pigments independent of its structural and/or its functional characteristics. However, to derive such a rule, one has to first decode the 'five-sites' rule, for it already accounts for variation in λ_{max} of the majority of the M/LWS pigments.

By studying the structural model and the spectral tuning mechanisms of monkey red- and green-sensitive visual pigments, an 'OH-site' rule that is shown to be site specific and not pigment specific (Sekharan et al. 2012b) was proposed. According to the 'OH-site' rule, if an amino acid contains an OH group and is present near the β-ionone ring part of the retinal in opsin, it will stabilize the excited state relative to the ground state and will induce a red shift in the λ_{max}. In contrast, if a hydroxyl-bearing amino acid is present near the SB site, it will stabilize the ground state relative to the excited state (Sekharan et al. 2012b) and will induce a blue shift in the λ_{max}. In other words, interaction of retinyl chromophore with dipolar residues near the β-ionone ring will increase the electron delocalization and decrease the bond length alternation (BLA) leading to red shifts, whereas, interaction with dipolar residues near the SB site will decrease the electron delocalization and increase the BLA, leading to blue shifts (Sekharan et al. 2012b). This principle was demonstrated using amino acid substitutions at sites 164, 261, and 269 (A164S, F261Y, and A269T) in the monkey M/LWS pigments, at site 292 (A292S, A292T, A292Y) in bovine RH1 pigment, and at site 111 (Y111) in squid RH1 pigment (Sekharan et al. 2012b). The findings confirmed that the 'OH-site' rule is indeed site specific and not pigment specific.

To explore the color tuning mechanism of butterfly pigments, A_3- and A_4 retinals were theoretically incorporated into the QM/MM optimized homology model of butterfly rhodopsin built using the QM/MM optimized squid rhodopsin model as a template (Sekharan et al. 2011). Contrary to general perception, the theoretical findings ruled out the formation of hydrogen bond between the hydroxyl-bearing β-ionone ring part of A_3- and A_4 retinals and opsin (Sekharan et al. 2011). Compared with the A_1 pigments, the color tuning mechanism of A_3 and A_4 pigments was found to involve an increase in BLA in conjunction with interactions of OH-bearing amino acids S116 and T185 and breaking of the hydrogen bond between sites E180 and T185 (Sekharan et al. 2011).

In the case of mouse melanopsin, the color tuning mechanism was attributed to the N87Q mutation and difference in the water-mediated hydrogen-bonding interactions near the SB region. The findings were found to be analogous to the spectral changes observed for the G89Q mutant in bovine rhodopsin, which suggests that single site mutations can convert visual pigments into non-visual sensory photoreceptors (Sekharan et al. 2012a).

In the case of UV pigments, upon F86Y mutation, both the USB11 and PSB11 forms of the chromophore are found to co-exist at physiological pH (Fasick et al. 2002). The origin of this intriguing equilibrium remained poorly understood at the molecular level. Therefore, homology structural models of the Siberian Hamster UV cone pigment were built using the X-ray structure of bovine rhodopsin as a template. To explore the structural rearrangements that stabilize USB11 over PSB11, structural models of the wild-type F86Y, S90A, and S90C mutants were compared. The study showed that formation of PSB11 upon F86Y mutation is stabilized by an 'inter-helical lock' (IHL) established by hydrogen-bonding networks between TM helices TM6, TM2, and TM3 (including water w2c and amino acid residues Y265, F86Y, G117, S118, A114, and E113) (Sekharan et al. 2013). The findings implicated the involvement of the IHL in constraining the displacement of TM6, an essential component of the activation of rhodopsin (Farrens et al. 1996; Sheikh et al. 1996; Hubbell et al. 2003; Altenbach et al. 2008), in the spectral tuning of UV pigments.

In summary, insights into the molecular mechanism of color tuning in the vertebrate (bovine, monkey, mouse) and invertebrate (squid, butterfly) species as well as into the non-visual pigment of a mammalian melanopsin are gathered and presented in this review. It is shown that by far the largest effect is exerted by the counter-ion and therefore, the origin of spectral tuning of all visual pigments lies with the counter-ion. Since the protein environment provides and stabilizes the chromophore distortion necessary for the selective and ultrafast photoisomerization (Sekharan and Morokuma 2011a), influence of the neutral polar/non-polar amino acids on the spectral tuning mechanism of type-2 rhodopsins is negligible (Sekharan et al. 2007a, b). Therefore, the neutral/uncharged residues cannot counterbalance the effect (blue shift of ~120 nm) induced by the negatively charged counter-ion as reported elsewhere (Cembran et al. 2005; Coto et al. 2006).

7.3 Microbial (Type-1) Rhodopsins

Except for some organisms, including human beings and higher plants, type-1 microbial retinylidene proteins are widely distributed through all three biological kingdoms, eukarya, bacteria, and archaea, indicating the biological significance of the retinal proteins (Fig. 7.6) (Inoue et al. 2014). The biological functions of the type-1 molecules are roughly divided into two basic functions: photo-induced ion transportation and photo-signal transduction (Fig. 7.7). Among ion transporters, some are light-driven active transporters, such as the proton pumps BR and archearhodopsin-3 (AR3) in haloarchaea, xanthorhodopsin in halophilic bacteria, proteorhodopsin in marine bacteria, Leptosphaeria rhodopsin in fungi, and as the anion pumps halorhodopsin (HR) in haloarchaea and in bacteria (Fig. 7.6a). These proteins generate an electrochemical membrane potential upon light activation, which is utilized by adenosine triphosphate (ATP) synthase to produce ATP (Mukohata et al. 1999; Lanyi 2004). Other ion-transporting retinylidene proteins are light-gated passive cation channels such as channel rhodopsins (ChRs)-1 (ChR1) and -2 (ChR2) in the eukaryotes *Chlamydomonas reinhardtii* and *Volvox carteri* (Fig. 7.6a) (Grote et al. 2014).

Microbial sensory rhodopsins (SRs) such as phototaxis receptors SRI and SRII in haloarchaea and eubacteria (Figs. 7.6a and 7.7) are responsible for signal transduction in the cells (Inoue et al. 2014). SRII, functioning as a negative phototaxis receptor with the cognate transducer protein HtrII, absorbs light of shorter wavelengths (>500 nm) than do ion-pumping rhodopsins, including BR, AR3, and HR

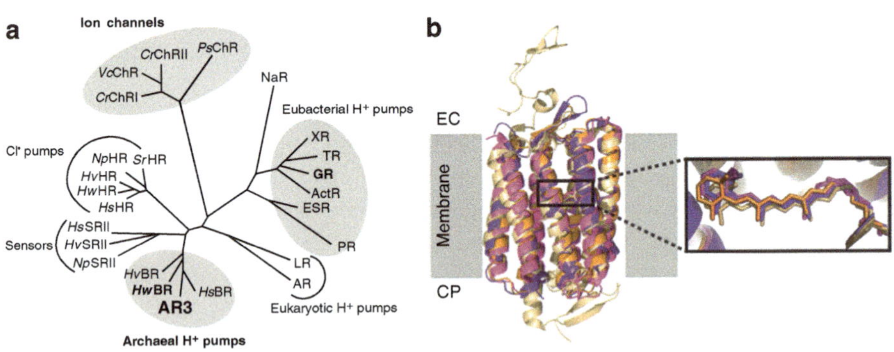

Fig. 7.6 *Phylogenetic tree of microbial type-1 retinylidene proteins and their crystal structures.* (**a**) The microbial type-1 retinylidene proteins widespread in archaea, eubacteria, and eukaryotes. The identity of their amino acid sequences among pumps, channels, and sensors is about 20 %. (**b**) Crystal structures of type-1 retinylidene proteins. Bacteriorhodopsin (BR) (Protein Data Bank [PDB] code: 1C3W) (Luecke et al. 1999), *Anabaena* sensory rhodopsin (SR) (PDB code: 1XIO) (Vogeley et al. 2004), SRII (PDB code: 1H68) (Royant et al. 2001), and channel rhodopsins (ChRs), also called CSR (PDB code: 3UG9) (Kato et al. 2012) are colored purple, magenta, orange, and yellow, respectively. The retinal chromophore covalently binds to a specific Lys residue located at the middle of the membrane via a protonated Schiff base linkage. *CP* cytoplasmic, *EC* extracellular

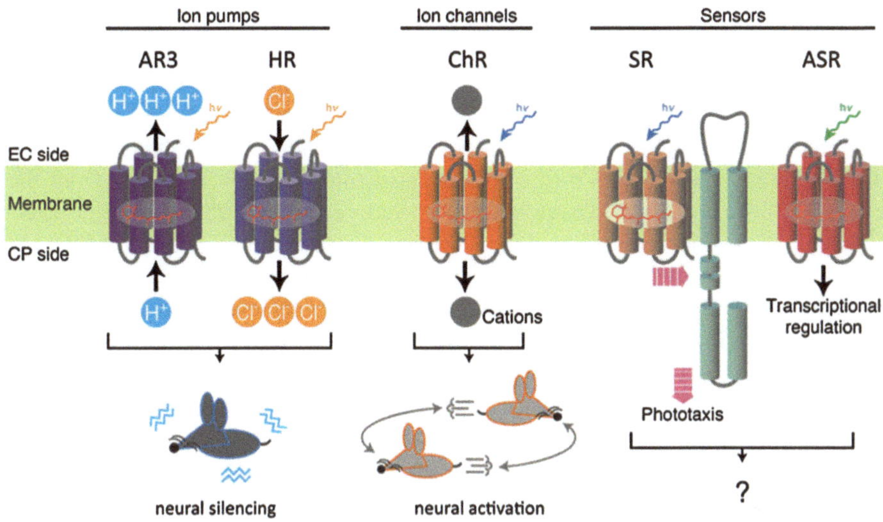

Fig. 7.7 *The biological function of type-1 microbial retinylidene proteins and their application for optogenetics technology.* The microbial type-1 proteins exhibit two basic biological functions: light-energy conversion (pumps and channels) and light-signal conversion (sensors). In the new technology 'optogenetics', ion-pumping rhodopsins, such as archearhodopsin-3 and halorhodopsin, are utilized for neural silencing through membrane hyperpolarization by the light-induced ion-pumping activity. On the other hand, channel rhodopsins (ChRs) mediate photo-induced membrane depolarization, leading to neural activation. Microbial sensory rhodopsins can regulate both kinase activity and transcription by light. Therefore, they could be novel tools for optogenetics

(Sudo et al. 2011a). This is reasonable because the light of shorter wavelengths contains harmful UV radiation. On the other hand, SRI mediates attractant motility responses to light of green/orange wavelengths (540–590 nm), which are also used by the ion pumps BR, AR3, and HR (Spudich and Bogomolni 1984). Thus, not only for type-2 animal rhodopsins, but also for type-1 microbial rhodopsins, it is essential to tune the color for their biological functions. Although all microbial retinylidene proteins are quite similar in their tertiary structures, especially in the chromophore-binding site (Fig. 7.6b), the λ_{max} value of SRII (~500 nm) is remarkably different from that of BR, HR, and SRI (560–590 nm). Therefore, it is obvious that the color tuning must be achieved by considerably small structural difference(s).

The comprehensive mutational studies have been performed to determine the key residue(s) contributing to the color change between SRII and other microbial retinylidene proteins such as BR and middle rhodopsin (Shimono et al. 2001, 2003; Sudo and Spudich 2006; Sudo et al. 2011a, 2013; Mori et al. 2013). These studies have revealed the importance of both the short-range (<5 Å) and the long-range (>5 Å) interaction between retinal chromophore and opsins. In particular, the importance of a specific threonine residue positioning at next to the SB lysine on spectral blue shift was universally confirmed in SRII, SRI, BR, and middle rhodopsin. The removal of the hydroxyl group may change the electrostatic field near the SB, which might cause the ground state to become more stable. Although the molecular origin for the long-range effect is still unclear, the long-range interaction (~25 Å) between the E–F loop in the eubacterium proton pump proteorhodopsin and its retinal chromophore has been reported by Kandori and coworkers (Yoshitsugu et al. 2008).

In addition to the threonine residue (Thr204 in SRII), the importance of the other residues, Met118 and Ser141, locating around the β-ionone ring of the retinal chromophore is revealed by the mutational experiments on a BR homolog from the archaeon *Haloquadrutum walsbyi* (*Hw*BR) (Sudo et al. 2011b, 2013). Free energetically optimized structural models of the native protein and the M118A/S141A/A215T triple mutant of BR from the archaeon *Halobacterium salinarum* corresponding to M126A/S149A/A223T in *Hw*BR effectively explained the observed large blue shift in the mutant. The polyene chain of the chromophore in the wild-type BR possesses a planer conformation, resulting in a long π conjugation. On the other hand, the β-ionone ring of the chromophore in the mutant is allowed to rotate about 130 degrees. It should be noted that a rotation of the β-ionone ring can, despite breaking the π conjugation, easily occur in isolated conditions due to the steric repulsions between the methyl groups of the β-ionone ring. It is noteworthy that the twisted conformation at the 6-S bond has also been found in the X-ray crystallographic structures of the visual cone pigments. However, the rotation of the β-ionone ring is prevented in the native protein with the side chains of Met118 and Ser141 by the steric repulsion of the methyl group at position 1 and the CH_2 moiety at position 2 of the β-ionone ring (Fig. 7.1). The replacement of those side chains with the smaller residues such as alanine removes this steric hindrance that imposes the planer conformation in the wild-type BR. Therefore, the breakage of the π conjugation due to the rotation of the β-ionone ring, in addition to electrostatic contributions of the A215T mutation, is likely to contribute largely to the blue shift of the photoabsorption of the chromophore (Sudo et al. 2013).

Comparative studies between SRI and SRII provided additional useful information about the color tuning mechanism in microbial rhodopsins. In 2008, Sudo and coworkers found a new SRI-like protein from the eubacterium *Salinibacter ruber* (*Sr*SRI), which has a λ_{max} of 557 nm, similar to SRI from the archaeon *H. salinarum* (*Hs*SRI) (Kitajima-Ihara et al. 2008). In addition to its high stability, the expression system utilizing *Escherichia coli* makes it possible to prepare large amounts of protein and enables studies of mutant proteins. The systematic mutational studies have revealed that Ala136 in *Sr*SRI is a novel color tuning residue (Sudo et al. 2011a). In the predicted structural model for *Hs*SRI, Ala136 orients its side chain toward the transmembrane region. It is characteristic of SRI among type-1 microbial rhodopsins. In fact, the side chains of the corresponding residues of BR and SRII in the X-ray crystal structures show outward-facing conformations, and the Y139A mutant of SRII shows only a 1-nm spectral shift (Sudo et al. 2011a). Interestingly, the A136Y mutant of *Sr*SRI affects not only the visible spectrum, but also the photoreaction cycling rate, whereas the inverse mutation of SRII (Y139A) affects neither the visible spectrum nor the reaction rate. This suggests that Ala136 of *Sr*SRI is involved in the structural rearrangement around the retinal chromophore because the decay of the intermediate corresponds to the reprotonation process of the SB nitrogen. Thus, in comparison with other type-1 retinylidene proteins, Ala136 in SRI would be slightly rotated, resulting in an inward-facing conformation of the 5th(E)-TM.

Another characteristic of SRI is the chloride binding to the opsin. Halophilic organisms including *H. salinarum* and *S. ruber* live in highly halophilic environments (~4 M NaCl), suggesting the possibility of the effect(s) of salts on the functions of the retinal proteins. BR does not bind any chloride ion under physiological conditions, whereas HR binds a chloride ion inside the protein moiety (Kolbe et al. 2000). This is quite reasonable since HR works as a light-driven chloride ion pump (Lanyi 2004). Some spectroscopic and structural studies clarified the binding position of the chloride ion as well as the local structure around the Schiff base in the dark state (Kolbe et al. 2000; Kouyama et al. 2010). The anion itself interacts with the proton of the protonated SB and the hydroxyl group of Ser130 in *Natronomonas pharaonis* HR (*Np*HR). The interaction consequently controls both the chloride pumping activity and the λ_{max} (Váró et al. 1995). In the case of *Sr*SRI, its high stability enabled the measurement of absorption spectra at various salt concentrations. Upon Cl^- binding, the absorption maximum is shifted from 542 to 556 nm with an affinity of 307 ± 56 mM (Suzuki et al. 2009). The binding of Cl^- to *Sr*SRI is much weaker than those of *Np*HR (dissociation constant, Kd = 1 mM). Interestingly, the anion binding induces a spectral blue shift in the case of *Np*HR (without and with Cl^-, $\lambda_{max} = 600$ and 577 nm, respectively) but a spectral red shift [$\Delta = +14$ nm] in the case of *Sr*SRI, which goes in the opposite direction. We estimated that a conserved residue among SRI-like proteins His131 located around the β-ionone ring region is involved in the anion binding in *Sr*SRI (Suzuki et al. 2009; Irieda et al. 2011; Reissig et al. 2012). Upon excitation, the positive charge on the SB nitrogen atom is delocalized along the π-conjugated double bonds and is centered around the β-ionone ring region, where it is stabilized by the chloride ion bound to His131 for *Sr*SRI and is concentrated on the SB region by chloride ion binding to Ser130 for *Np*HR, respectively. This property of *Sr*SRI leads to the stabilization of the excited state relative to the ground state, which, in turn leads to a red shift in the λ_{max}. Thus, effects of anion binding on the spectral tuning in *Sr*SRI and *Np*HR can be explained by the difference in binding site, and the relatively low binding affinity of *Sr*SRI would be caused by the hydrophobic environment around the β-ionone ring as compared with the SB region.

In summary, the color tuning mechanism of type-1 retinylidene proteins is independently achieved by a combination of a variety of effects such as polarity, steric hindrance, and ion binding.

7.4 Application for Optogenetics

Three types of microbial retinylidene proteins, proton-pumping rhodopsin archearhodopsin-3 (AR3), chloride-pumping rhodopsin HR from *N. pharaonis* (*Np*HR), and cation ChR2 have mainly been utilized as optogenetic tools for controlling neural activity (Fig. 7.7). The ion pumps or the ion channels induce a hyperpolarization used for neural silencing or a de-polarization used for neural activation, respectively. As for fluorescent proteins, the color variants of ion-transporting rhodopsins make it possible to use various wavelengths of light. AR3 and *Np*HR absorb light of longer wavelengths (550–580 nm), while ChR2 absorbs light of shorter wavelengths (~470 nm). Utilizing the difference in λ_{max}, Boyden and coworkers succeeded in the activation and inactivation of neurons by light of different colors (Boyden et al. 2005; Chow et al. 2010). Currently, many researchers are trying to modify the photochemical properties of ChR2, including its photoreaction, absorption maximum, etc. Spudich and coworkers have reported that the λ_{max} of AR3 and ChR2 containing the A_2 retinal (3,4-dehydroretinal), which is a natural retinoid that serves as the chromophore in red-shifted visual pigments of several lower aquatic animals (Sekharan et al. 2010b), are significantly shifted to longer wavelengths (red shift of ~30 nm) (Sineshchekov et al. 2012). For the blue-shifted variant, on the basis of the color tuning mechanism in microbial rhodopsins described above, we have produced a triple mutant of AR3, which has shorter absorption maxima (~500 nm), and which, therefore, overlap to an extent with that of wild-type ChR2 and coincide with that of E123T/T159C mutant (λ_{max} values for wild-type and E123T/T159C are ~470 and ~490 nm, respectively) (Sudo et al. 2013). The E123T/T159C mutant of ChR2 exhibits both increased photocurrents and faster kinetics than the wild-type, and therefore E123T/T159C has been called the ‘perfected ChR2’ (Pastrana 2011). Thus, now we can control both neural activation and silencing with the same wavelength of light at the same time. In contrast with type-1 microbial rhodopsins, type-2 animal rhodopsins are generally less effective as an optogenetic tool because of its non-cyclic photoreaction and chromophore configuration, in which 11-*cis* retinal is not stable in the cells except for visual cells. Lucas and coworkers have recently employed the jellyfish Gs-coupled opsin, which is bleach resistant and reusable, with the ubiquitously present all-*trans* retinal (Bailes et al. 2012). Because the jellyfish opsin can up-regulate the cyclic adenosine monophosphate (cAMP) level in the cells, the circadian entrainment has been artificially regulated by the light. Utilization of the type-2 rhodopsins would be expanded in the future.

In summary, the photophysical properties of type-1 retinal proteins are shown in Table 7.1. Today, the number of genes encoding retinylidene proteins identified is up to several tens of thousands and this continues to rise. In addition, the newly identified, optimized, and modified retinylidene proteins are applicable for controlling not only neural activity but also a variety of biological functions in the near future.

Table 7.1 A list of retinylidene proteins that can be utilized for optogenetics

Opsin type	Origin	λ_{max} [nm]	Function	Application
BR	*Halobacterium salinarum*	568	Proton pump	–
HR	*Natronomonas pharaonis*	575	Halide pump	Neural silencing
SRI	*Salinibacter ruber*	557	Photosensor	–
SRII	*Natronomonas pharaonis*	498	Photosensor	–
AR3	*Halorubrum sodomense*	552	Proton pump	Neural silencing
Blue-shifted AR3 (M128A/S151A/A225T)	*Halorubrum sodomense*	500	Proton pump	Neural silencing
ChR1	*Chlamydomonas reinhardtii*	500	Channel	Neural activation
ChR2	*Chlamydomonas reinhardtii*	470	Channel	Neural activation
Perfected ChR2 (E123T/T159C)	*Chlamydomonas reinhardtii*	490	Channel	Neural activation

References

Altenbach C, Kusnetzow AK, Ernst OP et al (2008) High-resolution distance mapping in rhodopsin reveals the pattern of helix movement due to activation. Proc Natl Acad Sci U S A 105:7439–7444

Andersen LH, Nielsen IB, Kristensen MB et al (2005) Absorption of Schiff-base retinal chromophore in Vacuo. J Am Chem Soc 127:12347–12350

Bailes HJ, Zhuang L-Y, Lucas RJ (2012) Reproducible and sustained regulation of Gαs signalling using a metazoan opsin as an optogenetic tool. PLoS One 7:e30774

Balashov SP, Imasheva ES, Boichenko VA (2005) Xanthorhodopsin: a proton pump with a light-harvesting carotenoid antenna. Science 309:2061–2064

Birge RR (1990) Photophysics and molecular electronic applications of the rhodopsins. Annu Rev Phys Chem 41:683–733

Boyden ES, Zhang F, Bamberg E et al (2005) Millisecond-timescale, genetically targeted optical control of neural activity. Nat Neurosci 8:1263–1268

Briggs WR, Spudich JL (2005) Handbook of photosensory receptors. Wiley-VCH, Weinheim

Cembran A, Luque RG, Altoè P et al (2005) Structure, spectroscopy, and spectral tuning of the gas-phase retinal chromophore: the beta-ionone "handle" and alkyl group effect. J Phys Chem A 109:6597–6605

Chow BY, Han X, Dobry AS et al (2010) High-performance genetically targetable optical neural silencing by light-driven proton pumps. Nature 463:98–102

Cornell WD, Cieplak P, Bayly CI et al (1995) A second generation force field for the simulation of proteins, nucleic acids, and organic molecules. J Am Chem Soc 117:5179–5197

Coto PB, Strambi A, Ferré N et al (2006) The color of rhodopsins at the ab initio multiconfigurational perturbation theory resolution. Proc Natl Acad Sci U S A 103:17154–17159

Farrens DL, Altenbach C, Yang K et al (1996) Requirement of rigid-body motion of transmembrane helices for light activation of rhodopsin. Science 274:768–770

Fasick JI, Applebury ML, Oprian DD (2002) Spectral tuning in the mammalian short-wavelength sensitive cone pigments. Biochemistry 41:6860–6865

Garczarek F, Gerwert K (2006) Functional waters in intraprotein proton transfer monitored by FTIR difference spectroscopy. Nature 439:109–112
Grote M, Engelhard M, Hegemann P (2014) Of ion pumps, sensors and channels – perspectives on microbial rhodopsins between science and history. Biochim Biophys Acta 1837:533–545
Hara T, Hara R, Takeuchi J (1967) Vision in octopus and squid: rhodopsin and retinochrome in the octopus retina. Nature 214:572–573
Hara-Nishimura I, Matsumoto T, Mori H et al (1993) Cloning and nucleotide sequence of cDNA for rhodopsin of the squid Todarodes pacificus. FEBS Lett 317:5–11
Hubbard R, St George RCC (1958) The rhodopsin system of the squid. J Gen Physiol 41:501–528
Hubbell WL, Altenbach C, Hubbell CM et al (2003) Rhodopsin structure, dynamics, and activation: a perspective from crystallography, site-directed spin labeling, sulfhydryl reactivity, and disulfide cross-linking. Adv Protein Chem 63:243–290
Inoue K, Tsukamoto T, Sudo Y (2014) Molecular and evolutionary aspects of microbial sensory rhodopsins. Biochim Biophys Acta 1837:562–577
Irieda H, Reissig L, Kawanabe A et al (2011) Structural characteristics around the β-ionone ring of the retinal chromophore in Salinibacter sensory rhodopsin I. Biochemistry 50:4912–4922
Irieda H, Morita T, Maki K et al (2012) Photo-induced regulation of the chromatic adaptive gene expression by *Anabaena* sensory rhodopsin. J Biol Chem 287:32485–32493
Jongejan A, Bruysters M, Ballesteros JA et al (2005) Linking agonist binding to histamine H1 receptor activation. Nat Chem Biol 1:98–103
Katayama K, Furutani Y, Imai H et al (2012) Protein-bound water molecules in primate red- and green-sensitive visual pigments. Biochemistry 51:1126–1133
Kato HE, Zhang F, Yizhar O et al (2012) Crystal structure of the channelrhodopsin light-gated cation channel. Nature 482:369–374
Kitajima-Ihara T, Furutani Y, Suzuki D et al (2008) Salinibacter sensory rhodopsin: sensory rhodopsin I-like protein from a eubacterium. J Biol Chem 283:23533–23541
Kito Y, Partridge JC, Seidou M et al (1992) The absorbance spectrum and photosensitivity of a new synthetic "visual pigment" based on 4-hydroxyretinal. Vision Res 32:3–10
Kolbe M, Besir H, Essen LO et al (2000) Structure of the light-driven chloride pump halorhodopsin at 1.8 Å resolution. Science 288:1390–1396
Kouyama T, Kanada S, Takeguchi Y et al (2010) Crystal structure of the light-driven chloride pump halorhodopsin from Natronomonas pharaonis. J Mol Biol 396:564–579
Koyanagi M, Terakita A (2008) G_q-coupled rhodopsin subfamily composed of invertebrate visual pigment and melanopsin. Photochem Photobiol 84:1024–1030
Lanyi JK (2004) Bacteriorhodopsin. Annu Rev Physiol 66:665–688
Luecke H, Schobert B, Richter HT et al (1999) Structure of bacteriorhodopsin at 1.55 A resolution. J Mol Biol 291:899–911
Luecke H, Schobert B, Stagno J et al (2008) Crystallographic structure of xanthorhodopsin, the light-driven proton pump with a dual chromophore. Proc Natl Acad Sci U S A 105:16561–16565
Matsui S, Seidou M, Uchiyama I et al (1988) 4-Hydroxyretinal, a new visual pigment chromophore found in the bioluminescent squid, *Watasenia scintillans*. Biochim Biophys Acta 966:370–374
Merbs SL, Nathans J (1992) Absorption spectra of human cone pigments. Nature 356:433–435
Michinomae M, Masuda H, Seidou M et al (1994) Structural basis for wavelength discrimination in the banked retina of the firefly squid, *Warasenia scintillans*. J Exp Biol 193:1–12
Mori A, Yagasaki J, Homma M et al (2013) Investigation of the chromophore binding cavity in the 11-*cis* acceptable microbial rhodopsin MR. Chem Phys 419:23–29
Mukohata Y, Ihara K, Tamura T et al (1999) Halobacterial rhodopsins. J Biochem 125:649–657
Murakami M, Kouyama T (2008) Crystal structure of squid rhodopsin. Nature 453:363–367
Nakanishi K (1991) 11-*cis*-retinal, a molecule uniquely suited for vision. Pure Appl Chem 63:161–170

Nathans J, Hogness DS (1983) Isolation, sequence analysis, and intron-exon arrangement of the gene encoding bovine rhodopsin. Cell 34:807–814
Nathans J, Thomas D, Hogness DS (1986) Molecular genetics of human color vision: the genes encoding blue, green, and red pigments. Science 232:193–202
Neese FA (2003) A spectroscopy oriented configuration interaction procedure. J Chem Phys 119:9428–9443
Nielsen IB, Lammich L, Andersen LH (2006) S1 and S2 excited states of gas-phase Schiff-base retinal chromophores. Phys Rev Lett 96:018304
Okada T, Fujiyoshi Y, Silow M et al (2002) Functional role of internal water molecules in rhodopsin revealed by x-ray crystallography. Proc Natl Acad Sci U S A 99:5982–5987
Okada T, Sugihara M, Bondar AN et al (2004) The retinal conformation and its environment in rhodopsin in light of a new 2.2 A crystal structure. J Mol Biol 342:571–583
Oprian DD, Asenjo AB, Lee N et al (1991) Design, chemical synthesis, and expression of genes for the three human color vision pigments. Biochemistry 30:11367–11372
Ota T, Furutani Y, Terakita A et al (2006) Structural changes in the Schiff base region of squid rhodopsin upon photoisomerization studied by low-temperature FTIR spectroscopy. Biochemistry 45:2845–2851
Pal R, Sekharan S, Batista VS (2013) Spectral tuning in Halorhodopsin: the chloride pump photoreceptor. J Am Chem Soc 135:9624–9627
Palczewski K, Kumasaka T, Hori T et al (2000) Crystal structure of rhodopsin: a G protein-coupled receptor. Science 289:739–745
Pardo L, Deupi X, Dölker N et al (2007) The role of internal water molecules in the structure and function of the rhodopsin family of G protein-coupled receptors. Chem Bio Chem 8:19–24
Pastrana E (2011) Perfecting ChR2. Nat Methods 8:447
Provencio I, Jiang G, De Grip WJ et al (1998) Melanopsin: an opsin in melanophores, brain, and eye. Proc Natl Acad Sci U S A 95:340–345
Provencio I, Rodriguez IR, Jiang G et al (2000) A novel human opsin in the inner retina. J Neurosci 20:600–605
Reissig L, Iwata T, Kikukawa T et al (2012) Influence of halide binding on the hydrogen bonding network in the active site of Salinibacter sensory rhodopsin I. Biochemistry 51:8802–8813
Royant A, Nollert P, Edman K et al (2001) X-ray structure of sensory rhodopsin II at 2.1-Å resolution. Proc Natl Acad Sci U S A 98:10131–10136
Saitou N, Nei M (1987) The neighbor-joining method: a new method for reconstructing phylogenetic trees. Mol Biol Evol 4:406–425
Seidou M, Sugahara M, Uchiyama H et al (1990) On the three visual pigments in the retina of the firefly squid, *Warasenia scintillans*. J Comp Physiol A 166:769–773
Sekharan S (2009) Water-mediated spectral shifts in rhodopsin and bathorhodopsin. Photochem Phobiol 85:517–520
Sekharan S, Buss V (2008) Glutamic acid 181 is uncharged in dark-adapted visual rhodopsin. J Am Chem Soc 130:17220–17221
Sekharan S, Morokuma K (2011a) QM/MM study of the structure, energy storage, and origin of the bathochromic shift in vertebrate and invertebrate bathorhodopsins. J Am Chem Soc 133:4734–4737
Sekharan S, Morokuma K (2011b) Why 11-*cis*-retinal? Why not 7-*cis*, 9-*cis* or 13-*cis*-retinal in the eye? J Am Chem Soc 133:19052–19055
Sekharan S, Weingart O, Buss V (2006) Ground and excited states of retinal Schiff base chromophores by multiconfigurational perturbation theory. Biophys J 91:L07–L09
Sekharan S, Sugihara M, Buss V (2007a) Origin of spectral tuning in rhodopsin-it is not the binding pocket. Angew Chem Int Ed Engl 46:269–271
Sekharan S, Sugihara M, Weingart O et al (2007b) Protein assistance in the photoisomerization of rhodopsin and 9-*cis*-rhodopsin—insights from experiment and theory. J Am Chem Soc 129:1052–1054

Sekharan S, Altun A, Morokuma K (2010a) Photochemistry of visual pigment in a G_q proton-coupled receptor (GPCR)-insights from structural and spectral tuning studies on squid rhodopsin. Chem Eur J 16:1744–1749
Sekharan S, Altun A, Morokuma K (2010b) QM/MM study of dehydro and dihydro β-ionone retinal analogues in squid and bovine rhodopsins: implications for vision in salamander rhodopsin. J Am Chem Soc 132:15856–15859
Sekharan S, Yokoyama S, Morokuma K (2011) Quantum mechanical/molecular mechanical structure, enantioselectivity, and spectroscopy of hydroxyretinals and insights into the evolution of color vision in small white butterflies. J Phys Chem B 115:15380–15388
Sekharan S, Wei JN, Batista VS et al (2012a) The active site of melanopsin: the biological clock photoreceptor. J Am Chem Soc 134:19536–19539
Sekharan S, Katayama K, Kandori H et al (2012b) Color vision: "OH-site" rule for seeing red and green. J Am Chem Soc 134:10706–10712
Sekharan S, Mooney VL, Rivalta I et al (2013) Spectral tuning of ultraviolet cone pigments: an interhelical lock mechanism. J Am Chem Soc 135:19064–19067
Sheikh SP, Zvyaga T, Lichtarge O et al (1996) Rhodopsin activation blocked by metal-ion-binding sites linking transmembrane helices C and F. Nature 383:347–350
Shichida Y, Imai H (1998) Visual pigment: G-protein-coupled receptor for light signals. Cell Mol Life Sci 54:1299–1315
Shimamura T, Hiraki K, Takahashi N et al (2008) Crystal structure of squid rhodopsin with intracellularly extended cytoplasmic region. J Biol Chem 283:17753–17756
Shimono K, Ikeura Y, Sudo Y et al (2001) Environment around the chromophore in pharaonis phoborhodopsin: mutation analysis of the retinal binding site. Biochim Biophys Acta 1515:92–100
Shimono K, Hayashi T, Ikeura Y et al (2003) Importance of the broad regional interaction for spectral tuning in Natronobacterium pharaonis phoborhodopsin (sensory rhodopsin II). J Biol Chem 278:23882–23889
Sineshchekov OA, Govorunova EG, Wang J et al (2012) Enhancement of the long-wavelength sensitivity of optogenetic microbial rhodopsins by 3,4-dehydroretinal. Biochemistry 51:4499–4506
Spudich JL, Bogomolni RA (1984) Mechanism of colour discrimination by a bacterial sensory rhodopsin. Nature 312:509–513
Spudich JL, Yang CS, Jung KH et al (2000) Retinylidene proteins: structures and functions from archaea to humans. Annu Rev Cell Dev Biol 16:365–392
Sudo Y, Spudich JL (2006) Three strategically placed hydrogen-bonding residues convert a proton pump into a sensory receptor. Proc Natl Acad Sci U S A 103:16129–16134
Sudo Y, Yuasa Y, Shibata J et al (2011a) Spectral tuning in sensory rhodopsin I from Salinibacter ruber. J Biol Chem 286:11328–11336
Sudo Y, Ihara K, Kobayashi S et al (2011b) A microbial rhodopsin with a unique retinal composition shows both sensory rhodopsin II and bacteriorhodopsin-like properties. J Biol Chem 286:5967–5976
Sudo Y, Okazaki A, Ono H et al (2013) A blue-shifted light-driven proton pump for neural silencing. J Biol Chem 288:20624–20632
Suzuki T, Makino-Tasaka M, Miyata S et al (1985) Competition between retinal and 3-dehydroretinal for opsin in the regeneration of visual pigment. Vision Res 25:149–154
Suzuki D, Furutani Y, Inoue K et al (2009) Effects of chloride ion binding on the photochemical properties of salinibacter sensory rhodopsin I. J Mol Biol 392:48–62
Terakita A, Tsukamoto H, Koyanagi M et al (2008) Expression and comparative characterization of G_q-coupled invertebrate visual pigments and melanopsin. J Neurochem 105:883–890
Váró G, Brown LS, Sasaki J et al (1995) Light-driven chloride ion transport by halorhodopsin from Natronobacterium pharaonis. 1. The photochemical cycle. Biochemistry 34:14490–14499
Vogeley L, Sineshchekov OA, Trivedi VD et al (2004) Anabaena sensory rhodopsin: a photochromic color sensor at 2.0 Å. Science 306:1390–1393

Vogt K (1983) Is the fly visual pigment a rhodopsin? Z Naturforsch Sect C Biosci 38:329–333
Wald G (1935) Carotenoids and the visual cycle. J Gen Physiol 19:351–371
Wald G (1936) Pigments of the retina. J Gen Physiol 20:45–56
Wald G (1937a) Visual purple system in fresh-water fishes. Nature 139:1017–1018
Wald G (1937b) Photo-labile pigments of the chicken retina. Nature 140:545–546
Wald G (1939) The porphyropsin visual system. J Gen Physiol 22:775–794
Williams SC, Deisseroth K (2013) Optogenetics. Proc Natl Acad Sci U S A 110:16287
Xu W, Campillo M, Pardo L et al (2005) The seventh transmembrane domains of the δ and κ opioid receptors have different accessibility patterns and interhelical intrercations. Biochemistry 44:16014–16025
Yan EC, Kazmi MA, Ganim Z et al (2003) Retinal couterion switch in the photoactivation of the G protein-coupled receptor rhodopsin. Proc Natl Acad Sci U S A 100:9262–9267
Yokoyama S (2008) Evolution of dim-light and color vision pigments. Annu Rev Genomics Hum Genet 9:259–282
Yokoyama S, Radlwimmer F (1998) The "five-sites" rule and the evolution of red and green color vision in mammals. Mol Biol Evol 15:560–567
Yoshitsugu M, Shibata M, Ikeda D et al (2008) Color change of proteorhodopsin by a single amino acid replacement at a distant cytoplasmic loop. Angew Chem Int Ed Engl 47:3923–3926

Part II
Optogenetics in Biological Systems

Chapter 8
General Description: Future Prospects of Optogenetics

Hiromu Yawo, Ryo Egawa, Shoko Hososhima, and Lei Wen

Abstract Recent optical neuro-research methods have several advantages over conventional techniques: high resolution in space and time, together with direct measurement and manipulation of cell functions by light. The use of fluorescent proteins, bioluminescence systems, and light-sensitive proteins facilitates the optical methods in combination with genetic engineering techniques. Techniques involving the application of light-sensitive proteins are collectively termed 'optogenetics' because they combine optics and genetics. Light-sensitive proteins, either natural or synthetic, are termed 'optogenetic molecular reagents' (OMRs) or 'optogenetic molecular tools' if they are used as tools for optogenetics. These techniques are receiving recognition as enabling breakthroughs in neuroscience research. In this chapter, the topics and the prospects of optogenetics are discussed from three view points: the optimization of OMRs, the optimization of gene-targeting methods, and the optimization of optics.

Keywords Optogenetics • Channelrhodopsin • Transporter-type rhodopsin • Flavoprotein • Phytochrome • Viral vector • Transgenic • Electroporation • Two-photon microscopy

8.1 Introduction

The brains of vertebrates and most invertebrates comprise vast numbers of neurons. For example, the human brain is assumed to have 10^{10}–10^{12} neurons. In addition, each neuron receives 100–1,000 synapses, connecting with other neurons to form

H. Yawo (✉) • R. Egawa • S. Hososhima
Tohoku University Graduate School of Life Sciences, Sendai, Japan
e-mail: yawo-hiromu@m.tohoku.ac.jp; r-egawa@m.tohoku.ac.jp; hososhima@dc.tohoku.ac.jp

L. Wen
Center for Functional Connectomics, Korea Institute of Science and Technology, Seoul 136-791, Republic of Korea

Lee Kong Chian School of Medicine, Nanyang Technological University, 50 Nanyang Drive, Research Techno Plaza, Singapore 637553, Singapore
e-mail: leiwen@ntu.edu.sg

H. Yawo et al. (eds.), *Optogenetics*, DOI 10.1007/978-4-431-55516-2_8

a complex network. Information exchange via the neuronal network underlies the brain's various functions, including perception, integration, and motor control. Clarifying the flow of signals in the neuronal network has been one of the main purposes of neuroscience for over a century since the idea was proposed by the Spanish neuro-anatomist, Santiago Ramón y Cajal (Albright et al. 2001). However, our knowledge about the brain remains fragmentary and bound by phenomenological descriptions. Furthermore, how such a complicated brain is formed from a single fertilized egg in the developmental stages of animals and how precise functions are acquired remain unclear. In the mature brain, the neuronal network is always being reconstituted, and individual synaptic weights vary. In addition, there are far more glial cells, which control neuronal functions, than neurons in the brain. What are the cause-and-effect relationships between the functions of these cells and synapses, and brain function? Through which processes does dysfunction cause neurological and/or psychiatric disorders? Thus, the brain is a vast and still relatively unknown region.

A neuronal network is a kind of computing system. It has been analyzed by measuring output as a result of various artificial inputs. With the development of electronics in the twentieth century, electrophysiological techniques in which a network is stimulated electrically and the results are recorded electrically have been the mainstream methods in research on neurons and neuronal networks. For example, in a series of studies in the 1930s, Penfield and colleagues created a detailed map of the motor cortex of the human brain by electrically stimulating various parts of the cerebral cortex and electrically measuring muscle contractions (Penfield and Rasmussen 1950). Meanwhile, electrophysiological studies generally use field electrodes or intracellular/patch electrodes. Stimulation and recording with field electrodes has the advantage of high temporal resolution but the disadvantage of low spatial resolution. Because an electric field is three-dimensional, it is difficult to identify neurons firing as a result of electric stimulation as well as which neurons are specifically being measured. On the other hand, techniques with intracellular/patch electrodes are suitable for stimulating and recording identified neurons but are unsuited to simultaneous application to multiple neurons. In contrast to these techniques, optical neuro-researching methods have the following advantages.

1. *Optical neuro-researching methods have excellent resolution in space and time*

Figure 8.1 compares the temporal and spatial resolution of various kinds of neuro-researching methods (Grinvald 2005). Electron microscopy has high spatial resolution but has the disadvantage of being inapplicable to living cells. Functional magnetic resonance imaging (fMRI) can measure the active brains of animals, including humans, but is unsuitable for investigating the functions of the network because of its low spatial and temporal resolution. However, optical microscopy can clearly distinguish individual neurons as well as the detailed connections between them.

2. *Optical neuro-researching methods allow the direct measurement and manipulation of cell functions by light*

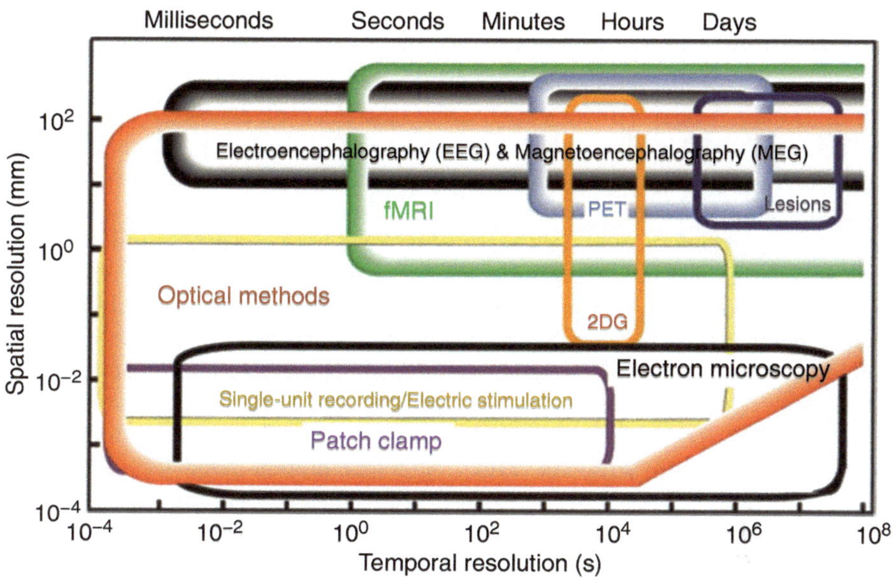

Fig. 8.1 Spatiotemporal resolution of available tools for studying brain and neuroscience. Copyright (2005) National Academy of Sciences, U.S.A.

Substances that reveal cell functions by light, i.e., optical molecular probes, allow the direct measurement of membrane potential, exocytosis, and intracellular messengers such as Ca^{2+} (Sakai et al. 2001; Miesenböck et al. 1998; Miyawaki et al. 1997). In addition, the interactions between substances can be visualized by applying phenomena such as Förster/fluorescence resonance energy transfer (FRET).

3. *Fluorescent proteins, bioluminescence systems, and light-sensitive proteins are available*

The development of optical molecular probes on the basis of proteins such as green fluorescent protein (GFP) enables the design of protein structures by DNA. By making such DNA work in a target cell, the cell creates optical molecular probes by itself. Such genetic engineering techniques have made it possible to measure the function of a specific neuron selectively, for example by localizing the probe to a specific region of a neuron (e.g., a synapse) to measure its function. Furthermore, even the brains of live animals can be studied by creating genetically modified animals into which optical molecular probes are incorporated.

Living organisms sense light by various means and control their behavior accordingly. The application of natural light-sensitive proteins makes it theoretically possible to optically manipulate the functions of neurons in which light-sensitive proteins are expressed. For example, Channelrhodopsin(ChR)-2 is a protein that is localized to the plasma membrane covering the eyespot of a green alga, *Chlamydomonas reinhardtii*; it is a member of the microbial/type I rhodopsin family and is involved in light-dependent behaviors such as the phototaxis of the alga

(Nagel et al. 2002, 2003). Upon absorption of blue light around 460 nm, ChR2 changes its conformation to conduct cations such as H^+, Na^+, and Ca^{2+}. Thus, blue light irradiation of neurons expressing ChR2 causes depolarization and the generation of action potentials (Boyden et al. 2005; Ishizuka et al. 2006).

The use of ChR2 is convenient because the firing of neurons in which the single molecule is genetically expressed can be controlled by light without further chemical modification or addition. In addition, the neurons can be stimulated by various light sources, including laser, light-emitting diode (LED), and the light source of an epifluorescence microscope with excellent temporal and spatial resolution. Techniques involving the application of such light-sensitive proteins are collectively termed 'optogenetics' because they combine optics and genetics. The light-sensitive proteins, either natural or synthetic, are termed 'optogentic molecular reagents' (OMRs) or 'optogenetic molecular tools' if they are used as tools for optogenetics (Boyden 2011). These techniques are receiving recognition as breakthroughs in neuroscience research (Pastrana 2011). Although protein-based techniques that optically measure functions are included in optogenetics in a broad sense, this article defines the light-dependent manipulations of cellular functions as optogenetics and reviews the potential applications and problems of optogenetics. However, optochemical genetics, which are techniques for optically controlling the functions of a protein to which a photo-switchable motif is chemically introduced, and thermogenetics, which are techniques involving an infrared-laser triggered temperature jump, are not mentioned in this article, although they are also promising for future developments and applications; please refer to other papers for more information (Fehrentz et al. 2011; Bernstein et al. 2012).

8.2 Increasing Variety of OMRs

8.2.1 Channelrhodopsins

At present, light-sensitive ion channels similar to ChR2 have been obtained from various living organisms. Furthermore, various options for the absorption spectra, light sensitivity, kinetics, conductance, and ion selectivity have been prepared with the search of natural variants, the production of chimeric ChRs and point mutations. Along with the development of structure–functional relationship studies (Wang et al. 2009) as well as the structural biology research (Kato et al. 2012) of ChR molecules, variations of the modified analogs are expected to increase; the era in which researchers can choose the most optimal ChR for a specific research purpose has come. Some of the modified analogs that may become widely used are reviewed here from the viewpoint of a user. For more details, please refer to other papers (Yizhar et al. 2011a; Zhang et al. 2011; Mattis et al. 2012).

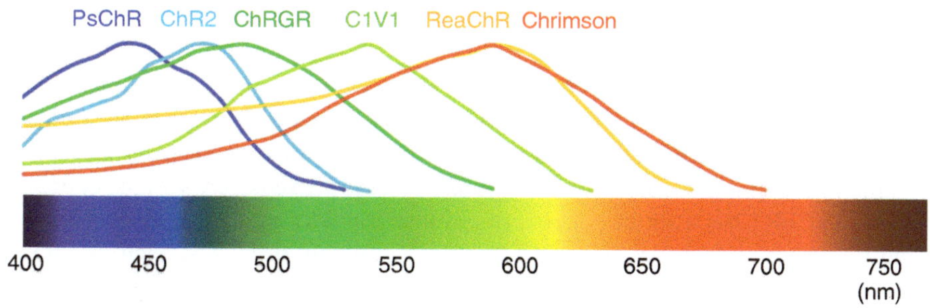

Fig. 8.2 Color variation of channelrhodopsins (ChRs)

8.2.1.1 Variety of Absorption Spectra

Discriminative stimuli using light of multiple wavelengths are expected. Green light is superior to blue light with respect to tissue transparency; the absorption of oxyhemoglobin is minimal at 500–520 nm. If red and near infrared light can be used, applications will be further expanded. At present, several ChRs with different spectral sensitivity almost cover the visible light band (Fig. 8.2): PsChR (derived from the alga *Platymonas* (*Tetraselmis*) *subcordiformis*) (400–490 nm), ChR2 and its variants (420–500 nm), ChR-green receiver (ChRGR) (440–540 nm), C1V1 and its derivatives (470–630 nm), ReaChR (470–630 nm), Chrimson (530–660 nm).

PsChR

PsChR has maximal spectral sensitivity at 445 nm (Govorunova et al. 2013). It exhibits higher unitary conductance and is more selective to Na^+ than ChR2.

ChRGR

ChR1, which is obtained from *C. reinhardtii* and was the first ChR identified, was considered to be unsuitable for application because of its low conductance (Nagel et al. 2005b). However, making a chimeric protein molecule in which certain domains are replaced by counterparts from ChR2 has made it possible to obtain enhanced photocurrent while maintaining the action spectrum of ChR1 with a peak absorption around 510 nm (blue-green light) (Wen et al. 2010). Because ChRGR shows minimal desensitization and fast on/off kinetics, it can be expected to have wide application.

ChR1/VChR1 Chimera (C1V1)

VChR1, which is obtained from the green alga *Volvox carteri*, has a maximum absorption range of around 540 nm, although it is impractical because of minimal trafficking to the membrane as well as poor expression in mammalian cells. However, a chimeric protein molecule in which certain domains are replaced with the counterparts from ChR1 has been developed; this chimera is referred to as C1V1 and has a large photocurrent (Yizhar et al. 2011b). Point mutants of C1V1, such as C1V1 (E162T) and C1V1 (E122T/E162T), are also expected to have wide application such as for two-photon scanning photoactivation (Prakash et al. 2012; Prigge et al. 2012).

ReaChR

ReaChR is one of the chimeric ChRs and consists of an *N*-terminal domain from ChR1, transmembrane (TM) domains of 1–5 and 7 from VChR1 and TM6 from VChR2 derived from *V. carteri* (Lin et al. 2013). The photocurrent amplitude is further enhanced by the introduction of L171I mutation. It is optimally excited with orange to red light (590–630 nm). When ReaChR was expressed in the deep brain neurons of mice, the neurons were photoactivated by red light irradiated through skin and bone as red light can better penetrate the tissue than blue or green light.

Chrimson (CnChR1)

Chrimson (CnChR1) is the first reported yellow-peaked ChR from an alga *Chlamydomonas noctigama* (Klapoetke et al. 2014). A relatively large photocurrent can be evoked even by far-red light as high as 660 nm. When adult flies of *Drosophila melanogaster* express Chrimson in sweet taste receptor cells, they even responded to 720 nm light with a behavioral response, proboscis extension reflex.

8.2.1.2 Light Sensitivity and Conductance

If large photocurrents can be obtained from relatively weak light power, various optical systems such as LEDs, projectors, and the light source of epifluorescence microscopes can be used in place of lasers. Furthermore, it would be possible to photostimulate the ChR-expressing cells deep in the tissue of the subject.

ChR2 (H134R) and ChR2 (T159C)

Both have high expression efficiency in the membrane and provide large photocurrents (Nagel et al. 2005a; Berndt et al. 2011). They show desensitization, although to a lesser extent than that of ChR2. It is noted that, in terms of the kinetics, their off time constants (τ_{OFF}) are almost twice as large as those of ChR2.

ChR-Wide Receiver (ChRWR, C1C2)

ChR-wide receiver (ChRWR), one of the chimeric ChRs between ChR1 and ChR2, is characterized by its high expression efficiency in the membrane and large photocurrent. In addition, it does not show prominent desensitization, which is observed in ChR2 (Wang et al. 2009; Umeda et al. 2013). Its τ_{OFF} is almost equal to that of ChR2. Because it has high sensitivity and a wide action spectrum, the optical stimulation efficiency can be enhanced by using a broadband light source.

8.2.1.3 Speed

ChRs with small τ_{OFF} are suitable for stimulating a neuron at high frequency. In addition, when ChRs have a larger τ_{OFF}, they can maintain depolarization longer and involve several action potentials even with a brief irradiation. High-speed ChRs such as ChR-fast receivers (ChRFRs) are also optimal for the combination of photostimulation and imaging of sensor molecules under single/two-photon scanning microscopy, as these ChRs are minimally activated by the scanning light (Egawa et al. 2013).

ChETA Series

Point mutants such as ChR2 (E123A), ChR2 (E123T), and ChR2 (E123T/T159C) are characterized by their small τ_{OFF} compared with that of ChR2 (Berndt et al. 2011). However, they are not versatile because they involve desensitization and have relatively low sensitivity to light.

ChRFR

ChRFR, a chimeric channelrhodopsin between ChR1 and ChR2, is characterized by its high expression efficiency in the membrane and large photocurrent. In addition, its desensitization is smaller than that of ChR2. The τ_{OFF} of the steady-state photocurrent is about half that of ChR2 and is among the fastest (Wang et al. 2009; Mattis et al. 2012).

ChIEF

ChIEF was obtained from a point mutation of a chimeric protein between ChR1 and ChR2. It has high expression efficiency in the membrane, high photocurrent, and almost negligible desensitization. The τ_{OFF} of the early photocurrent is among the fastest (Lin et al. 2009).

Chronos (ShChR)

Chronos (ShChR) is obtained from an alga *Stigeoclonium helveticum* and has fast on/off kinetics (Klapoetke et al. 2014). Importantly, Chronos displays about twice or higher effective photosensitivity than naturally occurring ChR2, and this huge photocurrent is not likely due to higher expression of apoprotein, but to improved channel conductance. Therefore, Chronos would be suitable for general use with blue-green light.

8.2.1.4 Step-Function Channelrhodopsin

Each ChR consists of a seven-path TM opsin apoprotein and a retinal, which is covalently bonded as the chromophore. During dark adaptation, a ChR is at the ground state with a retinal chromophore in the all-*trans* configuration. By absorbing the energy of a single photon, the retinal chromophore photoisomerizes to the 13-*cis* configuration. This photoisomerization triggers the activation of the ChR from the ground state via intermediates. Finally, the retinal chromophore returns to the all-*trans* configuration, and the ChR returns to the ground state (Fig. 8.3a). This photocycle is repeated during light irradiation (Stehfest and Hegemann 2010). Alternatively, the ChR transitions to the desensitized state and enters the desensitization photocycle. In the activated open state (i.e., P520 for ChR2), TM cation transfer generates photocurrents.

Step-Function Opsins (SFOs)

In the case of point mutants such as ChR2 (C128T), ChR2 (C128A), ChR2 (C128S), and ChR2 (D156A), the transition from an intermediate P520 to another intermediate P470 is retarded. Therefore, high photocurrent continues to flow in the order of tens of milliseconds without light absorption. On the other hand, the intermediates P390 and P520 absorb blue–violet and green light with peak wavelengths of 390 and 520 nm, respectively, and then return to the channel-closed, ground state (Berndt et al. 2009; Bamann et al. 2010). Consequently, during an experiment, the on/off status of the photocurrent can be made switchable by switching the color of the irradiating light. Similarly, chimeric ChRs such as ChRFR and ChRWR are made bistable by introducing a point mutation (Hososhima et al. 2015). For example, ChRFR(C167A) is turned on by absorbing blue or cyan light and turned off by yellow or orange light (Fig. 8.3b). These step-function opsins (SFOs) and their derivatives such as ChR2 (C128T, H134R) and ChR2 (C128A, H134R) are also applicable for photostimulation using single/two-photon scanning light (Prakash et al. 2012).

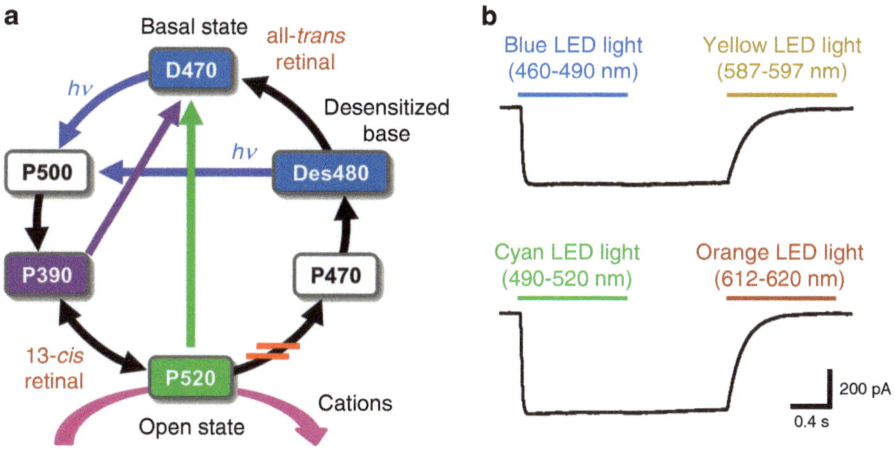

Fig. 8.3 Bi-stable channelrhodopsins (step-functiona opsins [SFOs]). (**a**) Channelrhodopsin (ChR)-2 photocycle. The stabilization of P520 (*red lines*) generates SFOs. (**b**) Wavelength dependence of on/off switching of ChR-fast receiver (ChRFR)–SFO photocurrents (From Yawo et al. (2014) Cell Technology Vol. 33 No. 3, 243–24 with permission)

Stable Step-Function Opsins (SSFOs)

In ChR2 (C128S/D156A), among others, the τ_{OFF} is prolonged in the order of tens of seconds to minutes. Therefore, even if light irradiation is weak, active intermediates accumulate with time and high photocurrent can be obtained (Yizhar et al. 2011b). Therefore, SFOs and stable SFOs (SSFOs) are suitable for stimulating cells deep within tissues where light hardly penetrates.

8.2.1.5 Ion Selectivity

ChR2 is a non-selective cation channel that conducts Ca^{2+} ions; the influence of Ca^{2+} influx cannot be ignored (Caldwell et al. 2008). The more Na^+-selective ChRs, such as PsChR and MChR1 from *Mesostigma viride*, would allow selective control of the membrane potentials (Govorunova et al. 2013). On the other hand, some ChR2 point mutants, such as CatCh and CatCh+, have relatively high Ca^{2+} conductance and are used as tools for optically controlling the intracellular Ca^{2+} levels (Kleinlogel et al. 2011; Prigge et al. 2012). However, they are permeable to other ions such as Na^+ and induce changes in membrane potentials; hence, they require further modification. Meanwhile, Cl^--selective ChRs and K^+-selective ChRs (if present) would be conducive to inducing hyperpolarization through light (Wietek et al. 2014; Berndt et al. 2014).

8.2.2 Transporter-Type Rhodopsins

Members of the microbial rhodopsin family have various functions and are expected to become effective tools in optogenetics (Zhang et al. 2011). Halorhodopsin (Halo/*Np*HR), a light-sensitive Cl^- transporter, absorbs yellow light with a peak wavelength of around 590 nm and transports Cl^- ions inwardly across the plasma membrane (Han and Boyden 2007; Zhang et al. 2007). Archaerhodopsin (Arch/aR-3, ArchT), a light-sensitive H^+ transporter, absorbs yellow light with a peak wavelength of around 580 nm and transports H^+ outwardly across the plasma membrane (Chow et al. 2010; Han et al. 2011). In addition, Mac, a blue–green light-sensitive H^+ transporter, and KR2, a green light-sensitive Na^+ transporter, have been found (Inoue et al. 2013; Kato et al. 2015). However, these proteins can transport only one ion along with the absorption of one photon. Therefore, relatively strong light is required to induce hyperpolarization sufficient to inhibit neuronal activity.

8.3 Control of Diverse Cell Functions

Cell functions are generated as a result of interactions among various molecules. Accordingly, direct optical control of these molecules may enable positive or negative control of cell functions. For example, Ca^{2+} plays a principal role in triggering various cellular functions, including secretion, motion, and transcription. CatCh and CatCh+may be used to increase intracellular Ca^{2+}. Meanwhile, it is conceivable to design light-sensitive Ca^{2+} chelators by using light-dependent molecular cleavage.

8.3.1 Flavoproteins

Flavoproteins (Fig. 8.4) are the main functional regulators in the cells of various living organisms (Hegemann 2008; Möglich et al. 2010; Christie et al. 2012). Photo-activated adenylyl cyclase (PAC obtained from *Euglena* and bacterial photoactivated adenylyl cyclase (bPAC; also called bacterial light-activated adenylyl cyclase, Blac) are proteins that have a blue light-using flavin adenine dinucleotide (BLUF) domain that binds flavins (Iseki et al. 2002; Stierl et al. 2011; Ryu et al. 2010). As its name implies, it is activated by the absorption of blue light, hydrolyzes adenosine triphosphate (ATP), and produces cyclic adenosine monophosphate (cAMP). Bacterial light-activated guanylyl cyclase (Blgc), a point mutant of Blac, produces cyclic guanosine monophosphate (cGMP) (Ryu et al. 2010). Light-sensitive proteins with flavin-binding light-oxygen-voltage-sensing (LOV) domains are widespread among plants, fungi, and archaebacteria. Many of them

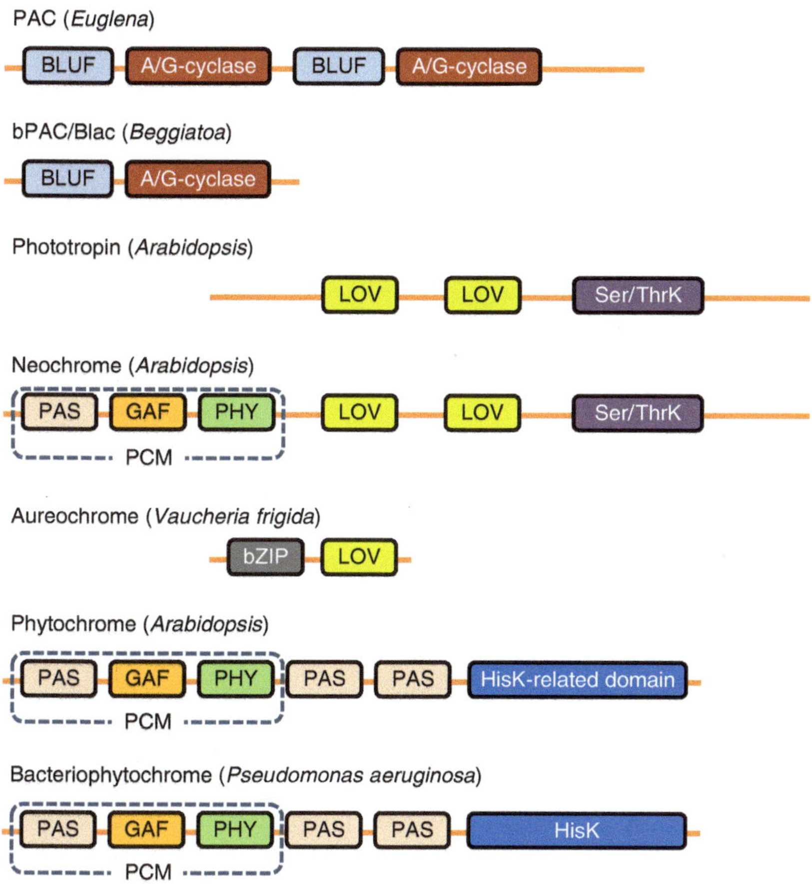

Fig. 8.4 Flavoproteins and phytochromes. *PAC* photo-activated adenylyl cyclase, *Blac* bacterial light-activated adenylyl cyclase, *BLUF* blue light-using flavin adenine dinucleotide (FAD) domain, *A/G-cyclase* adenylyl/guanylyl cyclase, *LOV* light-oxygen-voltage-sensing domain, *PAS* protein domain found in *Drosophila* period gene product (dPER), human Arnt, and *Drosophila* single-minded gene product (dSIM) and named after their initials, *GAF* protein domain found in cGMP-specific phosphodiesterase, adenylyl cyclase, and *Escherichia coli* FhlA (formate hydrogen lyase system activator) and named after their initials, *PHY* phytochrome domain, *PCM* photosensory core module, *Ser/ThrK* serine/threonine kinase domain, *bZIP* basic region/leucine zipper domain, *HisK* histidine kinase domain

act as photoswitches that switch functions on and off by light (Christie et al. 2012). For example, aureochrome, which is derived from yellow algae *Vaucheria frigida*, has one DNA-binding basic leucine zipper (bZIP) domain and one LOV domain. Because its affinity for DNA varies depending on light, aureochrome is expected to be used as a light-controllable transcriptional factor (Takahashi et al. 2007).

8.3.2 Phytochromes

Phytochromes (Fig. 8.4) were first found in higher plants and are involved in the regulation of seed germination and phototropism. They absorb red to near-infrared light through the photoisomerization of covalently attached bilin chromophores (Möglich et al. 2010). A bacterial protein with a similar phytochrome (PHY) domain has been reported. Phytochrome B (PhyB), which is found in *Arabidopsis*, absorbs 650 nm red light and changes its conformation to bind PHY-interaction factor 3 (PIF3). This conformational change is reversible; the absorption of 750 nm far-red light causes the dissociation of PIF3. On the basis of this mechanism, the activity of Rho-GTPase can be switched on or off (Levskaya et al. 2009) by targeting the Rho-GTPase activator to the membrane in a light-dependent manner.

8.3.3 Animal Rhodopsins

Animal rhodopsin has seven TM helices and is coupled with specific G-proteins via intracellular loops. Some studies have tried to drive some targeted signaling cascades in response to light by replacing the intracellular loops with those of other G-protein–coupled receptors (Kim et al. 2005; Li et al. 2005; Airan et al. 2009; Oh et al. 2010; Ye et al. 2011). In visual cells in the mammalian retina, rhodopsin is normally coupled with G_t, which is subsequently cleaved into $G_t\alpha$-GTP and the $\beta\gamma$-subunit. $G_t\alpha$-GTP activates cGMP phosphodiesterase. As a result, the cGMP concentration in the cell decreases and the cation channels are subsequently closed. When the intracellular loops of rhodopsin were replaced with those of β_2-adrenergic receptor, the rhodopsin became coupled with G_s, which activates adenylyl cyclase and subsequently increases the cAMP concentration. On the other hand, replacement with the intracellular loops of the α_1-adrenergic receptor led to rhodopsin coupling with G_q, which is involved in phospholipase activation, and increased the inositol triphosphate (IP_3) and diacylglycerol concentrations. Recently, the cytoplasmic loop of a microbial rhodopsin was replaced by a bovine rhodopsin to generate a chimera that activates G-proteins in a manner dependent on light (Sasaki et al. 2014). However, the specificity of coupling between the receptor and G-protein was not high. In addition, the localization in the membrane is important for signal transduction. Consequently, exogenously expressed chimeric protein receptors do not necessarily achieve the expected effect.

8.4 Introduction into Target Cells

8.4.1 Viral Vectors

Various methods have been devised to specifically express the above-mentioned OMRs in the cells to be controlled. One example is the use of adeno-associated virus (AAV) or lentivirus incorporating a target cell-specific promoter as a vector.

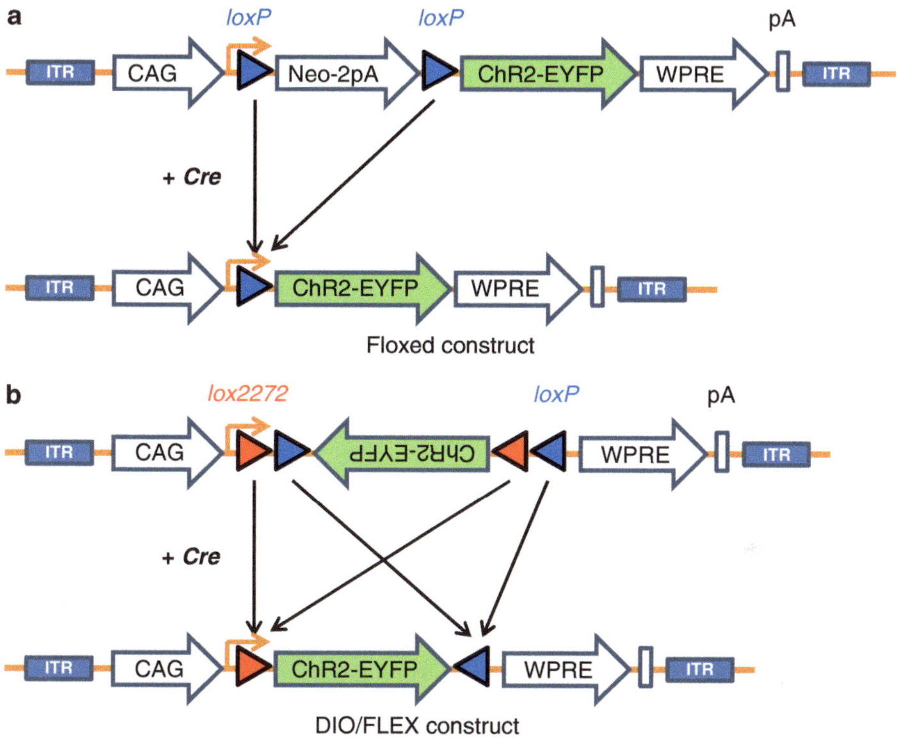

Fig. 8.5 Integrated constructs for viral vectors used for conditional expression systems. *ITR* inverted terminal repeat, *CAG* CAG promoter sequence, *pA* polyadenine sequence (stop codon), *Neo* neomycin-resistant gene sequence, *ChR2* channelrhodopsin-2 sequence, *EYFP* enhanced yellow fluorescent protein sequence, *WPRE* woodchuck hepatitis virus post-transcriptional regulatory element, *loxP, lox2272* bacteriophage P1-derived loxP sequence and its mutant, *Cre* bacteriophage P1-derived Cre-recombinase

However, contrary to expectations, the expression of the OMRs is frequently too low to effectively manipulate the cellular functions. In such cases, conditional expression systems are effective. These systems involve viral vectors incorporating the genes of OMRs into the downstream portion of relatively strong promoters (e.g., the CAG and EF1α promoters) as well as a floxed construct deactivated by loxP-STOP-loxP or a DIO/FLEX construct in which antiparallel DNA is inserted between two loxP sites (Fig. 8.5). These vectors are then injected into driver transgenic/knock-in animals that have a construct placing the Cre-recombinase gene in the downstream portion of the highly specific promoter (Yizhar et al. 2011a). Various Cre-transgenic/knock-in animals have been developed to date, and the variety will increase further. Meanwhile, methods in which viral vectors incorporating the Cre-recombinase into the downstream portion of the specific promoter are injected into reporter transgenic/knock-in animals incorporating a floxed construct or a FLEX construct of the light-sensitive protein gene also seem to be effective.

8.4.2 Genetically Engineered Animals

If both driver and reporter transgenic/knock-in animals are available, an experimental system using crossbred animals that can conditionally express OMRs is desirable because of good reproducibility. In addition to the above-mentioned Cre-loxP system, the tTA-*tet*O system is often used in mammals (Yawo et al. 2013). In the latter system, the on/off switching of reporter gene expression is regulated by the administration of a chemical substance, doxycycline (Tanaka et al. 2012). The tTA-*tet*O system has enabled the creation of mice in which ChR2 is specifically expressed in hippocampal neurons that have been active during fear learning (Liu et al. 2012). The memory was subsequently recalled upon photostimulation of the ChR2-expressing neurons.

Meanwhile, the Gal4-upstream activating sequence (UAS) system is often used in *Drosophila* (fruit fly) and zebrafish. Several Gal4 lines have been generated by enhancer trapping (Scott 2009). Of these Gal4 lines, those expressing Gal4 only in specific subsets of neurons were found by chance. Even if a gene is expected to be specifically expressed in a type of neuron of interest, it will often fail, because the specific promoter has not been identified in many cases. However, the contingent choice of the optimum line is independent of the identification of the promoter and is therefore a powerful and effective method. It is anticipated that an increase in the kinds of lines in which OMRs optimized for experiments are expressed downstream of the UAS promoter will promote studies using such animals (Umeda et al. 2013; Kimura et al. 2013).

8.4.3 Electroporation

Electroporation is a technique to introduce plasmid vector into target cells by electrical pulses. It is available for various targets such as the embryos in avian eggs (Odani et al. 2008) or in the uterus of rodents (Saito and Nakatsuji 2001) and ferrets (Kawasaki et al. 2012) (*in ovo/utero* electroporation), living organs (Matsuda and Cepko 2004) (*in vivo* electroporation) and isolated tissues (Kawabata et al. 2004) (*ex vivo* electroporation). Electroporation enables effective expression of the genes of interest in a brain-region-specific manner according to the position of electrodes (Momose et al. 1999; dal Maschio et al. 2012). Moreover, cell-type-specific expression with the Cre-loxP system (Nishiyama et al. 2012) and sparse expression using single-cell electroporation (Judkewitz et al. 2009) or Thy1s promoter (Ako et al. 2011) are promising for connectomics research.

Electroporation has advantages in simplicity and flexibility; it can be used as soon as the plasmid vector is prepared. In addition, co-expression of several genes of interest can be done easily by the injection of the mixture of plasmid vectors. These are great advantages when one wants to use a newly developed OMR and combine the OMR with additional genetic manipulations such as over-expression/knock-down of the target gene and expression of other genetically encoded tools. For example, an 'all-optical approach' became available with the combination of optical stimulation and Ca^{2+} imaging by co-electroporation of ChR

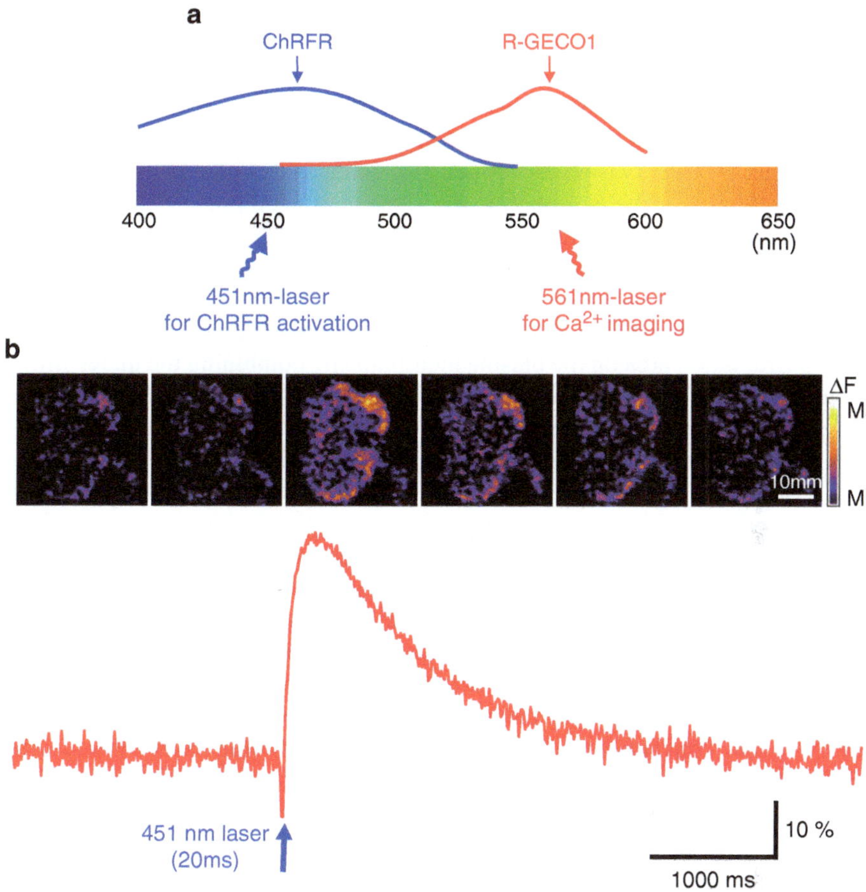

Fig. 8.6 All-optical approach by co-electroporation of ChR-fast receiver (ChRFR) and red fluorescent genetically-encoded Ca^{2+} indicators for optical imaging (R-GECO)-1. (**a**) Excitation spectra of ChRFR and red-fluorescent Ca^{2+} indicator. The interference was minimized by using 451-nm-laser for ChR activation and 561-nm-laser for Ca^{2+} imaging. (**b**) Ca^{2+} response evoked by ChRFR activation with 451-nm-laser pulse (frame interval, 500 ms). The lower trace is the change of average R-GECO1 fluorescence in the region of interest (ROI) covering a calyx as a whole (sampling rate, 120 fps)

and red-fluorescent Ca^{2+} indicator (Ohkura et al. 2012; Egawa et al. 2013) (Fig. 8.6). These advantages should be taken into consideration because optogenetics is a rapidly advancing field.

8.5 Two-Photon Optogenetics

Temporally precise manipulation of a subpopulation within genetically defined neurons mimicking the activity pattern from sparsely distributed neurons is a long-sought goal in neuroscience. Genetically specific targeting of OMRs on a defined neuronal population enables optical control of neuronal activity with high temporal resolution, but the constraints of spatial precision for photostimulation in terms of selected subpopulation, single-cell or subcellular compartment has hardly been achieved using conventional single-photon optics both *in vitro* and *in vivo* (Oron et al. 2012). The ability to confine the excitation narrowly along the *z* axis makes two-photon optics promising for fulfilling this experimental purpose. However the two-photon absorption has been proven rather ineffective for driving supra-threshold spikes due to the small single-channel conductance of ChR2 (~80 pS) (Feldbauer et al. 2009) and insufficient two-photon absorption volume (~2–5 μm^3) under the conventional diffraction-limited two-photon laser system (Papagiakoumou et al. 2010). Pioneer works have focused on enlarging the effective irradiation area on the plasma membrane by combining the under-filling back aperture of an objective lens with spiral scanning to allow the spatiotemporal accumulation of photocurrent (Rickgauer and Tank 2009). The compromised poor *z* axis resolution and long scanning duration (~32 ms) of the under-filling strategy was further overcome by using temporal focusing laser pulse (TEFO) (Andrasfalvy et al. 2010) or by combining generalized phase contrast (GPC) with TEFO. The action potentials were generated at the single-cell level with 2-ms temporal and 5-μm *z* axial resolution (Papagiakoumou et al. 2010). The key feature of the GPC-TEFO combination is that it allows simultaneous excitation of multiple neurons or neuronal compartments by shaping the two-photon region. Enlargement of the effective two-photon laser spot along *x*–*y* dimensions without using scanning modes requires elegant hardware innovations that are not really friendly to the conventional diffraction-limited two-photon laser system. The alternative strategy to instrumental innovation is digging mutagenesis of ChR. The idea is to extend the lifetime of a channel opening (τ_{OFF}) by slowing down the deactivation dynamics, so that the temporal accumulation of photocurrent would have a high chance of driving action potentials. Variants of C1V1 such as $C1V1_T$ (E162T) and $C1V1_{TT}$ (E122T/E162T) displayed moderately slow τ_{off} and had relatively high efficiency to drive action potentials upon raster scanning of a two-photon laser over the soma under a low-numerical aperture (NA) objective lens (Prakash et al. 2012). Taking advantage of the long τ_{off} of the engineered proton pump eArch3.0, two-photon inhibition of neuronal activity was also achieved by conventional raster scanning (Prakash et al. 2012). With these methods, even the subcellular compartments, such as dendritic spines, could be manipulated by a two-photon laser (Packer et al. 2012).

Considering the great precision of *z* axial resolution, a two-photon laser is not only capable of selectively stimulating a genetically identified target at the single-cell level, but also of drawing a functional mapping of neural circuits by interrogating selected neighboring cells with truly appreciable high spatial resolution. The technique is not limited to *in vitro* scenarios, but can also be used for *in vivo* optical manipulations. Therefore, three-dimensional two-photon mapping of functional neural circuits with single-cell resolution can be expected to be achieved in the near future.

8.6 Conclusion

One of the innovative techniques involving optical control is the reverse engineering of biological functions (O'Connor et al. 2009). Reverse engineering is a process of revealing the mechanisms, specifications, purposes, components, constituent technologies, etc. of software or hardware through disassembly or analysis. Conventional studies regard a neuronal network as a computing system and describe behavior on the basis of input/output relationships. On the other hand, reverse engineering studies attach significance to the functions of a neuron or a group of neurons on the basis of a change in the network performance caused by their direct manipulation. For example, neurons expressing Halo/*Np*HR or Arch/ArchT become hyperpolarized by yellow light, and their activity then declines. As a result, you could determine whether the neuron in question is necessary for the network performance of the behavioral expression. On the other hand, if light-sensitive cation channels such as ChR2 are expressed in a neuron, whether the neuron is sufficient for the network performance could be determined on the basis of light-dependent changes in the activity of the neuron. Applying both positive and negative manipulations together, it is possible to clarify the necessary and sufficient conditions.

So far, experiments involving regional brain lesions, local drug administration, knock-out animals, targeted expression tetanus toxin, or immunotoxin systems have been used to study necessary conditions. However, it is difficult to eliminate the possibility of the plastic changes in networks as a result of the long-term inhibition of functions. It is promising that optogenetics will enable studies that can determine the necessary and sufficient conditions with high temporal and spatial resolution, through positive or negative manipulation of the function of a neuron. But, to conclude that the neuron is responsible for the action is overly simplistic. In a network, an action is generated by the flow of information. Accordingly, the fact that a decline in the activity of a neuron is necessary for an action means that the decline in the activity of the neuron blocks the flow of information. On the other hand, the fact that the activity of a neuron is sufficient for an action means that the activity of the neuron only promotes the flow of information. In addition, an identical neuron may be involved in different actions.

Along with the increase and optimization of the kinds of OMRs and production of introduction–expression systems, optical control techniques are becoming increasingly accessible. The development of optical systems is expected to further accelerate the application. However, it is important to fully understand the limits of these systems and correctly interpret the results.

References

Airan RD, Thompson KR, Fenno LE et al (2009) Temporally precise *in vivo* control of intracellular signalling. Nature 458:1025–1029

Ako R, Wakimoto M, Ebisu H et al (2011) Simultaneous visualization of multiple neuronal properties with single-cell resolution in the living rodent brain. Mol Cell Neurosci 48:246–257
Albright TD, Jessell TM, Kandel ER et al (2001) Progress in the neural sciences in the century after Cajal (and the mysteries that remain). Ann N Y Acad Sci 929:11–40
Andrasfalvy BK, Zemelman BV, Tang J et al (2010) Two-photon single-cell optogenetic control of neuronal activity by sculpted light. Proc Natl Acad Sci U S A 107:11981–11986
Bamann C, Gueta R, Kleinlogel S et al (2010) Structural guidance of the photocycle of channelrhodopsin-2 by an interhelical hydrogen bond. Biochemistry 49:267–278
Berndt A, Yizhar O, Gunaydin LA et al (2009) Bi-stable neural state switches. Nat Neurosci 12:229–234
Berndt A, Schoenenberger P, Mattis J et al (2011) High-efficiency channelrhodopsins for fast neuronal stimulation at low light levels. Proc Natl Acad Sci U S A 108:7595–7600
Berndt A, Lee SY, Ramakrishnan C et al (2014) Structure-guided transformation of channelrhodopsin into a light-activated chloride channel. Science 344:420–424
Bernstein JG, Garrity PA, Boyden ES (2012) Optogenetics and thermogenetics: technologies for controlling the activity of targeted cells within intact neural circuits. Curr Opin Neurobiol 22:61–71
Boyden ES (2011) A history of optogenetics: the development of tools for controlling brain circuits with light. F1000 Biol Rep 3:11. doi:10.3410/B3-11
Boyden ES, Zhang F, Bamberg E et al (2005) Millisecond-timescale, genetically targeted optical control of neural activity. Nat Neurosci 8:1263–1268
Caldwell JH, Herin GA, Nagel G et al (2008) Increases in intracellular calcium triggered by channelrhodopsin-2 potentiate the response of metabotropic glutamate receptor mGluR7. J Biol Chem 283:24300–24307
Chow BY, Han X, Dobry AS et al (2010) High-performance genetically targetable optical neural silencing by light-driven proton pump. Nature 46:98–102
Christie JM, Gawthorneb J, Younga G et al (2012) LOV to BLUF: flavoprotein contributions to the optogenetic toolkit. Mol Plant 5:533–544
dal Maschio M, Ghezzi D, Bony G et al (2012) High-performance and site-directed *in utero* electroporation by a triple-electrode probe. Nat Commun 3:960. doi:10.1038/ncomms1961
Egawa R, Hososhima S, Hou X et al (2013) Optogenetic probing and manipulation of the calyx-type presynaptic terminal in the embryonic chick ciliary ganglion. PLoS One 8:e59179. doi:10.1371/journal.pone.0059179
Fehrentz T, Schönberger M, Trauner D (2011) Optochemical genetics. Angew Chem Int Ed Engl 50:12156–12182
Feldbauer K, Zimmermann D, Pintschovius V et al (2009) Channelrhodopsin-2 is a leaky proton pump. Proc Natl Acad Sci U S A 106:12317–12322
Govorunova EG, Sineshchekov OA, Li H et al (2013) Characterization of a highly efficient blue-shifted channelrhodopsin from the marine alga *Platymonas subcordiformis*. J Biol Chem 288:29911–29922
Grinvald A (2005) Imaging input and output dynamics of neocortical networks *in vivo*: exciting times ahead. Proc Natl Acad Sci U S A 102:14125–14126
Han X, Boyden ES (2007) Multiple-color optical activation, silencing, and desynchronization of neural activity, with single-spike temporal resolution. PLoS One 2:e299. doi:10.1371/journal.pone.0000299
Han X, Chow BY, Zhou H et al (2011) A high-light sensitivity optical neural silencer: development and application to optogenetic control of non-human primate cortex. Front Sys Neurosci 5:00018. doi:10.3389/fnsys.2011.00018
Hegemann P (2008) Algal sensory photoreceptors. Ann Rev Plant Biol 59:167–189
Hososhima S, Sakai S, Ishizuka T et al (2015) Kinetic evaluation of photosensitivity in bi-stable variants of chimeric channelrhodopsins. PLoS ONE in press
Inoue K, Ono H, Abe-Yoshizumi R et al (2013) A light-driven sodium ion pump in marine bacteria. Nat Commun 4:1678. doi:10.1038/ncomms2689

Iseki M, Matsunaga S, Murakami A et al (2002) A blue-light-activated adenylyl cyclase mediates photoavoidance in *Euglena gracilis*. Nature 415:1047–1051

Ishizuka T, Kakuda M, Araki R et al (2006) Kinetic evaluation of photosensitivity in genetically engineered neurons expressing green algae light-gated channels. Neurosci Res 54:85–94

Judkewitz B, Rizzi M, Kitamura K et al (2009) Targeted single-cell electroporation of mammalian neurons *in vivo*. Nat Protoc 4:862–869

Kato HE, Inoue K, Abe-Yoshizumi R et al (2015) Structural basis for Na^+ transport mechanism by a light-driven Na^+ pump. Nature in press

Kato HE, Zhang F, Yizhar O et al (2012) Crystal structure of the channelrhodopsin light-gated cation channel. Nature 482:369–374

Kawabata I, Umeda T, Yamamoto K et al (2004) Electroporation-mediated gene transfer system applied to cultured CNS neurons. Neuroreport 15:971–975

Kawasaki H, Iwai L, Tanno K (2012) Rapid and efficient genetic manipulation of gyrencephalic carnivores using *in utero* electroporation. Mol Brain 5:1–7

Kim JM, Hwa J, Garriga P et al (2005) Light-driven activation of β_2-adrenergic receptor signaling by a chimeric rhodopsin containing the β_2-adrenergic receptor cytoplasmic loops. Biochemistry 44:2284–2292

Kimura Y, Satou C, Fujioka S et al (2013) Hindbrain V2a neurons in the excitation of spinal locomotor circuits during zebrafish swimming. Curr Biol 23:843–849

Klapoetke NC, Murata Y, Kim SS et al (2014) Independent optical excitation of distinct neural populations. Nat Methods 11:338–346

Kleinlogel S, Feldbauer K, Dempski RE et al (2011) Ultra light-sensitive and fast neuronal activation with the Ca^{2+}-permeable channelrhodopsin CatCh. Nat Neurosci 14:513–518

Levskaya A, Weiner OD, Lim WA et al (2009) Spatiotemporal control of cell signalling using a light-switchable protein interaction. Nature 461:997–1001

Li X, Gutierrez DV, Hanson MG et al (2005) Fast noninvasive activation and inhibition of neural and network activity by vertebrate rhodopsin and green algae channelrhodopsin. Proc Natl Acad Sci U S A 102:17816–17821

Lin JY, Lin MZ, Steinbach P et al (2009) Characterization of engineered channelrhodopsin variants with improved properties and kinetics. Biophys J 96:1803–1814

Lin JY, Knutsen PM, Muller A et al (2013) ReaChR: a red-shifted variant of channelrhodopsin enables deep transcranial optogenetic excitation. Nat Neurosci 16:1499–1508

Liu X, Ramirez S, Pang PT et al (2012) Optogenetic stimulation of a hippocampal engram activates fear memory recall. Nature 484:381–385

Matsuda T, Cepko CL (2004) Electroporation and RNA interference in the rodent retina *in vivo* and *in vitro*. Proc Natl Acad Sci U S A 101:16–22

Mattis J, Tye KM, Ferenczi EA et al (2012) Principles for applying optogenetic tools derived from direct comparative analysis of microbial opsins. Nat Methods 9:159–172

Miesenböck G, De Angelis DA, Rothman JE (1998) Visualizing secretion and synaptic transmission with pH-sensitive green fluorescent proteins. Nature 394:192–195

Miyawaki A, Llopis J, Heim R et al (1997) Fluorescent indicators for Ca^{2+} based on green fluorescent proteins and calmodulin. Nature 388:882–887

Möglich A, Yang X, Ayers RA et al (2010) Structure and function of plant photoreceptors. Annu Rev Plant Biol 61:21–47

Momose T, Tonegawa A, Takeuchi J et al (1999) Efficient targeting of gene expression in chick embryos by microelectroporation. Dev Growth Differ 41:335–344

Nagel G, Ollig D, Fuhrmann M et al (2002) Channelrhodopsin-1: a light-gated proton channel in green algae. Science 296:2395–2398

Nagel G, Szellas T, Huhn W et al (2003) Channelrhodopsin-2, a directly light-gated cation-selective membrane channel. Proc Natl Acad Sci U S A 100:13940–13945

Nagel G, Brauner M, Liewald JF et al (2005a) Light activation of channelrhodopsin-2 in excitable cells of *Caenorhabditis elegans* triggers rapid behavioral responses. Curr Biol 15:2279–2284

Nagel G, Szellas T, Kateriya S et al (2005b) Channelrhodopsins: directly light-gated cation channels. Biochem Soc Trans 33:863–866

Nishiyama J, Hayashi Y, Nomura T et al (2012) Selective and regulated gene expression in murine Purkinje cells by *in utero* electroporation. Eur J Neurosci 36:2867–2876
O'Connor DH, Huber D, Svoboda K (2009) Reverse engineering the mouse brain. Nature 461:923–929
Odani N, Ito K, Nakamura H (2008) Electroporation as an efficient method of gene transfer. Dev Growth Differ 50:443–448
Oh E, Maejima T, Liu C et al (2010) Substitution of 5-HT_{1A} receptor signaling by a light-activated G protein-coupled receptor. J Biol Chem 285:30825–30836
Ohkura M, Sasaki T, Kobayashi C et al (2012) An improved genetically encoded red fluorescent Ca^{2+} indicator for detecting optically evoked action potentials. PLoS One 7:e39933. doi:10.1371/journal.pone.0039933
Oron D, Papagiakoumou E, Anselmi F et al (2012) Two-photon optogenetics. Prog Brain Res 196:119–143
Packer AM, Peterka DS, Hirtz JJ et al (2012) Two-photon optogenetics of dendritic spines and neural circuits. Nat Methods 9:1202–1205
Papagiakoumou E, Anselmi F, Begue A et al (2010) Scanless two-photon excitation of channelrhodopsin-2. Nat Methods 7:848–854
Pastrana E (2011) Optogenetics: controlling cell function with light. Nat Methods 8:24–25
Penfield W, Rasmussen T (1950) The cerebral cortex of man; a clinical study of localization of function. Macmillan, Oxford
Prakash R, Yizhar O, Grewe B et al (2012) Two-photon optogenetic toolbox for fast inhibition, excitation and bistable modulation. Nat Methods 9:1171–1179
Prigge M, Schneider F, Tsunoda SP et al (2012) Color-tuned channelrhodopsins for multiwavelength optogenetics. J Biol Chem 287:31804–31812
Rickgauer JP, Tank DW (2009) Two-photon excitation of channelrhodopsin-2 at saturation. Proc Natl Acad Sci U S A 106:15025–15030
Ryu MH, Moskvin OV, Siltberg-Liberles J et al (2010) Natural and engineered photoactivated nucleotidyl cyclases for optogenetic applications. J Biol Chem 285:41501–41508
Saito T, Nakatsuji N (2001) Efficient gene transfer into the embryonic mouse brain using *in vivo* electroporation. Dev Biol 240:237–246
Sakai R, Repunte-Canonigo V, Raj CD et al (2001) Design and characterization of a DNA-encoded voltage-sensitive fluorescent protein. Eur J Neurosci 13:2314–2318
Sasaki K, Yamashita T, Yoshida K et al (2014) Chimeric proton-pumping rhodopsins containing the cytoplasmic loop of bovine rhodopsin. PLoS One 9:e91323. doi:10.1371/journal.pone.0091323
Scott EK (2009) The Gal4/UAS toolbox in zebrafish: new approaches for defining behavioral circuits. J Neurochem 110:441–456
Stehfest K, Hegemann P (2010) Evolution of the channelrhodopsin photocycle model. ChemPhysChem 11:1120–1126
Stierl M, Stumpf P, Udwari D et al (2011) Light modulation of cellular cAMP by a small bacterial photoactivated adenylyl cyclase, bPAC, of the soil bacterium *Beggiatoa*. J Biol Chem 286:1181–1188
Takahashi F, Yamagata D, Ishikawa M et al (2007) AUREOCHROME, a photoreceptor required for photomorphogenesis in *stramenopiles*. Proc Natl Acad Sci U S A 104:19625–19630
Tanaka KF, Matsui K, Sasaki T et al (2012) Expanding the repertoire of optogenetically targeted cells with an enhanced gene expression system. Cell Rep 2:397–406
Umeda K, Shoji W, Sakai S et al (2013) Targeted expression of a chimeric channelrhodopsin in zebrafish under regulation of Gal4-UAS system. Neurosci Res 75:69–75
Wang H, Sugiyama Y, Hikima T et al (2009) Molecular determinants differentiating photocurrent properties of two channelrhodopsins from *Chlamydomonas*. J Biol Chem 284:5685–5696
Wen L, Wang H, Tanimoto S et al (2010) Opto-current-clamp actuation of cortical neurons using a strategically designed channelrhodopsin. PLoS One 5:e12893. doi:10.1371/journal.pone.0012893

Wietek J, Wiegert JS, Adeishvili N et al (2014) Conversion of channelrhodopsin into a light-gated chloride channel. Science 344:409–412
Yawo H, Asano T, Sakai S et al (2013) Optogenetic manipulation of neural and non-neural functions. Dev Growth Differ 55:474–490
Ye H, Daoud-El Baba M, Peng RW et al (2011) A synthetic optogenetic transcription device enhances blood-glucose homeostasis in mice. Science 332:1565–1568
Yizhar O, Fenno LE, Davidson TJ et al (2011a) Optogenetics in neural systems. Neuron 71:9–34
Yizhar O, Fenno LF, Prigge M et al (2011b) Neocortical excitation/inhibition balance in information processing and social dysfunction. Nature 477:171–178
Zhang F, Wang LP, Brauner M et al (2007) Multimodal fast optical interrogation of neural circuitry. Nature 446:633–639
Zhang F, Vieroch J, Yizhar O et al (2011) The microbial opsin family of optogenetic tools. Cell 147:1446–1457

Chapter 9
Optogenetic Manipulation and Probing

Masamichi Ohkura, Junko Sadakari, and Junichi Nakai

Abstract Controlling and monitoring the activities of defined cell populations should provide powerful methodologies for better understanding individual cellular functions in vivo. To enable the control and monitoring of cellular activities, 'photo-actuator molecules' and 'fluorescent probe molecules' have been generated, respectively. Photo-actuators are the motor molecules that can trigger cellular activities by photo-activation of specific intracellular molecules, and fluorescent probes are the molecules utilized to detect cellular activities by emitting fluorescence upon binding to their specific target structures of intracellular molecules. These actuators and probes are also known as 'optogenetic tools', and they can be expressed in specific cells or specific organelles for a long period, because they are genetically encoded. In recent years, the development and improvement of optogenetic tools has progressed rapidly. Researchers can now choose optogenetic tools that better suit their needs. In this review, we describe the history, species, and development of optogenetic tools, and future issues, limiting the definition of optogenetic tools to those based on proteins.

Keywords Optogenetic tools • Photo-actuator molecules • Fluorescent probes • Channelrhodopsin (ChR) variants • Halorhodopsin (*Np*HR/halo)/archaerhodopsin (Arch) variants • Light-driven G-protein-coupled receptor (opto-XR) variants • Ca^{2+} probes • Voltage probes • H^+ probes • Cl^- probes

9.1 Introduction

To understand how individual cells function in vivo, gain or loss of function experiments using live model animals should provide a powerful methodology. To control cellular functions, the following methods have often been used: (1) removal or ablation of specific cells of interest (anatomical method), (2) potentiation or suppression of cellular functions via drug application (pharmacological method), and (3) depolarization or hypopolarization of a single cell or a small number of cells via electric

M. Ohkura (✉) • J. Sadakari • J. Nakai
Saitama University Brain Science Institute, Saitama 338-8570, Japan
e-mail: mohkura@mail.saitama-u.ac.jp; s10dr001@mail.saitama-u.ac.jp; jnakai@mail.saitama-u.ac.jp

H. Yawo et al. (eds.), *Optogenetics*, DOI 10.1007/978-4-431-55516-2_9

stimulation (electrophysiological method). However, these methods carry some disadvantages. For example, it is not easy to selectively control cells of interest using the methods described in (1) or (2) without elaborating the methods. The method described in (3) allows us to selectively control cells, but the number of cells controlled at a time is limited. In addition, it is impossible to selectively control intracellular domains of interest by any of these methods.

Here, the cutting-edge methodologies for studying cellular functions in vivo by optogenetic manipulation and probing are featured. To perform optogenetic manipulation and probing, one has to express photo-actuator molecules and fluorescent probe molecules, respectively, in the target cells, then control and detect cellular functions by illuminating them. In recent years, development and improvement of photo-actuators and fluorescent probes have progressed rapidly. These molecules are now known as 'optogenetic tools'. Because optogenetic tools are genetically encoded, it becomes possible to express them in specific cells or specific organelles for a long period. Demands are now greater than before for the manipulation and probing of cellular functions in model animals in vivo. Optogenetic tools are suitable for performing such experiments.

In this review, we describe the history, species, and development of optogenetic tools, as well as future issues, limiting the definition of optogenetic tools to those based on proteins.

9.2 History of Photo-Actuator Molecules

Actuators are motors that are responsible for moving systems. Here, in the field of life science, photo-actuators as optogenetic tools are known as the motor molecules that can trigger cellular activities by photo-activation of specific intracellular molecules. The first protein-based photo-actuator was reported by Nagel's group in 2003 (Nagel et al. 2003). They isolated a non-selective cation (mainly Na^+) channel 'channelrhodopsin (ChR)-2' from *Chlamydomonas reinhardtii*, which can be activated by blue-light illumination, and cloned the gene coding ChR2. Further, in 2005, for the first time, Deisseroth's group applied ChR2 to neurons and found that blue-light illumination of the neurons expressing ChR2 could trigger neuronal firing (Boyden et al. 2005). Since then, ChR2 has been applied intensively in the field of life science, especially in the field of neuroscience, as a photo-actuator molecule.

In the following sections, we summarize representative photo-actuator molecules.

9.2.1 Channelrhodopsin (ChR) Variants

ChR variants such as ChR2 (Fig. 9.1a) are non-selective cation channels that permeate cations upon blue-light illumination; they are frequently used to trigger neuronal firing by light illumination at the arbitrary point of the time. ChR variants have seven transmembrane segments, in which the responsive sites for photo-activation

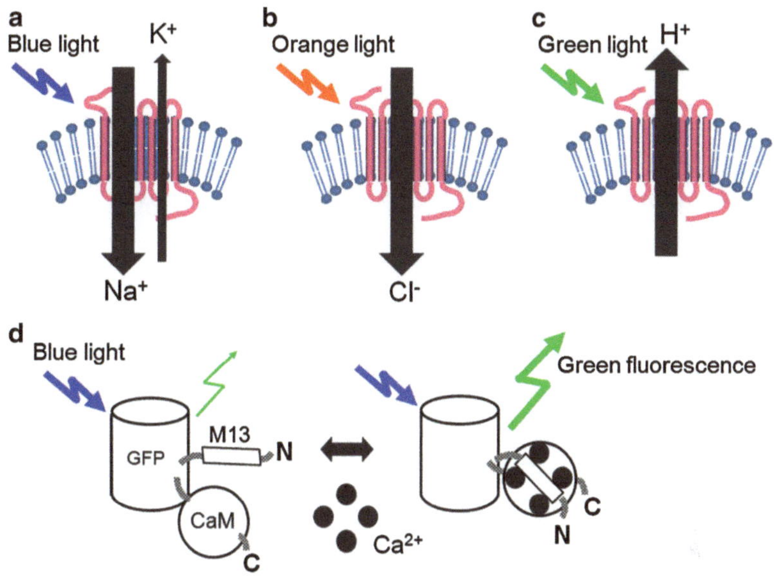

Fig. 9.1 *Operating principles of optogenetic tools.* (**a**) Channelrhodopsin (ChR)-2, a non-selective cation channel. This channel transports mainly Na^+ from the extracellular space to the cytoplasm upon blue-light illumination. (**b**) Halorhodopsin (HR) from *Natronomonas pharaonis* (*Np*HR/Halo), a light-driven Cl^- pump. This pump transports Cl^- from the extracellular space to the cytoplasm upon orange-light illumination. (**c**) Arch, a light-driven H^+ pump. This pump transports H^+ from the cytoplasm to the extracellular space upon orange-light illumination. (**d**) G-CaMP, a fluorescent Ca^{2+} probe. This probe emits bright green fluorescence upon Ca^{2+} binding

are located. The primary sequences of the transmembrane segments from ChR variants are similar to those from light-driven ion pumps of prokaryote rhodopsins.

Any ChR variant binds retinal (mainly *trans* form), an analog of vitamin A, as a substrate to its transmembrane segments, and forms the chromophore. Without retinal, either ChR variant cannot form the chromophore (remains apoprotein) and fails to be photo-activated. Thus, in order to photo-activate ChR variants, they must bind retinal and form chromophores (become holoproteins). Nonetheless, in the experiments using mammalian neurons, the external addition of retinal is not necessary. This is because mammalian cells can produce the necessary amounts of retinal by themselves.

Retinal bound to ChR variants is isomerized upon illumination at a wavelength peculiar to each ChR variant. In general, turning on the light induces isomerization of retinal and activates ChRs; turning off the light terminates this reaction. However, in recently developed ChR variants, termed step-function opsins (SFOs), e.g., ChR2(C128S) (Berndt et al. 2009), the isomerization state of retinal can be controlled with light at two different wavelengths (the SFOs can be activated with blue light and deactivated with yellow light). A favorable feature of SFOs is that once they are activated with blue light, their activation states last longer (τ_{off} ~100 s) than those of typical ChRs.

Nowadays, many ChR variants with different optical characteristics, including SFOs, have been generated. Researchers can select and use an appropriate ChR variant depending on the purposes of their experiments.

9.2.2 *Halorhodopsin (NpHR/Halo)/Archaerhodopsin (Arch) Variants*

Variants of halorhodopsin (HR) from *Natronomonas pharaonis* (*Np*HR/Halo) (Fig. 9.1b) are light-driven Cl^- pumps derived from archaea that transport Cl^- from extracellular space to the cytoplasm upon orange-light illumination (Han and Boyden 2007; Zhang et al. 2007); they are frequently used to suppress neuronal firing by the light illumination at the arbitrary point at the time. On the other hand, Arch variants (Fig. 9.1c) are light-driven H^+ pumps derived from archaea that transport H^+ from the cytoplasm to the extracellular space upon green-light illumination (Chow et al. 2010); they are also frequently used to suppress neuronal firing like *Np*HR/Halo variants. It has been reported that Arch variants can more strongly suppress neuronal firing than can *Np*HR/Halo variants, and that Arch variants are more suitable for performing repeated suppression or a lengthy suppression. Despite these advantages of Arch variants, it should be noted that photo-stimulation of Arch variants is likely to shift intracellular pH.

9.2.3 *Light-Driven G-Protein-Coupled Receptor (Opto-XR) Variants*

Opto-XR variants are chimeric G-protein-coupled receptors, of which agonist-binding sites are substituted with the sites derived from mammalian rhodopsin. Thus, Opto-XR variants activate their corresponding signal transduction cascades upon stimulation with light but not with agonists. Currently, three Opto-XR variants are available: opto-a1AR (a chimeric receptor between adrenergic alpha 1 receptor and rhodopsin: a variant for Gq cascade activation) (Airan et al. 2009), opto-b2AR (a chimeric receptor between adrenergic beta 2 receptor and rhodopsin: a variant for Gs cascade activation) (Airan et al. 2009), and Rh-CT5-HT1A (a chimeric receptor between serotonergic 5-HT1A receptor and rhodopsin: a variant for Gi/o cascade activation) (Oh et al. 2010).

9.3 History of Fluorescent Probe Molecules

In the life science field, fluorescent probe molecules are utilized to detect cellular activities by emitting fluorescence upon binding to their specific target structures of intracellular molecules. Before the 1990s, aequorin, which emits chemiluminescence upon Ca^{2+} binding, had been used as a probe to detect cellular activities. However, in recent years, aequorin is rarely used, because the chemiluminescence signal of aequorin is very weak and does not allow us to detect in a high spatiotemporal resolution. Instead, the use of genetically encoded fluorescent probes has become popular for detecting cellular activities. Since Miyawaki et al. reported such a Ca^{2+} probe, termed 'Cameleon' (Miyawaki et al. 1997) in 1997, and Miesenbock et al. reported a probe for H^+, termed 'pHluorin' (Miesenbock et al. 1998) in 1998, various kinds of high performance genetically encoded fluorescent probes have become available. Fluorescent probes are categorized into two types based on their structures: the Förster resonance energy transfer (FRET)-type probes (probes with two green fluorescent protein [GFP] variants with different colors that cause shifts in their fluorescence ratio following FRET), and the single GFP-type probes (probes with single GFP variants that cause changes in their fluorescence intensity following changes in probe structure).

In the following sections, we summarize representative fluorescent probe molecules.

9.3.1 Ca^{2+} Probes

Ca^{2+} is an intracellular second messenger that plays important roles in almost all cells, including neurons, muscle cells, and secretory cells. Thus, we generally use Ca^{2+} probes to monitor various cellular activities.

After the invention of Cameleon (Miyawaki et al. 1997), various types of Ca^{2+} probes were created, including Camgaroo (Baird et al. 1999), G-CaMP (Fig. 9.1d) (Nakai et al. 2001), and Pericam (Nagai et al. 2001). Among them, G-CaMP and its variants are currently most frequently used by researchers in the field of life science, especially in neuroscience, because their fluorescence changes are fairly large and easy to detect. As the G-CaMP variants are improved with more sophisticated functions, various cellular Ca^{2+} activities have become possible to detect, including Ca^{2+} activities of dendritic spines in neurons (Ohkura et al. 2005), Ca^{2+} activities in vivo in mice (37 °C) (Tallini et al. 2006), neuronal Ca^{2+} activities in response to single action potentials (Akerboom et al. 2012; Ohkura et al. 2012b; Tian et al. 2009), neuronal Ca^{2+} activities in response to optically evoked single action potentials by photo-stimulation of ChR2 (Ohkura et al. 2012a), and neuronal Ca^{2+} activities in response to excitatory synaptic inputs (Ohkura et al. 2012b).

As for other Ca^{2+} probes, Yellow Cameleon-Nano series (Horikawa et al. 2010), which are ultra-highly sensitive and suitable for detecting minute Ca^{2+} activities, and Troponeon series (Heim and Griesbeck 2004), which have fewer side effects in the cells, are also utilized. A large number of choices exist for Ca^{2+} probes, each with different advantages in terms of brightness, signal amplitude, Ca^{2+} sensitivity, kinetics, etc. Researchers can now choose Ca^{2+} probes that better suit their needs.

9.3.2 Voltage Probes

The change in membrane potential, which shows dynamic fluctuations in response to excitatory and inhibitory inputs, plays key roles in excitable cells such as neurons and muscle cells. In the brain, a myriad of neurons simultaneously communicate with each other, enabling a variety of information processing. In order to understand such information processing in the brain, real-time monitoring of neuronal communication is necessary. To date, recording the change in membrane potential has been popular for the purpose of detecting electrical activities of neurons. Here, electrophysiological approaches, including the micro-pipette method and the patch-clamp method allow us to record fast voltage changes at a sub-millisecond level (e.g., action potentials), whereas the number of simultaneously recordable neurons is limited. An alternative approach is the use of voltage-sensitive dyes, but the excitatory postsynaptic potentials (EPSP), the inhibitory postsynaptic potentials (IPSP), and even the action potentials are difficult to detect with currently available dyes. Another disadvantage of dyes is that, because they stain a large indefinite number of cells, it is necessary to devise means to identify the cell recorded with the dye after each experiment. To overcome these disadvantages, genetically encoded voltage probes have been developed.

In the development of the incipient voltage probes, including FlaSh (Siegel and Isacoff 1997), VSFP1 (Sakai et al. 2001), and SPARC (Ataka and Pieribone 2002), Na^+ or K^+ channels, the structures and functions of which have been well characterized, were used as the voltage-sensing domains. Although these incipient probes gave small fluorescence signals, they were able to convert structural changes of their voltage-sensing domains into fluorescence changes of their GFP domains. Recently, a voltage-sensitive enzyme, 'Ci-VSP' (Murata et al. 2005), in which the voltage-sensitive domain and the catalytic domain are separate, unlike ion channels, was first applied as a component of voltage probes. The improvement of Ci-VSP-based voltage probes is progressing (Mutoh et al. 2012).

On the other hand, Arch variants have been used not only as light-driven H^+ pumps, but also as voltage probes. Intriguingly, a probe termed Arch(D95N) (Kralj et al. 2011) shows voltage-dependent fluorescence change but no photocurrent, thereby causing no suppression of neuronal firings by the light illumination during the imaging session. Further improved versions of the Arch(D95N) probe, termed QuasARs (Hochbaum et al. 2014), with high voltage sensitivity and brightness have been reported more recently.

9.3.3 H^+ Probes

In neurons, pH inside the synaptic vesicles is kept low (pH ~5.7) because H^+ is actively pumped out by the H^+ pumps. When the synaptic transmission occurs, the synaptic vesicles fuse with the presynaptic membrane, and the inside space of

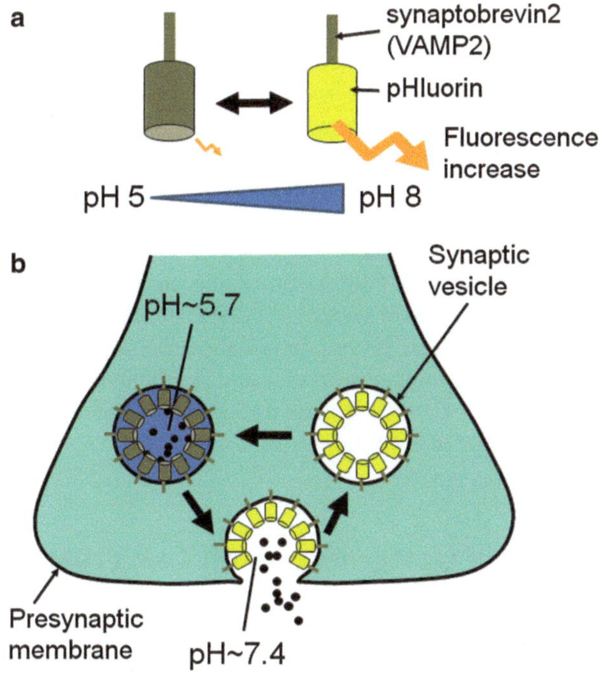

Fig. 9.2 *Detection principles for synaptic release activities using a fluorescent* H^+ *probe, Synapto-pHluorin.* (**a**) Schematic structure of Synapto-pHluorin, which is a fusion protein between pHluorin, a green fluorescent protein (GFP) variant sensitive to H^+ at pH 5–8, and synaptobrevin2 (VAMP2), a SNARE protein. Because pHluorin fused to the C-terminus of synaptobrevin2 faces the luminal side in the synaptic vesicle, luminal pH can be visualized as the fluorescence intensity of pHluorin. When the luminal pH increases, the fluorescence of pHluorin also increases. (**b**) Synapto-pHluorin is dim when the luminal pH of the synaptic vesicle is low (i.e., when the synaptic vesicle expressing the probe is unbound to the presynaptic membrane: pH ~ 5.7). During synaptic transmission, the probe turns bright with the pHluorin domain being exposed to extracellular pH (pH ~ 7.4) following the fusion of the synaptic vesicle to the presynaptic membrane. Afterwards, the probe turns dim again in response to acidification of the luminal space following withdrawal of the synaptic vesicle from the presynaptic membrane

synaptic vesicles is exposed to the same pH environment as the extracellular space (pH ~7.4). Thus, by expressing an H^+ probe inside the synaptic vesicles, the presynaptic release activities can be monitored. A popularly used H^+ probe is Synapto-pHluorin (Fig. 9.2) (Sankaranarayanan and Ryan 2001), which was made by fusing pHluorin (Miesenbock et al. 1998), a GFP variant sensitive to H^+ at pH 5–8, and synaptobrevin2 (VAMP2), a SNARE protein. A red H^+ probe has also recently been developed to detect presynaptic release activities (Li et al. 2011). This red H^+ probe is advantageous for neuroscientific applications, because it can be used simultaneously with photo-actuator molecules such as ChR2, *Np*HR/Halo, and Arch.

9.3.4 Cl^- Probes

As described above, Ca^{2+} probes have been widely used to monitor neuronal firing. Today, the need to monitor not only neuronal firing but also inhibitory inputs such as IPSP is increasing. Inhibitory neurotransmitters, including γ-aminobutyric acid (GABA) and glycine, cause transient increases in intracellular local Cl^- concentrations. Aiming to visualize IPSP in cells of interest, Cl^- probes have been developed.

In existing Cl^- probes such as Clomeleon (Kuner and Augustine 2000), yellow fluorescent protein (YFP), a yellow variant of GFP that binds halides, has been used as the Cl^--sensing domain. These probes do not allow the detection of minute intracellular Cl^- signals, because their sensitivities and responsivities are a little too low for the detection of physiological Cl^- concentrations. We expect that more improved Cl^- probes will be available in the future.

9.4 Development of Optogenetic Tools: Ca^{2+} Probes

In this section, we describe examples for the development of our latest G-CaMP-type Ca^{2+} probes.

9.4.1 Development of Green Fluorescent Ca^{2+} Probes, G-CaMP6/7/8, for Detecting Single Neuronal Spikes

As described above, the use of G-CaMP-type Ca^{2+} probes has been popular, because their fluorescence signals are greater than those obtained from other Ca^{2+} probes. Nonetheless, it was hard to detect minute Ca^{2+} signals such as those following neuronal single spikes with previous G-CaMP-type Ca^{2+} probes (subordinate G-CaMP variants in comparison with G-CaMP2 (Nakai et al. 2001; Ohkura et al. 2005; Tallini et al. 2006)).

To enable the detection of such minute neuronal Ca^{2+} signals, we started to develop novel G-CaMP-type Ca^{2+} probes with high sensitivity and high responsivity (Ohkura et al. 2012b). We first performed site-directed mutagenesis, aiming to enhance the Ca^{2+} sensitivity and improve the signal amplitude, against a prototype Ca^{2+} probe, G-CaMP2 (Tallini et al. 2006). Through in vitro screening and in a cell-based assay, we successfully found a G-CaMP2 variant with high sensitivity ($K_d = 158$ nM), termed G-CaMP6 (Fig. 9.3a). Next we performed random mutagenesis on G-CaMP6 by using an error-prone polymerase chain reaction (PCR), and were able to screen a G-CaMP6 variant with high responsivity, termed G-CaMP7, which differs from G-CaMP6 by a deletion of histidine in the RSET domain and an S205N mutation in the EGFP domain. The responsivity of G-CaMP7 was ~3-fold

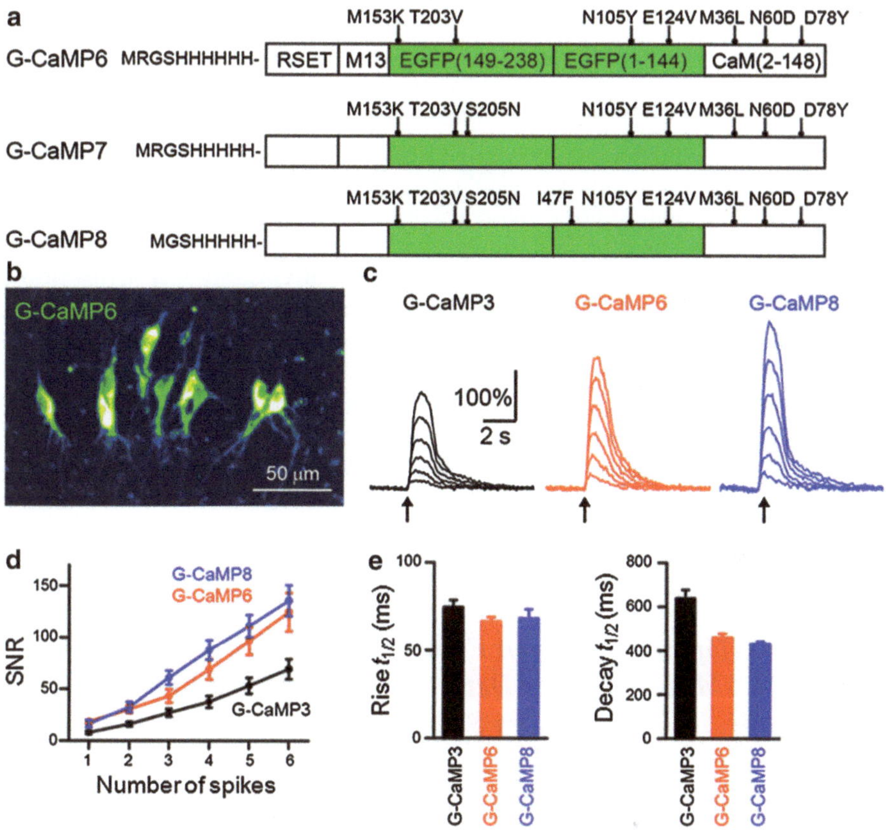

Fig. 9.3 *Comparisons for the characteristics of G-CaMP-type* Ca^{2+} *probes*. (**a**) Schematic structures of G-CaMP6/7/8. Mutations are indicated with respect to a prototype G-CaMP2. (**b**) Fluorescence image of G-CaMP6-expressing neurons. G-CaMP6 was expressed in CA3 pyramidal neurons of cultured hippocampal slices via targeted single-cell electroporation. (**c**) Respective traces of the Ca^{2+} fluorescence response ($\Delta F/F$) to trains of 1–6 spikes at a frequency of 50 Hz. (**d**) Mean signal-to-noise ratios (SNRs) plotted against the number of spikes. SNRs of G-CaMP6 and G-CaMP8 were consistently higher than those of G-CaMP3. (**e**) Rise and decay time constants for the responses to single spikes (Modified from Ohkura et al. 2012b)

greater than that of G-CaMP6, even though this variant showed a lower Ca^{2+} sensitivity (K_d=240 nM). By performing further random mutagenesis on G-CaMP7, we could obtain a more sensitive variant of G-CaMP7, termed G-CaMP8, bearing a deletion of arginine in the RSET domain and an I47F mutation in the EGFP domain. Because the Ca^{2+} sensitivity of G-CaMP8 (K_d=200 nM) was higher than that of G-CaMP7, despite that their responsivities were similar, the signal amplitude of G-CaMP8 was greater than that of G-CaMP7 in cell-based Ca^{2+} measurements.

We next evaluated the performance of G-CaMPs in pyramidal neurons in cultured rat hippocampal slices. Complementary DNAs (cDNAs) encoding G-CaMPs were subcloned into an expression plasmid with a CMV promoter, and the obtained plasmids were introduced in the cells of interest via targeted single-cell electroporation. At 24–48 h after electroporation, simultaneous patch-clamp recording and Ca^{2+} imaging with a Nipkow spinning-disc confocal microscope were performed on neurons expressing G-CaMPs (Fig. 9.3b). The neurons were current-injected to evoke spikes, and single spike-induced Ca^{2+} responses were successfully monitored from the neurons expressing G-CaMP6 and G-CaMP8 with 100 % detectability (Fig. 9.3c). The signal-to-noise ratios (SNRs) of Ca^{2+} responses evoked by 1–6 spikes are shown in Fig. 9.3d. Over the entire stimulus range, the SNRs of G-CaMP6 and G-CaMP8 were consistently higher than those of G-CaMP3. In terms of the kinetics, the rise time of the single spike-induced Ca^{2+} responses did not differ among G-CaMP3, G-CaMP6, and G-CaMP8, whereas the decay time for G-CaMP6 and G-CaMP8 was significantly faster than that for G-CaMP3 (Fig. 9.3e). The rapid kinetics contribute to an increased temporal resolution of the signals within spike trains. In fact, G-CaMP6 was shown to discrete fast individual spikes in burst-spike trains with a temporal resolution of up to 15–20 Hz. We have found that G-CaMP6/7/8 can detect neuronal Ca^{2+} responses in acute slices or in freely moving *Caenorhabditis elegans*.

9.4.2 Development of Green Fluorescent Ca^{2+} Probes, G-CaMP6-Actin, for Detecting Sub-threshold Input Activities at Excitatory Synapses

EPSP by excitatory synaptic inputs and IPSP by inhibitory synaptic inputs to a neuron are integrated, and when the integrated membrane potential exceeds a threshold membrane potential, an action potential occurs. Because EPSP and IPSP have not been successfully visualized, but are thought to be important factors in the control of synaptic plasticity, we are halfway down the road to better understanding their integration mechanism in a neuron. Visualization of EPSP and IPSP should enable us to evaluate long-term plasticity, which is thought to be an elementary component of learning and memory, so that our understanding of memory formation or engram will be facilitated.

Fluorescent probe molecules can be expressed in specific intracellular domains by fusion with another protein or signal peptide. To date, somatic Ca^{2+} activities evoked by neuronal spikes have been successfully detected, but spine Ca^{2+} activities evoked by sub-threshold synaptic inputs have still been difficult to detect. To solve this problem, we tried to develop a spine-targeted Ca^{2+} probe with high sensitivity (Ohkura et al. 2012b). For this purpose, we fused a high sensitivity Ca^{2+} probe G-CaMP6 with actin, a major cytoskeletal protein within spines, and yielded G-CaMP6-actin (Ohkura et al. 2012b). When expressed in hippocampal CA3 pyramidal neurons, G-CaMP6-actin was effectively localized to the spines (Fig. 9.4a).

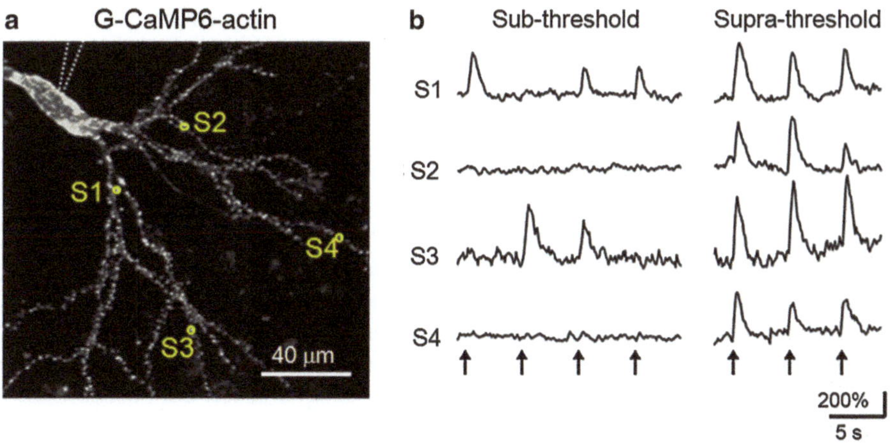

Fig. 9.4 *Detection of spine* Ca^{2+} *activity with G-CaMP6-actin*. (**a**) Fluorescence image of a representative neuron expressing G-CaMP6-actin. G-CaMP6-actin was expressed in a CA3 pyramidal neuron of a cultured hippocampal slice. Fluorescence was detected from virtually all of the spines in the imaged region. The position of the patch pipette is indicated by *dotted lines*. Four spines of interest (S1–S4) are indicated by *circles*. (**b**) Representative changes in Ca^{2+} fluorescence upon sub- or supra-threshold electrical stimulation. *Arrows* indicate the timing of stimulation. The sub-threshold stimulations triggered small fluorescence responses in a limited number of spines with a low probability in the active spines, whereas the supra-threshold stimulations triggered large fluorescence responses in virtually all of the spines with a 100 % probability (Modified from Ohkura et al. 2012b)

We then electrically stimulated the granule cells of the dentate gyrus, which innervate synapses in the CA3 region, with signals of two different strengths. Interestingly, the sub-threshold stimulations triggered small fluorescence responses in a limited number of spines, with a low response rate in the active spines. In contrast, the supra-threshold stimulations triggered large fluorescence responses in virtually all of the spines in the imaged region with a 100 % activity rate (Fig. 9.4b). Excitatory synaptic activity induces a Ca^{2+} transient in individual spines through the activation of voltage-sensitive Ca^{2+} channels and/or *N*-methyl-D-aspartate (NMDA) receptors. Although the exact mechanisms of spine Ca^{2+} responses remain unknown, it seems likely that sub-threshold stimulation triggers Ca^{2+} transients through postsynaptic NMDA receptors, while supra-threshold stimulation triggers Ca^{2+} transients at 100 % of the spines by the opening of voltage-gated Ca^{2+} channels through back-propagation of action potentials. In comparison with precedent studies using synthetic Ca^{2+} probes such as Oregon Green BAPTA-1 (Chen et al. 2011; Takahashi et al. 2012), G-CaMP6-actin seems to have a fairly high Ca^{2+} sensitivity. We expect that this spine-targeted Ca^{2+} probe should provide a powerful tool to detect spatiotemporal patterns of input activities at excitatory synapses.

9.4.3 Development of a Red Fluorescent Ca^{2+} Probe, R-CaMP1.07, for Detecting Action Potentials Evoked by Photo-Stimulation of ChR2

The majority of existing Ca^{2+} probes absorb blue (or green) light and emit green (or yellow) fluorescence. As described above, ChR2 is widely used as a tool to control neuronal excitability, but green (or yellow) fluorescent Ca^{2+} probes are not preferentially applied for combined use with ChR2, because their excitation light also activates ChR2, and this makes it difficult to independently manipulate ChR2 during the excitation of Ca^{2+} probes. Therefore, neuroscientists have been longing for a new Ca^{2+} probe that absorbs orange (or red) light and emits red fluorescence. Here, we describe an example for the development of a superior red Ca^{2+} probe that is excitable at a wavelength substantially distinct from the action spectra of ChR2 (Ohkura et al. 2012a).

To create such a red Ca^{2+} probe, we thought to improve R-GECO1 (Zhao et al. 2011), the first red Ca^{2+} probe in which a circularly permutated red fluorescent protein (RFP) is linked to calmodulin (CaM) and its target peptide M13 from myosin light chain kinase at its C- and N-termini, respectively, similar to G-CaMP. We first incorporated the RFP domain of R-GECO1 into a prototype molecule G-CaMP2 (Tallini et al. 2006), then introduced random mutagenesis into the RFP domain. The variants were screened in vitro and in HeLa cells, and the best variant R-CaMP1.01, bearing K47V and T49V mutations in the RFP domain, was identified. Since it was reported that the performance of G-CaMP2, our previous green fluorescent Ca^{2+} probe, was modified by peptide fusion, we further tested whether such fusion of several peptides with R-CaMP1.01 at the N- or C-terminus would modify the performance of R-CaMP1.01; we anticipated that some variants would be superior to R-CaMP1.01. Consequently, we found a variant of R-CaMP1.01, termed R-CaMP1.07, that had a self-cleaving peptide F2A at the C-terminus of R-CaMP1.01, exhibited the largest Ca^{2+} responses. Furthermore, we incidentally found that R-GECO1 and R-CaMP1.01 localized not only to the cytoplasm but also to the nucleus, whereas R-CaMP1.07 localized only to the cytoplasm in HeLa cells.

We next assessed the functionality of R-CaMP1.07 in neurons. R-CaMP1.07 was expressed in rat hippocampal CA3 pyramidal neurons in cultured slices. We induced trains of action potentials by current injection into the neurons, and tested whether R-CaMP1.07 could detect the action potential-induced Ca^{2+} transients. As a result, we found that R-CaMP1.07 exhibits an almost linear increase in the amplitudes of the Ca^{2+} responses evoked by up to six action potentials. We also tested the electrophysiological properties of neurons expressing R-CaMP1.07, and observed no significant differences in input resistance, membrane capacitance, resting potential, or amplitudes and frequency of spontaneous EPSCs between control (R-CaMP1.07-negative) and R-CaMP1.07-positive cell groups. These results suggest that R-CaMP1.07 expression per se does not cause abnormal changes in cellular electrophysiological properties or synaptic activity. R-CaMP1.07 was shown to be applicable for functional studies of neurons.

Because R-CaMP1.07 is hardly excited by the 400–500 nm light, which can effectively activate ChR2, it should be possible to use R-CaMP1.07 to monitor the neuronal activity evoked by photo-stimulation of ChR2 in an identical cell. To explore this hypothesis, we co-expressed R-CaMP1.07 and ChR2 in identical cells. Excitation and photo-stimulation wavelengths of 568 and 488 nm were used to monitor and manipulate neuronal action potentials, respectively. As expected, we successfully observed that photo-stimulation with 488 nm light triggered action potentials and transient increases in R-CaMP1.07 fluorescence. Thus far, in order to independently control ChR2 during the excitation of green Ca^{2+} probes, one must devise means such as using different laser powers for Ca^{2+} imaging and photo-stimulation. R-CaMP1.07 should provide a powerful tool for detection of optically evoked action potentials. As to the development of R-CaMP1.07, also see Chap. 10.

More recently, we reported an improved version of R-CaMP1.07, termed R-CaMP2, with ultra-high Ca^{2+} sensitivity and ultra-fast kinetics (Inoue et al. 2015). In the R-CaMP2 molecule, the M13 peptide is replaced with a peptide from CaMKK, which is known to bind to CaM at ultra-low Ca^{2+} concentrations.

9.5 Future Issues

Optogenetic tools have improved rapidly in the last few years. But several drawbacks remain with the use of these tools to manipulate and image cellular functions in in vivo model animals. We expect that the following points should be further improved in the future.

9.5.1 Photo-Actuator Molecules

1. Improvement for membrane trafficking. This improvement is likely to reduce a side effect of expressed probes, which is potentially caused by the probes being retained in the endoplasmic reticulum or in the Golgi apparatus.
2. Manipulation targeted to intracellular molecules. Recently, a light-driven actuator based on the Light-Oxygen-Voltage (LOV) domain, which is activated upon blue-light illumination, has been reported (Zoltowski et al. 2007). Further evolution of this kind of actuators is highly expected.

9.5.2 Fluorescent Probe Molecules

1. High brightness. This feature enables easier deep-tissue imaging.
2. High signal amplitude. This helps to enhance the efficiency of signal detection.
3. High sensitivity. This enables the probing of subtle signals.

4. Fast responsivity. Rapid kinetics contribute to discrete fast individual signals in burst signal trains with a high temporal resolution.

9.6 Concluding Remarks

Optogenetic manipulation and imaging are now becoming popular methodologies, not only in basic research but also in clinical research. For example, using model animals, trials for restoring impaired functioning of damaged tissues via photostimulation of transplanted cells expressing ChR2, or monitoring the functions of transplanted cells expressing G-CaMP, have already begun. Optogenetic tools are now applied for various types of cells, including induced pluripotent stem (iPS) cells. We expect that these tools will continue to be optimized for experimental purposes.

Together with the development and improvement of optogenetic tools, the performance of experimental devices used for optogenetic manipulation and probing are also improving. It is now easier than ever before to perform optogenetic experiments. We hope that this review helps readers choose actuators or fluorescent probes that are suitable for their own optogenetic experiments.

References

Airan RD, Thompson KR, Fenno LE et al (2009) Temporally precise in vivo control of intracellular signalling. Nature 458:1025–1029

Akerboom J, Chen TW, Wardill TJ et al (2012) Optimization of a GCaMP calcium indicator for neural activity imaging. J Neurosci 32:13819–13840

Ataka K, Pieribone VA (2002) A genetically targetable fluorescent probe of channel gating with rapid kinetics. Biophys J 82:509–516

Baird GS, Zacharias DA, Tsien RY (1999) Circular permutation and receptor insertion within green fluorescent proteins. Proc Natl Acad Sci U S A 96:11241–11246

Berndt A, Yizhar O, Gunaydin LA et al (2009) Bi-stable neural state switches. Nat Neurosci 12:229–234

Boyden ES, Zhang F, Bamberg E et al (2005) Millisecond-timescale, genetically targeted optical control of neural activity. Nat Neurosci 8:1263–1268

Chen X, Leischner U, Rochefort NL et al (2011) Functional mapping of single spines in cortical neurons *in vivo*. Nature 475:501–505

Chow BY, Han X, Dobry AS et al (2010) High-performance genetically targetable optical neural silencing by light-driven proton pumps. Nature 463:98–102

Han X, Boyden ES (2007) Multiple-color optical activation, silencing, and desynchronization of neural activity, with single-spike temporal resolution. PLoS One 2:e299. doi:10.1371/journal.pone.0000299

Heim N, Griesbeck O (2004) Genetically encoded indicators of cellular calcium dynamics based on troponin C and green fluorescent protein. J Biol Chem 279:14280–14286

Hochbaum DR, Zhao Y, Farhi SL et al (2014) All-optical electrophysiology in mammalian neurons using engineered microbial rhodopsins. Nat Methods 11:825–833

Horikawa K, Yamada Y, Matsuda T et al (2010) Spontaneous network activity visualized by ultrasensitive Ca^{2+} indicators, yellow Cameleon-Nano. Nat Methods 7:729–732
Inoue M, Takeuchi A, Horigane S et al (2015) Rational design of a high-affinity, fast, red calcium indicator R-CaMP2. Nat Methods 12:64–70
Kralj JM, Douglass AD, Hochbaum DR et al (2011) Optical recording of action potentials in mammalian neurons using a microbial rhodopsin. Nat Methods 9:90–95
Kuner T, Augustine GJ (2000) A genetically encoded ratiometric indicator for chloride: capturing chloride transients in cultured hippocampal neurons. Neuron 27:447–459
Li H, Foss SM, Dobryy YL et al (2011) Concurrent imaging of synaptic vesicle recycling and calcium dynamics. Front Mol Neurosci 4:34. doi:10.3389/fnmol.2011.00034
Miesenböck G, De Angelis DA, Rothman JE (1998) Visualizing secretion and synaptic transmission with pH-sensitive green fluorescent proteins. Nature 394:192–195
Miyawaki A, Llopis J, Heim R et al (1997) Fluorescent indicators for Ca^{2+} based on green fluorescent proteins and calmodulin. Nature 388:882–887
Murata Y, Iwasaki H, Sasaki M et al (2005) Phosphoinositide phosphatase activity coupled to an intrinsic voltage sensor. Nature 435:1239–1243
Mutoh H, Akemann W, Knöpfel T (2012) Genetically engineered fluorescent voltage reporters. ACS Chem Neurosci 3:585–592
Nagai T, Sawano A, Park ES et al (2001) Circularly permuted green fluorescent proteins engineered to sense Ca^{2+}. Proc Natl Acad Sci U S A 98:3197–3202
Nagel G, Szellas T, Huhn W et al (2003) Channelrhodopsin-2, a directly light-gated cation-selective membrane channel. Proc Natl Acad Sci U S A 100:13940–13945
Nakai J, Ohkura M, Imoto K (2001) A high signal-to-noise Ca^{2+} probe composed of a single green fluorescent protein. Nat Biotechnol 19:137–141
Oh E, Maejima T, Liu C et al (2010) Substitution of 5-HT_{1A} receptor signaling by a light-activated G protein-coupled receptor. J Biol Chem 285:30825–30836
Ohkura M, Matsuzaki M, Kasai H et al (2005) Genetically encoded bright Ca^{2+} probe applicable for dynamic Ca^{2+} imaging of dendritic spines. Anal Chem 77:5861–5869
Ohkura M, Sasaki T, Kobayashi C et al (2012a) An improved genetically encoded red fluorescent Ca^{2+} indicator for detecting optically evoked action potentials. PLoS One 7:e39933. doi:10.1371/journal.pone.0039933
Ohkura M, Sasaki T, Sadakari J et al (2012b) Genetically encoded green fluorescent Ca^{2+} indicators with improved detectability for neuronal Ca^{2+} signals. PLoS One 7:e51286. doi:10.1371/journal.pone.0051286
Sakai R, Repunte-Canonigo V, Raj CD et al (2001) Design and characterization of a DNA-encoded, voltage-sensitive fluorescent protein. Eur J Neurosci 13:2314–2318
Sankaranarayanan S, Ryan TA (2001) Calcium accelerates endocytosis of vSNAREs at hippocampal synapses. Nat Neurosci 4:129–136
Siegel MS, Isacoff EY (1997) A genetically encoded optical probe of membrane voltage. Neuron 19:735–741
Takahashi N, Kitamura K, Matsuo N et al (2012) Locally synchronized synaptic inputs. Science 335:353–356
Tallini YN, Ohkura M, Choi BR et al (2006) Imaging cellular signals in the heart *in vivo*: cardiac expression of the high-signal Ca^{2+} indicator GCaMP2. Proc Natl Acad Sci U S A 103:4753–4758
Tian L, Hires SA, Mao T et al (2009) Imaging neural activity in worms, flies and mice with improved GCaMP calcium indicators. Nat Methods 6:875–881
Zhang F, Wang LP, Brauner M et al (2007) Multimodal fast optical interrogation of neural circuitry. Nature 446:633–639
Zhao Y, Araki S, Wu J et al (2011) An expanded palette of genetically encoded Ca^{2+} indicators. Science 333:1888–1891
Zoltowski BD, Schwerdtfeger C, Widom J et al (2007) Conformational switching in the fungal light sensor Vivid. Science 316:1054–1057

Chapter 10
Probing Neuronal Activity Using Genetically Encoded Red Fluorescent Calcium Indicators

Takuya Sasaki

Abstract Understanding brain function requires experimental approaches that decipher the information coded by individual neurons. Calcium imaging with genetically encoded calcium indicators (GECIs) is a promising method that can visualize the spatiotemporal activity patterns of brain cells. Recent advances in protein engineering have greatly improved the properties of fluorescent GECIs, and they now have high flexibility for imaging defined cell populations over the course of months. The action spectra of single-wavelength GECIs have been extended by the development of color-shifted fluorescence probes, which increase the potential for multicolor imaging of different cell structures and, more importantly, can be integrated into optogenetics experiments with photoactivatable proteins. In particular, red-shifted GECIs are highly important for imaging deeper tissues because longer-wavelength light can reduce tissue scattering and background autofluorescence. This chapter mainly describes recent advances in the engineering of red GECIs, and highlights important neuroscience applications in optical monitoring and the manipulation of neuronal activity.

Keywords Calcium imaging • Genetically encoded calcium indicator • Red fluorescent protein • R-CaMP • ChR2 • Cortical neurons

10.1 Genetically Encoded Calcium Indicators (GECIs) for Monitoring the Spatiotemporal Dynamics of Neuronal Populations

Monitoring where, when, and how individual cells are active in the brain is a central requirement in neuroscience research. Over the past decades, electrophysiological recordings have been used extensively for capturing neuronal dynamics at a single-cell resolution both in vitro and in vivo. Advances in multi-channel unit recordings of animal behavior allows the simultaneous recording of up to hundreds of neurons,

T. Sasaki (✉)
Graduate School of Pharmaceutical Sciences, The University of Tokyo,
7-3-1 Bunkyo-ku, Hongo, Tokyo 113-0033, Japan
e-mail: tsasaki@mol.f.u-tokyo.ac.jp

H. Yawo et al. (eds.), *Optogenetics*, DOI 10.1007/978-4-431-55516-2_10

which have provided major contributions to the discovery of activity patterns in a variety of brain circuits (Buzsáki 2004). However, the drawbacks of microelectrode recording are the lack of sufficient spatial resolution to resolve the precise location of cells and the difficulty of tracking identified cells over consecutive recording sessions because of electrode drifting, cell death, and gliosis around the tips of electrodes.

Compared with electrophysiological techniques, optical imaging holds greater promise for probing the activity patterns of large groups of neurons with identified physical positions over extended periods of time (Gobel and Helmchen 2007; Takahashi et al. 2010b; Grienberger and Konnerth 2012). In particular, calcium fluorescent imaging has been used to obtain reliable measurement of neuronal activity with a high signal-to-noise ratio (SNR) because action potential firing is coupled tightly to large and rapid changes in the intracellular calcium concentration. Multicell calcium imaging is performed by loading cells with membrane-permeant forms of synthetic calcium indicators such as Fura-2, Oregon Green BAPTA (OGB)-1, and Fluo-4 (Regehr and Tank 1991; Yuste et al. 1992; Dombeck et al. 2007; Chen et al. 2011; Takahashi et al. 2010a). The major advantages of these organic probes are as follows: (i) they are bright, high-affinity, and have calcium-dependent fluorescence changes with large dynamic ranges; and (ii) the bulk loading procedure is relatively simple, i.e., simply incubating living tissues with dye-containing solution. Despite the widespread availability of synthetic indicators, they are generally unsuitable for labeling specific cell populations, which limits the examination of detailed circuit dynamics. Furthermore, it is technically challenging to visualize identical neuron populations over days and weeks due to the rapid clearance of indicators from the intracellular spaces.

Genetically encoded calcium indicators (GECIs) have the potential to overcome these technical limitations (Tian et al. 2012). GECIs are proteins; therefore, they can be expressed by delivering their genes into cells via viral vectors (Ziv et al. 2013; Osakada et al. 2011), electroporation techniques (Yamada et al. 2011; Ohkura et al. 2012b), or by constructing transgenic animal lines (Chen et al. 2012; Zariwala et al. 2012). GECIs have several remarkable advantages over synthetic indicators. First, GECIs can be targeted to specific cell types or specific subcellular compartments using cell-type specific promoters or cellular targeting sequences. Second, GECIs can be tracked repeatedly to analyze the neuronal activity of identical cell populations over the course of multiple recording sessions. Third, GECIs can be easily applied to mature neurons, whereas the loading efficiency of synthetic indicators generally declines as the age of neurons increases. These properties should be essential for unveiling the development, organization, and maintenance of neural circuit dynamics.

10.2 Recent Improvements in Green GECIs

Two important classes of GECIs have been engineered: (i) the Förster resonance energy transfer (FRET) type indicators such as a series of Cameleons (Nagai et al. 2004; Horikawa et al. 2010; Miyawaki et al. 1997), D3cpVenus (Palmer and Tsien

2006), and TN-XXL (Mank et al. 2008); and (ii) the single-wavelength type indicators, such as a series of GCaMPs (Nakai et al. 2001; Ohkura et al. 2005, 2012b; Tian et al. 2009, 2012; Chen et al. 2013) and flash-pericam (Nagai et al. 2001). Both types of GECIs comprise a calcium-binding domain such as calmodulin or troponin C and either one or two fluorescent proteins. These protein structures undergo changes in their fluorescence intensity, depending on the binding of calcium ions. Of these GECIs, GCaMPs are the most widely used sensors in a variety of brain regions and organisms, and they have been improved repeatedly to obtain faster kinetics and larger dynamic ranges of fluorescence changes. In particular, determinations of the X-ray crystal structures of GCaMPs have facilitated more rational improvements in the properties of probes. Recently, structure-based and random mutagenesis and semi-rational library screening have recently produced highly sensitive green calcium probes, including GCaMP3 (Tian et al. 2009), GCaMP4.1(Shindo et al. 2010), G-GECO (Zhao et al. 2011), and GCaMP5 (Akerboom et al. 2012). The latest version of GCaMPs is a family of GCaMP6 (Ohkura et al. 2012b; Chen et al. 2013), which have been developed by two independent groups (for further details, see Chap. 9 by Ohkura and Nakai). Such a new generation of calcium probes significantly outperforms other existing GECIs in the detection of neuronal activity at a sensitivity level similar to or better than those of synthetic calcium dyes. Green GECIs have been shown to be applicable to the long-term in vivo imaging of orientation-tuned activity in visual cortical neurons (Chen et al. 2013) and place-selective firing in hippocampal pyramidal neurons (Ziv et al. 2013) over days and weeks.

10.3 Development and Application of Red GECIs

10.3.1 Requirements for Red GECIs

Despite the ongoing efforts to improve GECIs, the majority of single-wavelength probes were still limited to green fluorescent proteins (GFPs) or yellow fluorescent proteins (YFPs) that emit fluorescence at a wavelength of 500–550 nm. This restricted emission spectrum prevents the use of GECIs in cells that have already been labeled with a green fluorescent probe. Furthermore, recent advances in optogenetic manipulation using light-modulated microbial opsins (Yizhar et al. 2011; Nagel et al. 2003; Chow et al. 2010; Zhang et al. 2007; Boyden et al. 2005) such as channelrhodopsin (ChR)-2 require the combinatorial use of GECIs and photoactivatable proteins in an identical cell population. However, the use of GECIs to visualize opsin-expressing cells without photoactivation is technically challenging because of the large overlap in their broad excitation spectra (GCaMPs 470–510 nm; ChR2 430–500 nm). Thus, GCaMP3 and ChR2 can only be combined in conditions where the laser power is lowered strictly to excite GCaMP3 alone but not ChR2 (Guo et al. 2009). The development of GECIs with different absorption and emission spectra from GFPs should resolve these

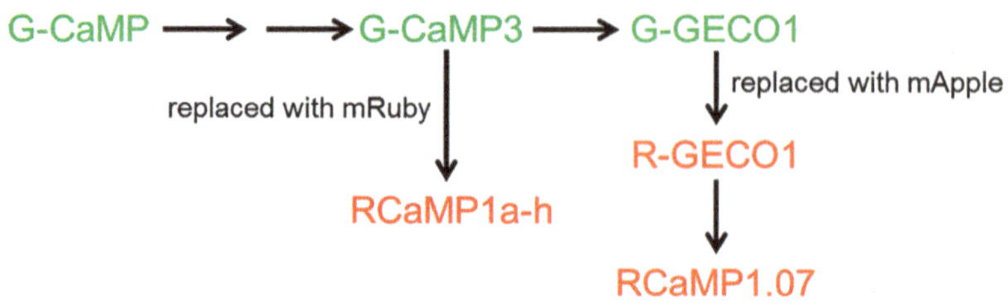

Fig. 10.1 Improvements of red genetically encoded calcium indicators (GECIs)

technical limitations and allow GECIs to be employed together with optogenetic tools. A number of fluorescent proteins have been developed, but few multicolor probes that exhibit clear fluorescent changes in response to calcium ions. Several research groups have recently engineered color-shifted GECIs by combining the fluorescent proteins and a calcium-binding protein (Zhao et al. 2011; Akerboom et al. 2012) (Fig. 10.1). Among these, red-shifted indicators are particularly desirable because longer-wavelength light is less susceptible to tissue scattering, blood absorption, phototoxicity, and background autofluorescence. These conditions are appropriate for in vivo live cell imaging from deeper tissues. This section describes the recent development and application of red-emitting GECIs in neuroscience research.

10.3.2 Repertoire of Red GECIs

10.3.2.1 R-GECO1

While neurophysiologists were awaiting new applications of GECIs, a pioneering study by Zhao et al. (2011) succeeded in creating color-shifted GECIs with unique excitation and emission spectra on the basis of a combination of single-wavelength fluorescent proteins. They used a variety of strategies to obtain efficient probes, including mutagenesis with an error-prone polymerase chain reaction (PCR) and digital fluorescence imaging of gene variants expressed in *Escherichia coli* colonies. Introducing targeted and random mutations into GCaMP3 yielded a blue calcium indicator, B-GECO1, and a series of high-sensitive green calcium indicators, G-GECOs. Part of the cfGFP in G-GECO1.1 was then replaced with an analogous cp version of a red fluorescent protein (mApple) (Shaner et al. 2008), which yielded red-shifted probes that are potentially calcium sensitive. Through the iterative evolution of variants created by error-prone PCR and randomization, the best optimized protein was finally identified as R-GECO1. The absorption and emission maxima of R-GECO1 were approximately 570 nm and 600 nm, respectively. On the basis of the calcium responses obtained from HeLa cells and cultured cells, the dynamic range of fluorescent changes in R-GECO1 appears to be almost similar to that in G-CaMP3.

10.3.2.2 R-CaMP1.07

The development of red-shifted GECIs by Zhao et al. (2011) motivated us to improve the probe further (Ohkura et al. 2012a). As an initial step, random mutations were introduced into a prototype molecule, R-GECO1, with the RSET tag at the N-terminus. Repeated rounds of in vitro screening yielded the best variant, R-CaMP1.01, which had mutations of K47V and T49V in the circularly permuted mApple domain. However, both the R-GECO1 and R-CaMP1.01 signals were found to be aggregated in the HeLa cell nucleus and the cytoplasm. This abnormal subcellular localization is probably related to the dysfunctioning of probes due to partial proteolysis (Tian et al. 2009). To prevent nuclear entry, R-CaMP1.01 was combined with a self-cleaving peptide, F2A, at the C-terminus. The resulting variant obtained, R-CaMP1.07, showed no signals of nuclear localization, and approximately two times larger fluorescent changes in response to adenosine triphosphate (ATP) application in HeLa cells compared with R-GECO1.

The performance of R-CaMP1.07 was also characterized in mammalian hippocampal CA3 pyramidal neurons in cultured slices. R-CaMP1.07 detected calcium transients induced by single spikes with 95 % probability. The signal amplitudes increased almost linearly up to six spikes with a frequency of 50 Hz (Fig. 10.2A). On average, R-CaMP1.07 yielded 1.5- to 2.0-fold larger dynamic responses ($\Delta F/F$) and higher SNRs than R-GECO1 (Fig. 10.2B). The rise and decay time constants of the spike-induced calcium transients of R-CaMP1.07 were approximately 120 ms and 900 ms, respectively, which are almost identical to those of R-GECO1. According to the temporal kinetics, this sensor could resolve individual spikes in bursts with a frequency of up to 5 Hz (Sasaki et al. 2008). However, R-CaMP1.07 has not yet reached the affinity and speed of the commonly used green GECIs. It may be possible to optimize the kinetics of sensors to further resolve the temporal patterns of spikes during burst firing.

10.3.2.3 RCaMPs

In parallel with the study of mApple-based GECIs, Akerboom et al. (2012) independently engineered a series of multi-color GECIs. To create color-shifted GECIs, they first introduced mutations into the GFP chromophore of GCaMP3 through the mutation sites that are crucial for conversion to color-shifted fluorescent proteins (Heim et al. 1994; Heim and Tsien 1996; Ormo et al. 1996). Additional systematic mutations of these color-shifted GCaMP3 yielded blue, cyan, and yellow probes, which displayed substantial fluorescent changes in response to calcium (i.e., BCaMP, CyCaMP, and YCaMP, respectively). However, this strategy failed to produce red variants of GCaMP3. As an alternative pathway, part of cpEGFP in GCaMP3 was replaced with a circularly permuted version of mRuby (Kredel et al. 2009), a red fluorescent protein. The addition of random mutations and subsequent structure-guided optimization produced red-emitting calcium indicators, RCaMP1a-h (note that these are different probes from R-CaMP1.07, which was developed by Ohkura

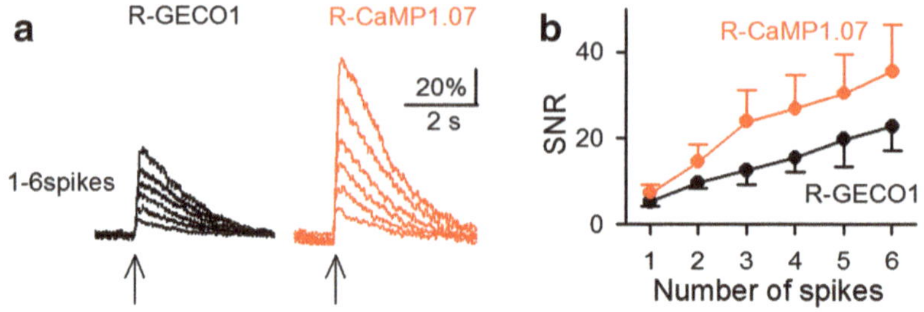

Fig. 10.2 *Detection of spiking activity using red genetically encoded calcium indicators (GECIs) in hippocampal pyramidal cells.* (**a**) Representative fluorescent traces ($\Delta F/F$) in response to spike trains with a frequency of 50 Hz. Imaging was performed at 50 frames per second (fps) using a Nipkow-disk confocal microscope. The *arrows* indicate the timing of spikes. (**b**) Mean signal-to-noise ratios (SNRs) plotted against the number of action potentials in R-GECO1 (*black*) and R-CaMP1.07 (*red*) (Copyright (2014) PLOS. From Ohkura M et al. (2012a) with permission)

et al. [2012a]). In cultured rat hippocampal neurons, RCaMP1d, 1e, 1f, and 1 h were brighter, and they exhibited smaller fluorescent changes in response to neuronal activity than did R-GECO1. The kinetics of these RCaMPs were slower than those of R-GECO1. These results suggest that R-GECO1 may be more suitable for detecting neuronal activity than R-CaMPs. However, RCaMPs have significant advantages over R-GECO1, i.e., RCaMPs are more tolerant to photobleeching and, as described later, they are never photoactivated by blue and green wavelength light, which is a prerequisite for their combined use with optogenetics tools. In addition, it has been demonstrated that RCaMPs can be used together with GCaMP5 for visualization of the calcium activity with multicolor fluorescence from different organelles in a single cell or different cell types in a single preparation.

10.3.3 Application of Red GECIs to Optogenetics

The excitation spectra of red-shifted GECIs are distinct from the action wavelengths used for photostimulating ChR2 (Fig. 10.3B). By exploiting their spectra properties, red GECIs can be used in optogenetics experiments for monitoring and manipulating neural activity with multiple light wavelengths. Ohkura et al. (2012a) demonstrated that R-CaMP1.07 fluorescence imaging can be used to detect spike-induced calcium events triggered via ChR2 photoactivation with blue light (Fig. 10.3C). A longer duration of photostimulation evoked larger numbers of spikes, which were linearly correlated with the amplitudes of the fluorescent changes. No membrane potential changes were induced during the imaging of R-CaMP1.07 fluorescence with 568 nm light, indicating that ChR2 was not activated in these imaging conditions. These results demonstrate that R-CaMP1.07 may be suitable for monitoring calcium transients in response to optically evoked spikes. However, the drawback of the imaging system by Ohkura et al. (2012a) is that blue light illumination causes the contamination of

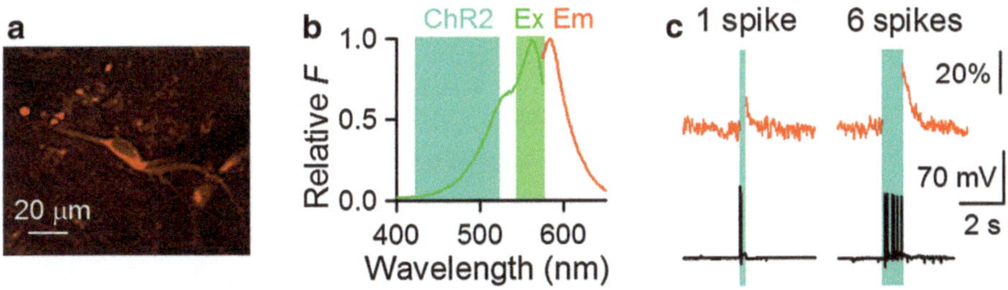

Fig. 10.3 *Optical monitoring and manipulation of neuronal activity*. (**a**) Hippocampal pyramidal cell co-expressing R-CaMP1.07 and channelrhodopsin (ChR)-2. (**b**) Excitation (*green*) and emission (*red*) spectra of R-CaMP1.07 superimposed on the excitation spectrum of ChR2 (*cyan*). The spectrum range of excitation filter used for RCaMP1.07 imaging is shown as the *green* region. (**c**) Imaging of ChR2-triggered-action potentials by R-CaMP1.07. Photostimulation (470-nm) is indicated by the *blue* region. Putative calcium increases during the photostimulation are represented by *broken lines*. The number of action potentials was recorded using the current clamp mode (Copyright (2014) PLOS. From Ohkura et al. (2012a) with permission)

photosignals, which prevents the direct recording of red fluorescent changes upon ChR2 photostimulation (as shown in the dashed line in Fig. 10.3A). Chang et al. (2012) overcame this issue by developing an optical system where a blue light pulse was applied during short intervals between image-acquisition events (Chang et al. 2012). This system allowed R-GECO calcium imaging without overlapping the photosignals generated by ChR2 photostimulation.

However, the results reported in these two previous studies have been questioned by the claim that mApple-based sensors, including R-GECO1 and R-CaMP1.07, inherit the property of photoreactivity to blue and green light; thus, implying that the ChR2-evoked fluorescent changes detected by these probes may be prone to artifacts. In contrast, RCaMPs produced by Akerboom et al. (2013) do not exhibit these photo-sensitive dynamics; therefore, it is possibly more suitable for use in artifact-free, integrated optogenetics experiments.

10.3.4 Note Regarding Red GECIs

It should be noted that the expression of GECIs alone may affect endogenous calcium-dependent signaling by buffering intracellular free calcium ions, thus leading to abnormal changes in cellular functions and behavioral phenotypes. For example, long-term and/or a high level of expression of GCaMP3 using in utero electroporation or viral infection caused nonfunctional indicators and abnormal physiology (Dombeck et al. 2010; Tian et al. 2009). In certain conditions, this concern may be excluded by demonstrating that GECI-positive cells in cultured slices do not show any changes in their intracellular and synaptic properties (Ohkura et al. 2012a, b), and GCaMP3-expressing transgenic mice generally display calcium transients in vivo without changing cellular functions (Chen et al. 2012). Overall, the most crucial factors are likely to be the accurate timing and magnitude of the expression of GECIs. Depending on the stability and affinity of sensors, the optimization of expression levels should be considered on a case-by-case basis. In addition, moderate levels of calcium indicators are crucial for obtaining the best performance with GECIs because excessively high concentrations of calcium indicators substantially affect the shapes of calcium transients with extended durations and smaller amplitudes (Helmchen et al. 1996).

The performance of red-shifted GECIs in rodent cells has been tested almost exclusively in slice preparations or cultured cells. However, little is known regarding whether these sensors may work in vivo with similar detection levels in rodent neurons. In general, the dynamic range of GECIs decreases in more intact preparations due to a variety of external factors such as pH, temperature, light scattering, and background autofluorescence (Tian et al. 2009; Ohkura et al. 2012b). In addition, the ability of detection by red GECIs in different neuron types has not yet been systematically examined. A number of physiological factors such as the electrical activity, the expression of calcium channels, and the endogenous calcium buffering capacity, vary considerably between neurons of different subtypes and among brain regions (Fierro and Llano 1996; Maeda et al. 1999). In other preparations, particularly in vivo rodents, the usefulness of each GECI should be confirmed and compared by individual end users. Further studies are needed to determine the optimal situations where each GECI shows the best sensitivity, dynamic range, and kinetics.

10.4 Concluding Remarks

Recent protein engineering has continuously improved the properties of GECIs in terms of their brightness, calcium affinity, response kinetics, and color expansion. In this trend, multicolor GECIs will facilitate new applications of optical imaging and activity manipulations, which have not been performed with existing experimental tools. To support these applications, gene delivery systems have been greatly improved to allow the expression of functional probes in a specific class of a neuronal population (Osakada et al. 2011; Chen et al. 2012; Zariwala et al. 2012). In addition, the microscopic instrumentation has improved recently, including multiphoton microscopy, high-speed scanning systems, Gallium Arsenide Phosphate (GaAsP) detectors, and portable integrated microscopy (Ghosh et al. 2011). This progress in biological and technological engineering should encourage neuroscientists to address a variety of fundamental questions regarding the neural mechanisms of behavior, learning, and diseases.

References

Akerboom J, Chen TW, Wardill TJ et al (2012) Optimization of a GCaMP calcium indicator for neural activity imaging. J Neurosci 32(40):13819–13840

Boyden ES, Zhang F, Bamberg E et al (2005) Millisecond-timescale, genetically targeted optical control of neural activity. Nat Neurosci 8(9):1263–1268

Buzsáki G (2004) Large-scale recording of neuronal ensembles. Nat Neurosci 7(5):446–451

Chang YF, Arai Y, Nagai T (2012) Optogenetic activation during detector "dead time" enables compatible real-time fluorescence imaging. Neurosci Res 73(4):341–347

Chen X, Leischner U, Rochefort NL et al (2011) Functional mapping of single spines in cortical neurons in vivo. Nature 475(7357):501–505

Chen Q, Cichon J, Wang W et al (2012) Imaging neural activity using Thy1-GCaMP transgenic mice. Neuron 76(2):297–308

Chen TW, Wardill TJ, Sun Y et al (2013) Ultrasensitive fluorescent proteins for imaging neuronal activity. Nature 499(7458):295–300

Chow BY, Han X, Dobry AS et al (2010) High-performance genetically targetable optical neural silencing by light-driven proton pumps. Nature 463(7277):98–102

Dombeck DA, Khabbaz AN, Collman F et al (2007) Imaging large-scale neural activity with cellular resolution in awake, mobile mice. Neuron 56(1):43–57

Dombeck DA, Harvey CD, Tian L et al (2010) Functional imaging of hippocampal place cells at cellular resolution during virtual navigation. Nat Neurosci 13(11):1433–1440

Fierro L, Llano I (1996) High endogenous calcium buffering in Purkinje cells from rat cerebellar slices. J Physiol 496(Pt 3):617–625

Ghosh KK, Burns LD, Cocker ED et al (2011) Miniaturized integration of a fluorescence microscope. Nat Methods 8(10):871–878

Gobel W, Helmchen F (2007) In vivo calcium imaging of neural network function. Physiology (Bethesda) 22:358–365

Grienberger C, Konnerth A (2012) Imaging calcium in neurons. Neuron 73(5):862–885

Guo ZV, Hart AC, Ramanathan S (2009) Optical interrogation of neural circuits in *Caenorhabditis elegans*. Nat Methods 6(12):891–896

Heim R, Tsien RY (1996) Engineering green fluorescent protein for improved brightness, longer wavelengths and fluorescence resonance energy transfer. Curr Biol 6(2):178–182

Heim R, Prasher DC, Tsien RY (1994) Wavelength mutations and posttranslational autoxidation of green fluorescent protein. Proc Natl Acad Sci U S A 91(26):12501–12504

Helmchen F, Imoto K, Sakmann B (1996) Ca^{2+} buffering and action potential-evoked Ca^{2+} signaling in dendrites of pyramidal neurons. Biophys J 70(2):1069–1081

Horikawa K, Yamada Y, Matsuda T et al (2010) Spontaneous network activity visualized by ultrasensitive Ca^{2+} indicators, yellow Cameleon-Nano. Nat Methods 7(9):729–732

Kredel S, Oswald F, Nienhaus K et al (2009) mRuby, a bright monomeric red fluorescent protein for labeling of subcellular structures. PLoS One 4(2):e4391. doi:10.1371/journal.pone.0004391

Maeda H, Ellis-Davies GC, Ito K et al (1999) Supralinear Ca^{2+} signaling by cooperative and mobile Ca^{2+} buffering in Purkinje neurons. Neuron 24(4):989–1002

Mank M, Santos AF, Direnberger S et al (2008) A genetically encoded calcium indicator for chronic in vivo two-photon imaging. Nat Methods 5(9):805–811

Miyawaki A, Llopis J, Heim R et al (1997) Fluorescent indicators for Ca^{2+} based on green fluorescent proteins and calmodulin. Nature 388(6645):882–887

Nagai T, Sawano A, Park ES et al (2001) Circularly permuted green fluorescent proteins engineered to sense Ca^{2+}. Proc Natl Acad Sci U S A 98(6):3197–3202

Nagai T, Yamada S, Tominaga T et al (2004) Expanded dynamic range of fluorescent indicators for Ca^{2+} by circularly permuted yellow fluorescent proteins. Proc Natl Acad Sci U S A 101(29):10554–10559

Nagel G, Szellas T, Huhn W et al (2003) Channelrhodopsin-2, a directly light-gated cation-selective membrane channel. Proc Natl Acad Sci U S A 100(24):13940–13945

Nakai J, Ohkura M, Imoto K (2001) A high signal-to-noise Ca^{2+} probe composed of a single green fluorescent protein. Nat Biotechnol 19(2):137–141

Ohkura M, Matsuzaki M, Kasai H et al (2005) Genetically encoded bright Ca^{2+} probe applicable for dynamic Ca^{2+} imaging of dendritic spines. Anal Chem 77(18):5861–5869

Ohkura M, Sasaki T, Kobayashi C et al (2012a) An improved genetically encoded red fluorescent Ca^{2+} indicator for detecting optically evoked action potentials. PLoS One 7(7):e39933. doi:10.1371/journal.pone.0039933
Ohkura M, Sasaki T, Sadakari J et al (2012b) Genetically encoded green fluorescent Ca^{2+} indicators with improved detectability for neuronal Ca^{2+} signals. PLoS One 7(12):e51286. doi:10.1371/journal.pone.0051286
Ormo M, Cubitt AB, Kallio K et al (1996) Crystal structure of the *Aequorea victoria* green fluorescent protein. Science 273(5280):1392–1395
Osakada F, Mori T, Cetin AH et al (2011) New rabies virus variants for monitoring and manipulating activity and gene expression in defined neural circuits. Neuron 71(4):617–631
Palmer AE, Tsien RY (2006) Measuring calcium signaling using genetically targetable fluorescent indicators. Nat Protoc 1(3):1057–1065
Regehr WG, Tank DW (1991) The maintenance of LTP at hippocampal mossy fiber synapses is independent of sustained presynaptic calcium. Neuron 7(3):451–459
Sasaki T, Takahashi N, Matsuki N et al (2008) Fast and accurate detection of action potentials from somatic calcium fluctuations. J Neurophysiol 100(3):1668–1676
Shaner NC, Lin MZ, McKeown MR et al (2008) Improving the photostability of bright monomeric orange and red fluorescent proteins. Nat Methods 5(6):545–551
Shindo A, Hara Y, Yamamoto TS et al (2010) Tissue-tissue interaction-triggered calcium elevation is required for cell polarization during *Xenopus* gastrulation. PLoS One 5(2):e8897. doi:10.1371/journal.pone.0008897
Takahashi N, Sasaki T, Matsumoto W et al (2010a) Circuit topology for synchronizing neurons in spontaneously active networks. Proc Natl Acad Sci U S A 107(22):10244–10249
Takahashi N, Takahara Y, Ishikawa D et al (2010b) Functional multineuron calcium imaging for systems pharmacology. Anal Bioanal Chem 398(1):211–218
Tian L, Hires SA, Mao T et al (2009) Imaging neural activity in worms, flies and mice with improved GCaMP calcium indicators. Nat Methods 6(12):875–881
Tian L, Hires SA, Looger LL (2012) Imaging neuronal activity with genetically encoded calcium indicators. Cold Spring Harb Protoc 2012(6):647–656
Yamada Y, Michikawa T, Hashimoto M et al (2011) Quantitative comparison of genetically encoded Ca indicators in cortical pyramidal cells and cerebellar Purkinje cells. Front Cell Neurosci 5:18. doi:10.3389/fncel.2011.00018
Yizhar O, Fenno LE, Davidson TJ et al (2011) Optogenetics in neural systems. Neuron 71(1):9–34
Yuste R, Peinado A, Katz LC (1992) Neuronal domains in developing neocortex. Science 257(5070):665–669
Zariwala HA, Borghuis BG, Hoogland TM et al (2012) A Cre-dependent GCaMP3 reporter mouse for neuronal imaging in vivo. J Neurosci 32(9):3131–3141
Zhang F, Wang LP, Brauner M et al (2007) Multimodal fast optical interrogation of neural circuitry. Nature 446(7136):633–639
Zhao Y, Araki S, Wu J et al (2011) An expanded palette of genetically encoded Ca^{2+} indicators. Science 333(6051):1888–1891
Ziv Y, Burns LD, Cocker ED et al (2013) Long-term dynamics of CA1 hippocampal place codes. Nat Neurosci 16(3):264–266

Chapter 11
Optogenetic Sensors for Monitoring Intracellular Chloride

Ken Berglund, Lei Wen, and George J. Augustine

Abstract Here we provide an overview of genetically encoded chloride indicators and their application in imaging intracellular chloride concentration ($[Cl^-]_i$). We first compare different chloride sensors and then describe the basic properties of two of these - Clomeleon and SuperClomeleon - in terms of their anion selectivity, pH sensitivity, and binding kinetics. We also describe several approaches that can be used to express Clomeleon in selected populations of neurons. Clomeleon imaging has been used to determine resting $[Cl^-]_i$ in a variety of neuron types and to determine the shifts in resting $[Cl^-]_i$ that occur during neuronal development and during pathological conditions. Clomeleon imaging has also been used to monitor dynamic changes in $[Cl^-]_i$ that are associated with activity of inhibitory synapses. Thus, optogenetic chloride imaging has provided a wide range of novel information about $[Cl^-]_i$ in neurons and should be generally useful for measuring $[Cl^-]_i$ within cells.

Keywords Chloride imaging • GABA receptors • Inhibitory synaptic transmission • Interneuron • Yellow fluorescent protein

---

K. Berglund
Departments of Neurosurgery and Anesthesiology,
Emory University School of Medicine, Atlanta, GA 30322, USA

L. Wen
Center for Functional Connectomics, Korea Institute of Science and Technology,
Seoul 136-791, Republic of Korea

Lee Kong Chian School of Medicine, Nanyang Technological University,
50 Nanyang Drive, Research Techno Plaza, Singapore 637553, Singapore

G.J. Augustine (✉)
Center for Functional Connectomics, Korea Institute of Science and Technology,
Seoul 136-791, Republic of Korea

Lee Kong Chian School of Medicine, Nanyang Technological University,
50 Nanyang Drive, Research Techno Plaza, Singapore 637553, Singapore

Institute of Molecular and Cell Biology, 61 Biopolis Drive, Singapore 138673, Singapore
e-mail: george.augustine@ntu.edu.sg

H. Yawo et al. (eds.), *Optogenetics*, DOI 10.1007/978-4-431-55516-2_11

11.1 Introduction

Optogenetic sensors allow quantitative monitoring of cellular functions with high spatiotemporal resolution. A number of optogenetic sensors have been developed in the last two decades to monitor protein–protein interactions and to report neural circuit activity (Adams et al. 1991; Miyawaki et al. 1997; Kuner and Augustine 2000; Miesenböck et al. 1998; Sakai et al. 2001). In this chapter, we review the development of genetically encoded chloride indicators and their application to the monitoring of intracellular chloride concentration.

In the nervous system, chloride is a fundamental anion involved in many physiological processes, such as synaptic transmission, stabilization of the resting membrane potential (RMP), regulation of cellular volume (Brochiero et al. 1995; Foskett 1990), fluid secretion (Coskun et al. 2002), and intracellular pH (Russell and Boron 1976; Kaila et al. 1989). The chloride gradient across the plasma membrane contributes to the RMP (Hinkle et al. 1971) and establishes the balance between excitation and inhibition via γ-aminobutyric acid (GABA)- and glycine-mediated synaptic inhibition (Buzsáki and Chrobak 1995; Paulsen and Moser 1998; Akerman and Cline 2006). Intracellular chloride concentration is selectively regulated in different types of cells by diverse types of chloride channels, transporters, and receptors (Deisz et al. 2011; Payne et al. 2003; Suzuki et al. 2006). The abnormal intracellular chloride concentration resulting from dysfunction of those chloride homeostatic mechanisms causes a number of serious diseases, such as cystic fibrosis (Kerem et al. 1989), epilepsy (Palma et al. 2006; Barmashenko et al. 2011), and other neurological disorders (De Koninck 2007). Therefore, the direct measurement of intracellular chloride concentration is important to establish a causal link between chloride and physiological function.

Here, we summarize the use of optogenetic sensors, such as Clomeleon and SuperClomeleon, for measurement of intracellular chloride concentration ($[Cl^-]_i$). We begin by briefly reviewing the evolution of chloride measurement techniques.

11.1.1 Chloride-Sensitive Microelectrodes

Although direct measurements of $[Cl^-]_i$ appear to be conceptually straightforward, it has been technically challenging to measure dynamic changes in $[Cl^-]_i$ because of the small transmembrane gradient for chloride ions (about 10-fold), resulting in a relatively weak driving force and small dynamic fluxes of chloride under physiological conditions (Bregestovski et al. 2009). In comparison, the gradient of calcium ions can be up to 10,000-fold, resulting in dynamic changes that can span several orders of magnitude on a much lower background signal.

Several approaches have been developed to overcome those difficulties. Direct measurement of $[Cl^-]_i$ began in the 1970s, when chloride-selective microelectrodes were developed. These allow direct measurement of resting concentration as well as dynamic changes in chloride with a time resolution in the order of minutes or less (Brown et al. 1970; Walker and Brown 1977). This technique is based on a micropipette that is filled with a liquid chloride ion exchanger for selective detection of chloride. However, application of this technique was limited by several technical difficulties, including the time-consuming fabrication of electrodes, and the slow kinetics of electrode responses to dynamic changes in $[Cl^-]_i$ (Aickin and Brading 1982). The most critical problem is that the large tip diameter of these electrodes limited their use to large cells, such as *Aplysia* neurons. Because of the low yield of information and potential damage to sampled cells from insertion of the electrode, less invasive techniques were subsequently developed.

11.1.2 Organic Indicator Dyes

Organic fluorescent dyes were initially applied to detect intracellular pH (Rink et al. 1982) and calcium (Tsien 1980; Grynkiecwicz et al. 1985), and later to detect intracellular chloride. A number of quinoline derivatives, including 6-methoxy-*N*-(3-sulfopropyl) quinolinium (SPQ) and *N*-(ethoxycarbonylmethyl)-6-methoxyquinolinium (MQAE), exhibit promising properties because chloride ions quench their fluorescence (Wolfbeis and Urbano 1983; Illsley and Verkman 1987; Krapf et al. 1988a; Verkman et al. 1989). A collisional mechanism, where halides are able to quench the fluorescence of heterocyclic organic compounds with quaternary nitrogen, has been proposed to explain the chloride sensitivity of these dyes (Verkman et al. 1989). The dynamic process of collisional quenching of an organic dye upon halide binding can be described by the Stern-Volmer equation (Krapf et al. 1988b):

$$\frac{F_0}{F} = 1 + K_D[Q]$$

where F_0 and F represent the fluorescence intensities in the absence and presence of halide quencher, respectively; K_D is the Stern-Volmer quenching constant and $[Q]$ is the concentration of the halide quencher.

Because of their relatively high sensitivity, rapid kinetics, and insensitivity to pH within the physiological range, SPQ and MQAE have been heavily used to measure chloride concentration in diverse preparations in vitro: isolated rabbit renal cortex (Chen et al. 1988), synaptoneurosomes (Engblom and Åkerman 1993), muscle cells (Heo et al. 2008; Koncz and Daugirdas 1994), different preparations of epithelial cells (Chao et al. 1990; Chu et al. 1992), cerebellar granule cells (Engblom et al. 1989, 1991; White et al. 1997), fibroblasts (Chao et al. 1989; Maglova et al. 1998), olfactory cells (Kaneko et al. 2001; Nakamura et al. 1997), and other kinds of neurons (Marandi et al. 2002; Song et al. 2005; Funk et al. 2008) in acute brain slices (Inglefield and Schwartz-Bloom 1997, 1999) and in vivo as well (Kovalchuk and Garaschuk 2012).

Despite the wide application of these organic dyes for monitoring chloride concentration, they have several disadvantages. All known organic dyes for detection of chloride were designed based on a quinolinium backbone and therefore are excited by ultraviolet (UV) light. Because UV light is relatively high in energy, dye excitation can cause phototoxicity during continuous imaging (Nakamura et al. 1997). Although MQAE has relatively good membrane permeability, it appears to be toxic to cells and brain slices during long-term exposure (Servetnyk and Roomans 2007). In addition, cells generally do not retain these fluorescent dyes very well, preventing long-term observation of chloride (Nakamura et al. 1997). Ideally, chloride indicators should have the following features: (1) long-lasting and stable fluorescence, (2) negligible cytotoxicity, (3) high selectivity for chloride, (4) chloride affinity near intracellular concentrations of interest (typically in the order of 5–10 mM for mature neurons); and (5) ability to detect chloride fluxes at the level of subcellular compartments.

11.2 Genetically Encoded Indicators

11.2.1 Yellow Fluorescent Protein and Its Variants

The green fluorescent protein (GFP) initially was used to report levels of gene expression, subcellular localization of proteins (Cheng et al. 1996; Zhang et al. 1996), and protein–protein interactions (Heim 1999). Subsequently, a great variety of GFP derivatives with diverse emission spectra have enriched the fluorescent protein family. Among these, yellow fluorescent proteins (YFPs) with red-shifted emission have been found to constitute a set of genetically encoded probes for reporting intracellular pH (Kneen et al. 1998; Llopis et al. 1998; Miesenböck et al. 1998). Taking advantages of fluorescence resonance energy transfer (FRET), fusion of a fluorescent protein pair with the addition of a calcium-binding motif from calmodulin has shown great ability to detect small changes in intracellular calcium concentration (Miyawaki et al. 1997, 1999).

Beyond those early applications, it was soon discovered that many YFP variants also respond rapidly and reversibly to changes in halide concentration (Wachter and Remington 1999; Jayaraman et al. 2000; Wachter et al. 2000; Kuner and Augustine 2000). This occurs over a wide range of pH and therefore means that YFPs constitute a set of genetically encoded chloride sensors. YFP (GFP/2V/S65G/V68L/S72A/T203Y; 10C in Ormö et al. 1996) initially was found to show a weak affinity for chloride at pH 7.5 (Wachter and Remington 1999). The affinity is strongly improved by lowering pH, implying an interaction between protons and chloride ions (Wachter et al. 2000). Crystallographic and energetic analysis revealed a chloride-binding pocket near the chromophore: chloride binding destabilizes the chromophore anion, shifts pK_a to a higher value, protonates the chromophore, and thereby quenches YFP fluorescence (Wachter et al. 1998, 2000). When expressed in transgenic mice under the control of the $K_V3.1$ promoter, EYFP yielded an apparent dissociation constant (K_{app} or IC_{50}) of 168 mM for chloride at pH 7.5 (Metzger et al. 2002). Cultured hippocampal neurons prepared from these transgenic mice have been used to study glutamate-evoked chloride increases (Slemmer et al. 2004).

An H148Q mutation in EYFP can boost K_{app} for chloride from 777 to 154 mM at pH 7.5. The anion selectivity of this mutant has been established as $ClO_3^- > I^- > SCN^- > NO3^- > Br^- > Cl^- >$ acetate $>$ sulphate $> F^-$ (Jayaraman et al. 2000; Wachter et al. 2000). Further improvement in EYFP/H148Q was made by random mutagenesis in the vicinity of the halide-binding site, with an additional V163S mutation lowering the half maximal inhibitory concentration (IC_{50}) for Cl^- down to 40 mM (Galietta et al. 2001). Although the H148Q mutation may improve chloride affinity, it also diminishes YFP fluorescence intensity (Elsliger et al. 1999), thus it does not necessarily improve the signal-to-noise ratio (SNR) of chloride measurements (Grimley et al. 2013).

Clomeleon, a ratiometric indicator (see below), utilizes a YFP variant called Topaz (GFP/2V/S65G/S72A/K79R/T203Y/H231L). When expressed by itself in vitro, Topaz has a K_d for chloride of 163 mM at pH 7.1 (Grimley et al. 2013). Topaz has the halide affinity sequence of $F^- > I^- > Br^- > Cl^-$. Screening of hundreds of Topaz mutants after structurally directed mutagenesis resulted in a Topaz variant with double mutations, Q69T/V163A, that has a higher affinity for chloride (K_d: 21 mM at pH 7.1). Indeed, some Topaz mutations yielded K_ds of even higher affinity, with the extreme being less than 1 mM (Grimley et al. 2013).

11.2.2 Ratiometric Indicators

A limitation with the use of these YFP mutants to quantify $[Cl^-]_i$ is that YFP fluorescence is dependent not only on chloride concentration but also on the protein concentration, optical path length, and excitation light intensity. A ratiometric indicator, meaning a fluorophore that changes its excitation or emission spectra, can circumvent this problem (Grynkiewicz et al. 1985; Bright et al. 1989; Kuner and Augustine 2000; Miyawaki et al. 2005; Bregestovski et al. 2009).

The first ratiometric chloride indicator protein was Clomeleon (Kuner and Augustine 2000). This probe links a variant of YFP with a chloride-insensitive cyan fluorescent protein (CFP). A 24-amino acid linker keeps the two fluorescent proteins in close proximity, permitting FRET between CFP and YFP. The Clomeleon emission spectrum has an isostilbic point around 515 nm. Clomeleon has been used to detect changes in chloride in a wide variety of organisms from plant (Lorenzen et al. 2004) to cultured neurons from the nervous systems of birds (Lindsly et al. 2014) and mammals (Kuner and Augustine 2000), in brain slices (Berglund et al. 2006; Duebel et al. 2006), and in vivo (Berglund et al. 2008). Recently, an improved version of Clomeleon was constructed with improved chloride affinity, FRET efficiency, and overall fluorescence intensity. This version, called SuperClomeleon, has been used to visualize synaptic inhibition in cultured hippocampal neurons (Grimley et al. 2013). In this chapter we focus on the properties of the Clomleon family of optogenetic indicators and their applications for neuroscience research.

A second type of FRET-based ratiometric chloride indicator protein was called Cl-sensor (Markova et al. 2008). Cl-sensor is also a fusion protein of CFP and YFP (H148Q/I152L/V163S), which has an isosbestic point at 465 nm in its excitation spectrum. Thus, the excitation ratio at 480 nm over 440 nm reports $[Cl^-]_i$ with a chloride affinity of ~30 mM at pH 7.3. This FRET pair has been used to map the distribution of intracellular chloride in different compartments of cultured hippocampal and spinal cord neurons. Cl-sensor demonstrated that chloride concentration was highest at the neuronal soma and had a tendency to decrease along dendritic branches, reaching a minimal value at distal dendrites (Waseem et al. 2010). Cl-sensor has also been used recently to detect elevation of chloride in neurons challenged with glioma-conditioned medium in response to glutamate receptor activation (Bertollini et al. 2012).

The major disadvantage of YFP-based chloride indicators such as Clomeleons and Cl-sensor is their pH sensitivity. This pH sensitivity is inescapable because it arises from the fundamental mechanism that couples chloride binding to quenching of YFP fluorescence. To make matters worse, chloride and pH in neurons often changes concomitantly. This problem can be solved by independent monitoring of intracellular pH. With this in mind, ClopHensor was developed to allow simultaneous determination of chloride and proton concentration (Arosio et al. 2010). ClopHensor is composed of a mutated GFP (E^2GFP; EGFP/T203Y), a 20-amino acid linker, and DsRed, a coral red fluorescent protein. Chloride concentration and pH are determined by ratiometric measurements obtained upon excitation with three different wavelengths: 488-nm Cl^- and pH-dependent GFP signal, 458-nm Cl^--dependent and pH-independent signal, and 543-nm Cl^-- and pH-independent DsRed signal. Although the chloride titration curve has pH dependence, the pH calibration curve is truly chloride independent, allowing dissociation of proton and chloride signals in the same measurement. A ClopHensor derivative, PalmPalm-ClopHensor, was developed for membrane targeting of the probe, while a double mutant, ClopHensor/H148G/V224L, exhibits better chloride affinity and its pH sensitivity is shifted to a more alkaline value (Mukhtarov et al. 2013). Another problem with the original ClopHensor was aggregation of the DsRed moiety, resulting in highly heterogeneous expression. To solve this issue, ClopHensorN has been constructed by replacing DsRed with tdTomato. The new fusion protein resulted in uniform and comparable expression of the two fluorescent proteins, allowing simultaneous monitoring of dynamic changes in intracellular chloride and pH in acute brain slices (Raimondo et al. 2013).

The properties of these various genetically encoded chloride indicators are compared in Table 11.1.

Table 11.1 Properties of genetically encoded Cl^- indicators

Indicator	Ratiometric properties	Isosbestic/isostilbic point (nm)	λ_{ex} (nm)	λ_{em} (nm)	*K*d (mM)	F_{max}/F_{min} or R_{max}/R_{min}	p*K*a	Halide selectivity
YFP(H148Q)[1]	NA	NA	480 ± 20	510 ± 20	100 (pH 7.5)	2.5	7.14 (0 mM Cl^-); 7.86 (150 mM Cl^-)	$ClO_4^- \sim I^- > SCN^- > NO3^- > Cl^- > Br^- >$ formate > acetate
Clomeleon[2]	Emission	515	440	485 ± 15 530 ± 15	167 (pH 7.4)	4.8	5.2 (0 mM Cl^-); 6.5 (150 mM Cl^-)	$F^- > I^- > Br^- > Cl^-$
Cl-sensor[3]	Excitation	465	480/440	>495	30 (pH 7.3)	2.4	7.1 (4 mM Cl^-); 8.0 (140 mM Cl^-)	NA
ClopHensor[4]	Excitation	458	458/488/543	500–550 and 580–700	49 (pH 7.25)	3.6	6.81 (0 mM Cl^-)	NA
SuperClomeleon[5]	Emission	NA	840	485/535	22 (pH 7.1)	4.0	6.4	$F^- > Cl^- > I^- > Br^-$

Reference
[1]Jayaraman et al. 2000; [2]Kuner and Augustine 2000; [3]Markova et al. 2008; [4]Arosio et al. 2010; [5]Grimley et al. 2013
NA: not applicable

11.3 Clomeleon and SuperClomeleon

11.3.1 Properties

11.3.1.1 FRET

When linked to CFP, YFP can act as a FRET acceptor for the CFP donor. In the absence of Cl^-, the FRET interaction is strong. However, increasing concentrations of Cl^- progressively quench YFP and eliminate YFP as an acceptor, thereby increasing CFP fluorescence. The ratio of YFP to CFP fluorescence emission therefore changes with the concentration of Cl^-, with the resulting titration curve described by the Hill equation. This allows the YFP/CFP emission ratio of Clomeleon to be directly translated into $[Cl^-]$ using an appropriate calibration curve, defining Clomeleon as a ratiometric indicator for $[Cl^-]$ (Kuner and Augustine 2000).

Clomeleon differs from most other genetically encoded FRET sensors (Hasan et al. 2004; Pologruto et al. 2004; Reiff et al. 2005), because the sensor function is provided by the YFP fluorescent protein itself and does not arise from a distinct sensing moiety. Apart from pHluorin, a pH-sensitive variant of GFP that can be used for detection of H^+ (Miesenböck et al. 1998), all other genetically encoded indicators are based on the principle that binding of a substrate results in changes in the distance and/or orientation between CFP and YFP. For example, Ca^{2+} binding to Cameleon results in a conformational change mediated by the M13 helix, pulling CFP and YFP closer together (Miyawaki et al. 1997). Because FRET efficiency changes with the sixth power of the distance between fluorophores, even minute conformational changes caused by substrate binding can change FRET efficiency significantly.

The linker between the donor and the acceptor fluorescent proteins is an important determinant of FRET efficiency. The original Clomeleon had a flexible linker of 24 amino-acid residues (Kuner and Augustine 2000). To improve FRET efficiency, the distance of the two fluorophores was shortened by replacing this linker with the Leu-Glu linker and further by removing 11 residues from the C terminus of CFP and five residues from the N terminus of YFP (Grimley et al. 2013). The updated version, SuperClomeleon, indeed exhibited increased FRET efficiency determined as a difference in FRET ratios (ΔR_{max}) observed in the absence of Cl^- and in the presence of saturating concentration of Cl^-. In addition to the linker, the brightness of donor and acceptor proteins was enhanced by introducing a point mutation, S30R, in both CFP and YFP, producing a further improvement in ΔR_{max}.

11.3.1.2 Chloride-Binding Affinity

The original Clomeleon, which utilized a chloride-sensitive YFP variant, Topaz, had chloride affinity of about 160 mM (Kuner and Augustine 2000). This is rather distant from the physiological range of $[Cl^-]_i$ for most cells of interest. Therefore, mutagenesis was required to improve the halide sensitivity of YFP.

Recently, high-throughput screening of mutations was developed based on cell-free, automated protein engineering (Grimley et al. 2013). By bypassing genetic cloning steps, this technique allows rapid synthesis and screening of the optical and biochemical properties of mutant YFPs. The performance of these mutants was evaluated using a metric that takes into account both fluorescence intensity and the SNR associated with a change in chloride concentration from 5 to 6 mM, which reflects a change in $[Cl^-]_i$ typically observed during inhibitory synaptic transmission in adult neurons. The most optimized mutant with double mutations (Q69T/V163A) improved K_d of Topaz from 163 to 21 mM. Together with the redesigned linker (see above), this Topaz variant achieved the highest SNR to date for detection of inhibitory synaptic transmission.

11.3.1.3 Ionic Selectivity

In addition to Cl^-, other halides such as F^-, I^-, and Br^- also bind to YFP and change the fluorescence emission of Clomeleon. Among these, Cl^- is the only physiologically relevant halide because most tissues and cells do not contain F^-, I^-, or Br^- at sufficient concentrations to bind to Clomeleon. However, we frequently use F^- to completely quench YFP during cellular calibration experiments (see below).

Other intracellular anions (such as glutamate, ATP, HCO_3^-, PO_4^{2-}, SO_4^{2-} and NO_3^-) do not affect Clomeleon at physiological concentrations. Further, anions commonly used in electrophysiology, such as gluconate and methylsulfate, do not affect the readout of Clomeleon. Hence, under physiological conditions, Cl^- is the only anion that is sensed by Clomeleon (Kuner and Augustine 2000).

11.3.1.4 Effects of pH

Changes in pH affect the apparent Cl^- affinity of Clomeleon and SuperClomeleon. For example, the IC_{50} for Clomeleon ranges from 15 mM at pH 6 to up to 900 mM at pH 8. The pK_a of the YFP variant used as a chloride sensor will influence the IC_{50} of Cl^- and define the extent of pH dependence. For example, Topaz (pK_a of ~6.4) has a larger IC_{50} but is less affected by physiological changes in pH (Kuner and Augustine 2000), while the EYFP/H148Q (pK_a of 7.14) has a smaller IC_{50} but is more susceptible to pH changes (Jayaraman et al. 2000).

In a cellular environment, pH is usually tightly regulated, although intracellular pH can change under certain conditions (Chesler and Kaila 1992). If such changes remain undetected, Clomeleon will report erroneous $[Cl^-]$. To minimize such errors, it is valuable to determine intracellular pH under the conditions in which Clomeleon is being used to measure $[Cl^-]$. However, for many experimental situations this may not be necessary, particularly when $[Cl^-]$ is less than ~25 mM. In this range, the readout error introduced by inadvertent changes in pH of ± 0.2 pH units will remain below ± 20 % and may be negligible depending on experimental paradigms (Kuner and Augustine 2000). However, when measuring $[Cl^-]$ higher than 25 mM, the measurement error will increase to 50 % for pH changes of ± 0.2 pH units. In such cases, pH must be measured independently in the same experimental conditions; this can be conveniently done using conventional, organic pH indicator dyes, such as SNARF-5f (Pond et al. 2006).

The influence of pH on the IC_{50} of Cl^- is well-defined and can be utilized to correct ratio readouts of Clomeleon for changes in pH. For example, if a calibration has been established in a certain cell type at a pH of 7.2, it can be still used to accurately determine $[Cl^-]$ in other cell types at different resting pH. The following relation can be used to calculate the appropriate IC_{50} for a given shift in pH (pH') from pH 7.2:

$$IC'_{50} = IC_{50} \times 10^{0.82 \times (pH' - 7.2)}$$

The corrected IC'_{50} can then be used to determine the correct $[Cl^-]$ (for details see Berglund et al. 2006; Pond et al. 2006). This approach could also be used to correct the $[Cl^-]$ readout dynamically in experiments where pH and $[Cl^-]$ are measured simultaneously. For example, a ratiometric pH dye, SNARF-5f, has an emission spectrum that is well separated from YFP and thus can be used to determine intracellular pH at the same time (Metzger et al. 2002).

In conclusion, the pH dependence of Clomeleon (and other YFP-based indicators) needs to be considered when measuring $[Cl^-]_i$, particularly when working under conditions of high $[Cl^-]_i$. However, in experimental conditions when large pH shifts are known to be absent and when $[Cl^-]_i$ is lower than approximately 25 mM, Clomeleon will accurately report $[Cl^-]_i$ without any correction. When pH changes do occur, corrections can be used to determine the appropriate $[Cl^-]_i$ under conditions of changing pH. Thus, Clomeleon can provide a completely reliable measure of $[Cl^-]_i$ regardless of pH or level of $[Cl^-]_i$.

11.3.1.5 Kinetics

The temporal resolution of Clomeleon is limited by the kinetics of Cl^- binding to YFP. Binding time constants of Clomeleon are temperature dependent, ranging from 500 to 600 ms at 20 °C down to 100–200 ms at 37 °C at physiologically relevant $[Cl^-]$. The slow binding of Cl^- is also evident in simultaneous imaging and electrophysiological recordings from hippocampal neurons (Kuner and Augustine 2000; Berglund et al. 2006).

11.3.2 Strategies for Expression

A rich variety of molecular genetic methods can be used to express Clomeleons in any cell type. Expression plasmids provide the simplest approach to express Clomeleons in cultured cells on a bulk scale using standard DNA lipofection protocols (Kuner and Augustine 2000; Grimley et al. 2013) or electroporation (Liedtke et al. 2013; Yeo et al. 2009). Electroporation can also be used to express Clomeleon in organotypic cultures of brain slice (Yeo et al. 2013). Viral transduction by recombinant adeno-associated virus (AAV) has also been used (Berglund et al. 2009).

Transgenic approaches provide a powerful means for targeted expression of Clomeleon in subpopulations of specific cell types in mice. The thy1 promoter can mediate strong, neuron-specific expression of transgenes within the brain (Caroni 1997). Pronounced positional variegation of this promoter results in striking line-to-line variations in neuronal expression, which can be used to advantage for fluorescence imaging of individual neuron types (Feng et al. 2000). We characterized seven thy1::Clomeleon mouse lines, with each line showing a unique pattern of neuronal Clomeleon expression (Berglund et al. 2006). Several studies of neuronal $[Cl^-]_i$ have been conducted with these thy1::Clomeleon transgenic mice (Berglund et al. 2006; Duebel et al. 2006; Pond et al. 2006, see below). Other transgenic strategies based on Cre recombinase are summarized in Berglund et al. (2009).

11.3.3 Microscopy

Similar to other FRET sensors, Clomeleon imaging has been conducted by observing either donor (i.e., CFP) fluorescence lifetime (Jose et al. 2007) or sensitized emission. Sensitized emission can be conveniently observed at two different wavelengths in either the excitation spectrum, such as 440 nm for CFP and 480 nm for YFP (Markova et al. 2008), or in the emission spectrum, such as 485 nm for CFP and 530 nm for YFP. Three modes of conventional microscopy have been successfully used in ratiometric emission imaging with Clomeleon: (1) wide-field epifluorescence microscopy, using either two photomultiplier (PMT) detectors (Kuner and Augustine 2000), a two-channel beam-splitter/filter-wheel and one charge-coupled device (CCD) camera (Berglund et al. 2006; Pond et al. 2006), or two CCD cameras; (2) confocal fluorescence imaging using a spinning disk microscope (Berglund et al. 2006); (3) two-photon imaging (Duebel et al. 2006).

Each of these approaches offers distinct advantages and could prove to be the optimal way to measure $[Cl^-]_i$, depending on the scientific question to be addressed and the experimental conditions available. For example, donor lifetime imaging microscopy is especially useful when imaging cells with low levels of Clomeleon expression. If Clomeleon expression is at relatively high levels, wide-field epifluorescence microscopy is the most convenient method to image primary neuronal cultures or acute brain slices. A wide-field microscope can collect fluorescence efficiently with relatively small loss, allowing imaging at a high temporal resolution of about 20 ms when combined with a highly sensitive camera. When a filter changer/wheel is employed, the time required for changing filters is the rate-limiting step. Imaging with a spinning disk confocal microscope offers the possibility of more precisely defining the volume from which the Clomeleon signal emerges, which is important in acute brain slices that may contain many neurons expressing high levels of Clomeleon. However, because of the relative inefficient photon throughput of spinning disk confocal microscopes, weak ratio signals are produced even in cells with high levels of Clomeleon expression. In practice, this limits the temporal resolution to about 200 ms per frame. Two-photon imaging of Clomeleon works well in cultured neurons, acute brain slices, whole-mounted retinas, and in brains in vivo. Unlike confocal microscopy, a two-photon microscope does not waste emitted photons, minimizing fluorophore excitation and photobleaching. Acquisition at several hertz to several tens of hertz can be achieved, depending on the signal levels. Clomeleon signals also can be obtained at a specified depth from the surface. This is especially useful to avoid damaged cells in acute brain slices, where cells at the surface are often compromised by preparation of the slice (Dzhala et al. 2012). Choice of excitation wavelength affects the emission spectrum and thus the absolute value and the dynamic range of the emission ratio (Berglund et al. 2011). Thus, calibration needs to be done for each microscope, being careful to use the same settings employed in the biological experiments.

11.3.3.1 Photobleaching and Photophysics

When using Clomeleon, it is important to know that YFP bleaches more rapidly than CFP. Interestingly, this bleaching of YFP has been found to be reversible over time (Dickson et al. 1997; Berglund et al. 2005; Sinnecker et al. 2005). Photobleaching of YFP becomes more prominent when it is excited by radiationless FRET through CFP than by direct excitation by 500-nm light (Friedel et al. 2013). Such differential bleaching will affect the fluorescence emission ratio of ratiometric chloride sensors, leading to an apparent time-dependent increase in $[Cl^-]_i$ (Berglund et al. 2005; Friedel et al. 2013).

To avoid this problem, the amount of excitation of Clomeleon should be reduced as much as practical. We have found that it is possible to reduce excitation light to levels that yield stable Clomeleon fluorescence with minimal bleaching. Considering this issue, the level of Clomeleon expression is one of the most important factors in Cl^- imaging experiments. The SNR will be ideal if Clomeleon expression is high to allow the excitation light to be dim. Because the levels of Clomeleon expression (typically micromolar) are orders of magnitude less than $[Cl^-]_i$ (typically millimolar), buffering of $[Cl^-]_i$ by Clomeleon is not an issue (Kuner and Augustine 2000). So the general goal should be to express as much Clomeleon in a cell as possible. In addition, to further increase the contrast between cells that express Clomeleon, ideally only a few cells in a given tissue should express Clomeleon.

11.3.3.2 Calibration

The calibration of $[Cl^-]$ measurements in intact cells is challenging, because cellular homeostasis mechanisms make it difficult to clamp $[Cl^-]_i$ at a defined concentration. Two main approaches have been used for this purpose: equilibration of $[Cl^-]_i$ via a whole-cell patch pipette or by treating the cell with a Cl^- antiporter to make the cell membrane permeable to Cl^-.

With the patch pipette calibration method, the cell is dialyzed with a solution containing a defined [Cl^-], and the Clomeleon emission ratio is measured at this defined [Cl^-]. This minimum ratio value (R_{min}) can be conveniently determined with 100 mM F^- in the pipette, which abolishes YFP fluorescence completely. Determining R_{max} with no $[Cl^-]_i$ is more difficult, because the extent to which the Cl^--free solution in the pipette can overwhelm cellular Cl^- homeostatic mechanisms and thereby maintain intracellular [Cl^-] close to 0 is unclear. The main limitation with this method is that each experiment can determine only one point on the Clomeleon calibration curve, meaning that a substantial number of experiments are needed to establish a reliable calibration curve.

With the antiporter approach, cells are exposed to tributyltin, a Cl^-/OH^- antiporter that forms pores in the cell membrane and allows external Cl^- to equilibrate with intracellular Cl^-. Because of the OH^- permeability of tributyltin, it is also necessary to clamp intracellular pH. For this purpose, the K^+/H^+ ionophore, nigericin, must also be applied to the cells at the same time. Because nigericin is also permeable to K^+, it can equilibrate and nullify the proton gradient only in the presence of high external [K^+]. Ideally, the pH of the calibration solutions should match pH_i of the intact cells of interest. Again, R_{min} values can be determined using an F^- solution, while values for the ratio at other levels of $[Cl^-]_i$, including the maximum ratio at low [Cl^-] (R_{max}) can be determined simply by substituting [Cl^-] with gluconate, an inert anion for Clomeleon. One convenient feature of this approach is that many points on the calibration curve can be determined from each cell, making determination of the calibration curve much faster and easier. However, in our experience, the ability of the tributyltin/nigericin treatment to control intracellular Cl^- can be variable. The precision of this approach critically depends on a number of factors, including the method of preparation and application of the antipoter/ionophore cocktail, the order of [Cl^-] titration, and external [K^+].

11.3.4 Applications

Clomeleon and SuperClomeleon have been employed in a variety of studies of neuronal $[Cl^-]_i$. In this section, we summarize all such applications that were published as of Spring 2014.

11.3.4.1 Resting $[Cl^-]_i$ in Mature Neurons

Many studies have shown that resting $[Cl^-]_i$ in mature neurons is maintained at a low level. This allows inhibitory neurotransmitters that work via Cl^- fluxes, such as GABA and glycine, to produce Cl^- efflux that hyperpolarizes the membrane potential. However, it has been difficult to determine the absolute level of resting $[Cl^-]_i$.

In thy1::Clomeleon transgenic mice, Clomeleon is expressed in a wide variety of neurons, allowing the first direct measurements of resting $[Cl^-]_i$ in mammalian neurons. Using these mice, resting $[Cl^-]_i$ has been measured in pyramidal cells in the hippocampus, pyramidal cells in the amygdala, granule cells in the cerebellum, and the premotor neurons in the superior colliculus (Berglund et al. 2006). Resting $[Cl^-]_i$ in these mature neurons was 5–10 mM, which is consistent with estimates made by other methods and can explain how inhibitory neurotransmitters hyperpolarize the membrane potential of mature neurons.

11.3.4.2 Shift in $[Cl^-]_i$ During Neuronal Development

Many studies have documented that, over the course of neuronal development, GABAergic transmission switches from depolarizing to hyperpolarizing (Ben-Ari et al. 1989; Cherubini et al. 1991; Owens et al. 1996). Although this is presumably due to a progressive drop in $[Cl^-]_i$, until recently this presumption had not been tested by direct measurements of neuronal $[Cl^-]_i$ at different postnatal stages.

Using Clomeleon, it has been possible to directly measure $[Cl^-]_i$ in developing neurons. In initial studies, $[Cl^-]_i$ was measured in cultured rat hippocampal neurons transfected with Clomeleon (Kuner and Augustine 2000). In these cultured cells, neuronal $[Cl^-]_i$ was higher in younger neurons and then declined over the next 2 weeks. Neuronal $[Cl^-]_i$ was estimated to be approximately 140 mM at embryonic day 18, decreasing to 60 mM at postnatal day 0 and then to approximately 20 mM by postnatal day 14. A follow-up study measured resting $[Cl^-]_i$ in CA1 pyramidal neurons in slices of hippocampus of thy1::Clomeleon transgenic mice (Berglund et al. 2006). The results were very similar to those found in the cultured rat neurons, although the absolute levels of $[Cl^-]_i$ were somewhat lower overall. At postnatal day 5, $[Cl^-]_i$ distributed widely in the range of 20–30 mM. At older ages, the distributions of $[Cl^-]_i$ became narrower, suggesting a more homogenous population of neurons at more mature stages. Mean $[Cl^-]_i$ was 17 mM at postnatal day 10 and further decreased to 6 mM at postnatal day 20. This corresponds to a drop in the Cl^- equilibrium potential of more than 30 mV between postnatal days 5 and 20.

These results are entirely consistent with the previously reported developmental changes in the polarity of GABA action on neurons. These developmental changes are associated with differential expression profiles of KCC2 and NKCC1 Cl^- transporters (Rivera et al. 1999; Stein et al. 2004), which presumably are responsible for the developmental drop in $[Cl^-]_i$. KCC2 expression is regulated through interaction of the REST transcriptional repressor complex and the RE-1 binding sites in the promoter of the *Kcc2* gene. Using embryonic rat neurons in culture transfected with Clomeleon, Yeo et al. (2009) revealed the function of a novel upstream RE-1 site. The novel RE-1 site provides strong repression for the *Kcc2* gene together with the well-characterized intronic RE-1 site during early development.

KCC2 expression is also affected by environmental factors. Yeo et al. (2013) studied the effect of bisphenol A (BPA), an additive commonly used during plastics manufacturing, on perinatal KCC2 regulation and found that BPA inhibits KCC2 upregulation and raises $[Cl^-]_i$ in neurons through an epigenetic mechanism. In addition to this GABA switch, KCC2 upregulation and subsequent $[Cl^-]_i$ downregulation signal interneuron precursors to halt migration during development (Bortone and Polleux 2009). Clomeleon-transfected interneuron precursors in organotypic cultures of embryonic brain slices had higher $[Cl^-]_i$ when cultured with BPA, and migration was increased (Yeo et al. 2013). Conversely, Liedtke et al. (2013) characterized the effects of carbon nanotubes on KCC2 expression and $[Cl^-]_i$ during development using Clomeleon-transfected neurons in culture and found that this highly electrically conductive nanomaterial can facilitate the Cl^- shift.

This conventional view of $[Cl^-]_i$ regulation by KCC2 has been challenged recently with the use of acute brain slices prepared from thy1::Clomeleon mice (Glykys et al. 2014). The study was prompted by an observation that an NKCC1-specific inhibitor, bumetanide, and a KCC2-specific inhibitor, VU0240551 (Delpire et al. 2009), had relatively small effects on $[Cl^-]_i$ in the hippocampal CA1 pyramidal cells of both neonatal and adult animals. In contrast, much larger effects on $[Cl^-]_i$ were observed when concentration of membrane-impermeant anions, such as intracellular gluconate and extracellular glycoproteins, were manipulated. These results indicate that the local gradient of impermeant anions across the cytoplasmic membrane is a more important determinant of $[Cl^-]_i$ than Cl^- cotransporters.

In conclusion, non-invasive optical imaging with Clomeleon provides direct measurements of $[Cl^-]_i$ that support the hypothesis that postnatal changes in the Cl^- electrochemical gradient underlie the functional switch in GABAergic transmission and in interneuron migration during development.

11.3.4.3 Compartmentalization of $[Cl^-]_i$ in Neurons

A great advantage of Clomeleon is its spatial resolution, which literally provides a new dimension of measurement that can reveal $[Cl^-]_i$ gradients between cell compartments.

The first hint of such gradients in neurons came from the observation that $[Cl^-]_i$ is higher in the dendrites than in the soma of cultured hippocampal neurons (Kuner and Augustine 2000). Measurements of $[Cl^-]_i$ in hippocampal slices have reiterated this conclusion: resting $[Cl^-]_i$ was 6–10 mM in apical dendrites of CA1 pyramidal cells and 8 mM in basal dendrites, whereas it was 5 mM in the cell bodies of these neurons (Berglund et al. 2006). $[Cl^-]_i$ gradients across cell compartments have also been reported in other neurons (Derdikman et al. 2003; Szabadics et al. 2006; Khirug et al. 2008).

Imaging resting $[Cl^-]_i$ in ON-type bipolar neurons in the retina has revealed a particularly interesting and important type of $[Cl^-]_i$ gradient. Although somewhat controversial (Satoh et al. 2001; Billups and Attwell 2002), it has been proposed that dendritic $[Cl^-]_i$ of these cells is high enough for GABA to depolarize the membrane potential (Miller and Dacheux 1983). To address this possibility, Duebel et al. (2006) used two-photon imaging to measure resting $[Cl^-]_i$ in ON-type bipolar neurons, employing retina from thy1::Clomeleon transgenic mice that express

Clomeleon in these cells (Haverkamp et al. 2005; Berglund et al. 2006). They found that resting $[Cl^-]_i$ is indeed very high in the dendrites of these cells (type 9 bipolar cells), averaging approximately 45 mM. Remarkably, resting $[Cl^-]_i$ in their somata is 20 mM lower on average, leading to steep standing gradients in resting $[Cl^-]_i$ between the soma and dendrite. Further, the effects of GABA on $[Cl^-]_i$ in these cells were measured. GABA caused a reduction in $[Cl^-]_i$ in the dendrites of these cells, showing that dendritic $[Cl^-]_i$ actually is high enough for GABA to cause a Cl^- efflux and subsequent depolarization of membrane potential. Although GABA is depolarizing in the dendrites of ON-type bipolar cells, GABA is inhibitory in the axon terminal of the same cell so that it can have a normal form of presynaptic inhibition. The ratiometric nature of Clomeleon imaging was key in providing the first direct demonstration of this Cl^- gradient within ON-type bipolar cells.

11.3.4.4 Imaging Cl^- Fluxes Associated with Synaptic Inhibition

One of the most promising applications of Clomeleon is to image the changes in association with synaptic inhibition in neurons. As mentioned above, the major forms of inhibition are due to inhibitory transmitters activating receptors that are coupled to Cl^- channels, resulting in transmembrane Cl^- fluxes that should change $[Cl^-]_i$.

Several experiments have evaluated the capacity of Clomeleon to detect synaptic inhibition by combining patch clamp recording and simultaneous Clomeleon imaging in hippocampal neurons. The first such experiments examined responses of cultured neurons to the application of GABA; Cl^- currents and subsequent $[Cl^-]_i$ changes associated with GABA receptor activation were simultaneously detected by whole-cell patch clamp recording and Clomeleon imaging (Kuner and Augustine 2000). Both the polarity and the magnitude of $[Cl^-]_i$ changes were well correlated with the GABA-induced Cl^- currents, regardless of whether Cl^- was entering or leaving a neuron. Changes in $[Cl^-]_i$ occurred with a delay of approximately 1 s, presumably limited by the kinetics of Clomeleon binding to Cl^-, as described above. These changes in $[Cl^-]_i$ returned to the baseline approximately 10 s after the Cl^- current ceased, indicating the relatively slow rate of Cl^- removal from these cells by transporters (Staley and Proctor 1999).

SuperClomeleon was also evaluated in a similar experimental paradigm (Grimley et al. 2013). Due to its higher sensitivity and wider dynamic range for Cl^-, SuperClomeleon improved the SNR six-fold, allowing detection of Cl^- changes by a train of synaptic responses at a single-cell level without voltage clamping.

GABA responses have also been measured in CA1 pyramidal neurons in hippocampal slices from thy1::Clomeleon transgenic mice (Berglund et al. 2006). A local inhibitory pathway (Freund and Buzsáki 1996; Somogyi and Klausberger 2005) was activated by an extracellular stimulating electrode that was placed at the distal part of the pyramidal cell apical dendrites while excitatory synaptic transmission was eliminated by treating the slices with a glutamate receptor antagonist. Under voltage clamp conditions, a train of electrical stimuli evoked summating inhibitory postsynaptic currents (IPSCs). These IPSCs elevated $[Cl^-]_i$ in the soma of individual pyramidal cells by as much as 4 mM; as was the case in cultured neurons, the $[Cl^-]_i$ changes correlated very well with the amount of Cl^- influx. As also observed in cultured neurons, the rise in $[Cl^-]_i$ lagged behind the current response by ~1 s and the change in $[Cl^-]_i$ lasted much longer than the duration of the Cl^- current.

Such results indicate that Clomeleon can report elevation of Cl^- upon activation of $GABA_A$ receptors when the driving force on Cl^- fluxes is kept relatively constant by the voltage clamp. However, in more physiological conditions, the changes in membrane potential produced by inhibitory postsynaptic potentials (IPSPs) will cause the driving force on Cl^- influx to decline during the IPSP, thereby reducing Cl^- flux. Because this effect could limit the ability to image inhibitory transmission, we next asked whether Clomeleon could image inhibitory synaptic transmission in neurons whose membrane potential was not kept constant by voltage clamp (Berglund et al. 2006).

Under these conditions, a single stimulus caused an IPSP that hyperpolarized the RMP from −71 to −76 mV. Trains of stimuli evoked summating IPSPs, and Clomeleon imaging revealed barely detectable changes in $[Cl^-]_i$ in the soma of the same cell (Berglund et al. 2006). The small magnitude of these responses is consistent with the expected reductions in Cl^- flux. When $[Cl^-]_i$ signals were collected and averaged from many somata, the SNR of the measurement was greatly improved, allowing responses to brief trains of IPSPs to be detected readily. Maximal changes in $[Cl^-]_i$ were 2 mM, which is smaller than voltage-clamped cells and is consistent with the reduced driving force on synaptic Cl^- fluxes in unclamped cells. These results indicate that Clomeleon can faithfully report the amount of inhibitory synaptic transmission in neurons not under voltage clamp. Averaging $[Cl^-]_i$ signals from 10 to 20 cells over four trials doubled the SNR, yielding a detection limit of a 90 μM increase in $[Cl^-]_i$. This corresponds to the response to approximately three IPSPs. The experiments above demonstrate that Clomeleon enables imaging of inhibitory transmission without any manipulation of the Cl^- driving force.

Such measurements can even be performed in the absence of any recording electrodes. In such conditions, interneuron stimulation still produces rises in $[Cl^-]_i$ in all layers of area CA1. The sensitivity of the method is sufficient to detect recruitment of presynaptic interneurons: increasing stimulus intensity, to activate larger numbers of interneurons, causes progressively larger changes in $[Cl^-]_i$. The magnitude of the spatially averaged $[Cl^-]_i$ change is proportional to stimulus intensity and is blocked by the $GABA_A$-specific antagonist, SR95531, indicating that the $[Cl^-]_i$ changes resulted from progressive activation of GABAergic interneurons by stronger stimuli (Berglund et al. 2006).

The spatiotemporal dynamics of inhibitory synaptic responses can be quantified by scanning a measurement line across different layers in the CA1 regions, which constitute different compartments of the pyramidal cells. While $[Cl^-]_i$ changes in the distal part of apical dendrites rise rapidly and reach a maximum at the end of the train of stimuli, $[Cl^-]_i$ rises more slowly in basal dendrites and in the proximal part of apical dendrites. Changes in $[Cl^-]_i$ rise most slowly in somata, probably due to slow diffusion of Cl^- from the dendrites and/or from the surface membrane of the cell bodies. These differences presumably reflect differences in the location of active inhibitory synapses on the pyramidal cells, in the surface-to-volume ratio of the different compartments and in the driving force on Cl^- (Staley and Proctor 1999).

In conclusion, these results document that Clomeleon and SuperClomeleon can detect changes in postsynaptic $[Cl^-]_i$ in hippocampal neurons that result from synaptic inhibition mediated by $GABA_A$ receptors. These results establish the utility of Clomeleon for imaging synaptic inhibition in the brain.

11.3.4.5 Pathological Changes of $[Cl^-]_i$

Abnormal regulation of neuronal $[Cl^-]_i$ contributes to the pathology of a multitude of diseases (Payne et al. 2003). For example, chronic pain is partly mediated by an increase in neuronal resting $[Cl^-]_i$, resulting in disinhibition of spinal cord neurons in the nociceptive pathway (Coull et al. 2003, 2005; Torsney and MacDermott 2005). Hypoxia and ischemia also induce pathological alterations in neuronal $[Cl^-]_i$ and GABAergic transmission, leading to excitotoxicity in the brain (Allen and Attwell 2004; Allen et al. 2004). Hippocampal slices from thy1::Clomeleon mice have been successfully used to track pathological increases of $[Cl^-]_i$ resulting from oxygen deprivation. The ability to measure these changes in $[Cl^-]_i$ also permitted identification of the source of the increased $[Cl^-]_i$, which is an influx of Cl^- resulting from activation of the NKCC1 transporter by phosphorylation (Pond et al. 2006).

Acute and organotypically cultured slices of the hippocampus from thy1::Clomeleon transgenic mice have also been used to study $[Cl^-]_i$ changes associated with ictal activity and epilepsy (Dzhala et al. 2010, 2012; Lillis et al. 2012).

11.3.4.6 In Vivo Imaging

The high expression of Clomeleon in thy1::Clomeleon transgenic mice also permits non-invasive brain imaging in vivo (Helmchen and Denk 2005). Two-photon microscopy has enabled the visualization of neocortical layer five pyramidal cells in living mice (Krieger et al. 2007; Berglund et al. 2008). This approach has been extended to image cerebellar granule cells in vivo (Berglund et al. 2009). This approach opens the possibility of imaging synaptic inhibition in the brains of intact, behaving mice.

11.3.4.7 In Vitro Screening

Clomeleon has been used for high-throughput screening of chemicals that affects Cl^- homeostasis in living cells. Utilizing a cell line stably transfected with Clomeleon, Gagnon et al. (2013) screened 100,000 compounds to identify a potent activator of KCC2. Such approaches will facilitate the discovery of novel therapeutics for neurological disorders caused by $[Cl^-]_i$ dysregulation.

11.4 Outlook

We hope this chapter makes clear that optogenetic chloride sensors, especially the initial and improved versions of Clomeleon, have provided an excellent range of possibilities for directly measuring $[Cl^-]_i$ in neurons and other cell types. Structurally directed mutagenesis has identified a YFP variant optimized for $[Cl^-]_i$ measurements in the nervous system. Future sensors should incorporate this variant to diversify optogenetic chloride sensors. For example, a FRET pair with Cl^--sensitive YFP and Cl^--insensitive red fluorescent protein would shift the emission spectrum so that it would penetrate tissue better. Another possibility is to combine this YFP with luciferase so that bioluminescence can be used for the detection of $[Cl^-]_i$ changes instead of fluorescence. For Ca^{2+}, there is already such an optogenetic sensor (Nano-lantern) consisting of *Renilla* luciferase linked to a YFP variant, Venus, through the Ca^{2+}-sensing calmodulin-M13 domain (Saito et al. 2012). We hope this chapter will inspire future efforts both to utilize current optogenetic chloride sensors and to engineer novel chloride sensors.

Acknowledgments Writing of this chapter was supported by a CRP grant from the National Research Foundation of Singapore and by the World Class Institute (WCI) Program of the National Research Foundation of Korea (NRF) funded by the Ministry of Education, Science and Technology of Korea (MEST) (NRF Grant Number: WCI 2009–003).

References

Adams SR, Harootunian AT, Buechler YJ et al (1991) Fluorescence ratio imaging of cyclic AMP in single cells. Nature 349(6311):694–697

Aickin CC, Brading AF (1982) Measurement of intracellular chloride in guinea-pig vas deferens by ion analysis, ^{36}chloride efflux and micro-electrodes. J Physiol Lond 326:139–154

Akerman CJ, Cline HT (2006) Depolarizing GABAergic conductances regulate the balance of excitation to inhibition in the developing retinotectal circuit in vivo. J Neurosci 26(19):5117–5130

Allen NJ, Attwell D (2004) The effect of simulated ischaemia on spontaneous GABA release in area CA1 of the juvenile rat hippocampus. J Physiol Lond 561(Pt 2):485–498

Allen NJ, Rossi DJ, Attwell D (2004) Sequential release of GABA by exocytosis and reversed uptake leads to neuronal swelling in simulated ischemia of hippocampal slices. J Neurosci 24(15):3837–3849

Arosio D, Ricci F, Marchetti L et al (2010) Simultaneous intracellular chloride and pH measurements using a GFP-based sensor. Nat Methods 7(7):516–518

Barmashenko G, Hefft S, Aertsen A et al (2011) Positive shifts of the $GABA_A$ receptor reversal potential due to altered chloride homeostasis is widespread after status epilepticus. Epilepsia 52(9):1570–1578

Ben-Ari Y, Cherubini E, Corradetti R et al (1989) Giant synaptic potentials in immature rat CA3 hippocampal neurones. J Physiol Lond 416:303–325
Berglund K, Dunbar RL, Lee P et al (2005) A practical guide: imaging synaptic inhibition with Clomeleon, a genetically encoded chloride indicator. In: Konnerth A, Yuste R (eds) Imaging in neuroscience and development: a laboratory manual. Cold Spring Harbor Laboratory Press, Cold Spring Harbor, pp 595–598
Berglund K, Schleich W, Krieger P et al (2006) Imaging synaptic inhibition in transgenic mice expressing the chloride indicator, Clomeleon. Brain Cell Biol 35(4–6):207–228
Berglund K, Schleich W, Wang H et al (2008) Imaging synaptic inhibition throughout the brain via genetically targeted Clomeleon. Brain Cell Biol 36(1–4):101–118
Berglund K, Kuner T, Augustine GJ (2009) Clomeleon, a genetically encoded chloride indicator. In: Alvarez-Leefmans FJ, Delpire E (eds) Physiology and pathology of chloride transporters and channels in the nervous system. Academic Press, London, pp 125–139
Berglund K, Kuner T, Feng G et al (2011) Imaging synaptic inhibition with the genetically encoded chloride indicator Clomeleon. Cold Spring Harb Protoc 2011(12):1492–1497
Bertollini C, Murana E, Mosca L et al (2012) Transient increase in neuronal chloride concentration by neuroactive amino acids released from glioma cells. Front Mol Neurosci 5. doi:10.3389/fnmol.2012.00100
Billups D, Attwell D (2002) Control of intracellular chloride concentration and GABA response polarity in rat retinal ON bipolar cells. J Physiol Lond 545(Pt 1):183–198
Bortone D, Polleux F (2009) KCC2 expression promotes the termination of cortical interneuron migration in a voltage-sensitive calcium-dependent manner. Neuron 62(1):53–71
Bregestovski P, Waseem T, Mukhtarov M (2009) Genetically encoded optical sensors for monitoring of intracellular chloride and chloride-selective channel activity. Front Mol Neurosci 2. doi:10.3389/neuro.02.015.2009
Bright GR, Fisher GW, Rogowska J et al (1989) Fluorescence ratio imaging microscopy. Methods Cell Biol 30:157–192
Brochiero E, Banderali U, Lindenthal S et al (1995) Basolateral membrane chloride permeability of A6 cells: implication in cell volume regulation. Pflugers Arch 431(1):32–45
Brown AM, Sutton RB, Walker JL Jr (1970) Increased chloride conductance as the proximate cause of hydrogen ion concentration effects in *Aplysia* neurons. J Gen Physiol 56(5):559–582
Buzsáki G, Chrobak JJ (1995) Temporal structure in spatially organized neuronal ensembles: a role for interneuronal networks. Curr Opin Neurobiol 5(4):504–510
Caroni P (1997) Overexpression of growth-associated proteins in the neurons of adult transgenic mice. J Neurosci Methods 71(1):3–9
Chao AC, Dix JA, Sellers MC et al (1989) Fluorescence measurement of chloride transport in monolayer cultured cells. Mechanisms of chloride transport in fibroblasts. Biophys J 56(6):1071–1081
Chao AC, Widdicombe JH, Verkman AS (1990) Chloride conductive and cotransport mechanisms in cultures of canine tracheal epithelial cells measured by an entrapped fluorescent indicator. J Membr Biol 113(3):193–202
Chen PY, Illsley NP, Verkman AS (1988) Renal brush-border chloride transport mechanisms characterized using a fluorescent indicator. Am J Physiol 254(1 Pt 2):F114–F120
Cheng L, Fu J, Tsukamoto A et al (1996) Use of green fluorescent protein variants to monitor gene transfer and expression in mammalian cells. Nat Biotechnol 14(5):606–609
Cherubini E, Gaiarsa JL, Ben-Ari Y (1991) GABA: an excitatory transmitter in early postnatal life. Trends Neurosci 14(12):515–519
Chesler M, Kaila K (1992) Modulation of pH by neuronal activity. Trends Neurosci 15(10):396–402
Chu TC, Socci RR, Coca-Prados M et al (1992) Comparative studies of furosemide effects on membrane potential and intracellular chloride activity in human and rabbit ciliary epithelium. Ophthalmic Res 24(2):83–91

Coskun T, Baumgartner HK, Chu S et al (2002) Coordinated regulation of gastric chloride secretion with both acid and alkali secretion. Am J Physiol Gastrointest Liver Physiol 283(5):G1147–G1155

Coull JA, Boudreau D, Bachand K et al (2003) Trans-synaptic shift in anion gradient in spinal lamina I neurons as a mechanism of neuropathic pain. Nature 424(6951):938–942

Coull JA, Beggs S, Boudreau D et al (2005) BDNF from microglia causes the shift in neuronal anion gradient underlying neuropathic pain. Nature 438(7070):1017–1021

De Koninck Y (2007) Altered chloride homeostasis in neurological disorders: a new target. Curr Opin Pharmacol 7(1):93–99

Deisz RA, Lehmann TN, Horn P et al (2011) Components of neuronal chloride transport in rat and human neocortex. J Physiol Lond 589(Pt 6):1317–1347

Delpire E, Days E, Lewis LM et al (2009) Small-molecule screen identifies inhibitors of the neuronal K-Cl cotransporter KCC2. Proc Natl Acad Sci U S A 106(13):5383–5388

Derdikman D, Hildesheim R, Ahissar E et al (2003) Imaging spatiotemporal dynamics of surround inhibition in the barrels somatosensory cortex. J Neurosci 23(8):3100–3105

Dickson RM, Cubitt AB, Tsien RY et al (1997) On/off blinking and switching behaviour of single molecules of green fluorescent protein. Nature 388(6640):355–358

Duebel J, Haverkamp S, Schleich W et al (2006) Two-photon imaging reveals somatodendritic chloride gradient in retinal ON-type bipolar cells expressing the biosensor Clomeleon. Neuron 49(1):81–94

Dzhala VI, Kuchibhotla KV, Glykys JC et al (2010) Progressive NKCC1-dependent neuronal chloride accumulation during neonatal seizures. J Neurosci 30(35):11745–11761

Dzhala V, Valeeva G, Glykys J et al (2012) Traumatic alterations in GABA signaling disrupt hippocampal network activity in the developing brain. J Neurosci 32(12):4017–4031

Elsliger M-A, Wachter RM, Hanson GT et al (1999) Structural and spectral response of green fluorescent protein variants to changes in pH. Biochemistry 38(17):5296–5301

Engblom AC, Åkerman KE (1993) Determination of the intracellular free chloride concentration in rat brain synaptoneurosomes using a chloride-sensitive fluorescent indicator. Biochim Biophys Acta 1153(2):262–266

Engblom AC, Holopainen I, Åkerman KE (1989) Determination of GABA receptor-linked Cl^- fluxes in rat cerebellar granule cells using a fluorescent probe SPQ. Neurosci Lett 104(3):326–330

Engblom AC, Holopainen I, Åkerman KE (1991) Ethanol-induced Cl^- flux in rat cerebellar granule cells as measured by a fluorescent probe. Brain Res 568(1–2):55–60

Feng G, Mellor RH, Bernstein M et al (2000) Imaging neuronal subsets in transgenic mice expressing multiple spectral variants of GFP. Neuron 28(1):41–51

Foskett JK (1990) $[Ca^{2+}]_i$ modulation of Cl^- content controls cell volume in single salivary acinar cells during fluid secretion. Am J Physiol 259(6 Pt 1):C998–C1004

Freund TF, Buzsáki G (1996) Interneurons of the hippocampus. Hippocampus 6(4):347–470

Friedel P, Bregestovski P, Medina I (2013) Improved method for efficient imaging of intracellular Cl^- with Cl-sensor using conventional fluorescence setup. Front Mol Neurosci 6. doi:10.3389/fnmol.2013.00007

Funk K, Woitecki A, Franjic-Würtz C et al (2008) Modulation of chloride homeostasis by inflammatory mediators in dorsal root ganglion neurons. Mol Pain 4:32

Gagnon M, Bergeron MJ, Lavertu G et al (2013) Chloride extrusion enhancers as novel therapeutics for neurological diseases. Nat Med 19(11):1524–1528

Galietta LJ, Haggie PM, Verkman AS (2001) Green fluorescent protein-based halide indicators with improved chloride and iodide affinities. FEBS Lett 499(3):220–224

Glykys J, Dzhala V, Egawa K et al (2014) Local impermeant anions establish the neuronal chloride concentration. Science 343(6171):670–675

Grimley JS, Li L, Wang W et al (2013) Visualization of synaptic inhibition with an optogenetic sensor developed by cell-free protein engineering automation. J Neurosci 33(41):16297–16309

Grynkiewicz G, Poenie M, Tsien RY (1985) A new generation of Ca^{2+} indicators with greatly improved fluorescence properties. J Biol Chem 260(6):3440–3450
Hasan MT, Friedrich RW, Euler T et al (2004) Functional fluorescent Ca^{2+} indicator proteins in transgenic mice under TET control. PLoS Biol 2(6):e163. doi:10.1371/journal.pbio.0020163
Haverkamp S, Wassle H, Duebel J et al (2005) The primordial, blue-cone color system of the mouse retina. J Neurosci 25(22):5438–5445
Heim R (1999) Green fluorescent protein forms for energy transfer. Methods Enzymol 302:408–423
Helmchen F, Denk W (2005) Deep tissue two-photon microscopy. Nat Methods 2(12):932–940
Heo KS, Ryoo SW, Kim L et al (2008) Cl^--channel is essential for LDL-induced cell proliferation via the activation of Erk1/2 and PI3k/Akt and the upregulation of Egr-1 in human aortic smooth muscle cells. Mol Cells 26(5):468–473
Hinkle M, Heller P, Van der Kloot W (1971) The influence of potassium and chloride ions on the membrane potential of single muscle fibers of the crayfish. Comp Biochem Physiol A 40(1):181–201
Illsley NP, Verkman AS (1987) Membrane chloride transport measured using a chloride-sensitive fluorescent probe. Biochemistry 26(5):1215–1219
Inglefield JR, Schwartz-Bloom RD (1997) Confocal imaging of intracellular chloride in living brain slices: measurement of $GABA_A$ receptor activity. J Neurosci Methods 75(2):127–135
Inglefield JR, Schwartz-Bloom RD (1999) Using confocal microscopy and the fluorescent indicator, 6-methoxy-*N*-ethylquinolinium iodide, to measure changes in intracellular chloride. Methods Enzymol 307:469–481
Jayaraman S, Haggie P, Wachter RM et al (2000) Mechanism and cellular applications of a green fluorescent protein-based halide sensor. J Biol Chem 275(9):6047–6050
Jose M, Nair DK, Reissner C et al (2007) Photophysics of Clomeleon by FLIM: discriminating excited state reactions along neuronal development. Biophys J 92(6):2237–2254
Kaila K, Pasternack M, Saarikoski J et al (1989) Influence of GABA-gated bicarbonate conductance on potential, current and intracellular chloride in crayfish muscle fibres. J Physiol Lond 416:161–181
Kaneko H, Nakamura T, Lindemann B (2001) Noninvasive measurement of chloride concentration in rat olfactory receptor cells with use of a fluorescent dye. Am J Physiol Cell Physiol 280(6):C1387–C1393
Kerem B, Rommens JM, Buchanan JA et al (1989) Identification of the cystic fibrosis gene: genetic analysis. Science 245(4922):1073–1080
Khirug S, Yamada J, Afzalov R et al (2008) GABAergic depolarization of the axon initial segment in cortical principal neurons is caused by the Na-K-2Cl cotransporter NKCC1. J Neurosci 28(18):4635–4639
Kneen M, Farinas J, Li Y et al (1998) Green fluorescent protein as a noninvasive intracellular pH indicator. Biophys J 74(3):1591–1599
Koncz C, Daugirdas JT (1994) Use of MQAE for measurement of intracellular $[Cl^-]$ in cultured aortic smooth muscle cells. Am J Physiol 267(6 Pt 2):H2114–H2123
Kovalchuk Y, Garaschuk O (2012) Two-photon chloride imaging using MQAE in vitro and in vivo. Cold Spring Harb Protoc 2012(7):778–785
Krapf R, Berry CA, Verkman AS (1988a) Estimation of intracellular chloride activity in isolated perfused rabbit proximal convoluted tubules using a fluorescent indicator. Biophys J 53(6):955–962
Krapf R, Illsley NP, Tseng HC et al (1988b) Structure-activity relationships of chloride-sensitive fluorescent indicators for biological application. Anal Biochem 169(1):142–150
Krieger P, Kuner T, Sakmann B (2007) Synaptic connections between layer 5B pyramidal neurons in mouse somatosensory cortex are independent of apical dendrite bundling. J Neurosci 27(43):11473–11482
Kuner T, Augustine GJ (2000) A genetically encoded ratiometric indicator for chloride: capturing chloride transients in cultured hippocampal neurons. Neuron 27(3):447–459

Liedtke W, Yeo M, Zhang H et al (2013) Highly conductive carbon nanotube matrix accelerates developmental chloride extrusion in central nervous system neurons by increased expression of chloride transporter KCC2. Small 9(7):1066–1075

Lillis KP, Kramer MA, Mertz J et al (2012) Pyramidal cells accumulate chloride at seizure onset. Neurobiol Dis 47(3):358–366

Lindsly C, Gonzalez-Islas C, Wenner P (2014) Activity blockade and $GABA_A$ receptor blockade produce synaptic scaling through chloride accumulation in embryonic spinal motoneurons and interneurons. PLoS One 9(4):e94559. doi:10.1371/journal.pone.0094559

Llopis J, McCaffery JM, Miyawaki A et al (1998) Measurement of cytosolic, mitochondrial, and Golgi pH in single living cells with green fluorescent proteins. Proc Natl Acad Sci U S A 95(12):6803–6808

Lorenzen I, Aberle T, Plieth C (2004) Salt stress-induced chloride flux: a study using transgenic *Arabidopsis* expressing a fluorescent anion probe. Plant J 38(3):539–544

Maglova LM, Crowe WE, Smith PR et al (1998) Na^+-K^+-Cl^- cotransport in human fibroblasts is inhibited by cytomegalovirus infection. Am J Physiol 275(5 Pt 1):C1330–C1341

Marandi N, Konnerth A, Garaschuk O (2002) Two-photon chloride imaging in neurons of brain slices. Pfluegers Arch/Eur J Physiol 445(3):357–365

Markova O, Mukhtarov M, Real E et al (2008) Genetically encoded chloride indicator with improved sensitivity. J Neurosci Methods 170(1):67–76

Metzger F, Repunte-Canonigo V, Matsushita S et al (2002) Transgenic mice expressing a pH and Cl^- sensing yellow-fluorescent protein under the control of a potassium channel promoter. Eur J Neurosci 15(1):40–50

Miesenböck G, De Angelis DA, Rothman JE (1998) Visualizing secretion and synaptic transmission with pH-sensitive green fluorescent proteins. Nature 394(6689):192–195

Miller RF, Dacheux RF (1983) Intracellular chloride in retinal neurons: measurement and meaning. Vision Res 23(4):399–411

Miyawaki A, Llopis J, Heim R et al (1997) Fluorescent indicators for Ca^{2+} based on green fluorescent proteins and calmodulin. Nature 388(6645):882–887

Miyawaki A, Griesbeck O, Heim R et al (1999) Dynamic and quantitative Ca^{2+} measurements using improved cameleons. Proc Natl Acad Sci U S A 96(5):2135–2140

Miyawaki A, Nagai T, Mizuno H (2005) Engineering fluorescent proteins. Adv Biochem Eng Biotechnol 95:1–15

Mukhtarov M, Liguori L, Waseem T et al (2013) Calibration and functional analysis of three genetically encoded Cl^-/pH sensors. Front Mol Neurosci 6. doi:10.3389/fnmol.2013.00009

Nakamura T, Kaneko H, Nishida N (1997) Direct measurement of the chloride concentration in newt olfactory receptors with the fluorescent probe. Neurosci Lett 237(1):5–8

Ormö M, Cubitt AB, Kallio K et al (1996) Crystal structure of the *Aequorea victoria* green fluorescent protein. Science 273(5280):1392–1395

Owens DF, Boyce LH, Davis MB et al (1996) Excitatory GABA responses in embryonic and neonatal cortical slices demonstrated by gramicidin perforated-patch recordings and calcium imaging. J Neurosci 16(20):6414–6423

Palma E, Amici M, Sobrero F et al (2006) Anomalous levels of Cl^- transporters in the hippocampal subiculum from temporal lobe epilepsy patients make GABA excitatory. Proc Natl Acad Sci 103(22):8465–8468

Paulsen O, Moser EI (1998) A model of hippocampal memory encoding and retrieval: GABAergic control of synaptic plasticity. Trends Neurosci 21(7):273–278

Payne JA, Rivera C, Voipio J et al (2003) Cation-chloride co-transporters in neuronal communication, development and trauma. Trends Neurosci 26(4):199–206

Pologruto TA, Yasuda R, Svoboda K (2004) Monitoring neural activity and [Ca^{2+}] with genetically encoded Ca^{2+} indicators. J Neurosci 24(43):9572–9579

Pond BB, Berglund K, Kuner T et al (2006) The chloride transporter Na^+-K^+-Cl^- cotransporter isoform-1 contributes to intracellular chloride increases after *in vitro* ischemia. J Neurosci 26(5):1396–1406

Raimondo JV, Joyce B, Kay L et al (2013) A genetically-encoded chloride and pH sensor for dissociating ion dynamics in the nervous system. Front Cell Neurosci 7. doi:10.3389/fncel.2013.00202
Reiff DF, Ihring A, Guerrero G et al (2005) *In vivo* performance of genetically encoded indicators of neural activity in flies. J Neurosci 25(19):4766–4778
Rink TJ, Tsien RY, Pozzan T (1982) Cytoplasmic pH and free Mg^{2+} in lymphocytes. J Cell Biol 95(1):189–196
Rivera C, Voipio J, Payne JA et al (1999) The K^+/Cl^- co-transporter KCC2 renders GABA hyperpolarizing during neuronal maturation. Nature 397(6716):251–255
Russell JM, Boron WF (1976) Role of chloride transport in regulation of intracellular *p*H. Nature 264(5581):73–74
Saito K, Chang YF, Horikawa K et al (2012) Luminescent proteins for high-speed single-cell and whole-body imaging. Nat Commun 3:1262. doi:10.1038/ncomms2248
Sakai R, Repunte-Canonigo V, Raj CD et al (2001) Design and characterization of a DNA-encoded, voltage-sensitive fluorescent protein. Eur J Neurosci 13(12):2314–2318
Satoh H, Kaneda M, Kaneko A (2001) Intracellular chloride concentration is higher in rod bipolar cells than in cone bipolar cells of the mouse retina. Neurosci Lett 310(2–3):161–164
Servetnyk Z, Roomans GM (2007) Chloride transport in NCL-SG3 sweat gland cells: channels involved. Exp Mol Pathol 83(1):47–53
Sinnecker D, Voigt P, Hellwig N et al (2005) Reversible photobleaching of enhanced green fluorescent proteins. Biochem Wash 44(18):7085–7094
Slemmer JE, Matsushita S, De Zeeuw CI et al (2004) Glutamate-induced elevations in intracellular chloride concentration in hippocampal cell cultures derived from EYFP-expressing mice. Eur J Neurosci 19(11):2915–2922
Somogyi P, Klausberger T (2005) Defined types of cortical interneurone structure space and spike timing in the hippocampus. J Physiol Lond 562(Pt 1):9–26
Song L, Seeger A, Santos-Sacchi J (2005) On membrane motor activity and chloride flux in the outer hair cell: lessons learned from the environmental toxin tributyltin. Biophys J 88(3):2350–2362
Staley KJ, Proctor WR (1999) Modulation of mammalian dendritic $GABA_A$ receptor function by the kinetics of Cl^- and HCO_3^- transport. J Physiol Lond 519(Pt 3):693–712
Stein V, Hermans-Borgmeyer I, Jentsch TJ et al (2004) Expression of the KCl cotransporter KCC2 parallels neuronal maturation and the emergence of low intracellular chloride. J Comp Neurol 468(1):57–64
Suzuki M, Morita T, Iwamoto T (2006) Diversity of Cl^- channels. Cell Mol Life Sci 63(1):12–24
Szabadics J, Varga C, Molnár G et al (2006) Excitatory effect of GABAergic axo-axonic cells in cortical microcircuits. Science 311(5758):233–235
Torsney C, MacDermott AB (2005) Neuroscience: a painful factor. Nature 438(7070):923–925
Tsien RY (1980) New calcium indicators and buffers with high selectivity against magnesium and protons: design, synthesis, and properties of prototype structures. Biochemistry 19(11):2396–2404
Verkman AS, Sellers MC, Chao AC et al (1989) Synthesis and characterization of improved chloride-sensitive fluorescent indicators for biological applications. Anal Biochem 178(2):355–361
Wachter RM, Remington SJ (1999) Sensitivity of the yellow variant of green fluorescent protein to halides and nitrate. Curr Biol 9(17):R628–R629
Wachter RM, Elsliger MA, Kallio K et al (1998) Structural basis of spectral shifts in the yellow-emission variants of green fluorescent protein. Structure 6(10):1267–1277
Wachter RM, Yarbrough D, Kallio K et al (2000) Crystallographic and energetic analysis of binding of selected anions to the yellow variants of green fluorescent protein. J Mol Biol 301(1):157–171
Walker JL, Brown HM (1977) Intracellular ionic activity measurements in nerve and muscle. Physiol Rev 57(4):729–778

Waseem T, Mukhtarov M, Buldakova S et al (2010) Genetically encoded Cl-Sensor as a tool for monitoring of Cl-dependent processes in small neuronal compartments. J Neurosci Methods 193(1):14–23

White HS, Brown SD, Woodhead JH et al (1997) Topiramate enhances GABA-mediated chloride flux and GABA-evoked chloride currents in murine brain neurons and increases seizure threshold. Epilepsy Res 28(3):167–179

Wolfbeis O, Urbano E (1983) Eine fluorimetrische, schwermetallfreie Methode zur Analyse von Chlor, Brom und Iod in organischen Materialien. Z Anal Chem 314(6):577–581

Yeo M, Berglund K, Augustine G et al (2009) Novel repression of Kcc2 transcription by REST-RE-1 controls developmental switch in neuronal chloride. J Neurosci 29(46):14652–14662

Yeo M, Berglund K, Hanna M et al (2013) Bisphenol A delays the perinatal chloride shift in cortical neurons by epigenetic effects on the Kcc2 promoter. Proc Natl Acad Sci U S A 110(11):4315–4320

Zhang G, Gurtu V, Kain SR (1996) An enhanced green fluorescent protein allows sensitive detection of gene transfer in mammalian cells. Biochem Biophys Res Commun 227(3):707–711

Chapter 12
Optogenetic Imaging of Protein Activity in the Synapse Using 2-Photon Fluorescence Lifetime Imaging Microscopy

Hideji Murakoshi and Akihiro C.E. Shibata

Abstract Elucidating the spatiotemporal dynamics of proteins in living cells, including protein–protein interactions and conformational changes, is essential for understanding the mechanisms of intracellular signal transduction and cellular functions such as synaptic plasticity, cell motility, and cell division. One of the best ways to visualize protein activity with high spatiotemporal resolution is to utilize optogenetic probes, such as green fluorescent protein, in combination with fluorescence resonance energy transfer (FRET) techniques, which enable us to measure the distance between donor and acceptor fluorescent proteins fused to the signaling proteins. Among the various FRET detection methods, 2-photon fluorescence lifetime imaging microscopy (2pFLIM) is the ideal method to monitor FRET in subcellular compartments of living cells located deep within tissues, such as in brain slices. This review introduces the principle of the 2pFLIM-FRET for monitoring intracellular protein activities and protein–protein interactions using two examples: detecting small GTPase activity and monitoring actin polymerization in dendrites and synapses of hippocampal neurons in brain slices.

Keywords 2pFLIM • FRET • GFP • Small GTPase • Cdc42 • RhoA • Actin • Synapse • Neuron

H. Murakoshi (✉)
Supportive Center for Brain Research, National Institute for Physiological Science, Okazaki, Japan

Department of Physiological Sciences, Graduate University for Advanced Studies, Okazaki, Aichi 444-8585, Japan

Japan Science and Technology Agency (JST), Kawaguchi, Saitama 332-0012, Japan
e-mail: murakosh@nips.ac.jp

A.C.E. Shibata
Supportive Center for Brain Research, National Institute for Physiological Science, Okazaki, Japan

Okazaki Institute for Integrative Bioscience, Okazaki, Aichi 444-8787, Japan

H. Yawo et al. (eds.), *Optogenetics*, DOI 10.1007/978-4-431-55516-2_12

12.1 Introduction

Inter-neuronal connections occur at synapses, where electrical signals in the presynaptic neuron trigger the release of chemical transmitter signals into the synaptic cleft, which then produces an electrical signal in the postsynaptic neuron (Lynch 2004; Vitureira et al. 2012). In the central nervous system, excitatory synapses are generally formed on postsynaptic structures called spines, which are approximately 300 nm in diameter and often have a mushroom-like shape (Harris and Stevens 1989) working as the container of postsynaptic components (Fig. 12.1). One of the characteristics of spines is their plasticity. The two main types of plasticity are long-term potentiation (LTP) and long-term depression (LTD). In hippocampal neurons, the mechanisms of LTP have been studied extensively via electrophysiological and biochemical methods. This has led to a deep understanding of plasticity, particularly the finding that AMPA receptor recruitment to the post-synaptic density (PSD) is critical for increased sensitivity to the neurotransmitter glutamate from the presynaptic neuron (Fig. 12.1) (Lynch 2004). The recent development and popularization of 2-photon fluorescence microscopy has also allowed us to investigate the mechanisms of LTP at the single-spine level in living neurons, both in cultured brain slices and in the intact brain of live animals (Svoboda and Yasuda 2006). Combining this powerful microscopy technique with electrophysiological recordings and 2-photon uncaging of caged compounds, such as MNI-glutamate, at the single-spine level have contributed even further to our understanding of LTP and LTD (Matsuzaki et al. 2004; Hayama et al. 2013). These studies showed that glutamate binding to *N*-methyl-D-aspartate (NMDA) receptors at the synapse leads to an influx of Ca^{2+} into the spine (Sabatini et al.

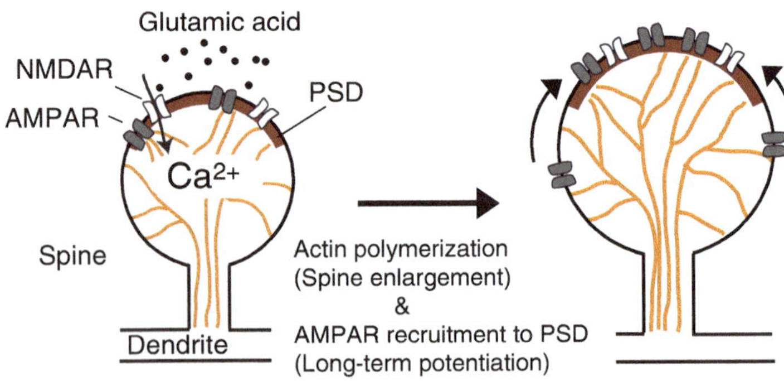

Fig. 12.1 *Schematic drawing of a dendritic spine and actin filaments.* The shape (mushroom-like structure) of a dendritic spine is maintained by actin filaments (*orange*). The binding of glutamate to *N*-methyl-D-aspartate (NMDA) receptors induces Ca^{2+} influx into the spine, which triggers the activation of the various signaling proteins, leading to spine enlargement due to actin polymerization and the recruitment of the AMPA receptors to post-synaptic density (PSD)

2002; Noguchi et al. 2005). The Ca^{2+} influx triggers activation of signaling molecules that produce subsequent plasticity-related events including spine enlargement, AMPA receptor recruitment to the PSD, and nuclear gene transcription (Lynch 2004; Matsuzaki et al. 2004; Zhai et al. 2013). However, the spatiotemporal activity patterns of these signaling molecules remained unknown because a good technique was lacking for the direct imaging of the activity of the specific signaling molecules within their cellular microcompartments, such as synapses, particularly in thick tissue sections with many interconnected neurons forming complex synaptic networks. Recently, the advance of combining 2-photon fluorescence microscopy with fluorescence lifetime measurement (2pFLIM) with fluorescence resonance energy transfer (FRET) has enabled us to visualize biochemical reactions such as protein–protein interactions and protein conformational changes at the single-synapse level, even deep within tissues (Fig. 12.2) (Yasuda et al. 2006; Harvey et al. 2008; Lee et al. 2009; Murakoshi et al. 2011). In this review, we describe the principle of 2pFLIM-FRET and its application to detect Cdc42 and RhoA activity and actin polymerization in dendritic spines of neurons in brain sections.

12.2 Fluorescence Lifetime

Fluorescent molecules, including synthetic organic fluorescent dyes and fluorescent proteins, have their own unique fluorescence lifetime, which is distinct from photobleaching and is defined as the time lag between fluorophore excitation and photon emission (i.e., the time spent in an excited state) (Lakowicz 2006). This physical parameter is affected by the local environment such as temperature, pH, ionic strength, oxygen, and FRET. Using green fluorescent protein (GFP) as an example, exposure to blue laser light excites a GFP fluorophore. This excited state lasts approximately 2.6 ns, which is the fluorescence lifetime, after which GFP emits a photon as green fluorescence and transitions back to the ground state (Fig. 12.3a, right). If one continuously excites a GFP molecule, the excitation and emission processes cycle continuously until photobleaching occurs due to the destruction of fluorophore. In the case of GFP, photobleaching occurs after 10^5 cycles of the excitation/emission process (Kubitscheck et al. 2000). Since fluorescence lifetime is affected by the local environment, it is very useful for the measurement of the intracellular state. One of the main uses of fluorescence lifetime is to monitor FRET using the Time-Correlated Single Photon Counting (TCSPC) (time domain) or phase-delay method (frequency domain) (Lakowicz 2006). The TCSPC method is suitable for imaging neuronal activities deep within tissues because it is highly compatible with 2-photon fluorescence microscopes equipped with deep penetrable long-wavelength pulse lasers, such as Ti:sapphire laser (Becker 2005; Yasuda 2006; Lleres et al. 2007).

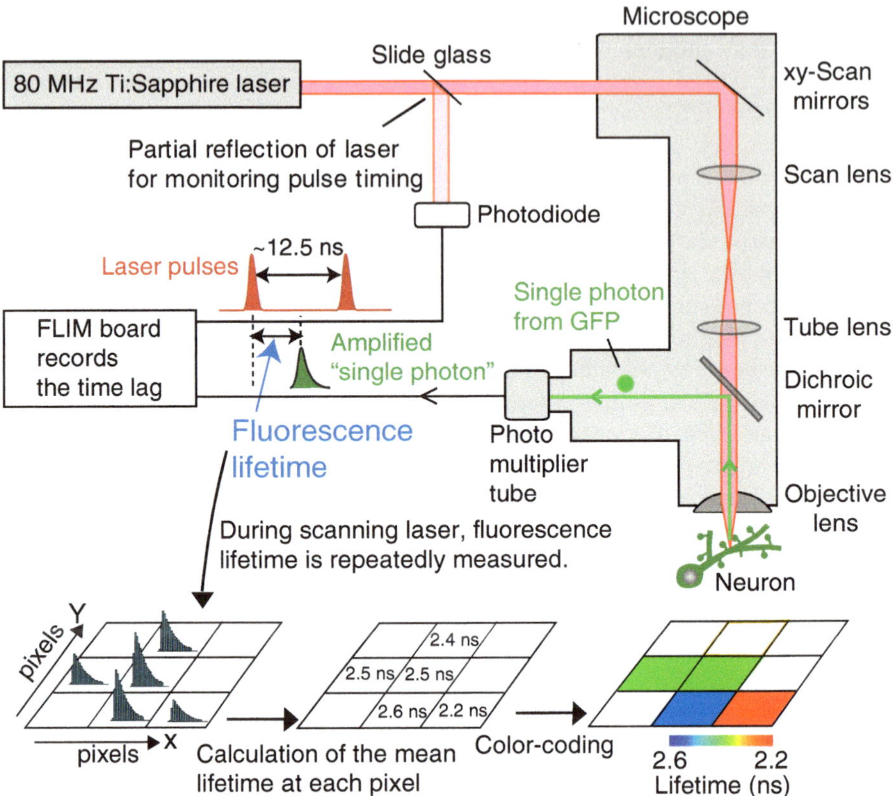

$$\tau_m = \frac{\sum_i t_i F(t_i)}{\sum_i F(t_i)}$$

Fig. 12.2 *Microscope setup for 2-photon fluorescence lifetime imaging.* An Olympus BX51WI microscope is often used as a base, and a hole is punctured through the microscope's side body near the fluorescence cube turret. Fluorescent signals are detected by the photomultiplier tube (PMT). The custom mirror cube is used to allow the side entry of green fluorescent protein (GFP) photons into PMT. A Ti:sapphire laser tuned to 920 nm is used for 2-photon GFP excitation with an xy-galvano scan mirror. This 2-photon microscope system is operated by Scanimage (Pologruto et al. 2003). For fluorescence lifetime imaging microscopy (FLIM) measurements, a minor fraction of laser light is split by a slide glass to take the reference pulses and to compare the arrival time of the fluorescence photon from the sample. The difference in arrival time between the laser pulse and the fluorescence photon is measured by a FLIM board (SPC-150, Becker and Hickle) (Becker 2005). After repeated lifetime measurements, the mean lifetime at each pixel is calculated and translated to the corresponding color with custom (Yasuda 2012) or commercially available software (Becker 2005)

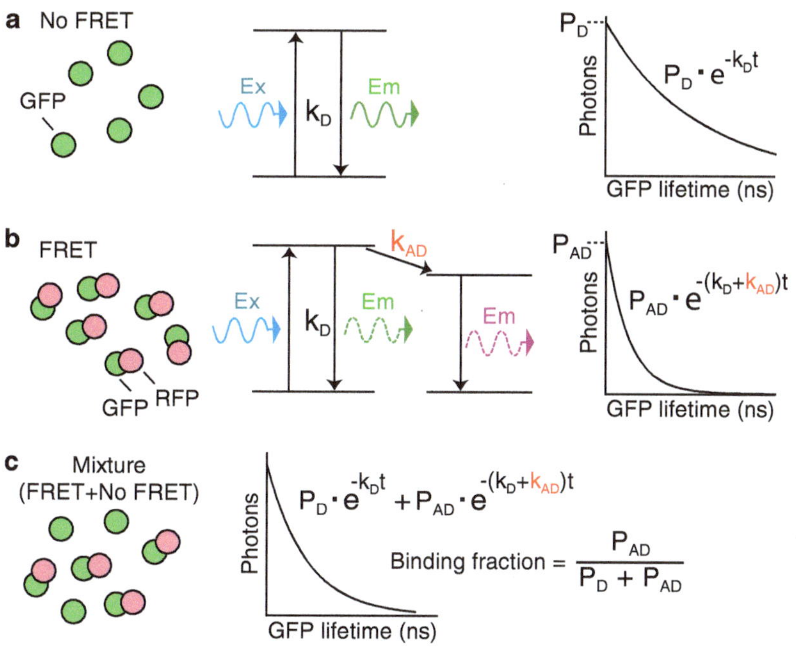

Fig. 12.3 *Fluorescence lifetime curve fitting.* (**a**) (*left*) Green fluorescent protein (GFP) molecules in the absence of fluorescence resonance energy transfer (FRET) acceptor molecules. (*middle*) The Jablonski diagram shows that, after the flash of light, a GFP molecule is excited, emits a photon, and then returns to ground state. (*right*) When a number of GFP molecules (P_D) are excited, the fluorescence decay is fit to a single exponential decay curve, since the photon emission occurs stochastically. (**b**) (*left*) All GFP molecules bind to red fluorescent protein (RFP) molecules (FRET acceptors). (*middle*) The Jablonski diagram shows that after a GFP molecule is excited by the flash of light, photon emission or FRET occurs with the rate constants k_D and k_{AD}, respectively. Since the photon emission and FRET occur competitively, the fluorescence decay is fit to a single exponential, similar to that of GFP lifetime in the absence of FRET acceptors. Note that the fluorescence lifetime is shorter than that of GFP in the absence of FRET acceptors because of the existence of two pathways (k_D and k_{AD}). See also Box 12.1. (**c**) (*left*) The mixture of free GFP and GFP bound to RFP with the ratio of P_D and P_{AD}. (*right*) Since the two populations coexist, the fluorescence lifetime of GFP is fit to two exponentials. The binding fraction is defined as *PAD/(PD + PAD)*

12.3 Fluorescence Resonance Energy Transfer (FRET)

FRET is a phenomenon that occurs when a donor fluorescent molecule is excited and an acceptor molecule is within the Förster distance (<10 nm). The donor molecule's energy excites the acceptor molecule by energy transfer, producing fluorescence emission from the acceptor molecule rather than the donor molecule (Miyawaki 2003; Lakowicz 2006; Lleres et al. 2007). For example, when GFP is excited, it emits green fluorescence. However, if red fluorescent protein (RFP) is located within 10 nm of the GFP molecule, the excited GFP transfers its energy to the RFP, causing RFP to emit red fluorescence (Fig. 12.3b). When FRET occurs,

GFP fluorescence intensity decreases whereas RFP fluorescence intensity increases. In addition, the fluorescence lifetime of GFP also decreases (See Box 12.1 for more detail). For FRET to occur, the following three criteria must be satisfied: (1) the emission spectrum of the donor fluorophore must overlap with the excitation spectrum of the acceptor fluorophore (energy overlap); (2) the donor and acceptor fluorophores must be in close proximity (<10 nm); and (3) the relative orientation of the donor and acceptor fluorophores must be appropriate, namely, the FRET efficiency is the maximum when the emission dipole moment of the donor and absorption dipole moment of the acceptor is parallel to each other.

Box 12.1: Why Does the Fluorescence Lifetime of a Molecule Shorten When FRET Occurs? An Explanation with an Analogy

As described in Sect. 12.3, when FRET occurs between a donor and an acceptor, the donor fluorescence lifetime is shortened. To help in the understanding of lifetime shortening, this Box 12.1 aims to explain it using an analogy. If a blue-excitation photon hits the fluorophore of a single fluorescent molecule, it sends the molecule into the excited state. Usually, a cell expresses many GFP molecules, so the laser excitation excites multiple GFP molecules at the same time. The analogy uses water molecules in a tank, where the individual water molecules represent the individual excited GFP molecules. The tank of water molecules is raised above the ground, and a hole is made in the bottom. All of the water molecules will fall down to the ground through the hole in 10 s (τ_D) (note that inverse of τ_D is k_D, which is proportional to the size of hole). This demonstrates that many of the fluorescent molecules in the excited state will return to the ground state with a rate constant k_D. In contrast, when FRET occurs, an additional rate constant k_{AD} appears, demonstrating the scenario that two holes are instantaneously made at the bottom of tank. In this situation, if the size of holes is equivalent, all molecules will fall down in 5 s (τ_{AD}) because two paths to the ground exist. Similarly, when FRET occurs (i.e., two pathways, k_D and k_{AD} exist), the lifetime of water in the tank shortens. In actuality, the TCSPC measurement described in Sect. 12.2 counts the fluorescence lifetime one by one.

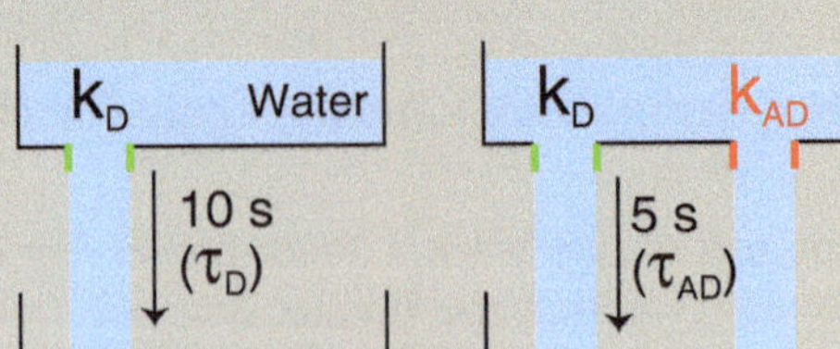

k_D, k_{AD}: the size of holes

$k_D = 1/\tau_D$

$k_D + k_{AD} = 1/\tau_{AD}$

$FRET_{eff} = k_{AD}/(k_{AD}+k_D)$

12.4 Fluorescence Lifetime Measurement and Curve Fitting

To monitor and quantify protein–protein interactions or protein conformational changes in living cells of tissue, TCSPC-based 2pFLIM is often used (Becker 2005). The typical setup of 2pFLIM is described in Fig. 12.2. With this technique, GFP is repeatedly excited by pulses of 2-photon excitation at a rate of 80 MHz. The laser intensity is adjusted so that only one GFP photon is detected by the photomultiplier tube in 100–1,000 excitation pulses. To determine the fluorescence lifetime, the time lag between the arrival time of GFP fluorescence photons and the excitation laser pulses are repeatedly measured using a Becker & Hickl FLIM board (Becker 2005) during two-dimensional image acquisition. After repeating this process approximately 20 times with the laser focused in the sample plane, the lifetime histogram at each pixel is constructed from the ensemble of measured time-lags (Fig. 12.2). Typically, a few seconds are required for single-plane image acquisition (128 × 128 pixels). For fluorescence lifetime calculation, the mean values of the constructed histograms are calculated and are used to generate a color-coded image (Fig. 12.2). For more details, see Yasuda et al. (2006) and Yasuda (2012).

For the quantification of free GFP and acceptor-bound GFP in a region of interest (ROI), a histogram is constructed from the ensemble of measured time-lags for each ROI and is fit with two exponentials. However, since the instrument response function (IRF) is Gaussian-shaped (Yasuda et al. 2006), mainly due to the Gaussian profile of the laser pulse, two exponentially modified Gaussian functions (Eq. (12.1)) are better for curve fitting than the two exponentials (Kalambet et al. 2011).

$$
\begin{aligned}
F(t) = {} & P_D \exp\left(\frac{\sigma_G^2}{2\tau_D^{\,2}} - \frac{t-t_0}{\tau_D}\right)\mathrm{erfc}\left(\frac{\sigma_G^2 - \tau_D(t-t_0)}{\sqrt{2}\tau_D\sigma_G}\right) \\
& + P_{AD} \exp\left(\frac{\sigma_G^2}{2\tau_{AD}^{\,2}} - \frac{t-t_0}{\tau_{AD}}\right)\mathrm{erfc}\left(\frac{\sigma_G^2 - \tau_{AD}(t-t_0)}{\sqrt{2}\tau_{AD}\sigma_G}\right)
\end{aligned}
\tag{12.1}
$$

In Eq. (12.1), erfc is complex error function, t_0 is the time offset, σ_G is the standard deviation of the IRF, and τ_D and τ_{AD} are the decay time constants of free GFP and acceptor-bound GFP, respectively. P_D and P_{AD} are the coefficients of the free GFP component and the acceptor-bound GFP component, respectively. Therefore, the binding fraction of GFP is calculated as $P_{AD}/(P_D + P_{AD})$. For more details, refer to Yasuda et al. (2006) and Yasuda (2012).

Since the aim of this review is to introduce the principle of 2pFLIM-FRET, a simplified form of equation (Eq. (12.2)) assumes that IRF is a delta function that may be used to fit a curve demonstrating its relation to the protein–protein interactions (Fig. 12.3).

$$
F(t) = P_D e^{-k_D t} + P_{AD} e^{-(k_D + k_{AD})t} \tag{12.2}
$$

In Eq. (12.2), k_D and k_{AD} are the decay rate of the lifetime curves of free GFP and acceptor-bound GFP, respectively ($k_D = 1/\tau_D$, $k_D + k_{AD} = 1/\tau_{AD}$). As the fluorescence lifetime of GFP expressed in a cell is measured, the lifetime histogram is fitted to a single-exponential function with decay rate of k_D (Fig. 12.3a). In contrast, when GFP and RFP are equally co-expressed, GFP binds to RFP with 1:1 stoichiometry, FRET occurs, and the fluorescence lifetime of GFP is fitted to a single-exponential function; however, the fluorescence lifetime of GFP is shortened due to additional decay rate k_{AD} (Fig. 12.3b, Box 12.1). Here, FRET efficiency is expressed as $k_{AD}/(k_D + k_{AD})$ or, in other words, $1\text{-}\tau_{AD}/\tau_D$. In reality, free GFP and RFP-bound GFP are usually both present in a cell. Thus, the fluorescence lifetime should be fitted to the two-exponential functions (Fig. 12.3c), where the binding fraction is estimated as $P_{AD}/(P_D + P_{AD})$.

The fitting method described above is quantitative when a strong enough signal is acquired. In contrast, when the signal is weak, for example, the signal emitted from a single synapse, the curve fitting occasionally fails due to relatively small signal-to-noise ratio. Therefore, the non-fitting method (Eq. (12.3)) is used for the estimation of the binding fraction, which is calculated by τ_D, τ_{AD}, and the mean lifetime τ_m (Yasuda 2012). Note that τ_D, τ_{AD} can be determined by independent experiments (Yasuda et al. 2006). Therefore, once the mean lifetime τ_m in the ROI is determined, P_{AD} is calculated directly without curve fitting.

$$P_{AD} = \frac{\tau_D (\tau_D - \tau_m)}{(\tau_D - \tau_{AD})(\tau_D + \tau_{AD} - \tau_m)} \tag{12.3}$$

12.5 Imaging and Estimation of Actin Polymerization with Optogenetic Probes

Actin is a main cytoskeletal component in cells and plays central roles in cell migration, cell division, and cell shape maintenance. Two states of actin, filamentous actin (F-actin) and globular actin (G-actin), co-exist in a cell. Local modulation of the equilibrium between F- and G-actin, namely actin polymerization–depolymerization, is essential to its function. In a neuron, actin is highly enriched in dendritic spines (Fig. 12.1) and regulates the shape of spines and filopodia during both structural plasticity and brain development (Krucker et al. 2000; Matus 2000; Okamoto et al. 2004; Sekino et al. 2007; Cingolani and Goda 2008; Honkura et al. 2008).

To visualize protein–protein interactions, the respective proteins must be fused to donor and acceptor fluorescent proteins. For example, monomeric enhanced GFP (mEGFP) as a donor molecule and super resonance energy-accepting chromoprotein (sREACh) as an acceptor molecule are both fused to the N-terminal of actin molecules, respectively (Fig. 12.4a). When these molecules are co-expressed and actin is polymerized, FRET will occur because the size of actin is 5.5 nm (Holmes et al. 1990), which brings the EGFP–sREACh pair within the Förster distance (Ganesan et al. 2006). In Fig. 12.4b, a neuron expressing mEGFP-actin/sREACh-

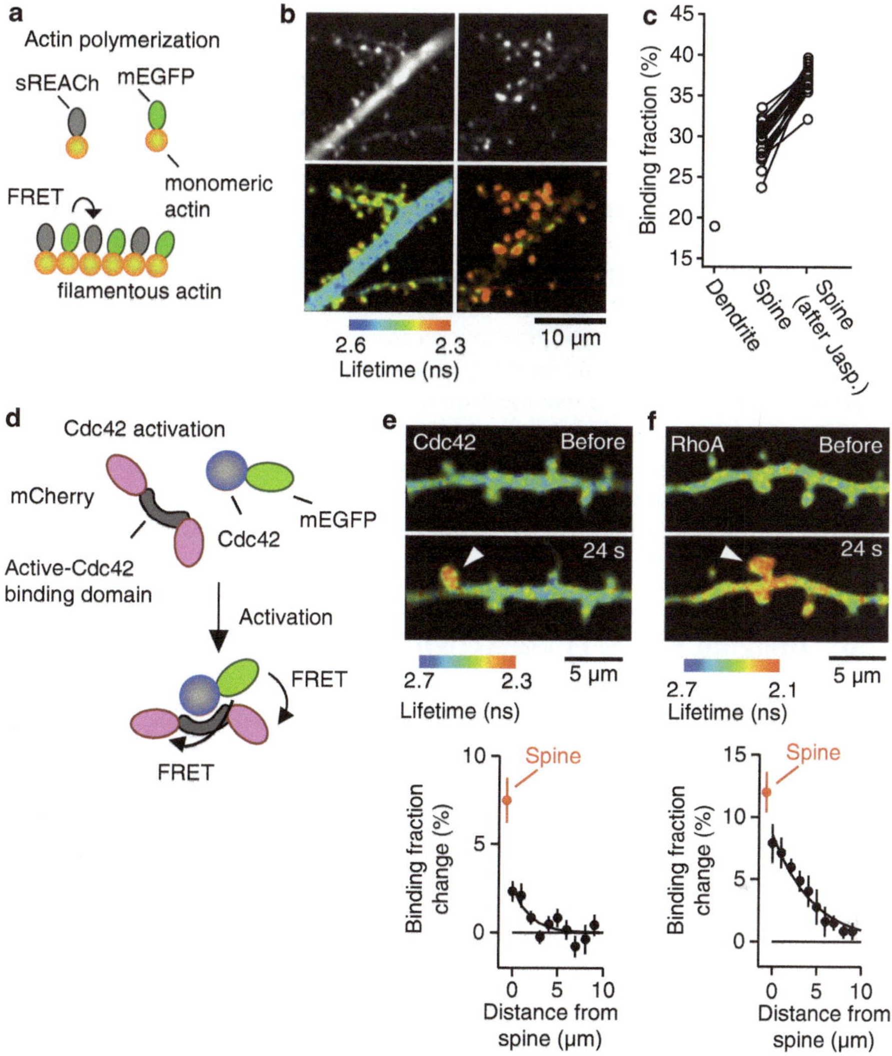

Fig. 12.4 *Imaging actin polymerization and Cdc42/RhoA activation.* (**a**) Actin fluorescence resonance energy transfer (FRET) sensor to detect actin polymerization. monomeric enhanced green fluorescent protein (mEGFP)-actin and super resonance energy-accepting chromoprotein (sREACh)-actin are used as the FRET donor and acceptor, respectively. These FRET sensors were sparsely transfected into a neuron of cultured hippocampal slices by ballistic gene transfer (McAllister 2000). (**b**) Imaging actin polymerization in dendritic spines (Murakoshi et al. 2008). (*Top left*) The fluorescence intensity image of mEGFP-actin. (*Bottom left*) The fluorescence lifetime image of mEGFP-actin. (*Top right*) The fluorescence intensity image after jasplakinolide treatment. Note that almost all F-actin molecules are localized in the spines after the treatment. (*Bottom left*) The fluorescence lifetime image after jasplakinolide treatment. (**c**) The binding fraction (%) in individual spines before and after jasplakinolide treatment. (**d**) Cdc42 FRET sensor to detect the activation of Cdc42. mEGFP-Cdc42 was used as the FRET donor, and Cdc42-binding domain fused to two mCherry molecules was used as the FRET acceptor. (**e, f**) The fluorescence lifetime images of mEGFP-Cdc42 (**e**) and RhoA (**f**). Cdc42 and RhoA are activated after single-spine glutamate uncaging stimulation with a frequency of 0.5 Hz for 60 s (*arrowheads*). The activity profiles of Cdc42 and RhoA in dendrites and spines are shown to the right of e and f (From Murakoshi et al. (2011) with permission)

actin was imaged by conventional 2-photon fluorescence microscopy and 2pFLIM. The expression pattern of mEGFP-actin, when imaged by 2-photon microscopy, shows the distribution of actin molecules in both the dendrite and the spine (Fig. 12.4b, top left), but G-actin cannot be discriminated from F-actin. In contrast, the 2pFLIM image shows the color-coded FRET signal, which is a readout of the level of polymerized F-actin (Fig. 12.4b, bottom left). These data clearly show that actin is highly polymerized in the spines but not in the dendrites. Treatment with jasplakinolide, which stabilizes and promotes actin polymerization, greatly increases the level of actin polymerization in spines (Fig. 12.4b, top right and bottom right). These data confirm that FRET levels (indicated by reduced fluorescence lifetime) are a quantitative measure of actin polymerization (Fig. 12.4b, c). Using Eq. (12.3), the binding fraction of mEGFP-actin in an individual spine can be quantified (Fig. 12.4c). The binding fraction in spines exhibits approximately a 7 % increase after jasplakinolide treatment. This shows that 2pFLIM is a sensitive, selective, and quantitative measure of protein–protein interactions.

12.6 Imaging Rho GTPase Activity with Optogenetic Probes

Rho GTPases, members of a family of small GTPase proteins, were originally identified as the regulators for the cytoskeleton. Their activity is spatiotemporally regulated and plays an important role in various signaling cascades (Machacek et al. 2009). Within the Rho GTPase family, the functions of Cdc42, RhoA, and Rac1 in neurons have been well studied by monitoring spine morphological changes in response to overexpression of constitutively active/inactive Rho GTPase mutants or small interfering RNA (siRNA)-mediated gene knockdowns (Lamprecht and LeDoux 2004). Furthermore, NMDA/Ca^{2+}-dependent activation of Rho GTPases has also been studied with biochemical methods. However, prior to the development of 2pFLIM-FRET, the spatiotemporal activity patterns of Rho GTPases after synaptic plasticity had not been studied because of the lack of the methods to image protein activity in brain slices with high spatiotemporal resolution.

To monitor the activity of Rho GTPases, FRET sensors optimized for 2pFLIM were developed for Cdc42 and RhoA (Fig. 12.4d) (Murakoshi et al. 2011). To image Cdc42 activity, mEGFP was fused to the N-terminal of Cdc42, and two mCherry fluorescent molecules were fused to the N-/C-terminus of Cdc42-binding domain (CBD) of Pak3 (Fig. 12.4d) (Murakoshi et al. 2011). Two mCherry molecules were fused to each CBD in order to increase the sensitivity of FRET. Since the maturation efficiency of mCherry is only 50 %, the fusion of two mCherrys to CBD is expected to increases the chances of FRET occurrence. Indeed, the introduction of two fluorescent proteins increases the sensitivity by ~30 % (Yasuda et al. 2006). In addition, the binding affinity between active Cdc42 and CBD was optimized to approximately 2 μM by introducing the mutations so that the CBD binding to an active Cdc42 does not inhibit the binding of endogenous downstream molecules such as GAP (GTPase activating protein) (Murakoshi et al. 2011). After successful development of this

Cdc42 FRET sensor, the FRET sensor plasmids were introduced into neurons by gene gun cotransfection (McAllister 2000). Since activated (GTP-bound) Rho GTPases have increased affinity for the CBD, the activation of mEGFP-Cdc42 leads to the binding of mCherry-CBD-mCherry, which results in FRET from mEGFP to mCherry. To image RhoA, mEGFP was fused to RhoA and two mCherry molecules were fused to the Rho-binding domain (RBD) of Rhotekin, similar to the strategy described above for Cdc42 (Murakoshi et al. 2011). The affinity between active RhoA and RBD was approximately 4 μM.

In Fig. 12.4e, focal stimulation of a single spine on a neuron expressing the Cdc42 FRET sensor was applied using 2-photon glutamate uncaging. After stimulation, the spine exhibited the enlargement with a concomitant increase of localized Cdc42 activation. In contrast, RhoA activation showed the spreading of activity into the dendrite (Fig. 12.4f). These examples clearly show that the combination of optogenetic FRET probes with 2pFLIM provides important spatiotemporal information in synapses, which cannot be measured by biochemical assays or by conventional overexpression experiments.

12.7 Conclusion

The optogenetic FRET sensors and the methods described in this chapter can be combined to provide important information about the spatiotemporal state of biochemical reactions, including protein–protein interactions, protein conformational changes, and protein activity in subcellular structures such as synapses that are located deep within tissues. In addition, the use of optogenetic manipulation tools such as photoactivatable fluorescent proteins or blue light-responsive proteins has gained strong momentum in recent years (Tye and Deisseroth 2012). The utilization of these tools in combination with 2pFLIM will allow us to gain more detailed information about synaptic plasticity with regard to spatiotemporally regulated signal transduction in the future.

Acknowledgments We thank R. Yasuda and H. Maebashi for discussion.

References

Becker W (2005) In: Castleman AW, Toennies JP, Zinth W (eds) Advanced time-correlated single photon counting techniques, vol 81. Springer, Berlin

Cingolani LA, Goda Y (2008) Actin in action: the interplay between the actin cytoskeleton and synaptic efficacy. Nat Rev Neurosci 9(5):344–356

Ganesan S, Ameer-Beg SM, Ng TT et al (2006) A dark yellow fluorescent protein (YFP)-based Resonance Energy-Accepting Chromoprotein (REACh) for Forster resonance energy transfer with GFP. Proc Natl Acad Sci U S A 103(11):4089–4094

Harris KM, Stevens JK (1989) Dendritic spines of CA1 pyramidal cells in the rat hippocampus: serial electron microscopy with reference to their biophysical characteristics. J Neurosci 9(8):2982–2997
Harvey CD, Yasuda R, Zhong H et al (2008) The spread of Ras activity triggered by activation of a single dendritic spine. Science 321(5885):136–140
Hayama T, Noguchi J, Watanabe S et al (2013) GABA promotes the competitive selection of dendritic spines by controlling local Ca^{2+} signaling. Nat Neurosci 16(10):1409–1416
Holmes KC, Popp D, Gebhard W et al (1990) Atomic model of the actin filament. Nature 347(6288):44–49
Honkura N, Matsuzaki M, Noguchi J et al (2008) The subspine organization of actin fibers regulates the structure and plasticity of dendritic spines. Neuron 57(5):719–729
Kalambet Y, Kozmin Y, Mikhailova K et al (2011) Reconstruction of chromatographic peaks using the exponentially modified Gaussian function. J Chemometrics 25(7):352–356
Krucker T, Siggins GR, Halpain S (2000) Dynamic actin filaments are required for stable long-term potentiation (LTP) in area CA1 of the hippocampus. Proc Natl Acad Sci U S A 97(12):6856–6861
Kubitscheck U, Kuckmann O, Kues T et al (2000) Imaging and tracking of single GFP molecules in solution. Biophys J 78(4):2170–2179
Lakowicz JR (2006) Principles of fluorescence spectroscopy, 3rd edn. Springer, Berlin
Lamprecht R, LeDoux J (2004) Structural plasticity and memory. Nat Rev Neurosci 5(1):45–54
Lee SJ, Escobedo-Lozoya Y, Szatmari EM et al (2009) Activation of CaMKII in single dendritic spines during long-term potentiation. Nature 458(7236):299–304
Lleres D, Swift S, Lamond AI (2007) Detecting protein-protein interactions *in vivo* with FRET using multiphoton fluorescence lifetime imaging microscopy (FLIM). Curr Protoc Cytom Chapter 12:Unit12 10
Lynch MA (2004) Long-term potentiation and memory. Physiol Rev 84(1):87–136
Machacek M, Hodgson L, Welch C et al (2009) Coordination of Rho GTPase activities during cell protrusion. Nature 461(7260):99–103
Matsuzaki M, Honkura N, Ellis-Davies GC et al (2004) Structural basis of long-term potentiation in single dendritic spines. Nature 429(6993):761–766
Matus A (2000) Actin-based plasticity in dendritic spines. Science 290(5492):754–758
McAllister AK (2000) Biolistic transfection of neurons. Sci STKE 2000(51):pl1
Miyawaki A (2003) Visualization of the spatial and temporal dynamics of intracellular signaling. Dev Cell 4(3):295–305
Murakoshi H, Lee SJ, Yasuda R (2008) Highly sensitive and quantitative FRET-FLIM imaging in single dendritic spines using improved non-radiative YFP. Brain Cell Biol 36(1–4):31–42
Murakoshi H, Wang H, Yasuda R (2011) Local, persistent activation of Rho GTPases during plasticity of single dendritic spines. Nature 472(7341):100–104
Noguchi J, Matsuzaki M, Ellis-Davies GC et al (2005) Spine-neck geometry determines NMDA receptor-dependent Ca^{2+} signaling in dendrites. Neuron 46(4):609–622
Okamoto K, Nagai T, Miyawaki A et al (2004) Rapid and persistent modulation of actin dynamics regulates postsynaptic reorganization underlying bidirectional plasticity. Nat Neurosci 7(10):1104–1112
Pologruto TA, Sabatini BL, Svoboda K (2003) ScanImage: flexible software for operating laser scanning microscopes. Biomed Eng Online 2:13
Sabatini BL, Oertner TG, Svoboda K (2002) The life cycle of Ca^{2+} ions in dendritic spines. Neuron 33(3):439–452
Sekino Y, Kojima N, Shirao T (2007) Role of actin cytoskeleton in dendritic spine morphogenesis. Neurochem Int 51(2–4):92–104
Svoboda K, Yasuda R (2006) Principles of two-photon excitation microscopy and its applications to neuroscience. Neuron 50(6):823–839
Tye KM, Deisseroth K (2012) Optogenetic investigation of neural circuits underlying brain disease in animal models. Nat Rev Neurosci 13(4):251–266

Vitureira N, Letellier M, Goda Y (2012) Homeostatic synaptic plasticity: from single synapses to neural circuits. Curr Opin Neurobiol 22(3):516–521

Yasuda R (2006) Imaging spatiotemporal dynamics of neuronal signaling using fluorescence resonance energy transfer and fluorescence lifetime imaging microscopy. Curr Opin Neurobiol 16(5):551–561

Yasuda R (2012) Studying signal transduction in single dendritic spines. Cold Spring Harb Perspect Biol 4(10):a005611

Yasuda R, Harvey CD, Zhong H et al (2006) Supersensitive Ras activation in dendrites and spines revealed by two-photon fluorescence lifetime imaging. Nat Neurosci 9(2):283–291

Zhai S, Ark ED, Parra-Bueno P et al (2013) Long-distance integration of nuclear ERK signaling triggered by activation of a few dendritic spines. Science 342(6162):1107–1111

Chapter 13
Optogenetics in *Drosophila*

Hiroshi Kohsaka and Akinao Nose

Abstract The fruit fly *Drosophila melanogaster*, an insect 5 mm long, has served as the experimental subject in a wide range of biological research, including neuroscience. In this chapter, we briefly introduce optogenetic applications in *Drosophila* neuroscience research. First, we describe the development of *Drosophila*, from egg to adult. In fly neuroscience, temperature-controlled perturbation of neural activity, sometimes called 'thermogenetics' has been an invaluable tool that pre-dates the advent of optogenetics. After briefly introducing this perturbation technique, we describe the process of generating transgenic flies that express optogenetic probes in a specific group of cells. Transgenic techniques are crucial in the application of optogenetics in *Drosophila* neuroscience; here we introduce the transposon P elements and ϕ*C31* integrase methods. As for cell-specific gene expression techniques, the binary expression systems utilizing Gal4-UAS and LexA-LexAop are described. We also present a short and basic optogenetic experiment with *Drosophila* larvae as a practical example. Finally, we review a few recent (as of 2014) studies in *Drosophila* neuroscience that made use of optogenetics. In this overview of fly development, transgenic methods, and applications of optogenetics, we present an introductory background to optogenetics in *Drosophila*.

Keywords *Drosophila* • Thermogenetics • Optogentics • Transgenic • P-element • ϕ*C31* integrase • Larvae • Adults • Locomotion • Motor neurons

---

H. Kohsaka (✉)
Department of Complexity Science and Engineering, University of Tokyo,
Kashiwa, Chiba 277-8561, Japan
e-mail: kohsaka@neuro.k.u-tokyo.ac.jp

A. Nose
Department of Complexity Science and Engineering, University of Tokyo,
Kashiwa, Chiba 277-8561, Japan

Department of Physics, Graduate School of Science, University of Tokyo,
Hongo, Bunkyo-ku, Tokyo 113-0033, Japan

H. Yawo et al. (eds.), *Optogenetics*, DOI 10.1007/978-4-431-55516-2_13

13.1 *Drosophila*

Drosophila melanogaster, a member of the guild of 'cosmopolitan' fruit flies, is a useful model organism in a wide range of biology research, from genetics to ecology. *Drosophila* lives in many regions of the world. With their small body size (~5 mm), rapid reproduction rate, and no special food requirements, they are easy to rear in laboratories and to use as model organisms in biology research. The use of *Drosophila* in neuroscience has been key to the large amount of accumulated knowledge and powerful techniques in genetics and developmental biology over the past century (Bellen et al. 2010). We begin this review with a brief introduction to the development of *Drosophila melanogaster* ('*Drosophila*', hereafter).

13.1.1 Overview of Drosophila *Development*

A single female fly lays about 1,000 eggs in a lifetime. Embryogenesis is completed in a day, then the first instar larva hatches. It takes 1 day for the first instar, 1 day for the second instar, and 2 days for the third instar (they molt between instars). After 5 days of the pupal stage, an adult fly emerges from the pupa case (eclosion) (Fig. 13.1).

Drosophila are holometabolous insects, which means they go through a pupal stage. Although larvae (before pupariation) and adults (after eclosion) are both the same individual animals, the behavioral repertoires are quite different between the two stages. Larvae spend a significant amount of time eating (body lengthens 10-fold in the 4 days after hatching), so feeding and locomotion behaviors are prominent. In the adult stage, reproduction is of paramount importance. Adult flies search

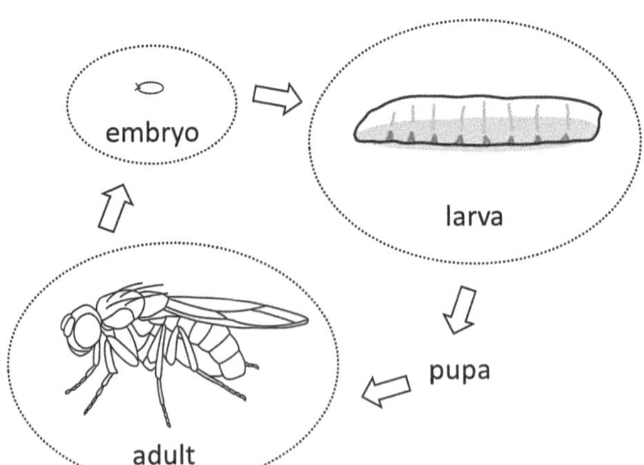

Fig. 13.1 Life cycle of *Drosophila*. Embryos are about 0.5 mm long; larvae and adults are about 4 mm long

for and recognize potential mates, and then execute stereotypical courtship behavior. Both larvae and adults have sensory systems that detect environmental information, including visual (object and circadian recognition), chemical (odor and taste), and physical (vibration, gravity, and touch) cues. Based on environmental information combined with individual experiences stored in the brain as memories, they exhibit adaptive behaviors. Using a vast amount of accumulated knowledge and techniques in genetic and developmental research, *Drosophila* researchers have untangled neural networks at the cellular and molecular level for a variety of behaviors (Venken et al. 2011).

13.1.2 Manipulation of Neural Activity in Drosophila

Optogenetics quickly spread throughout neuroscience research, starting in 2005. In *Drosophila* neuroscience, temperature-sensitive proteins had been used to manipulate neural activity since before the advent of optogenetics. Here we briefly introduce this method, which is sometimes called thermogenetics. Shibire^{ts} protein is a temperature-sensitive mutant of the dynamin protein. The *shibire* gene was cloned (van der Bliek and Meyerowitz 1991; Chen et al. 1991) by the analyses of a mutant (Grigliatti et al. 1973) that is paralyzed at high temperatures ('shibire' means paralysis in Japanese). Shibire^{ts} blocks endocytosis at temperatures higher than 30 °C ('restrictive temperatures'), but has no effect on endocytosis at lower temperatures ('permissive temperatures') (Kitamoto 2001). This change is reversible (See Thum et al. 2006). Since the protein works in a dominant-negative manner, one can block neurotransmission by overexpressing this gene in neurons. Shibire^{ts} in restrictive temperatures blocks secretion of synaptic vesicles (Kosaka and Ikeda 1983), so it is used as temporal blocker of neural activity. In application, one should keep in mind that Shibire^{ts} cannot block electrical synapses, because the target of Shibire^{ts} is dynamin-dependent vesicular neurotransmission, not gap junction communication.

While Shibire^{ts} is used to block neural activity, dTRPA1, a member of the TRP channel family, is an ion channel activated at >25 °C and can be used to activate neurons by temperature control (Hamada et al. 2008; Pulver et al. 2009). TRPM8 is a cation channel that remains open below 25 °C and also can be used as a neuron exciter (Peabody et al. 2009; McKemy et al. 2002). In *Drosophila* neuroscience, these temperature-controlled neuronal activity manipulations (thermogenetics) can be used in conjunction with optogenetics, depending on the experimental design.

13.2 Transgenic Methods in *Drosophila*

Optogenetics requires cell-specific gene expression to manipulate neural activity in a specific group of neurons. In *Drosophila*, transgenic methods are widely used for this purpose. There are two major methods of generating transgenic animals: using

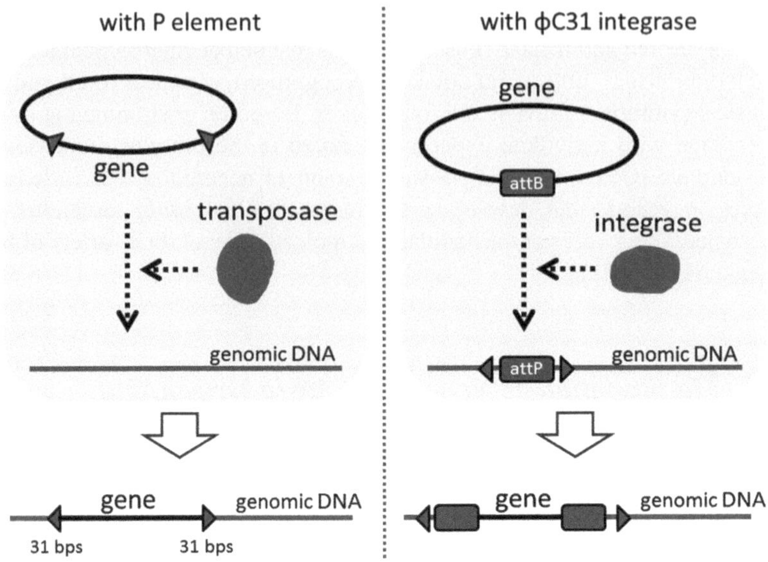

Fig. 13.2 Methods of transgenic fly generation. Vectors are used to insert genes into genomic DNA by enzymes (transposase or integrase)

a transposon P element or using ϕC31 integrase. As for exogenous gene expression techniques, binary expression systems utilizing Gal4-UAS and LexA-LexAop are commonly used. These methods are summarized in Fig. 13.2.

13.2.1 Making Transgenic Flies Using P-Element Transposons

A P-element is a *Drosophila*-specific transposon that is used for mutagenesis, as it possesses inverted repeats of 31 base pairs at both ends. It also contains a DNA sequence that encodes the transposase enzyme, which recognizes the 31-pair sequences and can incise and insert the DNA in between the ends into a region of the genome. Using this system, one can insert any DNA sequence into the *Drosophila* genome through the following procedure. First, one prepares a vector including a DNA sequence to be inserted ('exogenous gene') sandwiched between the 31 base-pair inverted repeats. Then one injects the vector with another vector encoding the transposase enzyme ("helper DNA") into fertilized eggs of a strain lacking the endogenous P-element transposon. In the embryos, the exogenous DNA is inserted into the genome of some cells in a random manner, by the function of transposase, which is transcribed and translated from the helper DNA. In early stages of

Drosophila embryogenesis, the embryo undergoes nuclear division without cell division. So, DNA injected into embryos just after egg laying (within 1 h) diffuses throughout the whole embryo and can be integrated in any nuclei, including those of the presumptive germ cells. Thus, transgenic animals that stably possess the exogenous genes in all cells can be obtained in the progeny of the germ cells that underwent the integration process (Rubin and Spradling 1982). Note that the exogenous gene no longer remobilizes because the progeny lacks transposase.

During the above procedures, the P-element is inserted into random positions in the chromosome (but with a tendency to be inserted into regions near the promoter of genes (Bellen et al. 2011)). This property bears advantages and disadvantages. The expression pattern of the exogenously introduced gene depends on the location of P-element insertion (called 'position effect' (Levis et al. 1985)). Accordingly, one can obtain transgenic lines with a range of expression levels of the exogenously introduced gene, and, in particular, those with a high expression level. However, the disadvantage is that expression of the exogenous gene may be influenced by the regulatory elements of nearby genes, potentially yielding unwanted ectopic expression. Moreover, P-element insertion may affect the expression and function of the endogenous genes near the insertion site. Accordingly, one must be careful about these caveats and perform appropriate controlled experiments when assessing the phenotype of the transgenic animals.

13.2.2 Making Transgenic Flies with *ϕC31* Integrase

Whereas an exogenous gene is inserted at random positions in the genome in the P-element system, the ϕC31 integrase system allows one to control the location of insertion (site-directed insertion). ϕC31 is the name of a phage that infects bacteria. Integrase, an enzyme produced by this phage, recognizes both a phage attachment site (*attP* DNA sequence) and a bacterial attachment site (*attB* DNA sequence) and induces DNA recombination between them. As a result, phage DNA can be inserted into the genome of host bacteria ("lysogenization"). This phenomenon has been used to generate transgenic lines of *Drosophila* (Groth et al. 2004). A collection of *attP* lines, in which a single *attP* sequence is inserted in various locations in the genome, has been generated. Since the location of the *attP* insertion sites in these lines is determined, one can insert an exogenous gene in a desirable position in the genome by cloning the gene in a vector with an *attB* sequence and injecting the vector with messenger RNA encoding integrase into embryos of the appropriate *attP* line. Thus, we can design transgenic animals with an exogenous gene inserted in a genomic location determined in advance (Markstein et al. 2008). Both the site-directed transgene insertion and the P-element method have been widely used to generate transgenic animals.

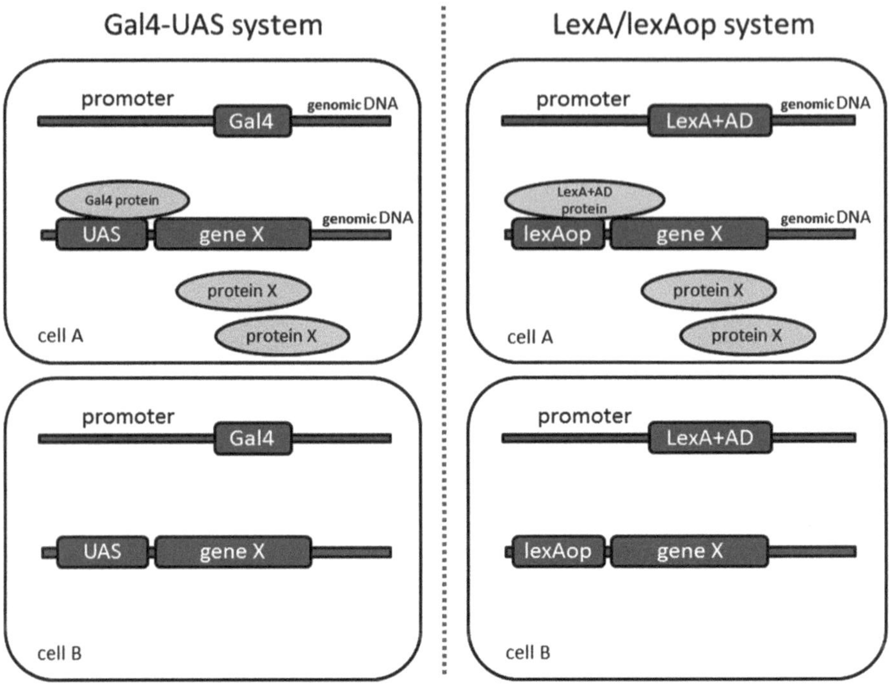

Fig. 13.3 Gene expression systems. In cell A, Gal4 or lexA+AD is expressed by the promoter, then in turn gene X is expressed by Gal4 or lexA+AD proteins. In cell B, since neither Gal4 nor lexA+AD is expressed, gene X is not expressed. *AD* activation domain

13.2.3 Gal4-UAS System and LexA/LexAop System

Cell-specific expression of exogenous genes is required in optogenetics. In *Drosophila*, the Gal4-UAS system is widely used for artificial gene expression (Brand and Perrimon 1993) (Fig. 13.3). The yeast transcription factor Gal4 binds to a DNA sequence known as the upstream activation sequence (UAS) and drives expression of a gene downstream of the UAS. By utilizing this mechanism, expression of genes can be induced in specific cells in *Drosophila*. A collection of Gal4 lines, in which *Gal4* is expressed in specific cells, and UAS lines, which can be used to express various reporters and optogenetic tools, have been established. Both Gal4 lines and UAS lines are made by the P-element or ϕC31 integrase method, described above. Gal4 lines are generated by (1) inserting a vector containing Gal4 gene sequence and the enhancer element that drives expression of Gal4 in specific cells (enhancer-Gal4 lines) or (2) screening P-element-mediated Gal4 insertions in random positions in the genome for those in which Gal4 is expressed, by influence of the nearby enhancers, in the cells of interest (enhancer-trap lines). By crossing a Gal4 line and a UAS line, one can obtain progeny that express an exogenous gene in specific cells. For example, by crossing 'motoneuron-Gal4' with 'UAS-ChR2', one can obtain animals expressing channelrhodopsin (ChR)-2 in motor neurons. Currently, thousands of Gal4 lines have been established and are publically available (Pfeiffer et al. 2008). Several UAS lines for expression of optogenetic probes have also been generated (Shearin et al. 2013).

The LexA-LexAop system has also been developed (Lai and Lee 2006) (Fig. 13.3). The *Escherichia coli* protein LexA works as a repressor by binding to the *lexA* operator (*lexAop*). A fusion protein of LexA and a transcription activator (such as the activator domain of Gal4) can induce expression of the gene downstream of the *lexAop* sequence. By using LexA transgenic lines, in which LexA is expressed in specific cells, exogenous genes can be expressed in specific cells with the LexA-LexAop system. Combinatorial use of the Gal4-UAS system and the LexA-LexAop system allows one to express different genes in distinct subsets of cells in the same animals and widens the range of optogenetic applications. For example, by using the Gal4-UAS system to express ChR2 in cells A and LexA-LexAop system to express GCaMP in cells B, one can study the effects of activating cells A on the activity of cells B.

13.2.4 Generate/Obtain Transgenic Lines

To establish new transgenic lines, one can design and generate a transgene vector as described above, inject the vector into fertilized eggs, select the animals that integrate the transgene and keep them as stocks. Solution containing the transgene is injected into embryos using a micro glass pipette. The injection process can be outsourced to some companies. Transgenic lines that are reported in published papers are, in principle, shared among fly researchers. They can be obtained by requesting them from individual researchers or from the stock centers in the USA, Europe, and Japan that maintain and supply valuable transgenic lines.

13.3 Example Optogenetics Experiment in *Drosophila* Larvae

We have described how to generate transgenic animals in *Drosophila*. Next, we describe an example of an optogenetic experiment using *Drosophila* larvae (Fig. 13.4). We expressed ChR2 in motor neurons, expecting muscles that are innervated by those motor neurons to contract when excited with blue light. (*Drosophila* motor neurons form excitatory synapses on muscle.)

First, we crossed a '*motoneuron-Gal4*' line that drives gene expression in motor neurons (*OK6-Gal4*; (Aberle et al. 2002)) with a '*UAS-ChR2*' line for expression of ChR2 (Pulver et al. 2009). This generates larvae that express ChR2 in motor neurons. However, blue-light illumination on the larvae (*OK6-Gal4*+*UAS-ChR2*)

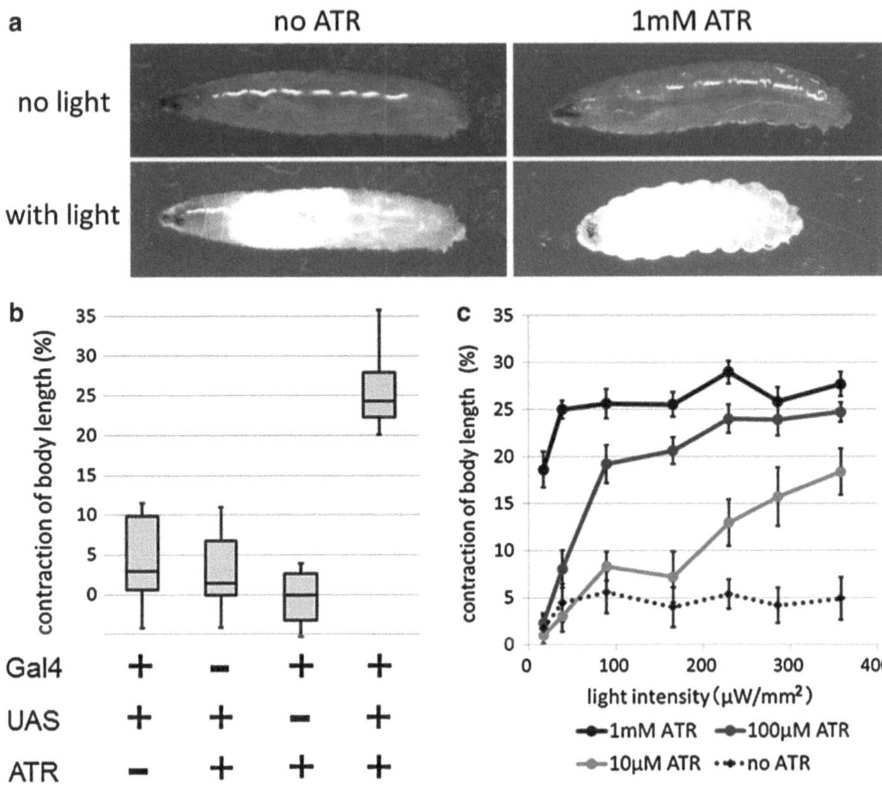

Fig. 13.4 Optogenetic activation of larval motor neurons. (**a**) Snapshots of optogenetic activation of larval motor neurons with 285 μW/mm^2 blue light. (**b**) Quantification of body contraction of experimental and control groups. (**c**) Dependence of efficiency of optogenetic stimulation on amounts of ATR and intensities of stimulation light. *ATR* all-*trans* retinal

cannot activate motor neurons because all-*trans* retinal (ATR) is missing. Unfortunately, *Drosophila* do not synthesize sufficient amounts of ATR for optogenetics, so it is necessary to feed larvae with ATR. As *Drosophila* larvae have a large appetite, a sufficient amount of ATR can be supplied by feeding a mixture of normal food and ATR. We simply spread yeast paste containing ATR (final concentration 1 mM) on agar gel, and reared the larvae overnight on the gel – the motor neurons were ready to be activated the next day (Matsunaga et al. 2013). The excitation light for imaging green fluorescent protein (GFP) in fluorescence microscopy is suitable for the excitation of ChR2. Upon illuminating free-running larvae on an agar plate with blue light, bodywall muscles of the larvae contract simultaneously.

Can we conclude from this observation that activating motor neurons with ChR2 leads to muscle contraction? Not without control groups. The experiment has three components: motoneurons-Gal4, *UAS-ChR2*, and ATR, and we tested whether blue-light illumination induces muscle contraction. Thus, the required control groups are *motoneuron-Gal4*+*UAS-ChR2* (lacking ATR), *UAS-ChR2*+ATR (lacking *motoneurons-Gal4*), and *UAS-ChR2*+ATR (lacking *UAS-ChR2*). None of these larvae showed muscle contraction as was seen in the experimental group, so we can conclude that muscles contracted as a result of ChR2 activation of motor neurons. In these three control groups, *UAS-ChR2*+ATR is especially important, since UAS lines sometimes induce gene expression even without Gal4 protein due to a position effect (called 'leak' (Pfeiffer et al. 2010)). To avoid misinterpretation of the results, a 'no Gal4' control must be included.

Efficacy of neural activation with ChR2 depends on several factors, including the expression level of ChR2, the amount of ATR, and the intensity of the light stimulation. We tested how the amount of ATR and light intensity affect the efficacy of activation of larval motor neurons. We reared larvae with three levels of ATR concentration (10 μM, 100 μM, and 1 mM) overnight, and then illuminated them with various intensities of blue light (20–360 $\mu W/mm^2$) and measured body contractions (Fig. 13.4). The higher the ATR concentration we fed the larvae, the more they exhibited contraction. Further, in a low-ATR concentration condition, the stronger the light we applied to larvae, the more the larvae showed body contractions. Although this observation seems to suggest that we should use blue light at maximum power, in practice we should pay attention to the innate photophobic behavior of larvae. Larvae have eyes on the head (actually, they also have photoreceptors on all surfaces of the body (Xiang et al. 2010)), and they exhibit photophobic response behavior upon being illuminated with strong blue light. To reduce the photophobic response in optogenetic experiments, the light intensity for stimulation should be minimized. Accordingly, the light intensity is determined by a trade-off between the efficiency of ChR2 excitation and the larval photophobic response. Larvae are less sensitive to longer wavelengths of light. Accordingly, red-shifted ChR2 (Lin et al. 2013; Klapoetke et al. 2014) can be used to activate larval neurons while avoiding innate photophobic responses.

13.4 Examples of Optogenetic Research in *Drosophila*

Finally, we briefly review some recent optogenetics applications in *Drosophila*, which have been growing quickly over the past several years. *Drosophila* provide a valuable model system for many areas of neuroscience, including synaptogenesis, sensory systems, motor circuits, and learning.

The larval neuromuscular junction (NMJ) is one of the major model systems in synapse formation studies. In the fly NMJ, the relationship between neural activity and synapse morphology can be examined in great resolution. Ataman et al. (2008) showed that optogenetic excitation of motor neurons in intact larvae induces formation of undifferentiated presynaptic structures ('ghost boutons') and potentiation of spontaneous synaptic vesicle release frequency. With presynaptic and/or postsynaptic cell-specific expression of ChR2, Ljaschenko et al. (2013) showed that the Hebbian plasticity rule governs the distribution of glutamate receptors in postsynaptic muscles. Through the application of optogenetics, NMJs have become a useful model system for further synapse formation research.

Larvae sense environmental cues through sensory neurons, and activation of sensory neurons in turn elicits adaptive behavior. To examine whether activation of specific neurons induces specific behaviors, one should activate candidate neurons specifically. One class of sensory neurons in the bodywall (class IV multidendritic neurons) was thought to be responsible for larval 'rolling' – an escape behavior. Dr. Daniel Tracey's group showed that optogenetic activation of the class IV sensory neurons induces the rolling behavior (Hwang et al. 2007; Honjo et al. 2012). In addition to nociceptive stimuli, Optogenetics has been used to activate specific sensory neurons in intact animals, and behavioral responses are then observed.

In typical larval behavior, as shown in Fig. 13.5, forward peristaltic locomotion is generated by propagation of body segment contraction from posterior to anterior (Kohsaka et al. 2012). This propagative activity is one example of axial locomotion, a form of animal motion including swimming, crawling, and walking. By local laser illumination of specific segments in the nervous system, we can perturb local neural activity during ongoing activity propagation and assess the spatiotemporal dynamics of motor circuits (Matsunaga et al. 2013). By silencing neuronal activity locally with halorhodopsin, we showed that motor neurons are responsible not only for muscle contraction but also for intersegmental progression of neural activity within the nervous system (Fig. 13.6) (Inada et al. 2011; Matsunaga et al. 2013). Operability

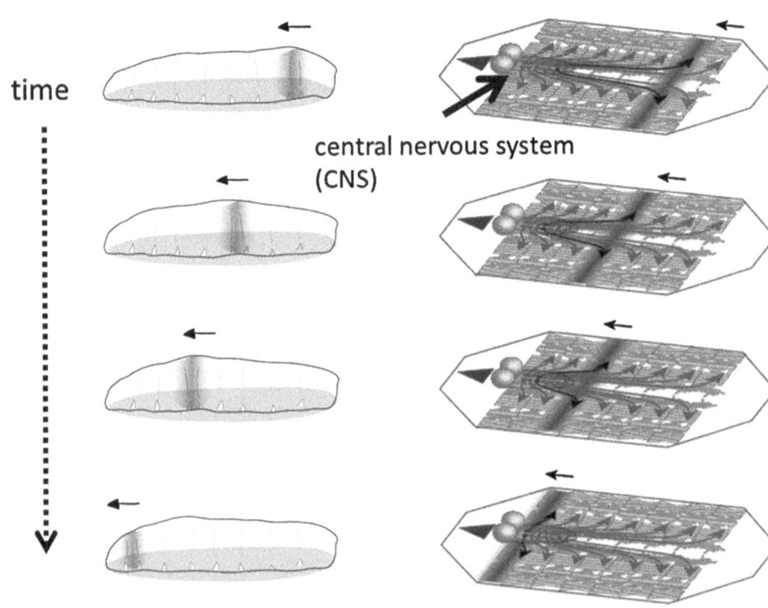

Fig. 13.5 Peristaltic locomotion of intact (*left*) or dissected (*right*) larvae. In forward locomotion, local segmental contraction is propagated from the posterior (*right*) to the anterior segment (*left*). *Arrows* in *left panels* show motor neuronal projection to muscles (From Kohsaka et al. (2012))

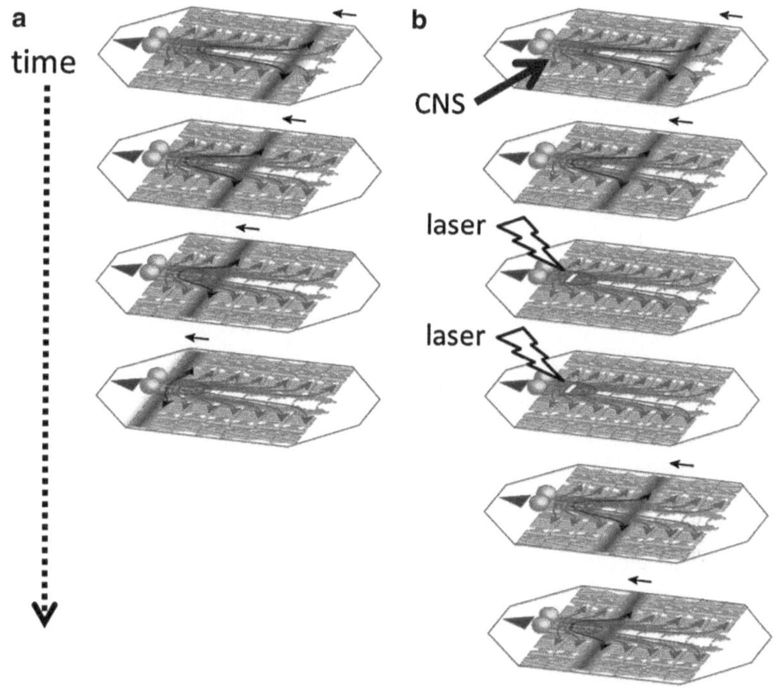

Fig. 13.6 Local optogenetic stimulation in larval motor circuits. (**a**) Normal peristaltic motion in a dissected larva. (**b**) Local inhibition of motor neuronal activities with *Np*HR during peristaltic motion. After the end of the local inhibition, peristaltic motion resumed from the inhibited region. This observation suggests that the peristaltic wave cannot progress without sufficient activation of motor neurons (Inada et al. 2011). *Np*HR halorhodopsin from *Natronomonas pharaonis*

of spatiotemporal activation with optogenetics allows us to study circuit activity dynamics, which is especially critical in motor function.

Larvae can also learn, and optogenetic activation can be used to dissect neural circuits for learning and memory. Activity of octopamine neurons (an invertebrate counterpart of adrenergic ligands) is an appetitive signal for *Drosophila* larvae. By combining optogenetic activation of octopamine neurons and an odor stimulus, it was shown that larvae can form an appetitive memory by Pavlovian classical conditioning (Schroll et al. 2006). Conversely, activity of dopamine neurons is an aversive signal, and optogenetic activation of dopaminergic neurons can substitute for the unconditional stimulus in aversive memory formation (Schroll et al. 2006).

Optogenetics can also be applied to adult flies. Pioneering work in optogenetics used caged-ATP and the P2×2 receptor to show that escape behavior of adult flies can be elicited by activation of the giant fiber system (Lima and Miesenbock 2005). More recently, optogenetics with ChR2 has been applied to the study of the adult fly sensory system, including olfactory (Suh et al. 2007), gustatory (Zhang et al. 2007; Gordon and Scott 2009; Inagaki et al. 2012), and visual (de Vries and Clandinin 2012) circuits. As red light is more penetrating than blue light to the cuticle of adult flies, Inagaki et al. (2014) used red-shifted ChR2 to activate neurons in free-running adult flies. By virtue of the high temporal resolution in optogenetics, the authors showed that the male courtship song can be divided into two components, probabilistic and deterministic. This improved optogenetic probe has opened the opportunity to examine neural circuits for natural behaviors such as courtship behavior in free-running adult flies.

13.5 Conclusion

A history of powerful genetic approaches in *Drosophila* neuroscience laid a foundation for widespread application of optogenetics in fly research. Comprehensive developments of Gal4 lines or LexA lines, which target specific cells, are ongoing in several institutes. UAS lines or LexAop lines that encode optogenetic or Ca imaging probes are also being established in parallel with improvements of the probes. Implementation of complicated but accessible neuronal circuits encoding multiple neural functions, including sensory processing and motor controls, combined with established methods for gene expression, make *Drosophila* an invaluable model system in untangling the functions of neural circuits.

Acknowledgment We thank Mrs. Kasumi Shibahara for help with preparing figures.

References

Aberle H, Haghighi AP, Fetter RD et al (2002) Wishful thinking encodes a BMP type II receptor that regulates synaptic growth in *Drosophila*. Neuron 33(4):545–558
Ataman B, Ashley J, Gorczyca M et al (2008) Rapid activity-dependent modifications in synaptic structure and function require bidirectional Wnt signaling. Neuron 57(5):705–718
Bellen HJ, Tong C, Tsuda H (2010) 100 years of *Drosophila* research and its impact on vertebrate neuroscience: a history lesson for the future. Nat Rev Neurosci 11(7):514–522
Bellen HJ, Levis RW, He Y et al (2011) The *Drosophila* gene disruption project: progress using transposons with distinctive site specificities. Genetics 188(3):731–743
Brand AH, Perrimon N (1993) Targeted gene expression as a means of altering cell fates and generating dominant phenotypes. Development 118(2):401–415
Chen MS, Obar RA, Schroeder CC et al (1991) Multiple forms of dynamin are encoded by shibire, a *Drosophila* gene involved in endocytosis. Nature 351(6327):583–586
de Vries SE, Clandinin TR (2012) Loom-sensitive neurons link computation to action in the *Drosophila* visual system. Curr Biol 22(5):353–362
Gordon MD, Scott K (2009) Motor control in a *Drosophila* taste circuit. Neuron 61(3):373–384
Grigliatti TA, Hall L, Rosenbluth R et al (1973) Temperature-sensitive mutations in *Drosophila* melanogaster. XIV. A selection of immobile adults. Mol Gen Genet 120(2):107–114
Groth AC, Fish M, Nusse R et al (2004) Construction of transgenic *Drosophila* by using the site-specific integrase from phage phiC31. Genetics 166(4):1775–1782
Hamada FN, Rosenzweig M, Kang K et al (2008) An internal thermal sensor controlling temperature preference in *Drosophila*. Nature 454(7201):217–220

Honjo K, Hwang RY, Tracey WD Jr (2012) Optogenetic manipulation of neural circuits and behavior in *Drosophila* larvae. Nat Protoc 7(8):1470–1478

Hwang RY, Zhong L, Xu Y et al (2007) Nociceptive neurons protect *Drosophila* larvae from parasitoid wasps. Curr Biol 17(24):2105–2116

Inada K, Kohsaka H, Takasu E et al (2011) Optical dissection of neural circuits responsible for *Drosophila* larval locomotion with halorhodopsin. PLoS One 6(12):e29019. doi:10.1371/journal.pone.0029019

Inagaki HK, Ben-Tabou de Leon S, Wong AM et al (2012) Visualizing neuromodulation in vivo: TANGO-mapping of dopamine signaling reveals appetite control of sugar sensing. Cell 148(3):583–595

Inagaki HK, Jung Y, Hoopfer ED et al (2014) Optogenetic control of *Drosophila* using a red-shifted channelrhodopsin reveals experience-dependent influences on courtship. Nat Methods 11(3):325–332

Kitamoto T (2001) Conditional modification of behavior in *Drosophila* by targeted expression of a temperature-sensitive shibire allele in defined neurons. J Neurobiol 47(2):81–92

Klapoetke NC, Murata Y, Kim SS et al (2014) Independent optical excitation of distinct neural populations. Nat Methods 11(3):338–346

Kohsaka H, Okusawa S, Itakura Y et al (2012) Development of larval motor circuits in *Drosophila*. Dev Growth Differ 54(3):408–419

Kosaka T, Ikeda K (1983) Possible temperature-dependent blockage of synaptic vesicle recycling induced by a single gene mutation in *Drosophila*. J Neurobiol 14(3):207–225

Lai SL, Lee T (2006) Genetic mosaic with dual binary transcriptional systems in *Drosophila*. Nat Neurosci 9(5):703–709

Levis R, Hazelrigg T, Rubin GM (1985) Effects of genomic position on the expression of transduced copies of the white gene of *Drosophila*. Science 229(4713):558–561

Lima SQ, Miesenbock G (2005) Remote control of behavior through genetically targeted photostimulation of neurons. Cell 121(1):141–152

Lin JY, Knutsen PM, Muller A et al (2013) ReaChR: a red-shifted variant of channelrhodopsin enables deep transcranial optogenetic excitation. Nat Neurosci 16(10):1499–1508

Ljaschenko D, Ehmann N, Kittel RJ (2013) Hebbian plasticity guides maturation of glutamate receptor fields in vivo. Cell Rep 3(5):1407–1413. doi:10.1016/j.celrep.2013.04.003

Markstein M, Pitsouli C, Villalta C et al (2008) Exploiting position effects and the gypsy retrovirus insulator to engineer precisely expressed transgenes. Nat Genet 40(4):476–483

Matsunaga T, Fushiki A, Nose A et al (2013) Optogenetic perturbation of neural activity with laser illumination in semi-intact *Drosophila* larvae in motion. J Vis Exp 77:e50513. doi:10.3791/50513

McKemy DD, Neuhausser WM, Julius D (2002) Identification of a cold receptor reveals a general role for TRP channels in thermosensation. Nature 416(6876):52–58

Peabody NC, Pohl JB, Diao F et al (2009) Characterization of the decision network for wing expansion in *Drosophila* using targeted expression of the TRPM8 channel. J Neurosci 29(11):3343–3353

Pfeiffer BD, Jenett A, Hammonds AS et al (2008) Tools for neuroanatomy and neurogenetics in *Drosophila*. Proc Natl Acad Sci U S A 105(28):9715–9720

Pfeiffer BD, Ngo TT, Hibbard KL et al (2010) Refinement of tools for targeted gene expression in *Drosophila*. Genetics 186(2):735–755

Pulver SR, Pashkovski SL, Hornstein NJ et al (2009) Temporal dynamics of neuronal activation by Channelrhodopsin-2 and TRPA1 determine behavioral output in *Drosophila* larvae. J Neurophysiol 101(6):3075–3088

Rubin GM, Spradling AC (1982) Genetic transformation of *Drosophila* with transposable element vectors. Science 218(4570):348–353

Schroll C, Riemensperger T, Bucher D et al (2006) Light-induced activation of distinct modulatory neurons triggers appetitive or aversive learning in *Drosophila* larvae. Curr Biol 16(17):1741–1747

Shearin HK, Dvarishkis AR, Kozeluh CD et al (2013) Expansion of the gateway multisite recombination cloning toolkit. PLoS One 8(10):e77724. doi:10.1371/journal.pone.0077724

Suh GS, Ben-Tabou de Leon S, Tanimoto H et al (2007) Light activation of an innate olfactory avoidance response in *Drosophila*. Curr Biol 17(10):905–908

Thum AS, Knapek S, Rister J et al (2006) Differential potencies of effector genes in adult *Drosophila*. J Comp Neurol 498(2):194–203

van der Bliek AM, Meyerowitz EM (1991) Dynamin-like protein encoded by the *Drosophila* shibire gene associated with vesicular traffic. Nature 351(6325):411–414

Venken KJ, Simpson JH, Bellen HJ (2011) Genetic manipulation of genes and cells in the nervous system of the fruit fly. Neuron 72(2):202–230

Xiang Y, Yuan Q, Vogt N et al (2010) Light-avoidance-mediating photoreceptors tile the *Drosophila* larval body wall. Nature 468(7326):921–926

Zhang W, Ge W, Wang Z (2007) A toolbox for light control of *Drosophila* behaviors through channelrhodopsin 2-mediated photoactivation of targeted neurons. Eur J Neurosci 26(9):2405–2416

Chapter 14
Optogenetics in *Caenorhabditis elegans*

Yuki Tsukada and Ikue Mori

Abstract With a compact neural circuit consisting of entirely mapped 302 neurons, *Caenorhabditis elegans* plays an important role in the development and application of optogenetics. Application of optogenetics in *C. elegans* accelerates the studies of neural circuits by offering the tools that drastically change experimental designs with increasing accessibility for neural activity. Combination with a different approach, such as electrophysiology, expands the potential of both optogenetics and other approaches by increasing resolution of elucidation. Moreover, the technologies specifically developed in combination with optogenetics, such as patterned illumination, provide new tools to interrogate neural functions. In this chapter, we first introduce the reasons to use optogenetics in *C. elegans* studies, and discuss the technical issues of optogenetics, especially for *C. elegans*. We then review early and recent milestone works using optogenetics to investigate neural and behavioral mechanisms. Note that, in this chapter, the term 'optogenetics' includes both imaging with fluorescence probes and neural activity manipulation using opsins.

Keywords *Caenorhabditis elegans* • *C. elegans* • Neural circuit • Behavioral analysis • Photo-electrophysiology • Tracking • Patterned illumination

14.1 Introduction: Reasons to Use *Caenorhabditis elegans*

Tiny nematode *Caenorhabditis elegans* (Fig. 14.1) has been used as an important model organism for various kinds of biological research for about 40 years (Brenner 1974). Abundant advantages for scientific studies such as a short life cycle, a transparent body, and the convenience of genetic analysis have been pushing various fields of research, including genomics, cell biology, development, aging, and neuroscience.

Y. Tsukada (✉) • I. Mori
Group of Molecular Neurobiology, Division of Biological Science,
Graduate School of Science, Nagoya University, Furo-cho, Chikusa-ku,
Nagoya 464-8602, Japan
e-mail: tsukada.yuki@nucc.cc.nagoya-u.ac.jp

H. Yawo et al. (eds.), *Optogenetics*, DOI 10.1007/978-4-431-55516-2_14

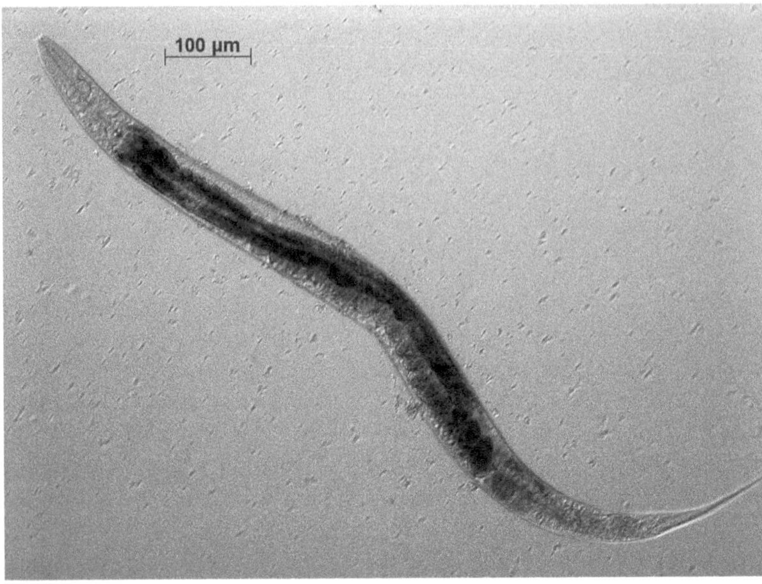

Fig. 14.1 Differential interference contrast (DIC) image of adult hermaphrodite *Caenorhabditis elegans*

Optogenetics is actively and naturally introduced into work with *C. elegans* from the beginning, not only because neuroscience is a prominent region in *C. elegans* research, but also because the characteristics of *C. elegans* are amenable to optogenetics. In this section, we overview the advantages of *C. elegans* as a platform for neuroscience research, especially for using optogenetics.

14.1.1 Transparent Body

Optical manipulation requires the conveyance of specific light stimulus to target molecules. The transparent body of *C. elegans* renders it suitable for optogenetic manipulation with the supply of enough light to the opsins or fluorescent proteins in a body. Thus, in vivo analysis with optogenetic tools is convenient in *C. elegans* because of its transparency and tiny body size. These characteristics are even more useful in optogenetics with behavioral assays in the case of freely migrating animals, since non-invasive and of a low power of photo stimulus is sufficient to drive optogenetic tools inside the body of target animals without altering their natural behavior.

14.1.2 Abundant Promoters

Specific expression of opsins or genetically encoded indicators (GEIs) is a key technology in optogenetics. There are abundant annotated promoters for *C. elegans*, so expression of optogenetic tools in most targeted neurons is promising. Convenient sites exist for searching expression patterns of annotated promoters (Wormbase: http://www.wormbase.org/) or services providing transgenic animals (*C. elegans* Genetic Center (CGC): http://www.cbs.umn.edu/cgc/) and plasmids (plasmid search: http://hokatsu-nou.biochem.s.u-tokyo.ac.jp/index.html [only in Japanese]). Genetic manipulation, such as the generation of transgenic animals, keeping strains as frozen stock, and crossbreeding, are established and the short life cycle helps to generate transgenic lines quickly. Testing a new optogenetic tool in body muscles using *myo-3* promoter is also practical in *C. elegans* because the apparent phenotype is measurable with muscle contraction (Zhang et al. 2007).

14.1.3 The Connectome Has Been Produced

Fully annotated anatomical information for *C. elegans*, including a neural circuit map, is beneficial when planning effective experiments to study neural circuits using optogenetics. Represented by the completely identified cell lineage, details of developmental and anatomical information about *C. elegans* are fully described and available on some web databases (wormatlas: http://www.wormatlas.org/, wormbook: http://www.wormbook.org/). Electron microscopic analysis elucidated whole connections among 302 neurons of the entire neural network in hermaphrodites (White et al. 1986) and 383 neurons in males (Jarrell et al. 2012). Connectomes including the positions of chemical synapses and gap junctions is mapped out and includes the strength of each connection. Furthermore, many visual online databases are available to help in the understanding of circuit properties in the compact neural system of *C. elegans* (*C. elegans* neural net: http://wormweb.org/neuralnet, worm wiring project: http://wormwiring.org/, Database of Synaptic Connectivity of *C. elegans* for computation: http://ims.dse.ibaraki.ac.jp/ccep-tool/).

Based on these annotated neural maps and information about the function of each neuron, one can conduct optogenetics to explore deep into the principles of neural circuits. Noninvasive perturbation and monitoring of each neural activity using opsins and genetically encoded calcium or voltage indicators takes advantage of the complete neural map and should shed light on its functional properties.

14.1.4 Preserving High-Order Functions with a Compact Neural Circuit

The simple neural circuit of *C. elegans* preserves high-order functions such as sensation of environmental stimuli, learning, and memory that regulate apparent quantifiable behavior like chemotaxis or thermotaxis. Therefore, the essential mechanisms of such functions can be investigated using abundant and powerful tools in *C. elegans* research. In addition to classical genetics or physiology, optogenetic tools offer the ability to dissect neural circuit properties by providing perturbation or monitoring tools for individual neurons.

Because of the above, *C. elegans* is regarded as an effective model organism for the study of neural circuits using optogenetics.

14.2 Protocols and Notes

This section describes how to introduce optogenetics into *C. elegans* work. The section includes both worm-specific tips and general topics in optogenetics.

14.2.1 Transfection of Optogenetic Genes

To conduct optogenetics with *C. elegans*, one must have transgenic animals. The published strains expressing optogenetic proteins may be available directly from researchers, or the CGC, which carries out the collection, maintenance, and distribution of *C. elegans* strains. Transgenic animals can be generated via conventional procedures with microinjection to gonads or other methods (Boulin and Hobert 2012). The online protocols, including the creation of transgenic animals, are available on wormbook (wormbook: http://www.wormbook.org/). Although there is no optimized protocol for the generation of transgenic animals expressing optogenetic genes in *C. elegans*, it would be better to have a high enough number of candidates for the transgenic lines, because expression of opsin protein itself does not guarantee it will drive neural manipulation. An appropriate quantity of expressed opsins is critical in some cases.

14.2.2 Supplying All-Trans Retinal (ATR)

Expressed opsin proteins such as channelrhodopsin from *Chlamydomonas reinhardtii* (ChR) and halorhodopsin from *Natronomonas pharaonis* (*Np*HR) need a cofactor, all-*trans* retinal (ATR), to work with photoactivation. Because nematodes do not generate ATR, exogenous supplementation with ATR in the nematode growth medium (NGM) agar or bacterial lawn with ethanol is necessary throughout the development. Note that ATR solution decomposes in about 10 days in room temperature; fresh NGM plates should be used for light-induced neural manipulation. Conversely, NGM plates without ATR serve as a good negative control for transgenic lines as mentioned in the later parts of this chapter.

14.2.3 Notes for Illumination

14.2.3.1 Illumination Devices and Light Paths

Illumination devices have a critical role in optogenetic experiments. Nevertheless, a standard epifluorescence microscope system (Fig. 14.2A) is usually sufficient for illuminating opsins expressed in worms. The requirements of an illumination system for opsins are similar to those for fluorescence probes: to effectively provide an appropriate wavelength, strength, and timing of light to an object. Yet, the purposes of both systems differ slightly in that one is needed for visualization of fluorescence and the other for stimulation of opsins. An illumination system for fluorescence probes focuses on collecting fluorescence light, while illumination for opsins mainly focuses on maximizing light power to supply enough energy to drive opsins.

A complex optogenetic illumination system is necessary when simultaneously monitoring and manipulating neural activity. The selection of dichroic mirrors and filters depends on the combination of absorption and emission spectrums of both fluorescence probes and opsins. A prospective example is the use of genetically encoded calcium indicators (GECIs) and ChR2 (Akerboom et al. 2013). Some approaches exist to avoid interventions between the excitation of florescence probes and the stimulation of opsins: dividing the given light stimuli based on wavelength, strength, location, and timing. Avoiding overlapping absorption wavelengths and/or using different light strengths of fluorescence probes and opsins is critical for the simultaneous monitoring and manipulation of neural activity. Specially designed

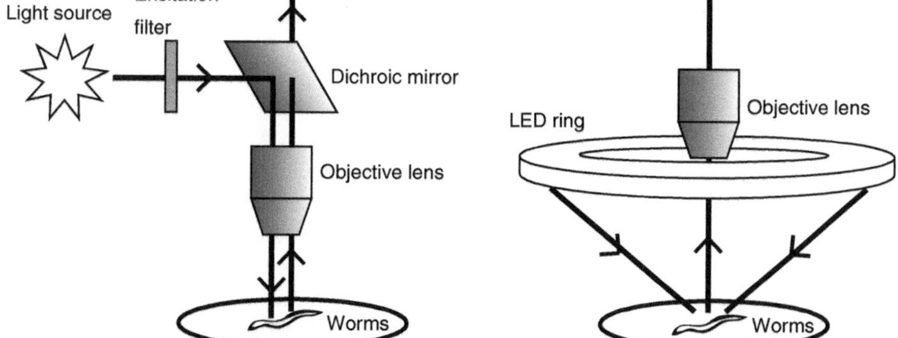

Fig. 14.2 Examples of optogenetic illumination applicable for *Caenorhabditis elegans*. (**a**) Schematic of a standard epifluorescence illumination system. The light source, excitation filter, and dichroic mirror are especially important for optogenetic photo stimulus. The choice of light source or filters is critical for optogenetic manipulation. (**b**) Light-emitting diode (LED) ring illumination: external illumination eliminates a limitation of epifluorescence illumination or even of microscopes. Of course, the combination of epifluorescence and the LED ring is feasible

devices for multi-color or multi-location illumination support simultaneous monitoring and manipulation with optogenetics. Spatial and temporal division of photostimulation using patterned illumination is described in later sections.

Another approach independently illuminates opsins without a microscope or epifluorescence system. Kawazoe et al. used a customized high-power light-emitting diode (LED) ring to regulate improved ChR, called channelrhodopsin green receiver (ChRGR) expressed in body wall muscle (Fig. 14.2b) (Kawazoe et al. 2012). An independent illumination device is free of the restrictions of epifluorescence and microscope systems. The application of such an illuminating device for opsins extends the possible experimental variations, for example, conditioning for the population of worms in specific timing with optogenetics. The application of optogenetics is thus expanded in combination with suitable illuminating devices.

14.2.3.2 Avoiding Heat from Illumination

Opsins often need several mW/mm^2 of high-power illumination to drive their functions in neurons. The heat caused by the illumination per se will be a problem for dealing with small volumes of *C. elegans*. In particular, it will be critical for behavioral experiments because a known behavior of worms is heat avoidance. Some approaches solve or reduce the effects of heat: shortening the illumination exposure time and/or using an external cooling device such as a metal plate or Peltier device are valuable to consider. To ensure that the heat generated by light stimulation does not induce any artifacts, it is useful to show whether the events could be induced only in the presence of ATR and not in the absence of ATR.

14.2.3.3 Light Toxicity in Short Wave Length

Wild-type *C. elegans* exhibits apparent avoidance behavior to ultraviolet (UV), and blue light. Moreover, continuous UV lighting even causes death to *C. elegans* in 10 min. For these reasons, it is important to discern optogenetic manipulation of neural activity from light-induced responses in *C. elegans*. An eight-transmembrane protein, LITE-1, is identified as a main regulator of the avoidance behavior evoked by light (Edwards et al. 2008), although *C. elegans* lacks a photoreceptive organ or eye. Therefore, *lite-1* mutant strains are commonly used for behavioral optogenetics, as the *lite-1* mutants reduce light-induced avoidance behavior. It should be noted that there is *lite-1*-independent light-evoked behavior, and the use of the *lite-1* background strain does not assure the complete suppression of light avoidance (Edwards et al. 2008). Also, similar to the heat problem, a negative control experiment without ATR is useful to judge whether observed phenomenon are caused by opsins or not.

14.2.3.4 Regulation of Illumination

The regulation of illumination is one of the most important factors for successful optogenetic experiments. Synchronization of the illumination device and camera is critical for fluorescence imaging, especially in combination with neural activity manipulation because the light power used in imaging and in optical manipulation is quite different. Also, precise control of illumination timing and power is necessary to find the optimum light stimulation for neural manipulation, and ON/OFF states of illumination timestamps validate the correlation between a given optogenetic manipulation and its outcome. In addition, the millisecond order control of illumination has the potential to improve the efficiency of long-term optogenetic manipulation of neural activity. *Np*HR shows a decrease in efficiency when continuous light stimulation is used to evoke neural inhibition (Zhang et al. 2007). However, millisecond order pulsed light restores (Zhang et al. 2007) or prevents (Kuhara et al. 2011) this decrease, and thus regulation of illumination increases the performance of opsins.

14.3 Applications of Optogenetics in *C. elegans* Research

This section introduces the applications of optogenetics in *C. elegans* studies, and discusses the present situation and prospects.

14.3.1 C. elegans *Optogenetics as Pioneering Works*

As mentioned in the beginning of this chapter, *C. elegans* provides many advantageous characteristics to investigate multicellular organisms, and many important pioneering works have emerged with *C. elegans*. The pioneering works with GFP, the whole genome project, and the complete connectome of neural circuits clearly illustrate the significance of *C. elegans* as a model animal serving as a platform for studies on living systems. In terms of the use of opsin proteins for neural manipulation, *C. elegans* was the first multicellular organism in which behavioral regulation by a ChR was investigated (Nagel et al. 2005). This was in the same year as the first neural manipulation using hippocampal cell culture (Boyden et al. 2005).

Nagel el al. expressed a gain-of-function mutant of ChR2, the histidine of which at position 134 was altered to an arginine residue (H134R) in order to have high depolarization by illumination, into the body wall and egg-laying muscle using *myo-3* promoter. Transgenic worms with ATR show muscle contraction at 500 ms of illumination, showing that exogenous ATR works on transgenic strains expressing ChR2. With the advantage of many available mutants, Nagel et al. also elucidated that L-type voltage-gated Ca^{2+}-channels (VGCCs) and ryanodine receptors (RyRs) regulate this ChR2-mediated muscle contraction. Further, light-evoked

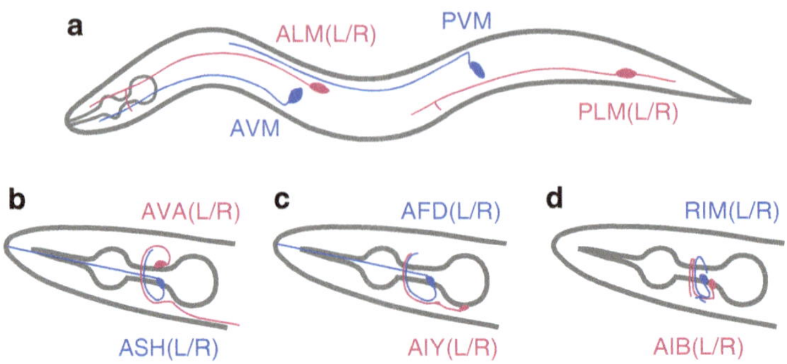

Fig. 14.3 Positions and projections of neurons. All figures show heads in left. *Blue* and *red* colors are used to discriminate closely located individual neurons. *(L/R)* denotes left and right pair of neurons; however, one is omitted for legibility. (**a**) Mechanosensitive nociceptive neurons labeled by *mec-4* promoter-regulated expression. (**b**) Nociceptive sensory neuron ASH and interneuron AVA. (**c**) Thermosensory neuron AFD and interneuron AIY. (**d**) Interneuron AIB and motor neuron RIM

expression of ChR2 (H134R) in mechanosensory neurons ALML, ALMR, PLML, PLMR, AVM, and PVM using *mec-4* promoter (Fig. 14.3a) induced withdrawal behavior in a similar way to mechanical stimulation.

This study clearly showed temporal- and cell-specific manipulation of neural or muscular activity in a living multicellular organism using optogenetics. This first optogenetic study of *C. elegans* symbolically shows the good compatibility of optogenetics with *C. elegans*. The study thoroughly used the merits of *C. elegans*, such as identifying the neurons required for withdrawal behavior, and promoters that regulate gene expression in the identified neurons and muscles. As shown here, accumulation of genetic, cellular, and behavioral levels of information and technologies enables the effective use of optogenetics.

14.3.2 Combination of Electrophysiology and Optogenetics

The complete identification of *C. elegans* neural circuit connectivity is pushing neuroscience research to the new challenge of a functional connectome. The combination of electrophysiology and optogenetics, 'photo-electrophysiology', provides a powerful toolbox to elucidate functional relationships among neurons on the complete neural map (Husson et al. 2013). Photo-electrophysiology extends electrophysiological dissection by arbitrary selection of a neuron to manipulate activity and thus examine the functional properties between target neural connections.

As an example of the target of photo-electrophysiology, ASH sensory neurons and AVA interneurons are known to work as part of the nociceptive neural network (Fig. 14.3b). The functional relationship between these neurons, i.e., how the sensory signal received at ASH transmits to AVA, was unclear. For this problem, to access multiple neurons for electrophysiology is technically challenging due to the difficulty of surgery with hard cuticle and the small size of the neurons in *C. elegans*.

Lindsay et al. expressed ChR2 in presynaptic ASH neurons using an *sra-6* promoter, and recorded postsynaptic AVA neural activity with electrodes during light-evoked ASH activation (Fig. 14.4a). The results showed gradual synaptic transmission between these two types of neurons and a graded probabilistic change of reversal behavior by photo-stimulation (Lindsay et al. 2011). The combination of electrophysiology, behavioral analysis, and molecular genetics delivered the conclusion that stronger aversive stimulus elicits a greater number of escape responses and depends on a glutamate signal. This approach clearly shows the power of photo-electrophysiology to dissect neural circuit mechanisms.

Graded synaptic transmission is not only found in severe aversive stimulus but also in relatively mild environmental sensory systems, such as thermosensation. A sensory neuron, AFD is known to work as a main thermosensory neuron and is connected to AIY interneurons; both types of neurons are known to play important roles in thermotaxis (Fig. 14.3c) (Mori and Ohshima 1995). Narayan et al. (2011) showed graded transmission between AFD and AIY by expressing ChR2 in AFD using *gcy-8* promoter, and they recorded postsynaptic current (PSC) and postsynaptic potential (PSP) in AIY (Fig. 14.4b). AFD thermosensory neurons sense temperature differences at least as small as 0.05 °C (Clark et al. 2006), and such faint signals

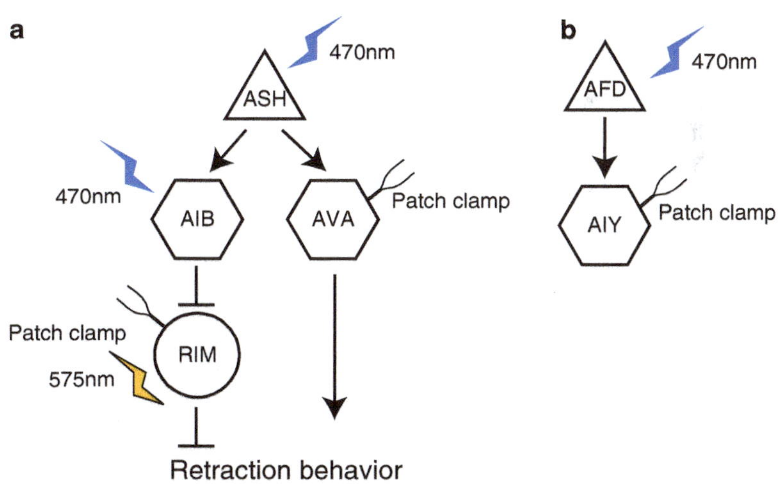

Fig. 14.4 Neural network diagrams and schemes of the photo-electrophysiologies. *Triangles*, *hexagons*, and *circles* indicate sensory, inter-, and motor-neurons, respectively. (**a**) Photo-electrophysiology for the nociceptive neural circuit. Photo-stimulation of upstream neurons (ASH and AIB) evokes activity from downstream neurons (AVA and RIM). Piggot et al. quantified reversal behavior when they stimulated RIM with photo-stimulation. (**b**) Photo-electrophysiology for the thermosensory neural circuit. The synaptic connection between AFD sensory neurons and AIY interneurons was measured by stimulating AFD with light, and conducting a whole cell patch clamp with AIY

received by AFD might be amplified in AFD. The graded signal transmission between AFD and AIY implies that the amplified signal in AFD is scaled in a quantitative manner. The graded transmissions in two different sensory circuits suggests the general scaling mechanism between sensory and inter neurons for information integration of different sensory inputs.

Photo-electrophysiology is also applicable to the output of neural information processing such as motor circuits. A backward regulatory neural circuit consists of AIB interneuron and RIM motorneuron under several sensory neurons (Fig. 14.3d). Piggot et al. expressed ChR2 and NpHR in AIB and RIM, respectively; they uncovered an inhibitory pathway between AIB and RIM using photo-manipulation of these neurons with tracking and photo-electrophysiology (Fig. 14.4a). This study also used a tracking system to analyze free behavioral animals as described in the next section.

14.3.3 Combination with Tracking System

The application of noninvasive manipulation of neural activity expands to studies of behavioral regulation when it is combined with a tracking microscope. The transparent body and slow migration speed of *C. elegans* are suitable characteristics for such experiments. However, microscopes are required to be able to effectively observe or illuminate the small body of *C. elegans*. Then, the tracking system is required to keep a target animal within a microscopic field during neural manipulation and observation. Accordingly, several laboratories have developed various tracking systems for *C. elegans* for dissecting neural circuits responsible for behavioral regulation (Faumont et al. 2011; Kuhara et al. 2011; Piggott et al. 2011; Kocabas et al. 2012).

Thermotaxis of *C. elegans* is an intriguing behavior that includes the mechanism of learning and memory: worms store cultivated temperature when they were previously cultivated for a few hours at a certain temperature with plenty of food, and these conditioned worms move to the cultivated temperature when they are on a thermal gradient without food. Although the neural circuit related to the thermotaxis is already identified (Mori and Ohshima 1995), how the information processing occurs inside the thermotaxis neural circuit is still unknown. Kuhara et al. attenuated the activity of the AFD thermosensory neuron by expressing *Np*HR in AFD, and then showed behavioral change that depended on the strength of AFD neuronal activity (Kuhara et al. 2011). Using the automated tracking system with feedback regulation of a motorized microscopic stage, Kuhara et al. analyzed the tracks of freely migrating *C. elegans* on an agar plate for 1 h (Fig. 14.5). This system enables the recording of behavioral status during free movement, and also allows for long periods of photo stimulation. The records of motorized-stage regulation provide traces of migrating worms as behavioral data. This approach is generally applicable to any other behavioral analysis.

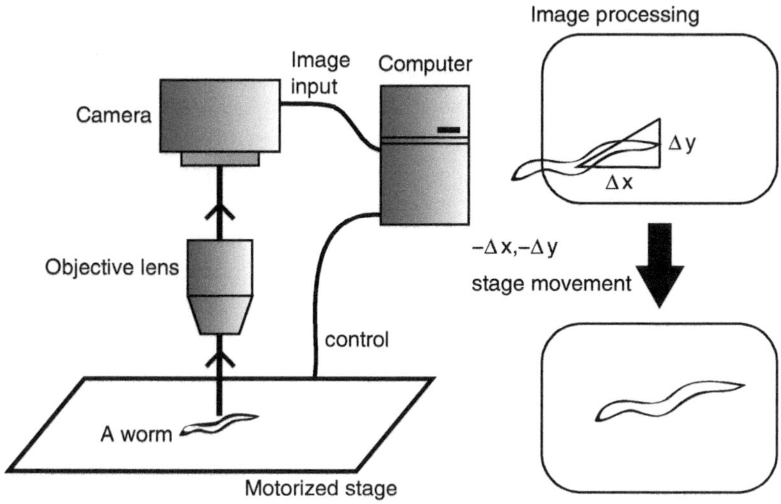

Fig. 14.5 An example of a tracking system. Real-time feedback regulation using a camera and a motorized stage enable the tracking of individual *Caenorhabditis elegans*. Coordinates of the center of mass for a binary image of the worm body or a target neuron is a convenient variable to keep the tracking animal in the microscopic field. A record of stage regulation denotes a trace of the migrating history of the tracking worm. Combination with a movie acquired by camera input helps to further behavioral analysis

Faumont et al. (2011) adopted a four-quadrant photomultiplier tube (PMT) as a tracking system instead of cameras. PMT reduces exposure, data reading, and image processing time, thereby enabling a tracking system to quickly track a moving worm under high-magnification objective lens. The tracking system captures fluorescence signals of a target neuron in a freely moving *C. elegans* using 63× objective lens. Photo-manipulation of ASH neuron with this tracking system showed imitating osmotic avoidance behavior caused by ASH activity. In addition, the tracking system enabled the quantification of the activities of reverse command neuron AVA and forward command neuron AVB by expressing ratiometric calcium sensors (TN-XL or YC3.60) in AVA and AVB. The combination of tracking systems and optogenetics should expand the ability of the methodology, and shed light on the neural mechanism related to behavioral regulation.

14.3.4 Calcium Imaging and Neural Manipulation

In the same way as optogenetic neural activity manipulation, GECIs are impacting on neuroscience studies. Förster resonance energy transfer (FRET)-based calcium sensors called cameleons (Miyawaki et al. 1997) and circular permutated-based GCaMPs (Nakai et al. 2001) are remarkable examples of genetically encoded neural activity sensors that are popular in neuroscience. *C. elegans* is compatible with these fluorescent probes as discussed in the beginning of this chapter. Because calcium imaging and optogenetic operation realize complete noninvasive monitoring and manipulation of neural activity, a method of combining these approaches is valuable for behavioral neuroscience. Moreover, such noninvasive approaches expand the targets to freely moving animals in combination with the tracking system described above. Although these approaches have technological or engineering difficulties that remain unsolved, they are unveiling functional meanings of neural activities.

A tracking and calcium imaging approach obtains quantitative data, and thus expands the scope of analysis to continuous phenomena. Kocabas et al. (2012) focused on the mechanism of navigation during chemotaxis, and found that AIY interneurons control gradual curves in navigation. They expressed ChR2 and GCaMP3 in AIY neurons, and confirmed the increase of AIY activity using photostimulation with calcium imaging. They then manipulated navigation behavior by activating AIY with ChR2. Although the experiments did not show the relationship between the intact activity of AIY and behavior, the results did show that the manipulated activity of AIY neurons is sufficient to navigate migrating *C. elegans*.

Representing a good combination of calcium imaging and optogenetic manipulation, some of the studies discussed in this chapter have also used calcium imaging with optogenetic manipulation (Faumont et al. 2011; Piggott et al. 2011). Since the combination of multiple technologies often provides solutions to overcome technical difficulties, the multi-technological approach serves as a driving force to elucidate remaining problems in modern neuroscience. Fluorescence imaging and optogenetic manipulation offers combinatorial components to monitor or manipulate neural activity by providing several options for probes and opsins that have different characteristics (Yizhar et al. 2011). The flourishing development of new probes and opsins should continue to expand their abilities.

14.3.5 Patterned Illumination Devices

Patterned illumination expands the application of optogenetics by illuminating multiple objects with different wavelengths in a specific spatio-temporal pattern. Fortunately, a hurdle to making pattern illumination systems has lowered because the popularization of digital projectors increased the performance of the components and decreased its cost. Digital micro-mirror devices (DMDs) and liquid crystal displays (LCDs) were used as a pattern illumination device for optogenetics (Guo et al. 2009; Leifer et al. 2011; Stirman et al. 2011). These devices enabled the simultaneous illumination of multiple neurons or muscles of spatially different positions, and thus the system complements the specificity of optogenetic manipulation beyond promoter-dependent regulation of gene localization.

Leifer et al. (2011) developed a system called Control Locomotion and Behavior in Real Time (CoLBeRT), which performs patterned light stimulus to a freely migrating worm with DMD. The worms expressing ChR2 and *Np*HR in body wall muscles, motor neurons, and mechanosensory neurons were used to validate the performance of the CoLBeRT system. Patterned illumination discriminates opsins in different locations and, consequently, enables to interrogate specific cells. Similarly to that described above, they used *mec-4* promoter to express ChR2 in ALM (L/R), PLM (L/R), AVM, and PVM neurons. Patterned illumination distinguished different roles in these neurons by selectively activating opsins in each neuron: PLM promotes forward movement, AVM and ALM promotes backward movement, and AVM and ALM adapt to stimulus as shown previously (Chalfie et al. 1985).

Stirman et al. (2011) used liquid crystal display (LCD) to provide patterned illumination in a relatively easy and less expensive manner. Similar to the study with CoLBeRT, *mec-4* promoter-driven opsin protein expression was dissected via patterned illumination. A mechanism of integration of localized neural input for mechanosensory avoidance behavior was thus examined. Moreover, Stirman et al. established the simultaneous activation of different opsins such as ChR2 and Mac, that have different excitation peaks, with simultaneous illumination of different wavelengths at different places. This study provides a good example of optogenetic neural manipulation in multiple colors and positions. Thus, the combination of optogenetics and these light manipulation technologies promotes studies for a functional connectome.

14.4 Conclusion

Studies using the convenient characteristics of *C. elegans* have greatly contributed to the establishment of optogenetics. At the same time, optogenetics has become one of the general tools in *C. elegans* neuroscience. Synergistic effects accelerate the development of upcoming new technologies and consolidate the basis of optogenetic methodologies. These methodological progresses continue to support pioneering research in neural circuits and behavioral regulation.

References

Akerboom J, Carreras Calderón N, Tian L et al (2013) Genetically encoded calcium indicators for multi-color neural activity imaging and combination with optogenetics. Front Mol Neurosci 6:2. doi:10.3389/fnmol.2013.00002

Boulin T, Hobert O (2012) From genes to function: the *C. elegans* genetic toolbox. WIREs Dev Biol 1:114–137

Boyden ES, Zhang F, Bamberg E et al (2005) Millisecond-timescale, genetically targeted optical control of neural activity. Nat Neurosci 8:1263–1268

Brenner S (1974) The genetics of *Caenorhabditis elegans*. Genetics 77:71–94

Chalfie M, Sulston JE, White JG et al (1985) The neural circuit for touch sensitivity in *Caenorhabditis elegans*. J Neurosci 5:956–964

Clark DA, Biron D, Sengupta P et al (2006) The AFD sensory neurons encode multiple functions underlying thermotactic behavior in *Caenorhabditis elegans*. J Neurosci 26:7444–7451
Edwards SL, Charlie NK, Milfort MC et al (2008) A novel molecular solution for ultraviolet light detection in *Caenorhabditis elegans*. PLoS Biol. doi:10.1371/journal.pbio.0060198
Faumont S, Rondeau G, Thiele TR et al (2011) An image-free opto-mechanical system for creating virtual environments and imaging neuronal activity in freely moving *Caenorhabditis elegans*. PLoS One 6:e24666. doi:10.1371/journal.pone.0024666
Guo ZV, Hart AC, Ramanathan S (2009) Optical interrogation of neural circuits in *Caenorhabditis elegans*. Nat Methods 6:891–896
Husson SJ, Gottschalk A, Leifer AM (2013) Optogenetic manipulation of neural activity in *C. elegans*: from synapse to circuits and behaviour. Biol Cell 105:235–250
Jarrell T a, Wang Y, Bloniarz AE et al (2012) The connectome of a decision-making neural network. Science 337:437–444
Kawazoe Y, Yawo H, Kimura KD (2012) A simple optogenetic system for behavioral analysis of freely moving small animals. Neurosci Res 75:65–68
Kocabas A, Shen C-H, Guo ZV et al (2012) Controlling interneuron activity in *Caenorhabditis elegans* to evoke chemotactic behaviour. Nature 490:273–277
Kuhara A, Ohnishi N, Shimowada T et al (2011) Neural coding in a single sensory neuron controlling opposite seeking behaviours in *Caenorhabditis elegans*. Nat Commun 2:355. doi:10.1038/ncomms1352
Leifer AM, Fang-Yen C, Gershow M et al (2011) Optogenetic manipulation of neural activity in freely moving *Caenorhabditis elegans*. Nat Methods 8:147–152
Lindsay TH, Thiele TR, Lockery SR (2011) Optogenetic analysis of synaptic transmission in the central nervous system of the nematode *Caenorhabditis elegans*. Nat Commun 2:306. doi:10.1038/ncomms1304
Miyawaki A, Llopis J, Heim R et al (1997) Fluorescent indicators for Ca^{2+} based on green fluorescent proteins and calmodulin. Nature 388:882–887
Mori I, Ohshima Y (1995) Neural regulation of thermotaxis in *Caenorhabditis elegans*. Nature 376:344–348
Nagel G, Brauner M, Liewald JF et al (2005) Light activation of channelrhodopsin-2 in excitable cells of *Caenorhabditis elegans* triggers rapid behavioral responses. Curr Biol 15:2279–2284
Nakai J, Ohkura M, Imoto K (2001) A high signal-to-noise Ca(2+) probe composed of a single green fluorescent protein. Nat Biotechnol 19:137–141
Narayan A, Laurent G, Sternberg PW (2011) Transfer characteristics of a thermosensory synapse in *Caenorhabditis elegans*. Proc Natl Acad Sci U S A 108:9667–9672. doi:10.1073/pnas.1106617108
Piggott BJ, Liu J, Feng Z et al (2011) The neural circuits and synaptic mechanisms underlying motor initiation in *C. elegans*. Cell 147:922–933
Stirman JN, Crane MM, Husson SJ et al (2011) Real-time multimodal optical control of neurons and muscles in freely behaving *Caenorhabditis elegans*. Nat Methods 8:153–158
White JG, Southgate E, Thomson JN et al (1986) The structure of the nervous system of the nematode *Caenorhabditis elegans*. Philos Trans R Soc Lond B Biol Sci 314:1–340
Yizhar O, Fenno LE, Davidson TJ et al (2011) Primer optogenetics in neural systems. Neuron 71:9–34
Zhang F, Wang L-P, Brauner M et al (2007) Multimodal fast optical interrogation of neural circuitry. Nature 446:633–639

Chapter 15
Optogenetics Research Using the Mouse as a Model System

Kenji F. Tanaka

Abstract Smooth progress in the field of optogenetic research can be, as the suffix '-genetics' indicates, expected using biological models that provide conditions suitable for genetic control or gene manipulation. The artificial expression of opsin molecules is expected to be carried out smoothly in mice compared with rats or non-human primates because of established inheritable genetic modification technologies in this species. However, when compared with the nematode or the fly, there are still many difficulties in the expression of opsin in mice if flexible gene manipulation is attempted.

Keywords Mouse genetics • Tet-system • Cre-loxP system • BAC transgenic • KENGE-tet • β-actin gene

15.1 Introduction

There are difficulties in obtaining the expression of appreciable amounts of opsin in specifically targeted cell populations in the mouse compared with in the nematode or the fly. Why is it necessary to manipulate specifically targeted cells in mice despite the existence of such difficulties? In the first half of this article, I present the historical evolution of how early neuroscientists progressed research in brain science. In the process, I also show the reason for the necessity of manipulating activities of specific populations of cells. In the latter half of this article, I provide a list of currently available genetically modified mice that express opsin molecules.

K.F. Tanaka (✉)
Department of Neuropsychiatry, Keio University School of Medicine, Tokyo, Japan
e-mail: kftanaka@a8.keio.jp

H. Yawo et al. (eds.), *Optogenetics*, DOI 10.1007/978-4-431-55516-2_15

15.2 Why Do Neuroscientists Focus on Cell-Type-Specific Expression of Target Molecules?

What type of organ is the brain? Gross observation was the only way to answer this question in the pre-microscopy days. By this method, neuroscientists applied the terms gray matter or white matter to particular areas of the brain. Under gross observation of the brain, areas rich in dendrites or synapses appeared gray, while myelinated axons appeared white because of a high lipid composition in myelin. Similarly, the region rich in dopamine neurons appeared dark and was thus termed the substantia nigra; the region rich in noradrenergic neurons appeared blue and was named the locus coeruleus.

After the establishment of microscopy, neuroscientists were able to observe the micro-world in addition to the macroscopic field. However, when magnifying tiny or thin objects using a microscope, the objects appeared pale, colorless, or transparent. It was almost impossible to distinguish the differences between objects being observed. To approach this problem, staining techniques were conceived and applied. Nissl and Golgi staining were introduced as methods for staining neuronal cells in the late nineteenth century (Fig. 15.1). Nissl staining dyes ribonucleic acids,

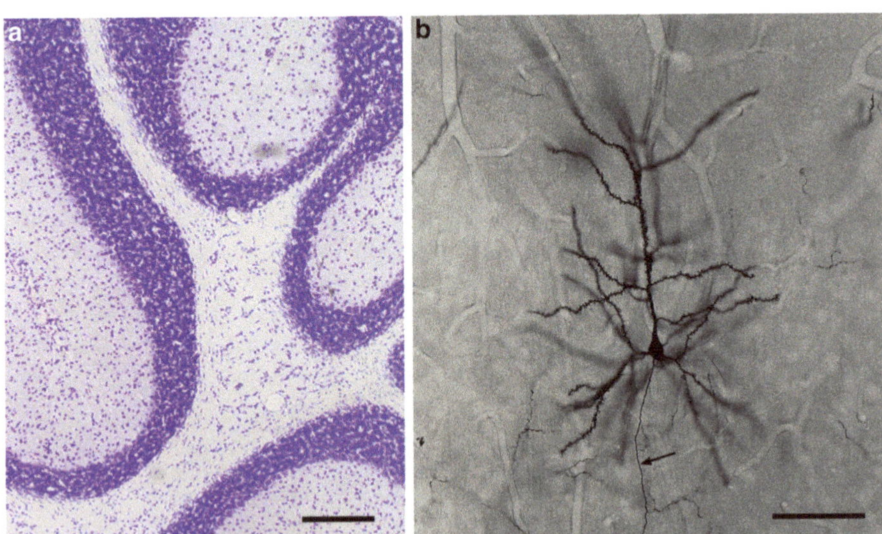

Fig. 15.1 Methods of identification of brain cells. (**a**) Nissl staining of the mouse brain. Each spot (5–20 μm in diameter) represents a single cell body. Zonal congestion of cell bodies is found in the cerebellum. Nissl staining can show the location of cell bodies but fails to indicate their shapes. Bar: 200 μm. (**b**) Golgi staining of the mouse cerebral cortex (Courtesy of Dr. Kiyoshi Hama, Professor emeritus, National Institute for Physiological Sciences). Only one cortical pyramidal cell was stained by this method. The *arrow* indicates the axon. All the other projections are dendrites. As shown in this picture, Golgi staining can stain only a small proportion of cells, but can reveal the shapes of cells in detail once stained. Bar: 100 μm

allowing for clear observation of cell bodies and of the position of cells. This staining method reveals the cell body of each cell clearly but cannot distinguish shapes of the cells. For example, imagine two people sitting side by side on a bench. Nissl stain reveals only the cell bodies, in this metaphor that would be the hearts of the two people. The image you can see is merely two hearts floating 70 cm apart on the bench. You may have a vague idea of two old people spending time in the sun or perhaps two happy young lovers holding hands. However, you cannot see which is right. The Golgi staining method, on the other hand, strongly stains cytoskeletons, allowing the shapes of cells to be distinguished clearly. However, Golgi staining can stain only one cell in every several hundred cells. Therefore, the alignment of cells or shapes of neighboring cells cannot be distinguished. Imagine that the person sitting on the left side of the bench is looking at their right knee in the above metaphor. They seem to be holding something in their left hand and their right forefinger is pointing to it. It looks as if this person is explaining something to the other person about the object in his/her left hand; it may be a picture or an iPad.

As seen in the above historical background, by the late nineteenth century to the early twentieth century, scientists needed to find a way to observe how cells are shaped in specific regions and how cells are arranged relative to each other in the brain. Santiago Ramón y Cajal, known as the father of neuroanatomy, left a huge catalog of drawings of brain cells for future generations. He indicated that a wide variety of cell types existed in the brain and showed that these cells could be categorized to some extent by their characteristic shapes. He forecasted that each group of cells plays a specific role, depending on their type.

15.3 Shapes, Functions, and Gene Expression of Brain Cells

Cell shapes are formed to perform their specific functions. Specific genes expressed in cells encode functional expression and morphogenesis of the cells. In other words, cells expressing a specific gene can perform a specified function with a specified shape. Based on this philosophy, approaches have been attempted from approximately 1980 to determine distributions and shapes of brain cells expressing specific genes. In early studies, the expression of specific genes (messenger RNA [mRNA]) and their encoded proteins were investigated via in situ hybridization and immunohistochemistry (or immuno-staining), respectively. For example, neurons that release dopamine were determined to distribute to the substantia nigra by in situ hybridization of the gene *tyrosine hydroxylase*, which is essential for dopamine synthesis. Furthermore, extensions of neuronal dendrites and neurites were found via immunohistochemical staining. Moreover, these techniques enabled anatomical information about specific neuronal cell populations that expresses specific genes or gene products to be obtained. Although it was apparent that these techniques were superior to Nissl or Golgi staining, neuroscientists were still not satisfied with the situation: these methods obtained information about no more than dead neuronal cells.

15.4 Understanding the Activities of Living Neuronal Cells

Neurons communicate with each other using electrical signals. Different electrical signals convey different messages to the other cells. One of the major goals of neuroscience is to understand the relationships between the mode of signal transmission and changes in activities of neuronal cells. Again, imagine the above metaphor of a couple sitting on a bench. If you see a still picture that shows a woman's hand stopping close to the man's cheek, it is impossible to read whether she is about to slap him on his cheek or if she is cleaning lint off his cheek. However, if the movement of her right arm can be recorded dynamically, it will be easy to distinguish whether or not the movement of the women's hand was a slap.

Real-time monitoring of nervous cell activities can be achieved using electrophysiological techniques. This is particularly easy if monitoring is not targeted to a specific cell population. However, if any neuronal cell populations that express specific genes are targeted, monitoring the activities of such cells will be technically much more difficult.

The introduction of green fluorescent protein (GFP) to neuroscience has made it much easier to study the electrical activity of particular neuronal cell populations. By expressing GFP under the control of a specific gene promoter, GFP will be expressed only in cell populations that express that gene. Thus, living cells can be identified by fluorescence of the expressed GFP, distribution of the particular type of cell population can be determined, and the electrical activity can be measured.

Expression of GFP under specific promoters was introduced to neuroscience in the 1990s. Today, a mouse resource is available (GENSAT) that organizes information on the expression of GFP under the control of a specific gene promoter (Gong et al. 2010). By entering the name of a gene into the query in the GENSAT home page (http://www.gensat.org/), the distribution and pattern of the gene-expressing cells can be found with one click. Moreover, GFP mouse lines shown in the database can be purchased without any restrictions. This indicates that by using GFP-expressing mice, everybody has an opportunity to monitor the activities of particular types of cells via GFP fluorescence.

15.5 Once Neuronal Cell Activities Are Detected, We Scientists Tend to Manipulate Them

By using gene promoters, it will be possible to understand how specific neuronal cell populations are positioned and how they extend their neuronal processes in a living state. Activities of such specific cell populations can be monitored in case-by-case situations of activities such as anger, laughing, or running. To draw an analogy, in an orchestra, beautiful music is produced by the harmony of different instruments in the ensemble. For example, in a heroic piece of music, the generation of the feeling of heroism may be understood by analyzing each tone of every instrument. This process is similar to the above-mentioned "analyzing activities of particular neuronal cell populations." Imagine that analysis of the music reveals that the pianist played weakly for lower pitches. Does it mean that the piano part was the major component for generating the feeling of heroism? That is not necessarily the case. Unfortunately, only analyzing single tones cannot explain an entire musical piece.

To address this question empirically, the playing style or keystroke of the pianist should be modified. By changing how a specific instrument is played, the role for each note of that instrument in the entire ensemble may be understood. For instance, it may be found that a stronger piano keystroke results in calmness and lesser feeling of bravery in the entire musical piece, or perhaps playing octave tones results in cruder music.

After any relationship has been clarified between activities of a particular population of neuronal cells and behavior of the animal, then we scientists tend to be interested in more detailed causal association; whether or not the activity of the particular neuronal cell population directly controls the behavior of the animal. To elucidate the causal relationships between neuronal cell activity and animal behavior, it is necessary for the investigator to manipulate the activities of neural cell populations. As in the above metaphor of playing a piano in an orchestra, the presence or absence of a causal relationship can be indicated only after manipulation of method or activity.

Before optogenetics was developed, no appropriate technology was available to manipulate activities of neural cell populations, even after the cell activity was revealed. Thus, causal associations could not be identified from the apparent relationship between the activities of particular neuronal cell populations and animal behavior. Optogenetics was the long-awaited investigational tool that made manipulation of neuronal cell activity feasible.

15.6 Optogenetics in Mice; a Challenging Subject

Karl Deisseroth first reported optogenetics in 2005 using cultured neuronal cells (Boyden et al. 2005). Optogenetic research using living mice was first reported in 2007. In this study, the *Thy1* promoter was used to express a target molecule in a particular subset of neuronal cells and channelrhodopsin (ChR)-2 transgenic mice were generated (Arenkiel et al. 2007). The *Thy1* promoter was a promoter of choice because of its strong expression and wide distribution in many types of neuronal cells, but not specifically to a particular type of cell population. Since ChR2 molecules are expressed in neuronal cell populations that control the activity of the motor area, manipulation of movement was capable by illuminating the cells with guided light. This treatment induced a prompt motor action in the mouse. This was recorded in a movie and was accessible via YouTube. This experiment clearly indicated the possibility of "cell-type-specific manipulation of neuronal cells" using the *Thy1* promoter and ChR2 transgenic mice.

In the academic field of neuroscience, mice expressing GFP in a cell-type-specific manner have been generated for research purposes. Thus, it was expected that obtaining ChR2-expressing mice would be simple by exchanging GFP and ChR2 molecules. If a useful ChR2 transgenic mouse is generated, studies can be carried out conveniently in any laboratory, since mice will be freely available through mating. Furthermore, by various measures to illuminate light on the mouse, studies of "cell-type-specific manipulation of neuronal cells" are feasible. Many neuroscientists dreamed this delusional idea.

To be honest, I was one of those dreamers. As a psychiatrist, I was interested in understanding the activity of the serotonergic nervous system using optogenetics. Manipulation of serotonergic neurons would answer a wide range of important questions in the field.

15.7 Facing a Delusional Idea and Reality, Then Breakthrough

The delusional idea I fell into was that I believed suitable mice could be prepared without much difficulty. ChR2 conductance was reported to be 50–250 fs (Lin et al. 2009), indicating a smaller conductance than those of most ionic channels reported until then. This fact suggested that a "more than a moderate level of expression" of ChR2 would be needed to control the membrane potential. Although this information had been recognized early, scientists were not clear as to how much ChR2 should be expressed on the cell membrane to successfully control the membrane potential. A higher level of expression was considered to be better than less expression. However, aggregation of ChR2 molecules was a concern for having too much expression. Scientists were confused by the variety of information. Furthermore, it was suggested that ChR2 molecules stayed in the endoplasmic reticulum and did not reach the cell membrane, causing failure of detection of photocurrents. To evaluate this possibility, expression of a ChR2-GFP fusion protein was attempted. This made the already difficult situation more difficult. Expression of a ChR2-GFP fusion protein was a challenging subject when compared with expression of ChR2 molecules alone.

It was surprising to hear an experiment that attempted to express ChR2 in the *ROSA* locus, which was known to be able to express exogenous genes in any cell type, failed. In this experiment, Richard Axel, a Nobel Prize winner in medicine for his discovery of odorant receptors, attempted to express ChR2 molecules specifically in odorant receptor-expressing cells by using a combined system of the *ROSA* locus with Cre-loxP system. However, they made an unsuccessful attempt because ChR2 expression was too low to be used in their study. Until that time, the *ROSA* locus had been considered a proven promoter for expression of GFP or YFP molecules. In fact, direct fluorescence of YFP was utilized in electrophysiological experiments using living cells. However, the expression of ChR2-GFP fusion protein was too low, and direct fluorescence could not be detected. Later it became clear that at least tenfold induction power was needed to express GFP-ChR2 fusion protein at similar levels of light (direct fluorescence) as free GFP.

To establish both cell-type-specific expression and sufficient amounts of expression simultaneously, I have adopted an expression induction system using a tetracycline-controllable gene expression system (tet system) (Gossen and Bujard 1992) for expression of a sufficient amount of ChR2 molecules in a cell-type-specific manner. The tet system is a bipartite system (Fig. 15.2). The first mouse line expresses the gene encoding tetracycline transactivator (tTA) protein under the control of a cell-type-specific promoter. In the second mouse line, ChR2 complementary DNA (cDNA) is connected downstream of the tetO promoter, which is activated only by the presence of tTA protein. Cell-type-specific expression of ChR2 can be expected only in tTA-tetO double transgenic mice. In the prototype mouse generated by this strategy, ChR2-expressing cells were only sparsely observed. This line was usable in experiments with acute slices of the brain since the amount of expression reached functional levels in expressing cells (Chuhma et al. 2011). However, because of the low efficiency of expression, many users remained unsatisfied.

We then attempted to improve the tet system. Why did the tetO promoter fail to induce ChR2 expression despite the fact that expression of tTA was obtained? For

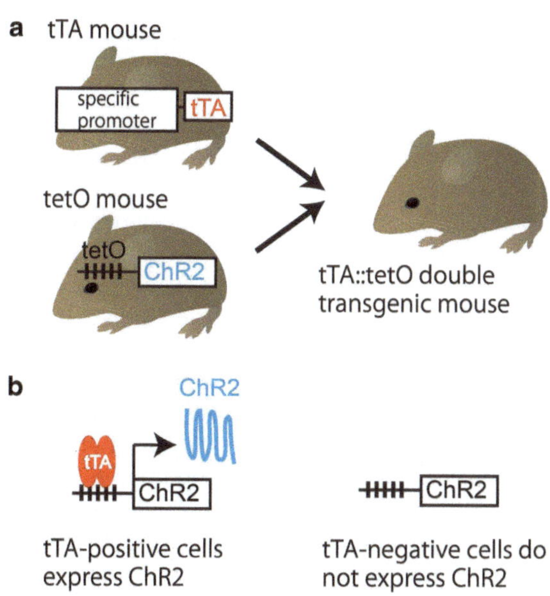

Fig. 15.2 The tet system. (**a**) Bipartite system using tTA mice and tetO-ChR2 mice. Mice expressing gene-encoding tTA under the control of a cell-type-specific promoter and mice with tetO promoter–ChR2 are mated and the obtained double transgenic mice are used in the experiment. (**b**) Cell-type-specific expression of ChR2 by the tet system. tTA binds to the tetO promoter and initiates transcription in tTA-expressing cells. ChR2 expresses in these cells (*left*). Since the tetO promoter cannot be activated in tTA non-expressing cells, ChR2 expression cannot be achieved. *ChR* channelrhodopsin

this question, we got ideas from our own experience (Tanaka et al. 2010). In past experiments, we established tetO mice via microinjection of transgenic DNA into the fertilized eggs, resulting in poor tTA-mediated gene induction. However, we found that the efficiency of tTA-mediated gene induction was improved dramatically when tetO mice were generated via a homologous recombination technique through embryonic stem (ES) cells. From this, we generated the tetO-ChR2 mouse, not by transgenics but by a knocking-in approach called the Knockin-mediated ENhanced Gene Expression (KENGE)-tet (Tanaka et al. 2012). The house-keeping *β*–actin gene locus was used as the knock-in site and the tetO-ChR2 cassette was inserted. This strategy was a major success and tremendously high levels of ChR2 expression was achieved compared with the previous tetO lines.

There was another advantage in the duplex composition of KENGE-tet system. Once this system worked successfully, the tetO-ChR2 mouse line was mated with a variety of other tTA lines. By this, it was expected that a variety of bigenic lines that express sufficient amounts of ChR2 molecules in different kinds of cells could be prepared easily. If the tetO line had insertions of proteins other than ChR2 protein, it would be possible to express high amounts of functional proteins in specifically targeted cell populations.

15.8 Mice Expressing ChR2 in a Specific Type of Cell Population

Optogenetic manipulation of the mouse can be possible only after obtaining ChR2-expressing mice. First, researchers need to prepare mice that express sufficient amounts of functional ChR2 in a particular type of cell population. Currently, three choices exist for this purpose: KENGE-tet mice, Ai mice using Cre-loxP (Madisen et al. 2012), and BAC transgenic mice (Zhao et al. 2011). These three candidates are very useful research tools and provide a wide variety of possibilities.

Previously, I have mentioned an example of failure of the *ROSA* promoter in optogenetics. Another Cre-loxP system is the technical basis for generation of Ai mice (Nagy 2000). The Cre-loxP system is also a bipartite system (Fig. 15.3). The first mouse line expresses Cre recombinant enzyme under the control of a cell-type-specific promoter, while in the second transgenic mouse line, a ubiquitous promoter–loxP-STOP-loxP-ChR2 gene is inserted. In the double transgenic mouse generated by mating these two mouse lines, ChR2 is expressed under a ubiquitous promoter only in Cre recombinant enzyme-expressing cells after excision of the STOP sequence between the two loxP sequences. Cells that do not express Cre recombinase do not express ChR2. It is suggested that the failure indicated in the above example was caused by the expression of an insufficient amount of ChR2 by the native *ROSA* promoter.

Hongkui Zeng of the Allen Institute generated a wide variety of mouse reporter lines by insertion of the CAG promoter, a strong promoter obtained by combining

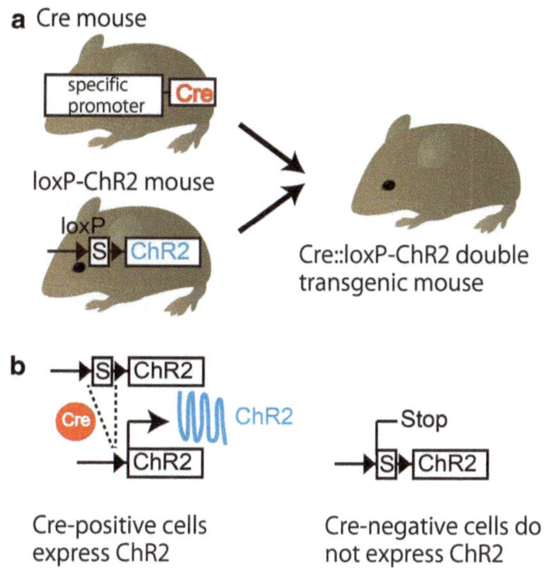

Fig. 15.3 Cre-loxP system. (**a**) Bipartite system using Cre mice and loxP-ChR2 mice. Mice expressing Cre recombinase under the control of a cell-type-specific promoter, and mice with a ubiquitous promoter–loxP (*black triangle*)-STOP (S)-loxP-ChR2 are mated, and the obtained double transgenic mice are used in the experiment. (**b**) Cell-type-specific expression of ChR2 by the Cre-loxP system. The region between loxP sites is excised by recombination, and ChR2 expresses under the ubiquitous promoter (*left*) in cells expressing Cre recombinase. In cells not expressing Cre recombinase, the ubiquitous promoter fails to induce downstream genes because of the presence of the STOP(S) sequence. *ChR* channelrhodopsin

the cytomegalovirus early enhancer with the chicken β-actin promoter, into the *ROSA* locus (Madisen et al. 2010). One of these lines is the Ai32 line, in which the *ROSA*-CAG-loxP-STOP-loxP-ChR2 is inserted. Ai stands for Allen Institute. It became possible to express functional ChR2 with this powerful tool. There are many mouse lines in the Ai series. Among them, Arch-expressing Ai35 mice and e*Np*HR-expressing Ai39 mice are considered to be useful lines relevant to optogenetics.

Which mouse line is better, KENGE-tet or Ai? In regards to ChR2 expression, it would seem nearly equal between these two lines because both lines provide a sufficient level of ChR2 expression to manipulate membrane potential. As for the variety of mouse lines with assured expression of ChR2 in specifically targeted cell populations, many more Cre mouse lines are available than tTA lines. As for availability, KENGE-tet lines can be better since lines will be purchased at a low cost from RIKEN BioResource Center (http://www.brc.riken.jp/lab/animal/). Ai mice can be obtained from the Jackson Laboratory (http://jaxmice.jax.org).

Aside from the duplex composition mouse lines such as KENGE-tet or Ai lines, ChR2 BAC transgenic mice were generated using only one mouse line. This mouse line was generated by a simple replacement of GFP with ChR2 in GFP-expressing mice (Gene Expression Nervous System Atlas [GENSAT]). This strategy is successful only when the gene promoters used have sufficiently strong power. The advantage of this method is that it is not necessary to mate two different mouse lines. BAC transgenic mice are also available from the Jackson Laboratory.

15.9 Conclusion

Manipulation of nervous cells in a cell-type-specific manner was the investigational strategy that brain researchers have been looking for. The mouse expressing ChR2 protein in specifically targeted cell populations is the basis of this investigation, and obtaining this mouse line is the first step to successful optogenetic research. A wide variety of the mouse lines were established within the past several years (Beppu et al. 2014; Miyazaki et al. 2014; Ohmura et al. 2014; Tsunematsu et al. 2013, 2014; Yamazaki et al. 2014), and these mice are easily obtained. Even if the desired cell populations are not targeted in the present list of mouse lines, it is possible to establish such lines via the combination of a virus and genetically modified mouse lines. Please refer to other publications for these needs (Ting and Feng 2013; Yizhar et al. 2011).

References

Arenkiel BR, Peca J, Davison IG et al (2007) In vivo light-induced activation of neural circuitry in transgenic mice expressing channelrhodopsin-2. Neuron 54:205–218

Beppu K, Sasaki T, Tanaka KF et al (2014) Optogenetic countering of glial acidosis suppresses glial glutamate release and ischemic brain damage. Neuron 81:314–320

Boyden ES, Zhang F, Bamberg E et al (2005) Millisecond-timescale, genetically targeted optical control of neural activity. Nat Neurosci 8:1263–1268

Chuhma N, Tanaka KF, Hen R et al (2011) Functional connectome of the striatal medium spiny neuron. J Neurosci 31:1183–1192

Gong S, Kus L, Heintz N (2010) Rapid bacterial artificial chromosome modification for large-scale mouse transgenesis. Nat Protoc 5:1678–1696

Gossen M, Bujard H (1992) Tight control of gene expression in mammalian cells by tetracycline-responsive promoters. Proc Natl Acad Sci U S A 89:5547–5551

Lin JY, Lin MZ, Steinbach P et al (2009) Characterization of engineered channelrhodopsin variants with improved properties and kinetics. Biophys J 96:1803–1814

Madisen L, Zwingman TA, Sunkin SM et al (2010) A robust and high-throughput Cre reporting and characterization system for the whole mouse brain. Nat Neurosci 13:133–140

Madisen L, Mao T, Koch H et al (2012) A toolbox of Cre-dependent optogenetic transgenic mice for light-induced activation and silencing. Nat Neurosci 15:793–802

Miyazaki KW, Miyazaki K, Tanaka KF et al (2014) Optogenetic activation of dorsal raphe Serotonin neurons enhances patience for future rewards. Curr Biol. doi:10.1016/j.cub.2014.07.041

Nagy A (2000) Cre recombinase: the universal reagent for genome tailoring. Genesis 26:99–109

Ohmura Y, Tanaka KF, Tsunematsu T et al (2014) Optogenetic activation of serotonergic neurons enhances anxiety-like behaviour in mice. Int J Neuropsychopharmacol 21:1–7

Tanaka KF, Ahmari SE, Leonardo ED et al (2010) Flexible Accelerated STOP Tetracycline operator-knockin (FAST): a versatile and efficient new gene modulating system. Biol Psychiatry 67:770–773

Tanaka KF, Matsui K, Sasaki T et al (2012) Expanding the repertoire of optogenetically targeted cells with an enhanced gene expression system. Cell Rep 2:397–406

Ting JT, Feng G (2013) Development of transgenic animals for optogenetic manipulation of mammalian nervous system function: progress and prospects for behavioral neuroscience. Behav Brain Res 255:3–18

Tsunematsu T, Tabuchi S, Tanaka KF et al (2013) Long-lasting silencing of orexin/hypocretin neurons using archaerhodopsin induces slow-wave sleep in mice. Behav Brain Res 255:64–74

Tsunematsu T, Ueno T, Tabuchi S et al (2014) Optogenetic manipulation of activity and temporally controlled cell-specific ablation reveal a role for MCH neurons in sleep/wake regulation. J Neurosci 34:6896–6909

Yamazaki Y, Fujiwara H, Kaneko K et al (2014) Short- and long-term functional plasticity of white matter induced by oligodendrocyte depolarization in the hippocampus. Glia 62:1299–1312

Yizhar O, Fenno LE, Davidson TJ et al (2011) Optogenetics in neural systems. Neuron 71:9–34

Zhao S, Ting JT, Atallah HE et al (2011) Cell type–specific channelrhodopsin-2 transgenic mice for optogenetic dissection of neural circuitry function. Nat Methods 8:745–752

Part III
Optogenetics in Neuroscience

Chapter 16
Neuroscientific Frontline of Optogenetics

Akihiro Yamanaka

Abstract Optogenetics is a recently developed experimental technique to control the activity of neurons using light. Optogenetics shows its power to reveal the physiological role of specific neural circuits in the brain. In particular, the manipulation of specific types of neurons using optogenetics with high-accuracy timing enables us to analyze causality between neural activity and initiation of animal behaviors. However, to manipulate the activity of specific neurons in vivo, two steps need to be fulfilled to succeed in the manipulation of neural activity and control of the behavior of individual animals. Step 1: an adequate number of molecules of light-activated protein must be expressed in the cell membrane of the neurons of interest. Step 2: the optical system must illuminate the targeted neurons with enough intensity to activate the light-activated protein. In this chapter, I illuminate the tricks to succeeding in the manipulation of targeted neurons in vivo using optogenetics to control animal behaviors.

Keywords Behavior • Virus vector • Transgenic animal • Cre/loxP • tTA/TetO • LED • Laser

16.1 Aspects of Optogenetics (Compared with Conventional Techniques)

Imagine obtaining a precise mechanical device that contains a vast number of intricate parts. Two techniques are used to learn how that device operates, i.e. the mechanism by which it functions. These techniques are 'observation' and 'manipulation'. Observation provides an understanding of the shape and function of individual parts and it reveals what they look like when functioning normally. If certain parts were to function and others were to stop functioning, then the other technique, manipulation, reveals how nearby parts will behave as a result of the on/off of each part of interest. Manipulation also reveals how certain parts affect the device's operation as

A. Yamanaka (✉)
Research Institute of Environmental Medicine, Nagoya University, Nagoya, Japan
e-mail: yamank@riem.nagoya-u.ac.jp

H. Yawo et al. (eds.), *Optogenetics*, DOI 10.1007/978-4-431-55516-2_16

a whole. Manipulation of certain intricate parts is initially made possible using fine tools like tweezers. Optogenetics is a technology that is perhaps akin to a pair of tweezers. The central nervous system functions via an exceedingly vast number of neurons that form complex and elaborate pathways. Before optogenetics first emerged in 2005, there were no technologies to artificially manipulate the activity of nerves with certain properties with very precise timing. Before the advent of optogenetics, the primary techniques to manipulate neural circuits were electrical stimulation or local application of a drug. Electrical stimulation involves implanting microelectrodes in the brain regions that one wishes to activate and then passing a current between those electrodes. This allows the activation of neurons in a relatively limited area with precise timing. However, the soma of neurons near the electrode are non-specifically activated as are nerve fibers passing through near the electrode, so neurons are activated with no specificity. In addition, the electrical stimulation can up-regulate the activity of the targeted neuron, whereas it cannot be down-regulated in the same manner. The second technique to manipulate neural circuits, local application of a drug, involves the use of agonists and antagonists of the receptors present in neurons. Glutamate and its analogs are locally injected to up-regulate neural activity, while substances such as muscimol, a γ-aminobutyric acid (GABA) receptor agonist, are locally injected to down-regulate it. Injecting drugs into areas of the brain results in either activation or inhibition of neurons in those regions. However, like electrical stimulation, neurons are activated with no specificity, and precision timing is exceedingly difficult since a drug disperses once it is injected. In contrast to these conventional techniques, optogenetics uses light, which has good penetrating capability and is minimally invasive. Optogenetics involves expressing protein molecules that sense light and that are capable of affecting membrane potential and cell function (light-activated molecules) in certain neurons with specific properties. Thus, the activity of these neurons can be manipulated with light. Optogenetics allows the regulation of the activity of certain target neurons with very precise timing for the first time in human history. Channelrhodopsin (ChR)-2, which up-regulates neural activity, is switched on by blue light while halorhodopsin (HaloR), which down-regulates neural activity, is switched on by orange light. Thus, neurons can instantly be switched from activation to inhibition or *vice versa* by changing the light wavelength (blue to orange). This capability to activate certain neurons while inhibiting others at the same time is a prominent aspect of optogenetics.

16.2 From Manipulating Neural Activity to Controlling Behavior with Light

Optogenetics was first used to manipulate neural activity with light in primary cultured neurons. A viral vector was used to introduce the ChR2 gene in cultured neurons and express that gene thereafter. A cultured neuron was illuminated with blue light while recording its electrical activity using the patch clamp technique.

As a result, an action potential was induced with a frequency matching that of the pulse of blue light (Boyden et al. 2005). This is because ChR2 is a non-selective cation channel that opens when it absorbs blue light. Na^+ ions flow into a neuron from extracellular regions, causing the cell's membrane potential to be depolarized. When the membrane potential is depolarized to exceed the threshold of triggering voltage-gated sodium channels in the membrane, it is robustly depolarized as a result of opening sodium channels with the generation of an action potential (see Fig. 16.1). Similarly, when cultured neurons expressing HaloR are illuminated with orange light, their membrane potential is hyperpolarized, allowing the generation of an action potential to be inhibited, because HaloR is a Cl^- ion pump that is operated by orange light (see Fig. 16.1). This is explained as follows. When Cl^- ion pumps start working, they transport Cl^- ions from outside to inside the neuron to hyperpolarize its membrane potential (because Cl^- is negatively charged). Consequently, the membrane potential does not reach the threshold of a voltage-gated sodium channel, even in the presence of excitatory inputs. Once ChR2 and HaloR were characterized in vitro, scientists began to use these molecules to manipulate specific neural activity and control behavior in vivo in individual animals. This allowed the direct analysis of neural activity that governs certain animal behaviors, and promoted major advances in elucidating mechanisms regulating physiological phenomena that are evident only in an animal as a whole (rather than individual cells). Research on *Caenorhabditis elegans* first indicated that light could be used to manipulate neural

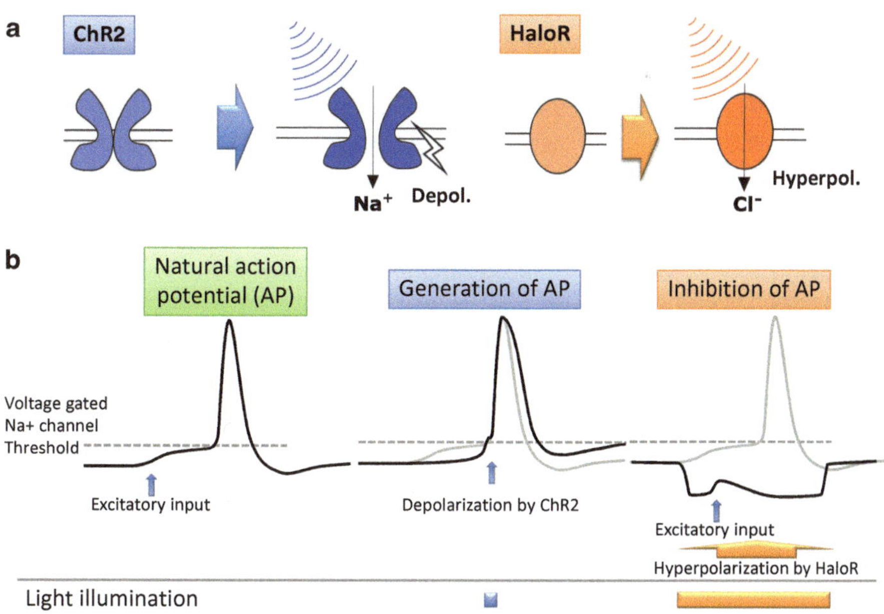

Fig. 16.1 (**a**) Principle backgrounds on the action of the light-activated proteins channelrhodopsin (ChR)-2 and halorhodopsin (HaloR). (**b**) Manipulation of neural activity using ChR2 and HaloR

activity and control behavior (Nagel et al. 2005). Light was then successively used on various animal species ranging from flies to primates, elucidating the role of neural circuits. The following are typical studies using mice. We and Karl Deisseroth's group independently reported that an animal's arousal from sleep can be regulated by the manipulation of orexin-producing neural activity in the hypothalamus (Adamantidis et al. 2007; Tsunematsu et al. 2011). Kravitz et al. (2010) manipulated neurons expressing dopamine D1 receptor or D2 receptor in order to clarify the role of direct and indirect pathways in the basal ganglia. Liu et al. (2012) found that memories are formed in neurons of the hippocampus. Two conditions need to be fulfilled for any research to succeed in the manipulation of the neural activity and control of the behavior of individual animals: (1) an adequate number of molecules of light-activated protein that must be expressed in the cell membrane of the neurons of interest, and (2) the optical system that illuminates the targeted neurons with enough light intensity to activate the light-activated protein in question. The first condition is the focus of the following sections.

16.3 Expressing Light-Activated Protein in Specific Neurons

16.3.1 Approaches Using a Viral Vector

Induced expression with a viral vector is one of the simplest and quickest ways to express a light-activated protein like ChR2 or HaloR in specific neurons. Lentivirus (Adamantidis et al. 2007) and adeno-associated virus (AAV) (Zhang et al. 2010) are often used as viral vectors to express light-activated proteins. There are limits on the length of a gene that can be packaged in a viral vector (about 9 kb for lentivirus, about 5 kb for AAV). This makes it difficult to fit a specific promoter sequence that limits expression of a gene to certain cells in a viral vector in its entirety. Thus, an approach using Cre/loxP or Flp/FRT, which combines a viral vector and genetically modified mice is often used. In such instances, cell specificity is accomplished with genetically modified mice expressing a recombinase such as Cre or Flp only in specific types of neurons, while the light-activated protein one wishes to express is usually introduced via the viral vector. A STOP cassette flanked by two loxP sequences or FRT sequences (e.g. loxP-STOP-loxP or FRT-STOP-FRT) is removed, or a gene to express is inverted and allowed to return to its correct orientation (FLEX), by the action of Cre or Flp (see Fig. 16.2). This allows a powerful promoter to induce gene expression. This powerful promoter is usually unselective to cell type and is used to cause a strong expression of the downstream gene in whatever cells are to be targeted. However, when a STOP cassette is present, it stops translation immediately, which prevents the expression of light-activated molecules. Since a gene is inverted in the FLEX system, its translation should be kept from proceeding or its product should be non-functional. As a result, the light-activated proteins are expected to be expressed specifically only in the neurons that express Cre or Flp with sufficient quantity (number of molecules). The manipulation by light could

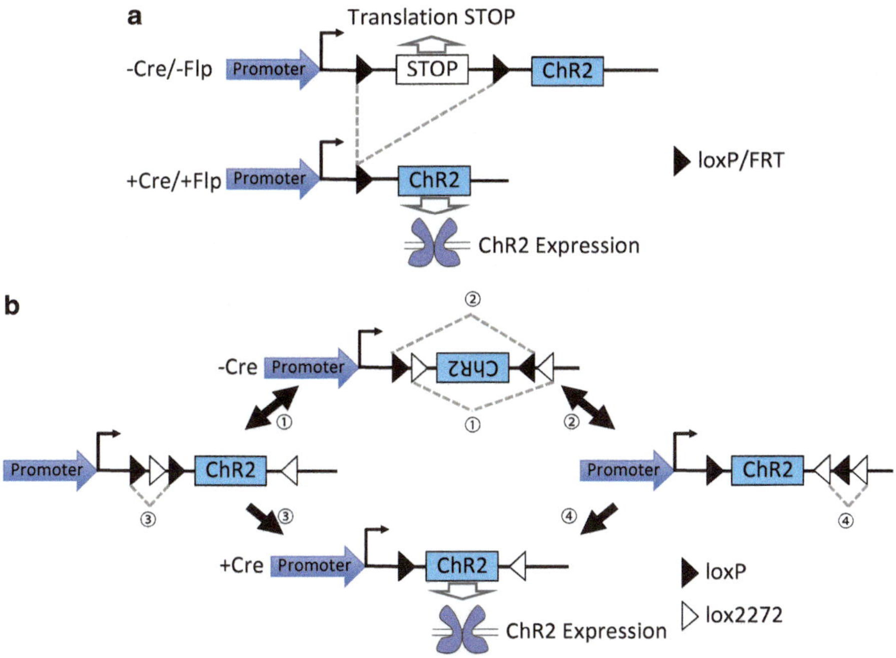

Fig. 16.2 (**a**) Regulation of gene expression by removal of a STOP cassette using Cre/loxP or Flp/FRT. (**b**) Regulation of gene expression using the FLEX switch system

be further refined by the localization of viral transfection as well as by regional illumination. A great deal of versatility is prospected in these approaches, as any new light-activated molecules, either through molecular modification or gene mining, would be incorporated in the targeted neurons simply by preparing or purchasing appropriate viral vectors.

16.3.2 Approaches Using Genetically Modified Mice

A number of approaches to express a light-activated protein in specific neurons have been tried using genetically modified mice. These are categorized in the following three ways: (1) A non-specific promoter is used to express a light-activated protein in a rather extensive area. (2) A specific promoter is used to express a light-activated protein only in specific neurons. (3) A Cre/loxP system, Flp/FRT system, or a tetracycline gene (Tet) expression system is used to express a light-activated protein in specific neurons by cross-breeding two types of genetically modified mice (one line determines the type of cell that will express the protein, the other line determines the type of light-activated protein that will be expressed). Examples of category (1) are mice (LeChasseur et al. 2011) and rats (Ji et al. 2012) that have been bred to have

Thy1 promoter. An example of category (2) is a mouse line in which HaloR was expressed under regulation by the human prepro-orexin promoter (Tsunematsu et al. 2011). In this mouse, HaloR was specifically expressed in orexin-producing neurons in the hypothalamus, which are crucial to regulate arousal from sleep. Mice that have been genetically modified using bacterial artificial chromosome (BAC) are also categorized as (2). BAC can carry fragments of DNA of several-hundred kbases, large enough to contain promoters that have yet to be identified. Examples are promoters of the choline acetyltransferase (ChAT) gene, which is expressed specifically in cholinergic neurons, and the tryptophan hydroxylase 2 (Tph2) gene, which is expressed specifically in serotonergic neurons (Zhao et al. 2011). Each of them worked as expected and expressed light-activated proteins selectively in the cholinergic neurons or in the serotonergic neurons. In these experiments, a promoter with sufficient activity must be chosen since the number of molecules of light-activated protein that are expressed depends on the strength of the promoter. Accordingly, relatively powerful promoters such as neurotransmitter-synthesizing enzymes and neuropeptides are chosen. Light-activated proteins are often expressed as a fusion protein with fluorescent protein such as green fluorescent protein or red fluorescent protein. Optogenetic light manipulation of neural activity often fails when fluorescence from a fluorescent protein is not evident in live tissue under a fluorescence microscope although the presence of fluorescence is ultimately confirmed by fluorescence enhancement using specific antibodies. Examples of category (3) include approaches using Cre/loxP or Flp/FRT (Madisen et al. 2012; Imayoshi et al. 2013) and approaches using tetracycline transactivator (tTA)/TetO (Tanaka et al. 2012) (Fig. 16.3). In both of these approaches, two types of mice had to be cross-bred to create bigenic mice. With the former, mice expressing Cre or Flp in specific type of neurons were cross-bred with mice expressing a light-activated protein in a manner dependent on Cre or Flp. In contrast, in the Tet system, tTA mice expressing tTA in specific neurons were cross-bred with TetO mice expressing a light-activated protein in a manner dependent on tTA. Several advantages of using such genetically modified mice are that a light-activated protein can be expressed in

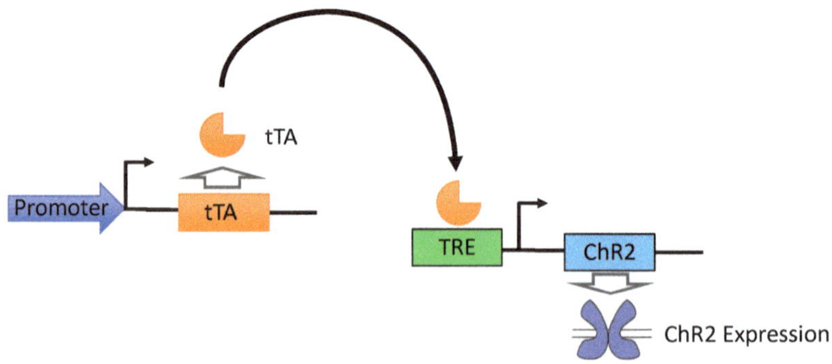

Fig. 16.3 Regulation of gene expression using the tTA/TetO system

a wide area, there is no variability among individual animals since they all have the same pattern of expression, and that gene expression can be induced in a manner that is not invasive at all.

16.4 Prospects for the Future

Numerous ChR2 variants have been created. Aspects have been improved, such as improved sensitivity to light and improved translocation to the cell membrane. In addition, numerous studies have described new molecules and variants to facilitate activation by light at longer wavelengths and allow more frequent stimulation and more prolonged stimulation. This means that the number of useful tools is increasing so as to be optimized to the individual experiments. However, current users are having greater difficulty selecting the molecules and laboratory equipment that are best suited to their own experiments.

Over the past few years, new techniques to manipulate specific neural activity, such as designer receptors exclusively activated by designer drugs (DREADD) (Armbruster et al. 2007) and pharmacologically selective actuator modules/pharmacologically selective effector molecules (PSAMs/PSEMs) (Magnus et al. 2011) have been reported. All of these techniques express a receptor that recognizes a special compound in specific neurons. The compound is administered intraperitoneally and binds to the receptor, allowing neural activity to be manipulated. Both activation and inhibition are achieved by choosing the type of receptor and downstream effector. These approaches to manipulation of neural activity offer highly imprecise timing compared with optogenetics, but they do not require equipment to provide light illumination, so physical constraints and issues of invasiveness (like those with fiber optics) are eliminated. In addition, these approaches allow the manipulation of neural activity for a relatively prolonged period. These approaches are a highly effective way to manipulate cell function found in vast areas and cells that are not readily illuminated. Nowadays, optogenetics is not a special technique but is instead becoming a common research technique that is essential for research on the role of neural circuits. Future research on the role of neural circuits will develop to intricately manipulate neural circuits by combining other new techniques and optogenetics.

References

Adamantidis AR, Zhang F, Aravanis AM et al (2007) Neural substrates of awakening probed with optogenetic control of hypocretin neurons. Nature 450(7168):420–424

Armbruster BN, Li X, Pausch MH et al (2007) Evolving the lock to fit the key to create a family of G protein-coupled receptors potently activated by an inert ligand. Proc Natl Acad Sci U S A 104(12):5163–5168

Boyden ES, Zhang F, Bamberg E et al (2005) Millisecond-timescale, genetically targeted optical control of neural activity. Nat Neurosci 8(9):1263–1268

Imayoshi I, Tabuchi S, Hirano K et al (2013) Light-induced silencing of neural activity in Rosa26 knock-in mice conditionally expressing the microbial halorhodopsin eNpHR2.0. Neurosci Res 75(1):53–58

Ji ZG, Ito S, Honjoh T et al (2012) Light-evoked somatosensory perception of transgenic rats that express channelrhodopsin-2 in dorsal root ganglion cells. PLoS One 7(3):e32699

Kravitz AV, Freeze BS, Parker PR et al (2010) Regulation of parkinsonian motor behaviours by optogenetic control of basal ganglia circuitry. Nature 466(7306):622–626

LeChasseur Y, Dufour S, Lavertu G et al (2011) A microprobe for parallel optical and electrical recordings from single neurons in vivo. Nat Methods 8(4):319–325

Liu X, Ramirez S, Pang PT et al (2012) Optogenetic stimulation of a hippocampal engram activates fear memory recall. Nature 484(7394):381–385

Madisen L, Mao T, Koch H et al (2012) A toolbox of Cre-dependent optogenetic transgenic mice for light-induced activation and silencing. Nat Neurosci 15(5):793–802

Magnus CJ, Lee PH, Atasoy D et al (2011) Chemical and genetic engineering of selective ion channel-ligand interactions. Science 333(6047):1292–1296

Nagel G, Brauner M, Liewald JF et al (2005) Light activation of channelrhodopsin-2 in excitable cells of *Caenorhabditis elegans* triggers rapid behavioral responses. Curr Biol 15(24):2279–2284

Tanaka KF, Matsui K, Sasaki T et al (2012) Expanding the repertoire of optogenetically targeted cells with an enhanced gene expression system. Cell Rep 2(2):397–406

Tsunematsu T, Kilduff TS, Boyden ES et al (2011) Acute optogenetic silencing of orexin/hypocretin neurons induces slow-wave sleep in mice. J Neurosci 31(29):10529–10539

Zhang F, Gradinaru V, Adamantidis AR et al (2010) Optogenetic interrogation of neural circuits: technology for probing mammalian brain structures. Nat Protoc 5(3):439–456

Zhao S, Ting JT, Atallah HE et al (2011) Cell type-specific channelrhodopsin-2 transgenic mice for optogenetic dissection of neural circuitry function. Nat Methods 8(9):745–752

Chapter 17
Elucidation of Neuronal Circuitry Involved in the Regulation of Sleep/Wakefulness Using Optogenetics

Tomomi Tsunematsu and Akihiro Yamanaka

Abstract Although sleep is an absolutely essential physiological phenomenon to maintain normal health in animals, little is known about the function and regulatory mechanism of sleep so far. In this section, we introduce how optogenetics was applied to freely behaving animals to elucidate neuronal circuits involved in the regulation of sleep/wakefulness. When optogenetics was applied to the specific type of neurons involved in sleep/wakefulness regulation, we could control the sleep/wakefulness state, changes among wakefulness, non-rapid eye movement (NREM) sleep and REM sleep state. Selective activation of orexin neurons and noradrenergic neurons in the locus coeruleus (LC) using channelrhodopsin-2 and melanopsin induced the transition from sleep to wakefulness. In contrast, suppression of these using halorhodopsin and archaerhodopsin induced the transition from wakefulness to NREM sleep and increased the time spent in NREM sleep. Selective activation of melanin-concentrating hormone (MCH) neurons induced the transition from NREM sleep to REM sleep and prolonged the time spent in REM sleep accompanied by a decrease in time spent in NREM sleep. These studies help to answer how this specific type of neural activity contributes to the regulation of sleep/wakefulness.

Keywords Orexin • Melanin-concentrating hormone (MCH) • Locus coeruleus (LC) • Wakefulness • Non-rapid eye movement (NREM) sleep • Rapid eye movement (REM) sleep • Melanopsin/OPN4 • Channelrhodopsin-2 • Archaerhodopsin • Halorhodopsin

---

T. Tsunematsu (✉)
Research Institute of Environmental Medicine, Nagoya University, Nagoya, Japan

Strathclyde Institute of Pharmacy and Biomedical Science, University of Strathclyde, Glasgow, UK
e-mail: tomomi.tsunematsu@strath.ac.uk

A. Yamanaka
Research Institute of Environmental Medicine, Nagoya University, Nagoya, Japan

H. Yawo et al. (eds.), *Optogenetics*, DOI 10.1007/978-4-431-55516-2_17

17.1 Introduction to Sleep/Wakefulness, Wakefulness, Non-rapid Eye Movement (NREM) Sleep, and REM Sleep

We first introduce the concepts of sleep and wakefulness before discussing the regulation of sleep/wakefulness using optogenetics. Sleeping is one of the most instinctive behaviours and is a universal physiological phenomenon conserved in insects and vertebrates. If we sleep for 8 h per day, we will spend one-third of our lifetime sleeping. When animals such as rats or dogs are deprived of sleep, body weight reduces notably while food intake increases. Animals will finally die if they are kept in such a sleep-deprived state for 2 weeks due to the development of opportunistic infections (Rechtschaffen et al. 1983). These reports suggest that sleep is an absolutely essential physiological phenomenon that maintains normal health in animals.

There are two completely different states of sleep: rapid eye movement (REM) sleep and non-REM (NREM) sleep. To determine these distinct states (wakefulness, NREM sleep, and REM sleep), recordings such as electroencephalogram (EEG) and electromyogram (EMG) are typically used. There are also functional differences in the autonomic activity among these states. During wakefulness, an EEG shows low-amplitude and an EMG shows large-amplitude patterns due to muscle contraction (Fig. 17.1). When NREM sleep starts, an EEG shows slow-wave patterns with high-amplitude, and an EMG becomes low-amplitude compared with patterns observed during wakefulness (Fig. 17.1). On the other hand, during REM sleep, an EEG shows a relatively low-amplitude pattern that looks similar to the wakefulness EEG at first glance. However, the component of EEG frequency in this case is predominantly theta waves (6–10 Hz). EMG activity is barely observed due to the relaxation of voluntary muscles (Fig. 17.1). If we look at the EEG power spectra that is calculated by fast Fourier transformation, it is easy to visually understand what kind of EEG frequency composes each vigilance state, and it greatly helps us to determine the vigilance state (Fig. 17.1).

17.2 Flip-Flop Circuit that Regulates the Timing of Sleep/Wakefulness: Sleep Center and Arousal Center

What controls the sleep/wakefulness state in the brain, and how is it generated? It has been generally believed that sleep is a passive phenomenon to alleviate tiredness that is caused by continuous wakefulness. However, this might be a significant misunderstanding. A part of the brain actively promotes quiescence of the entire brain and body, i.e., a 'sleep center' in the brain. This sleep center is located in the anterior part of the hypothalamus called the ventrolateral preoptic (VLPO) area, which contains γ-aminobutyric acid (GABA)-ergic neurons. These GABAergic neurons are activated during sleep. The longer the time spent in wakefulness, the greater increase

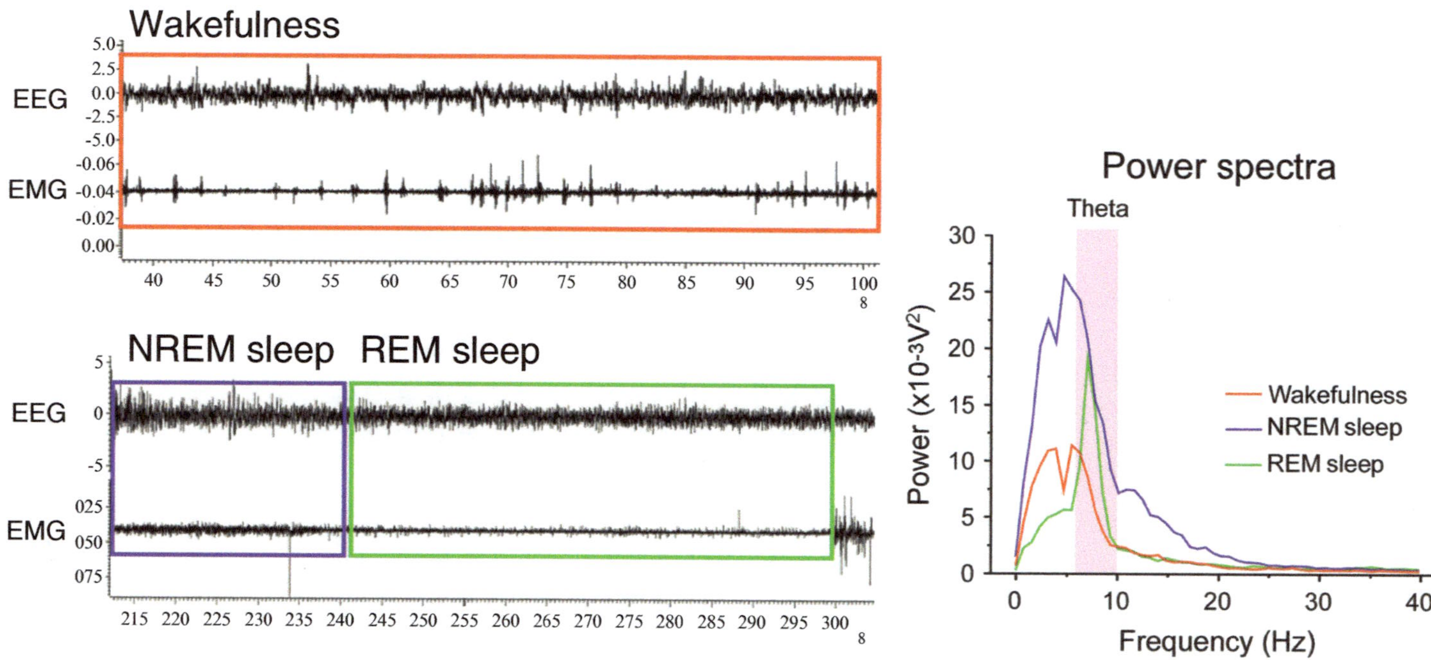

Fig. 17.1 Electroencephalogram (EEG) and electromyogram (EMG) during wakefulness, non-rapid eye movement (NREM) sleep, and REM sleep. Line graph shows EEG power spectra in each vigilance state. X-axis indicates frequency of EEG (Hz). Y-axis indicates power value (x 10^{-3} V^2)

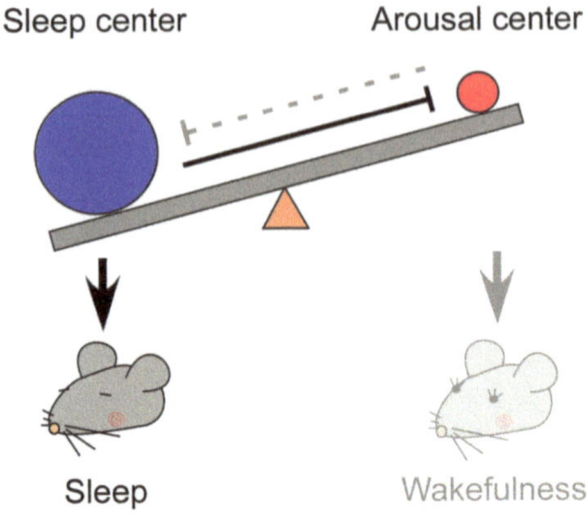

Fig. 17.2 Flip-flop circuit that regulates the timing of sleep/wakefulness

in sleep pressure (sleepiness). As a result, sleep pressure increases the activity of GABAergic neurons in the VLPO.

In comparison, the ‘arousal centers’, which induce wakefulness, are located in the brain stem and posterior part of the hypothalamus. Monoaminergic neurons, such as histaminergic neurons in the tuberomammillary nucleus (TMN), serotonergic neurons in the raphe nucleus, and noradrenergic neurons in the locus coeruleus (LC), are the major players as arousal centers in the generation of arousal. The activity of monoaminergic neurons is synchronized and strongly associated with sleep/wakefulness states. These neurons exhibit tonic firing during wakefulness, decrease firing during NREM sleep, and are electrically almost quiescent during REM sleep. Monoaminergic neurons project axons broadly throughout the brain, not only in the brain stem and posterior part of the hypothalamus, but also into the cerebral cortex to activate entire brain regions. This activation results in the promotion of arousal.

Sleep and arousal centers share mutually inhibitory circuitry. That is, sleep centers inhibit arousal centers, and *vice versa*. Sleep and arousal centers function like a seesaw relationship, which is known as a flip-flop circuit (Fig. 17.2). In other words, sleep or wakefulness is determined by the relationship between the balance of activity between sleep and arousal centers. For example, if the sleep center activity overcomes arousal center activity, sleep is initiated (Fig. 17.2). The sensation that promotes sleep after an all-nighter might imply the sudden flipping of a switch within the flip-flop circuit.

17.3 Orexin/Hypocretin and Narcolepsy

Narcolepsy is one type of sleep disorder. It typically begins in adolescence with the primary symptom of excessive daytime sleepiness. Patients tend to fall asleep even at inappropriate times and places. In the context of a boring lecture, even a healthy person may fall asleep (as they might not make an effort to maintain wakefulness).

However, narcolepsy patients fall asleep even if they are doing critical things such as giving a lecture. Cataplexy is a specific symptom of narcolepsy. It is the sudden weakness of muscle triggered by strong emotions such as surprise, anger, and humor. In addition to sleepiness and cataplexy, narcoleptic patients exhibit a direct transition from wakefulness to REM sleep although normal subjects display a marked transit from wakefulness to REM sleep via NREM sleep. As a result, narcoleptic patients frequently experience sleep paralysis.

Narcolepsy was a disease of unknown etiology for a long time. The discovery of the neuropeptide orexin (also known as hypocretin) in 1998 opened a door to understanding the etiology of narcolepsy. Orexin-producing neurons (orexin neurons) are specifically and sparsely located in the lateral hypothalamic area, which is a well-known feeding center, and project their axons to almost all regions of the brain (Fig. 17.3). Orexin actually consists of two peptides, orexin A and B, which are enzymatically processed from the common precursor peptide, prepro-orexin. Orexin acts physiologically by binding to two kinds of G-protein-coupled receptors (GPCRs), orexin 1 receptor (OX1R) and orexin 2 receptor (OX2R) (Fig. 17.3). Orexin was initially recognized as a regulator of feeding behavior because intracerebroventricular injections of orexin induce feeding behavior in rats and mice. Subsequently, *prepro-orexin* knock-out mice, *OX2R* knock-out mice, and orexin neuron-ablated mice displayed a narcolepsy-like phenotype, specifically a fragmentation of sleep/wakefulness (or a frequent repetition of sleep and wakefulness). These animals cannot maintain wakefulness and thus behavior is commonly arrested. In addition, a post mortem study of human narcolepsy patients revealed a degeneration of orexin neurons in the hypothalamus. These facts support the idea that degeneration of orexin neurons by unknown reasons (possibly autoimmune disease) leads to narcolepsy.

While orexin neurons project their axons to almost all regions of the brain, they are densely projecting to arousal centers, such as the TMN, raphe nucleus, and LC. Orexin neurons activate these monoaminergic neurons. Orexin neurons are thought to be phasically active during wakefulness and almost silent during NREM and REM sleep. Therefore, it seems that orexin neurons should contribute to stabilization of the flip-flop circuit as an activator of arousal centers, especially for the maintenance of wakefulness (Fig. 17.4). In contrast, in narcolepsy patients, the balance between sleep and arousal centers should be unstable because of the specific degeneration of orexin neurons. This would cause the fragmentation of sleep/wakefulness states as a result of frequent transition within the flip-flop circuit (Fig. 17.4). It is thus presumable that orexin neurons have a crucial role in the stabilization of arousal states.

We have previously presented evidence supporting this hypothesis using genetically modified mice. However, these mice likely compensate for the genetic modifications with activity of other neurons during development since the regulation of sleep/wakefulness is essential to the physiological maintenance of life. Thus, the question remains, how does orexin neuron activity drive behavior, particularly sleep/wakefulness regulation? Optogenetics could provide a powerful tool to address this question.

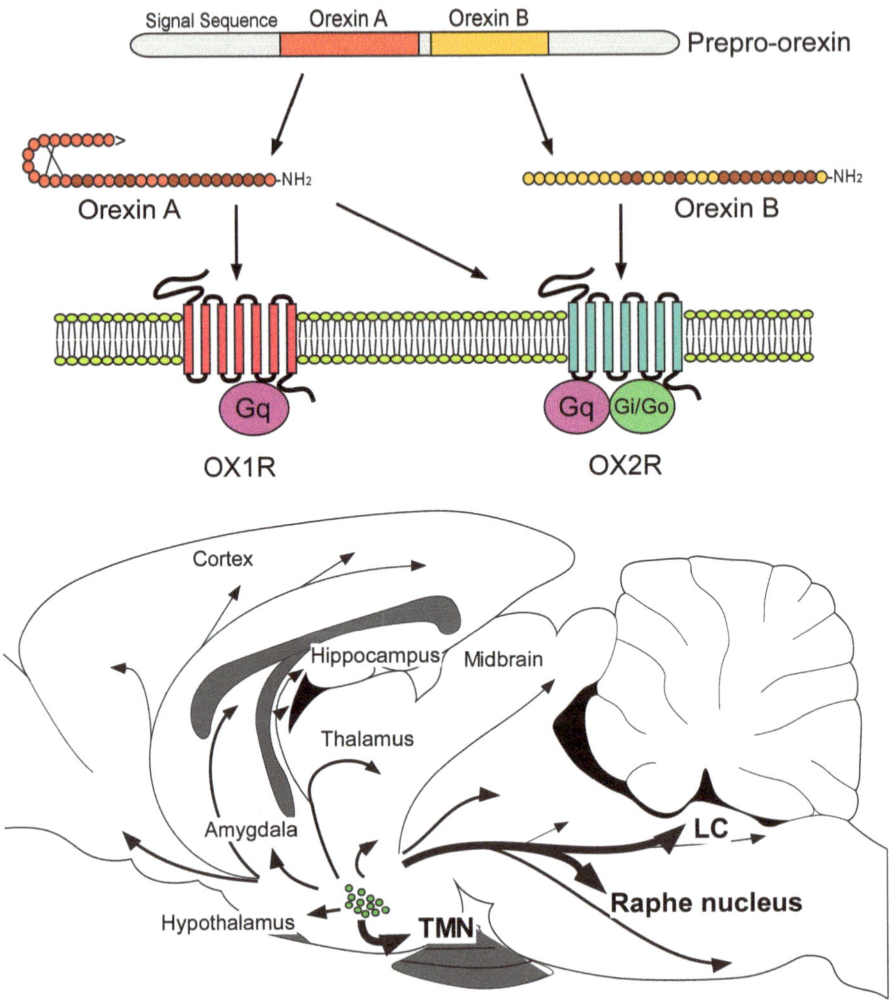

Fig. 17.3 Orexin and orexin receptors

17.4 The Optogenetics Wired System for Freely Moving Mice

Here, I introduce the optogenetics wired system for freely moving mice, which is used by our group at Nagoya University. To determine sleep/wakefulness states in mice, we have to record EEG and EMG. In addition, we have to simultaneously illuminate light onto orexin neurons, which are located in the lateral hypothalamic area. After much trial and error over a few years, we have successfully designed such a wired system that fulfills the above requirements (Fig. 17.5) (Yamanaka 2012).

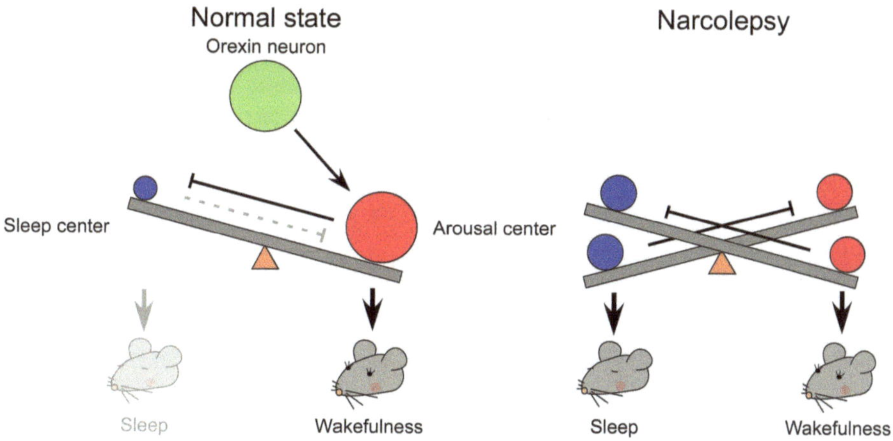

Fig. 17.4 Orexin and narcolepsy

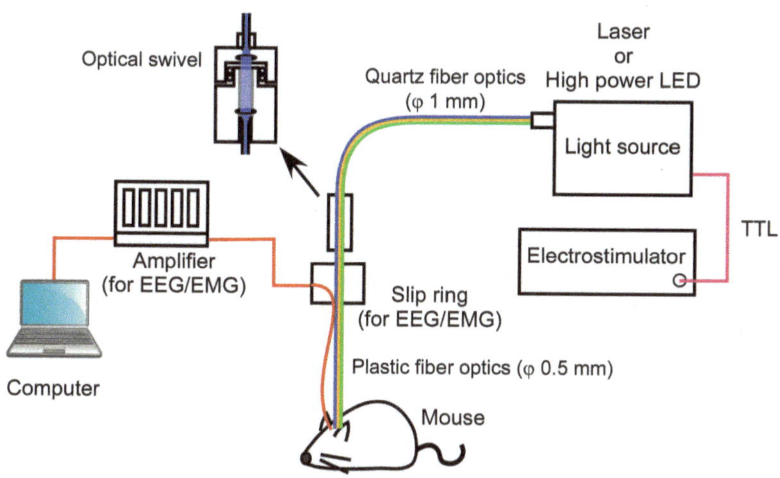

Fig. 17.5 Schematic of the optogenetics wired system for freely moving mice

A high-power light-emitting diode (LED) or laser is used as a light source. On and off switching of the light source is regulated by a transistor-transistor logic (TTL) signal connected to an electrical stimulator (or a pulse generator). The light is guided through quartz fiber optics (at a diameter of 1 mm), which is connected to a light source. Flexible plastic fiber optics (at a diameter of 0.5 mm) are inserted bilaterally into the brain. Tips of fibers are located 1 mm above the targeted neurons. Quartz and plastic fibers are connected via a optical swivel, which is a rotatable device that maintains a high light-coupling efficacy. The device enables mice to move freely without twisting the optical fibers. The position of inserted fiber optics is determined by the location of target cells, the three-dimensional range of their distribution, and the numerical aperture (NA) of the fiber optics. In our case, it was

1 mm above the location of target cells with an NA of 0.5 and with the assumption that orexin neurons are approximately distributed within a range of 1 mm^3 cube.

To determine sleep/wakefulness states at the same time as light illumination, EEG and EMG are simultaneously recorded. An EEG was recorded from the surface of the cortex using screw electrodes and an EMG was recorded from neck muscles using wire electrodes. The EEG/EMG signal is continuously recorded via a slip ring. The signal is amplified and then data are acquired and analyzed via computer.

17.5 Induction of Wakefulness via Activation of Orexin Neurons

In 2007, the group of Deisseroth and de Lecea at Stanford University applied optogenetics to freely behaving animals and successfully controlled sleep/wakefulness by targeting orexin neurons in the hypothalamus (Adamantidis et al. 2007).To selectively activate orexin neurons with light, channelrhodopsin (ChR)-2 was exclusively expressed in them using a viral vector. They generated mice in which orexin neurons specifically expressed ChR2 by using a lenti-virus vector containing the 3.1-kb upstream region of the mice *prepro-orexin* gene as a promoter. To confirm the function of ChR2 expression in orexin neurons, slice patch clamp recordings were performed. Blue light illumination induced depolarization of the ChR2-expressing orexin neurons. It was discovered that both continuous and pulse light illumination could induce depolarization and increases in firing frequency.

As we mentioned before, orexin neurons phasically fire during wakefulness and are almost silent during sleep. One question arises from this. Can we artificially promote wakefulness in mice via activation of orexin neurons during sleep? They were the first to provide the answer. Blue light illumination was applied in vivo in freely moving mice that displayed spontaneous sleep or wakefulness states. During spontaneous sleep, continuous or pulsative light illumination of greater than 5 Hz in frequency induced a transition to wakefulness after a delay of approximately 30 s from the initiation of light illumination. This effect was observed not only during NREM sleep but also during REM sleep. These facts suggest that the up-regulation of orexin neuronal activity is sufficient to promote the initiation of arousal (Adamantidis et al. 2007). This study was the first successful report linking the activation of specific types of neurons in mice to behavior by using optogenetics. In addition, this study established that optogenetics is a powerful tool to regulate neuronal activity. Since this report, optogenetic approaches have been applied to the study of many neuronal populations and various behaviors.

Another challenge was made independently using transgenic mice in which orexin neurons specifically expressed melanopsin (OPN4) under a tet-off system (*Orexin-tTA; BitetO human OPN4/mCherry* bigenic mice) (Tsunematsu et al. 2013b). OPN4 is blue light-driven Gq-coupling GPCR normally expressed in retinal ganglion cells. Blue light illumination depolarizes OPN4-expressing neurons via activation of the Gq signal transduction cascade. The advantage of using OPN4 is that even applying short light pulse induces long-lasting depolarization since GPCR signaling is relatively long lasting. In addition, the other advantage in our study is using the bigenic mouse strain, which led to reliable and reproducible expression of OPN4 in the orexin neurons compared with using a viral vector. The authors also confirmed a continuous light-evoked depolarization after blue light cessation using slice patch clamp. Again, blue light pulses induced the transition from NREM sleep to wakefulness in bigenic mice.

17.6 Induction of NREM Sleep via Acute Inhibition of Orexin Neurons

The reports using ChR2 and OPN4 revealed that the acute activation of orexin neuronal activity induces the transition from sleep to wakefulness in mice. The next question is whether the inhibition of orexin neuronal activity induces sleep and/or a narcolepsy-like phenotype in the opposite manner.

To inhibit orexin neuronal activity, we employed an orange light-driven chloride pump, halorhodopsin (HaloR). To express HaloR in orexin neurons, the 3.2-kb upstream sequence of the human *prepro-orexin* gene was used as a promoter to generate new transgenic mice (*orexin/HaloR* transgenic mice). Immunohistochemical analyses revealed that 94 % of orexin neurons expressed HaloR in the *orexin/HaloR* mice.

At first, to confirm the function of HaloR expressed in orexin neurons in the transgenic mouse brain, slice patch clamp analysis was performed using brain slice preparations from *orexin/HaloR* mice. Orange light illumination immediately hyperpolarized HaloR-expressing orexin neurons and completely inhibited the generation of action potentials. In addition, we determined how long orexin neuronal activity could be inhibited. Firing was completely inhibited for approximately 1 min from the initiation of orange light illumination. However, firing gradually recovered, probably due to the desensitization of HaloR or the accelerated outward transport of chloride ions. From this in vitro evidence, we attempted to silence orexin neurons acutely for 1 min in freely moving *orexin/HaloR* mice.

Using the optogenetics wired system for freely moving mice as mentioned before, orexin neuronal activity was inhibited for 1 min via continuous orange light illumination. Light illumination was applied when mice were in a state of arousal. To determine sleep/wakefulness states, EEG and EMG were simultaneously recorded. Mice maintained a wakefulness state for a few seconds after the initiation of orange light illumination. EMG power gradually decreased and the slow-wave component of EEG increased. These facts indicated an induction of NREM sleep. It is suggested that the acute inhibition of orexin neuronal activity is sufficient to promote the initiation of NREM sleep or that the activation of orexin neurons is necessary to maintain an arousal state. However, a narcolepsy-like phenotype, such as cataplexy and a direct transition from wakefulness to REM sleep was not observed.

To further study the physiological significance of orexin neuronal activity in sleep/wakefulness regulation in conjunction with other arousal centers, EEG and EMG activity and the activity of dorsal raphe (DR) serotonergic neurons were simultaneously recorded. In vivo single unit extracellular recording was performed using conscious, head-fixed *orexin/HaloR* mice. The firing of serotonergic neurons in the DR was recorded when orexin neuronal activity was inhibited by light illumination via fiber optics located in the hypothalamus. Serotonergic neurons in the DR are densely innervated by orexin neurons and are activated by orexin either directly or indirectly. During orange light illumination, the firing frequency of serotonergic neurons in the DR gradually decreased in conjunction with the EEG delta power increase (Fig. 17.6). This evidence suggests that acute silencing of orexin neuronal activity by illuminating orange light into the hypothalamus decreased the firing frequency of serotonergic neurons in the DR and induced NREM sleep in mice. This study provided the first example that the selective inhibition of orexin neurons is sufficient to induce a transition from wakefulness to NREM sleep (Tsunematsu et al. 2011).

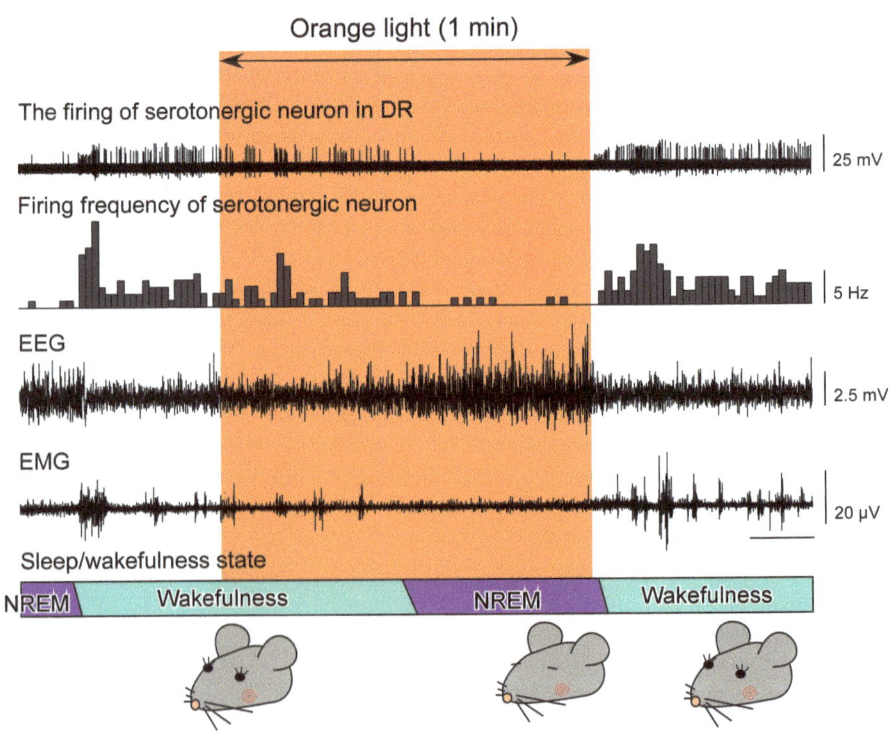

Fig. 17.6 Transition from wakefulness to non-rapid eye movement (NREM) sleep and inhibition of serotonergic neuronal activity by light illumination

17.7 Increase in Total Time in NREM Sleep via Long-Lasting Inhibition of Orexin Neurons

The study using *orexin/HaloR* mice revealed that acute inhibition of orexin neural activity functions as the switch to initiate NREM sleep. However, to completely understand the regulatory circuitry and the mechanisms of sleep/wakefulness under physiological conditions, silencing for more than 1 min is essential. However, as described above, it is difficult to inhibit neuronal activity for a longer duration, at least in orexin neurons, probably due to the desensitization of HaloR or the accelerated outward transport of chloride ions. To address this issue, we employed a green light-driven proton pump, Archaerhodopsin-3 (ArchR) and Archaerhodopsin TP009 (ArchT) (Tsunematsu et al. 2013a). The authors generated two transgenic mice in which orexin neurons specifically express ArchR or ArchT using the human prepro-orexin promoter in the case of ArchR *(orexin/ArchR* mice) and using the tet-off system in the case of ArchT (*orexin-tTA; TetO ArchT* mice). The expression rate of ArchR and ArchT in orexin neurons in *Orexin/ArchR* mice and *orexin-tTA; TetO ArchT* mice was 80 % and 72 %, respectively.

Light-induced responses of HaloR-expressing, ArchR-expressing, or ArchT-expressing orexin neurons were evaluated using slice patch clamp. As shown in Fig. 17.7, ArchR and ArchT induced stronger inhibition of activity in terms of light-induced current, hyperpolarization, and available duration of inhibition compared with HaloR (at least in orexin neurons). It seems that ArchR and ArchT are suitable for longer-duration inhibition due to limited desensitization. In fact, not only by using slice patch clamp but also via immunohistochemical measurement of c-Fos, a marker of activated neurons, we were able to detect a significant inhibition of orexin neurons over at least 1 h.

Using novel transgenic mice lines, we examined the effects of 1-h light illumination on the sleep/wakefulness state during both the light and the dark period. Inhibition of orexin neurons by applying continuous green light illumination induced a significant increase in total time of NREM sleep during the dark period (the active period for nocturnal mice) (Fig. 17.8). Mice repeated ~10 min bouts of NREM sleep several times and they could not maintain wakefulness but continued to sleep during light illumination. This can be interpreted as a fragmentation of sleep/wakefulness, such as seen in the narcolepsy phenotype. In contrast, light illumination had little effect on sleep/wakefulness patterns during the light period (the inactive period for nocturnal mice). These results suggest that orexin neuronal activity contributes to the maintenance of wakefulness during the active phase (Tsunematsu et al. 2013a).

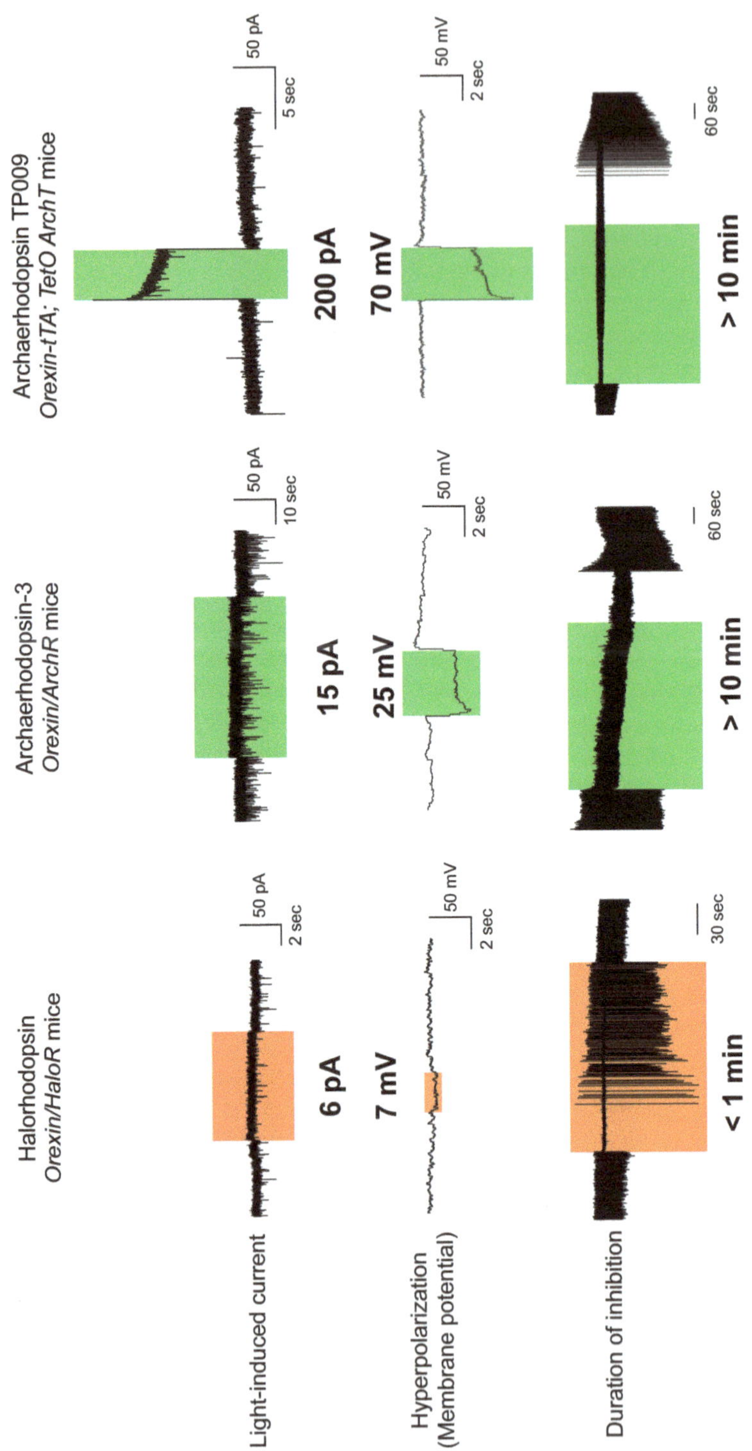

Fig. 17.7 Comparison of light-evoked responses via light-driven pumps in orexin neurons

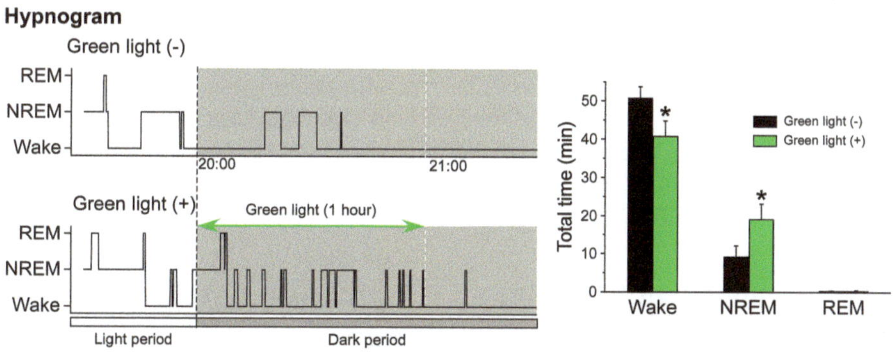

Fig. 17.8 Long-lasting silencing of orexin neurons induces increases in total time in non-rapid eye movement (NREM) sleep

17.8 Control of Sleep/Wakefulness via Regulation of LC Noradrenergic Neuron Activity

We have so far introduced the concept of sleep/wakefulness regulation via orexin neuronal activity. Orexin neurons contribute to the stabilizing of flip-flop circuitry, which is composed of mutual inhibitory circuits between sleep centers and arousal centers in the brain. In addition to optogenetic control of orexin neurons, optogenetic strategies directly applied to neurons within sleep and arousal centers would be expected to control sleep/wakefulness states.

In 2010, the group of Deisseroth and de Lecea at Stanford University reported the optogenetic control of one of the primary arousal centers, noradrenergic neurons of the LC (Carter et al. 2010). Adeno-associated virus (AAV) was injected into transgenic mice in which tyrosine hydroxylase-expressing cells exclusively expressed Cre recombinase. Noradrenergic neurons in these mice specifically expressed ChR2 or HaloR. Light-induced responses in ChR2-expressing noradrenergic neurons and HaloR-expressing noradrenergic neurons were confirmed using slice patch clamp.

Is it possible to induce the transition from sleep to wakefulness when noradrenergic neurons in the LC are specifically activated by light illumination? A 5 Hz frequency of blue light pulses over 5 s was applied to ChR2-expressing mice during the light period. Light illumination induced transitions from either NREM sleep or REM sleep to wakefulness within 5 s from the initiation of light exposure. This fact suggests that the activation of noradrenergic neurons in the LC acts as a transition switch from a sleep to an arousal state. For comparison, they also examined the effects of a 1-h light illumination on the sleep/wakefulness cycle using HaloR-expressing, freely moving mice. Yellow light illumination significantly increased total time in NREM sleep and decreased total time in wakefulness during the dark (active) period. However, light illumination had little effect on sleep/wakefulness states during the light (inactive) period. This result is in agreement with inhibition of orexin neuronal activity. Furthermore, long-lasting activation via ChR2 was also performed. A 3-Hz frequency of blue light pulses for 1 h induced significant increases in total time in wakefulness, decreases in total time in NREM sleep, and increases in locomotor activity. These results indicate that the activity of noradrenergic neurons in the LC plays an important role in the maintenance of wakefulness (Carter et al. 2010), in addition to the documented role of orexin.

17.9 Induction of REM Sleep via Activation of Melanin-Concentrating Hormone Neurons

We have mainly focused on the transition from wakefulness to NREM sleep. REM sleep typically appears after NREM sleep; however, little is known about what regulates the transition from NREM sleep to REM sleep.

Melanin-concentrating hormone (MCH) neurons are distributed in the lateral hypothalamic area. This is similar to the orexin neuron distribution. However, a completely different population of neurons produce MCH. Previous reports revealed that MCH neurons play an important role in the regulation of feeding behavior, energy metabolism, and locomotor activity. In comparison, in vivo extracellular recording studies using rats have revealed that MCH neurons are silent during wakefulness, discharge during sleep, and are maximally active during REM sleep. This firing pattern is reciprocal to that of orexin neurons, implying that the MCH neurons could be involved in REM sleep regulation.

The group of Adamantidis at McGill University in 2013 and our group in 2014 independently published studies that applied optogenetics to the study of MCH neurons. Jego et al. (2013) injected AAV in transgenic mice in which MCH neurons expressed Cre recombinase. In the brains of these mice, ChR2, HaloR, or ArchR was expressed in MCH neurons. Conversely, we generated transgenic mice in which MCH neurons specifically expressed ChR2 or ArchT using the tet-off system (*MCH-tTA; TetO ChR2* bigenic mice and *MCH-tTA; TetO ArchT* bigenic mice).

In the above study, they performed in vivo optogenetic activation of MCH neurons using freely moving mice. They reported that optogenetic stimulation of MCH neurons at a 20 Hz frequency at the onset of NREM sleep had little effect on NREM sleep duration, although it increased the probability of NREM to REM sleep transitions. Comparatively, at the onset of REM sleep, activation of MCH neurons significantly extended REM sleep duration. Regarding inhibition of MCH neurons, optogenetic silencing of MCH neurons during REM sleep did not reduce the duration, but induced a shift in the dominant theta wave peak frequencies toward slower oscillations (Jego et al. 2013). Our group described an optogenetic manipulation of MCH neurons using transgenic mice. Optical activation of MCH neurons by a 10 Hz frequency of blue light pulses for 3 h using *MCH-tTA; TetO ChR2* resulted in an increase in time spent in REM sleep accompanied by a decrease in time spent in NREM sleep without affecting the total time of wakefulness. In addition, an acute 1-min activation of MCH neurons at a 10 Hz frequency during the NREM sleep state induced a transition to REM sleep. In comparison, silencing MCH neurons via continuous green light illumination for 3 h using *MCH-tTA; TetO ArchT* mice had little effect on the vigilance state (Tsunematsu et al. 2014). These results from two reports suggest that MCH neuronal activity is sufficient to induce REM sleep, but is not necessary for REM sleep to occur. This conclusion supports the notion that other neurons are responsible for the generation of REM sleep.

17.10 Perspective

The optical control of neuronal activity involved in the regulation of sleep/wakefulness, including orexin neuron activity in the hypothalamus and noradrenergic neuron activity in the LC, indicates that sleep/wakefulness states can be controlled in vivo. These studies have helped answer how specific neuronal activity contributes to the regulation of sleep/wakefulness in neuronal circuits driving these behaviors. Additionally, the application of optogenetic strategies to the study of MCH neurons has revealed that these neurons also play a physiological role in the regulation of REM sleep, providing a giant step toward the complete understanding of the neuronal circuitry of REM sleep regulation.

In the future, we would like to reveal not only the neuronal circuits involved in sleep/wakefulness but also to determine the ultimate functions of sleep and the precise definitions of sleep and wakefulness using optogenetics.

References

Adamantidis AR, Zhang F, Aravanis AM et al (2007) Neural substrates of awakening probed with optogenetic control of hypocretin neurons. Nature 450:420–424

Carter ME, Yizhar O, Chikahisa S et al (2010) Tuning arousal with optogenetic modulation of locus coeruleus neurons. Nat Neurosci 13:1526–1533

Jego S, Glasgow SD, Herrera CG et al (2013) Optogenetic identification of a rapid eye movement sleep modulatory circuit in the hypothalamus. Nat Neurosci 11:1637–1643

Rechtschaffen A, Gilliland MA, Bergmann BM et al (1983) Physiological correlates of prolonged sleep deprivation in rats. Science 221:182–184

Tsunematsu T, Kilduff TS, Boyden ES et al (2011) Acute optogenetic silencing of orexin/hypocretin neurons induces slow-wave sleep in mice. J Neurosci 31:10529–10539

Tsunematsu T, Tabuchi S, Tanaka KF et al (2013a) Long-lasting silencing of orexin/hypocretin neurons using archaerhodopsin induces slow-wave sleep in mice. Behav Brain Res 255:64–74

Tsunematsu T, Tanaka KF, Yamanaka A et al (2013b) Ectopic expression of melanopsin in orexin/hypocretin neurons enables control of wakefulness of mice in vivo by blue light. Neurosci Res 75:23–28

Tsunematsu T, Ueno T, Tabuchi S et al (2014) Optogenetic manipulation of activity and temporally controlled cell-specific ablation reveal a role for MCH neurons in sleep/wake regulation. J Neurosci 34:6896–6909

Yamanaka A (2012) Optogenetical approach to control the activity of specific types of neurons in vivo. Nihon Yakurigaku Zasshi 140:280–284

Chapter 18
Optogenetic Analysis of Striatal Connections to Determine Functional Connectomes

Nao Chuhma

Abstract Neural circuit function is determined not only by anatomical connections but also by the strength and nature of the connections, that is functional connectivity. To elucidate functional connectivity, selective stimulation of presynaptic terminals of an identified neuronal population is crucial. However in the central nervous system (CNS), intermingled input fibers make selective electrical stimulation impossible. With optogenetics this becomes possible, and enables the comprehensive study of functional synaptic connections between an identified population of neurons and defined postsynaptic targets to determine the functional connectome. By stimulating convergent synaptic inputs impinging on individual postsynaptic neurons, low frequency and small amplitude synaptic connections can be detected. Further, the optogenetic approach enables measurement of cotransmission and its relative strength. In this chapter, optogenetic studies in the striatum (Str) are introduced to demonstrate the functional connectome approach. For spiny projection neurons, this has revealed cell-type specific intra-striatal connections as well as striatal output connections. Cholinergic interneurons in the ventral striatum have been shown to use glutamate as a cotransmitter. Examining striatal afferents from the ventral midbrain has identified fast, direct dopaminergic connection onto cholinergic interneurons and preferential connections of γ-aminobutyric acid (GABA) neurons to cholinergic interneurons in the nucleus accumbens. Further, it has revealed regionally heterogeneous glutamate and GABA cotransmission of dopamine neurons in the Str. These connections can be quite plastic, revealing a new vista of connectivity that is likely to be important in understanding the circuitry of neuropsychiatric disorders.

Keywords Nucleus accumbens • Spiny projection neurons • Channelrhodopsin • Dopamine neurons • Ventral tegmental area • Substantia nigra • Cholinergic interneurons

N. Chuhma (✉)
Department of Psychiatry, Columbia University, New York, NY, USA

Department of Molecular Therapeutics, New York State Psychiatric Institute, New York, NY, USA
e-mail: nc2027@columbia.edu

H. Yawo et al. (eds.), *Optogenetics*, DOI 10.1007/978-4-431-55516-2_18

18.1 Introduction

The mammalian brain is made from billions of cells and the majority of its functions are determined by connectivity of neurons. Since the identification of synapses as the point of communication between neurons by Sherrington (Shepherd 1994), determining the wiring diagram of a brain has been a central interest of neuroscientists in order to understand brain functions. Actual neuronal connectivity includes two levels: the physical wiring diagram (anatomical connectivity) and the strength and nature of the connection (physiological or functional connectivity). Anatomical connectivity is typically determined using retrograde and anterograde axonal tracers, including viruses with reporter proteins. Though these tracers are powerful tools to elucidate the wiring diagram, they are not suitable to examine synaptic connections. Synaptic connections can be identified by trans-synaptic tracers (e.g., pseudo-rabies virus), green fluorescent protein (GFP) reconstitution across synaptic partners (GRASP) (Feinberg et al. 2008) or observation with electron microscopy. However, the functional strength of a particular synaptic connection is difficult to discern with these methods. Thus, direct measurement of synaptic response between identified populations is crucial to define physiological or functional connectivity.

18.1.1 Functional Connectome

We introduced the concept of the *functional connectome* to describe comprehensive studies of functional connectivity (Chuhma et al. 2011). The word *connectome* was used by Sporns in 2005 to describe the wiring diagram of the human brain (Sporns et al. 2005). The current definition of *connectome* is the comprehensive mapping of wiring in a brain (not limited to human) or a region of brain (Lichtman and Sanes 2008; Seung 2009). In contrast to *connectome*, defined as the comprehensive study of anatomical connections of neurons, *functional connectome* can be defined as the comprehensive study of the physiological strength and nature (e.g., kinds of transmitters) of monosynaptic connections between two identified populations of neurons. It should be noted that the concept of functional connectivity used in the *functional connectome* approach is different from the *functional connectivity* used in brain imaging, which refers to temporally associated activities of two or more brain regions during a certain brain state (Friston et al. 1993). Functional connectivity in brain imaging is not necessarily relevant to direct synaptic connections among recorded regions, while *functional connectome* is restricted to *direct synaptic connections* from an identified population of neurons to their defined postsynaptic targets.

18.1.2 Why the Optogenetic Approach Is Suitable for Functional Connectome Studies

For a functional connectome study, selective and reliable stimulation of an identified population of neurons and recordings from defined neurons are crucial. Although identification of postsynaptic neurons is possible in vivo, intracellular recording from brain slices has advantages. In contrast to relatively easy identification of postsynaptic neurons, selective stimulation of identified presynaptic neurons has been hard to achieve. Particularly in the basal ganglia, where cell architecture is not layered and input fibers from multiple structures are intermingled, population-selective stimulation with electrical stimulation is almost impossible. The advent of optogenetics has enabled selective activation of identified neuronal populations by genetically limited expression of photoactivatable proteins. Optogenetic activation causes convergent input stimulation to a single neuron and is suitable to detect low-frequency small-amplitude responses. The optogenetic approach also reveals existence and strength of cotransmission. Immunohistochemistry may indicate the potential for cotransmission; however, the functional contribution of any cotransmitter is difficult to discern without physiological measurement.

For functional connectome studies, fast-on fast-off excitation current generation is necessary. The suitable photoactivatable proteins are non-selective cation channel channelrhodopsin (ChR)-2 and its mutants with faster kinetics and/or higher light sensitivity (Yizhar et al. 2011). Because of its membrane-targeted nature, ChR2 can also be used as a genetically controlled axonal tracer. Indeed, a study using ChR2 as an axonal tracer revealed novel projections from pontine tegmental nuclei to the striatum (Str) (Dautan et al. 2014). Although this makes coordinated anatomical studies feasible, this chapter focuses on excitation of genetically identified axon terminals to reveal functional connectivity. In the following section, I focus on the Str and its intrinsic and output connectivity. The Str is one of the most suitable locations to apply the optogenetic approach to the functional connectome, because of intermingled inputs, existence of genetic markers suitable for conditional expression of ChR2, and lack of thorough studies of functional strength of synaptic connections in spite of its importance both in normal functions of a brain and in the pathophysiology of neuropsychiatric disorders.

18.2 Functional Connectome of the Striatum with Optogenetics

18.2.1 Striatal Make-up and Inputs

The rodent Str comprises the nucleus accumbens (NAc) and the dorsal or neostriatum (dStr). The NAc corresponds to the ventral Str in primates. Although the striatal region corresponding to the primate caudate or putamen is not clear in rodents, the ventromedial dStr is regarded as roughly homologous to the caudate and the dorsolateral dStr as roughly homologous to the putamen (Graybiel 2008). Both the NAc and the dStr share similar cytoarchitectures; about 95 % γ-aminobutyric acid (GABA)-ergic spiny projection neurons (SPNs) and about 5 % interneurons (Wilson 2004). SPNs are the sole population of output neurons, and their major projection sites are the globus pallidus (GP)/ventral pallidum (VP) and the ventral midbrain (Wilson 2004). Striatal interneurons comprise cholinergic and GABAergic interneurons. Cholinergic interneurons (ChIs) are a single population and make up about 1 % of striatal neurons, while GABAergic interneurons comprise several different types, namely fast spiking interneurons (FSIs), low-threshold spike interneurons (LTSIs), and neurogliaform interneurons (Tepper et al. 2010; Kreitzer 2009). Although interneurons make up small minority populations, they exert strong control of SPN excitability and axo-axonal modulation of dopamine neuron afferents (Tepper et al. 2004; Oldenburg and Ding 2011; Exley and Cragg 2008).

Both the Str and the NAc receive inputs from the prefrontal cortex (PFC), the thalamus (Thal) and the ventral midbrain (Yin and Knowlton 2006). The NAc is distinguished by inputs from limbic structures: the hippocampus and the basolateral amygdala (BLA) (Pennartz et al. 2009; Belujon and Grace 2011). Those inputs show a topographic projection pattern. Ventral midbrain dopamine (DA) neuron fibers from the medial ventral tegmental area (VTA) project to the NAc medial shell, namely the most ventromedial subregion of the striatal complex, and more lateral DA neurons in the substantia nigra pars compacta (SNc) project to dorsolateral subregion of the Str (Haber et al. 2000; Ikemoto 2007). These connections modulate cortex-basal ganglia-thalamus loops that run in parallel from the more ventral 'limbic loop' to the dorsal 'sensorimotor loop' (Pennartz et al. 2009; Yin and Knowlton 2006).

18.2.2 Output from the Striatum/Accumbens

SPNs are divided into two populations based on their projections to the output nuclei of the basal ganglia directly (direct pathway) or via the GP/VP (indirect pathway). The output nuclei comprise the VTA/SN and the entopeduncular nucleus (the rodent counterpart of the internal segment of the GP in primates) and send projections to the Thal (Gerfen and Surmeier 2011). Direct pathway SPNs (dSPNs) express DA D1 receptors and enkephalin, while indirect pathway SPNs (iSPNs) express DA D2 receptors and substance P (Gerfen and Surmeier 2011). Those two populations of SPNs are segregated almost completely in the dStr and NAc core in adult rodents (Bertran-Gonzalez et al. 2008). In the NAc shell, 17–38 % of SPNs express both D1 and D2 receptors (Bertran-Gonzalez et al. 2008; Gangarossa et al. 2013), although it is not clear whether SPNs co-expressing both receptors belong to dSPNs or iSPNs.

18.2.2.1 Direct Pathway

Retrograde tracer studies have shown that dSPNs send axons to both DA neurons in the SNc and to GABA neurons in the SN pars reticulata (SNr) (Gerfen 1985; Fujiyama et al. 2011). Synaptic connections of dSPNs with both populations in the SN were confirmed by pseudo-rabies virus tracer (trans-synaptic retrograde tracer) injected into the ventral midbrain (Watabe-Uchida et al. 2012). However, these anatomical studies revealed neither the strength of the connections nor the responsible transmitters.

The first functional connectome study with optogenetics used transgenic mice expressing ChR2 selectively in SPNs with a tetracycline transactivator (tTA)-tetracycline operator (tetO) strategy (Chuhma et al. 2011). In these mice, tTA is expressed selectively in SPNs driven by the alpha-calcium-calmodulin-kinase II (αCaMKII) promoter (Mayford et al. 1996). tTA binds and activates tetO driving ChR2 expression. Whole cell recording from SN neurons and activating SPN terminals with wide-field photostimulation revealed that dSPNs make strong connections to SNr GABA neurons via $GABA_A$ receptors, but made no detectable connections to SNc DA neurons (Chuhma et al. 2011) (Fig. 18.1a). However, only 10 % of SPNs express ChR2 in αCaMKII-tTA::tetO-ChR2 mice. Although the distribution of ChR2 expression in SPNs appears to be random, as it does not distinguish between SPNs in patch and matrix striatal regions or between direct or indirect pathway SPNs, it is possible that SPN inputs to SNc DA neurons were too weak to be detected or that the SPNs expressing ChR2 did not include SPNs connecting to DA neurons. Similarly, a functional connectome study of accumbal SPNs with ChR2-viral vector injection found only photo-evoked $GABA_A$ responses in non-DA neurons, but not DA neurons in the VTA (Xia et al. 2011). In a subsequent study using viral transfection, weak SPN GABAergic connections to VTA DA neurons were seen; however, the GABAergic input to VTA GABA neurons was overwhelmingly stronger (Bocklisch et al. 2013). Taken together, these optogenetic functional connectome studies reveal that dSPNs make preferential connections to non-DA neurons in the VTA/SN and that connections to DA neurons are extremely weak (Fig. 18.1a).

Most dSPNs also send collaterals to the GP (Fujiyama et al. 2011), which account for about one-third of striatal inputs to the region (Kita 2007). In vivo activation of the dSPN collaterals in the GP inhibits GP neuron firing (Cazorla et al. 2014). Although the inhibition is likely mediated by direct collateral connections activating $GABA_A$ receptors, inhibitory synaptic responses were not evident in GP neurons (N. Chuhma, unpublished observations). The functional connectome of dSPNs in the GP remains to be determined (Fig. 18.1a, dashed line).

18.2.2.2 Indirect Pathway

The major projections of iSPNs are to the GP/VP. In αCaMKII-tTA::tetO-ChR2 mice, photo-activation of striatal inputs in the GP evokes $GABA_A$ responses (Chuhma et al. 2011). Almost all GP neurons are GABAergic projection neurons;

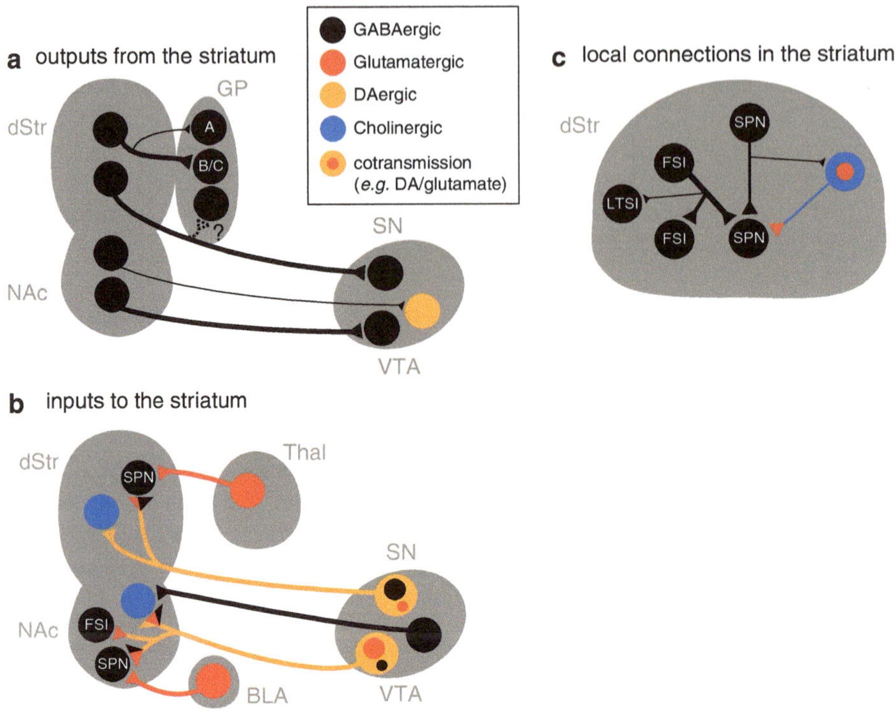

Fig. 18.1 Optogenetically identified fast direct synaptic connections are shown for striatal outputs (**a**), inputs (**b**), and local connectivity (**c**). Within cells, transmitters are color coded. Cotransmission is indicated by smaller circles in cells using the cotransmitters. Axon terminals are color coded accordingly. Significant differences in the strength of connections are indicated by the thickness of axons. In *panel A*, the dashed line indicates a connection suggested by in vivo recording, but the direct synaptic connection has not been confirmed. In *panel B*, observations in NAc shell and core are merged. *dStr* dorsal striatum, *NAc* nucleus accumbens, *GP* globus pallidus, *SN* substantia nigra, *VTA* ventral tegmental area, *Thal* thalamus, *BLA* basolateral amygdala, *A* Type A-like neurons, *B/C* Type B/C-like neurons, *SPN* spiny projection neuron, *FSI* fast spiking interneuron, *LTSI* low-threshold spiking interneuron

however, there has not been a clear consensus about classification of these neurons. We observed two types of electrophysiologically distinguishable populations; slower-firing high-input impedance neurons and faster-firing low-input impedance neurons (Chuhma et al. 2011). The former population shared the nature of Cooper-Stanford type A neurons (Cooper and Stanford 2000) and Lim homeobox 6 (Lhx6) positive neurons (Mastro et al. 2014). The latter shared some characteristics of Cooper-Stanford type B and C neurons (Cooper and Stanford 2000) and parvalbumin (PV)-positive neurons (Mastro et al. 2014). Striatal $GABA_A$ input was stronger in type B/C-like neurons and very weak in type A-like neurons (Chuhma et al. 2011) (Fig. 18.1a). Since ChR2 was expressed in both dSPNs and iSPNs in αCaMKII-tTA::tetO-ChR2 mice, the $GABA_A$ responses were combined responses of iSPN inputs and dSPN collateral inputs. However, when ChR2 was selectively expressed in dSPNs by viral vector injection, small $GABA_A$ response was evoked and the relative connectivity strength to the two populations was the same (Chuhma, unpublished observation), suggesting that the preferential connectivity to type B/C-like neurons was mostly determined by iSPNs.

18.2.3 *Inputs to the Striatum/Accumbens*

The Str receives spatially segregated subcortical and cortical inputs. Intermingled subcortical inputs enter the Str from the caudal side, so electrical stimulation activates multiple inputs, while cortical inputs enter the Str from the rostral side, so electrical stimulation activates these inputs selectively. Thus, functional connectome studies have focused on the subcortical inputs.

18.2.3.1 Ventral Midbrain Dopamine Neurons

The most characteristic projection to the Str is the one from DA neurons in the ventral midbrain. This is the densest DA projection in the brain, and it has a range of functions depending on projection sites in the Str, e.g., motivation and instrumental learning in the NAc shell, cognitive learning in the ventromedial Str, and motor control and habit formation in the dorsolateral Str (Haber et al. 2000; Graybiel 2008). Ventral midbrain DA neuron inputs are particularly suitable for optogenetic studies, because of intermingled inputs, expression of selective genetic markers, and controversial cotransmission. Optogenetic studies have revealed two interesting features of this DA neuron transmission besides DA release and its modulatory action: cotransmission involving two fast transmitters and fast direct DA transmission through D2 receptors.

Glutamate cotransmission by ventral midbrain DA neurons is an excellent example to show how powerful optogenetics is in functional connectome studies. DA neuron glutamate cotransmission was reported in DA neuron culture (Sulzer et al. 1998) and in slices (Chuhma et al. 2004). However, its existence in a mature intact brain was questioned, because the stimulation in slice preparation was not selective to DA neurons, and there are purely glutamatergic neurons in the VTA that possibly project to the NAc (Yamaguchi et al. 2007). The existence of functional cotransmission and its strength is hard to discern with anatomical methods. Selective stimulation of DA neuron terminals in the NAc with ChR2 solves these problems. Optogenetic stimulation of DA neuron terminals generated fast glutamatergic responses in all recorded SPNs in the NAc medial shell (Stuber et al. 2010; Tecuapetla et al. 2010; Mingote et al. 2012; Chuhma et al. 2014), and most recorded SPNs in the NAc core (Mingote et al. 2012). In the NAc medial shell, photo-evoked glutamate responses were also observed in all recorded ChIs and FSIs. Responses in ChIs were significantly larger (Chuhma et al. 2014). Therefore, it can be concluded that all the NAc medial shell neurons receive glutamate cotransmission from VTA DA neurons and preferential connections are to ChIs (Fig 18.1b). In the dStr, glutamate cotransmission to SPNs was observed by one group (Tritsch et al. 2012), but not by others (Stuber et al. 2010; Mingote et al. 2012). This discrepancy could be due to finer regional heterogeneity of DA neuron input in the dStr and bears further study.

Finding GABA cotransmission with VTA/SNc DA neurons is another triumph of optogenetic studies of synaptic transmission in the basal ganglia (Tritsch et al. 2012). This GABAergic cotransmission has a very unique nature, as it does not require glutamic acid decarboxylase (GAD) or vesicular GABA transporters, which have been thought to be critical for GABAergic transmission. Therefore, anatomical identification of cotransmission (e.g., co-localization of DA and GABA markers) may not detect this non-canonical cotransmission. The DA neuron GABA cotransmission was observed in both dSPNs and iSPNs in the dStr (Tritsch et al. 2012), both dSPNs and iSPNs in the NAc (Tritsch et al. 2014), and ChIs in the NAc core (Chuhma et al. 2014), indicating that GABA cotransmission is not limited to nigro-striatal projections in the dStr (Fig. 18.1b). However, DA neurons do not make significant GABA connections to ChIs in the most medial part of the NAc shell (Chuhma et al. 2014), suggesting that there is a regional difference in the strength of GABAergic cotransmission within the NAc.

Selective activation of DA neuron terminals with ChR2 revealed fast D2-mediated DA transmission in ChIs in the medial dStr (Chuhma et al. 2014) (Fig. 18.1b). Fast D2-mediated responses are observed in ventral midbrain DA neuron through dendritic release (Beckstead et al. 2004; Gantz et al. 2013); however it was not believed that DA has fast direct transmission in projection areas. These D2-mediated fast inhibitory responses were observed in all recorded ChIs in medial dStr (Chuhma et al. 2014). Since D2-mediated fast response has not been reported in iSPNs in the dStr, this direct DA transmission appears to be specific to ChIs.

18.2.3.2 Ventral Tegmental Area GABA Neurons

The VTA contains GABA neurons that project to the NAc (Van Bockstaele and Pickel 1995). However, the direct synaptic connectivity was not studied. Since the basal ganglia have GABA-dominant circuits, it has been difficult to stimulate only a particular subset of GABAergic inputs. ChR2 virus injection to the ventral midbrain of GAD-Cre mice enabled selective activation of VTA-GABA neurons (Brown et al. 2013). Large-amplitude photoactivated GABAergic responses were observed in all recorded ChIs, while modest connections were seen in SPNs in the NAc (Brown et al. 2013) (Fig. 18.1b). In NAc FSIs, 20 % of recorded neurons showed responses but the amplitude of the response was extremely small (Brown et al. 2013). This study revealed another preferential ventral midbrain input to ChIs.

18.2.3.3 Other Inputs

Compared with projections from the ventral midbrain, functional connectome studies of other inputs have not been conducted in detail. Thalamic inputs were stimulated with a viral vector of ChR2 injected to the intralaminar nuclei (Ellender et al. 2013) or the parafascicular nucleus (Ding et al. 2012), and strong glutamatergic connections were observed both in dSPNs and iSPNs in the dStr (Fig. 18.1b). Photoactivation of ChR2-expressing BLA terminals generated strong glutamatergic responses in SPNs in the NAc (Stuber et al. 2011) (Fig. 18.1b). However, these studies were conducted to confirm proper activation of synaptic connections with ChR2, and a functional connectome study has not yet been carried out.

18.2.4 Intrinsic Striatal Connections

In addition to synaptic inputs from other structures, striatal neurons make synaptic connections with each other; there are not only connections from interneurons to SPNs, but also connections amongst SPNs (Tepper et al. 2004). Because of physical proximity, paired recordings are possible to study synaptic connections between identified neurons. However, paired recordings will miss weak or lower-probability connections. Optogenetic stimulation of an identified neuronal population activates convergent inputs, and is more likely to detect low-probability synaptic connections.

18.2.4.1 SPN Connections to Striatal Neurons

Inter-SPN connections have been studied with paired whole-cell recordings (Taverna et al. 2008). However, the probability of connections is low (10–37 %), response amplitude is small (Tepper et al. 2004), and connections are difficult to detect. Introducing ChR2 to a subset of SPNs with the tetO-tTA strategy (Chuhma et al. 2011) and recording from non-ChR2-expressing SPNs provides information about convergent inputs from other SPNs. Under these conditions, about 60 % of recorded SPNs show GABAergic connections (Chuhma et al. 2011). Considering that only 10 % of SPNs expressed ChR2 in these mice, inter-SPN connections are stronger and more frequent than was estimated from paired recordings.

SPNs also make synaptic connections to ChIs, though the frequency is lower than connections to other SPNs (Chuhma et al. 2011) (Fig. 18.1c). Since the frequency and amplitude of this SPN–ChI connection is low, it is hard to identify this connection without optogenetics. On the other hand, SPNs do not make recognizable connections to FSIs (Chuhma et al. 2011). Considering the strong input from FSI to SPN (Tepper et al. 2004; Szydlowski et al. 2013), the connection between FSI and SPN is unidirectional (Fig. 18.1c).

18.2.4.2 Interneuron Connection

Most interneuron-to-SPN connection studies were conducted with simultaneous whole-cell recording, since interneurons have relatively strong and high-incidence connections to SPNs. However, for a comprehensive connectivity study, optogenetic stimulation of interneurons is necessary. ChR2 expression in PV-positive FSIs revealed inter-FSI connections, strong connections to SPNs, and weak and low-frequency connections to LTSIs, while there were no detectable connections to ChIs (Szydlowski et al. 2013) (Fig. 18.1c).

Striatal ChIs express vesicular glutamate transporter 3 (VGLUT3) and have been thought to be capable of glutamatergic transmission (Fremeau et al. 2002). However, it has been difficult to detect physiologically, since the response is not strong enough to be detected by simultaneous recording. Population activation of ChIs with ChR2 made the recording of convergent synaptic inputs possible, and fast glutamate cotransmission onto SPNs was revealed (Higley et al. 2011) (Fig. 18.1c). This shows two advantages of optogenetic approaches: stimulation of convergent inputs and identifying cotransmission.

18.3 Discussion

The advent of optogenetics has made possible selective stimulation of identified neuronal populations and has enabled functional connectome studies in the Str. Optogenetic activation of presynaptic terminals stimulates convergent inputs to single postsynaptic neurons. This enables detection of low-probability and small-amplitude synaptic connections. The optogenetic approach is also a powerful tool to identify cotransmission and its relative strength. A functional connectome approach revealed cell-type-specific connections of striatal output and preferential connections among different cell types in the local intra-striatal circuits. A regionally heterogeneous blend of cotransmitters of DA neuron projection and its preferential connection to particular cell types in the Str were also shown. The optogenetic approach also revealed new synaptic connections: VTA GABA neuron connection to ChIs and fast D2 response in ChIs.

ChR2 can be introduced to identified populations of neurons in two ways: conditional viral vector injection and transgenic mice. Viral vector injection provides strong expression with multiple copies of ChR2 and expression selectivity by both conditional expression and injection location and timing. However, ChR2 expression varies among animals and may not reach all cells when the targeted structure is large, e.g., the dStr, where this variation cannot be ignored in functional connectome studies. The virus used to deliver the ChR2 may also have an impact on synaptic properties (Jackman et al. 2014). The transgenic strategy gives reliable expression among animals, while expression is weaker than viral injection even using enhanced-expression techniques. This implies that there is a possibility of false-negative findings of synaptic connectivity in transgenic mice. These two approaches are complementary. Since there is no one method combining the advantages of both approaches, it will be important to compare the results obtained by these two approaches to achieve a more comprehensive functional connectome.

Several issues exist in assessing the fidelity of functional connectome studies with optogenetics. Because of penetration of blue light in tissue (Al-Juboori et al. 2013), it is not clear how much of the total synaptic inputs are activated with illumination. Light penetration can be improved by mutants of ChR2 with red-shifted excitation (Lin et al. 2013), since longer wavelength light penetrates further in brain tissue. Another possible issue is synchronous activation of terminals. Since ChR2 is activated by blue laser or high-power LED, it may activate terminals more synchronously than occurs in vivo under normal physiological conditions. However, in functional connectome studies, the aim is to obtain relative strength and capacity of transmission in different populations of postsynaptic neurons, and synchronous activation is preferable for estimation of relative strength. This information of basic functional connectivity is necessary for further studies of regional or synapse specific plasticity, which is likely to happen in vivo under both physiological and pathological conditions.

Though the anatomical connectome is important, anatomical information alone is not sufficient to elucidate brain function. Determining the connectivity of identified populations of neurons will be crucial for understanding how neuronal wiring diagrams correlate with function. Comprehensive mapping of the functional connectome is one of the ways to explore how our brain works, and optogenetics shines light to show the direction.

References

Al-Juboori SI, Dondzillo A, Stubblefield EA et al (2013) Light scattering properties vary across different regions of the adult mouse brain. PLoS One 8(7):e67626. doi:10.1371/journal.pne.0067626

Beckstead MJ, Grandy DK, Wickman K et al (2004) Vesicular dopamine release elicits an inhibitory postsynaptic current in midbrain dopamine neurons. Neuron 42(6):939–946

Belujon P, Grace AA (2011) Hippocampus, amygdala, and stress: Interacting systems that affect susceptibility to addiction. Annal NY Acad Sci 1216:114–121

Bertran-Gonzalez J, Bosch C, Maroteaux M et al (2008) Opposing patterns of signaling activation in dopamine D1 and D2 receptor-expressing striatal neurons in response to cocaine and haloperidol. J Neurosci 28(22):5671–5685

Bocklisch C, Pascoli V, Wong JCY et al (2013) Cocaine disinhibits dopamine neurons by potentiation of GABA transmission in the ventral tegmental area. Science 341(6153):1521–1525

Brown MTC, Tan KR, O'Connor EC et al (2013) Ventral tegmental area GABA projections pause accumbal cholinergic interneurons to enhance associative learning. Nature 492(7429):452–456

Cazorla M, de Carvalho FD, Chohan MO et al (2014) Dopamine D2 receptors regulate the anatomical and functional balance of basal ganglia circuitry. Neuron 81(1):153–164

Chuhma N, Zhang H, Masson J et al (2004) Dopamine neurons mediate a fast excitatory signal via their glutamatergic synapses. J Neurosci 24(4):972–981

Chuhma N, Tanaka KF, Hen R et al (2011) Functional connectome of the striatal medium spiny neuron. J Neurosci 31(4):1183–1192

Chuhma N, Mingote S, Moore H et al (2014) Dopamine neurons control striatal cholinergic neurons via regionally heterogeneous dopamine and glutamate signaling. Neuron 81(4):901–912
Cooper AJ, Stanford IM (2000) Electrophysiological and morphological characteristics of three subtypes of rat globus pallidus neurone in vitro. J Physiol 527(2):291–304
Dautan D, Huerta-Ocampo I, Witten IB et al (2014) A major external source of cholinergic innervation of the striatum and nucleus accumbens originates in the brainstem. J Neurosci 34(13):4509–4518
Ding JB, Oh W-J, Sabatini BL et al (2012) Semaphorin 3E-plexin-D1 signaling controls pathway-specific synapse formation in the striatum. Nature Neurosci 15(2):215–223
Ellender TJ, Harwood J, Kosillo P et al (2013) Heterogeneous properties of central lateral and parafascicular thalamic synapses in the striatum. J Physiol 591(1):257–272
Exley R, Cragg SJ (2008) Presynaptic nicotinic receptors: A dynamic and diverse cholinergic filter of striatal dopamine neurotransmission. Brit J Pharmacol 153(Suppl 1):S283–S297
Feinberg EH, Vanhoven MK, Bendesky A et al (2008) GFP reconstitution across synaptic partners (GRASP) defines cell contacts and synapses in living nervous systems. Neuron 57(3):353–363
Fremeau RT, Burman J, Qureshi T et al (2002) The identification of vesicular glutamate transporter 3 suggests novel modes of signaling by glutamate. Proc Nat Acad Sci 99(22):14488–14493
Friston KJ, Frith CD, Liddle PF et al (1993) Functional connectivity: The principal-component analysis of large (PET) data sets. J Cereb Blood Flow Metab 13(1):5–14
Fujiyama F, Sohn J, Nakano T et al (2011) Exclusive and common targets of neostriatofugal projections of rat striosome neurons: A single neuron-tracing study using a viral vector. Eur J Neurosci 33(4):668–677
Gangarossa G, Espallergues J, de Kerchove d'Exaerde A et al. (2013) Distribution and compartmental organization of gabaergic medium-sized spiny neurons in the mouse nucleus accumbens. Front Neural Circuits 7:22. doi: 10.3389/fncir.2013.00022
Gantz SC, Bunzow JR, Williams JT (2013) Spontaneous inhibitory synaptic currents mediated by a g protein-coupled receptor. Neuron 78(5):807–812
Gerfen CR (1985) The neostriatal mosaic. I Compartmental organization of projections from the striatum to the substantia nigra in the rat. J Comp Neurol 236(4):454–476
Gerfen CR, Surmeier DJ (2011) Modulation of striatal projection systems by dopamine. Ann Rev Neurosci 34:441–466
Graybiel AM (2008) Habits, rituals, and the evaluative brain. Ann Rev Neurosci 31:359–387
Haber SN, Fudge JL, McFarland NR (2000) Striatonigrostriatal pathways in primates form an ascending spiral from the shell to the dorsolateral striatum. J Neurosci 20(6):2369–2382
Higley MJ, Gittis AH, Oldenburg IA et al (2011) Cholinergic interneurons mediate fast VGLUT3-dependent glutamatergic transmission in the striatum. PLoS One 6(4):e19155. doi:10.1371/journal.pone.0019155
Ikemoto S (2007) Dopamine reward circuitry: Two projection systems from the ventral midbrain to the nucleus accumbens-olfactory tubercle complex. Brain Res Rev 56(1):27–78
Jackman SL, Beneduce BM, Drew IR et al (2014) Achieving high-frequency optical control of synaptic transmission. J Neurosci 34(22):7704–7714
Kita H (2007) Globus pallidus external segment. Prog Brain Res 160:111–133
Kreitzer AC (2009) Physiology and pharmacology of striatal neurons. Ann Rev Neurosci 32:127–147
Lichtman JW, Sanes JR (2008) Ome sweet ome: What can the genome tell us about the connectome? Curr Opin Neurobiol 18(3):346–353
Lin JY, Knutsen PM, Muller A et al (2013) ReaChR: a red-shifted variant of channelrhodopsin enables deep transcranial optogenetic excitation. Nat Neurosci 16(10):1499–1508
Mastro KJ, Bouchard RS, Holt HAK et al (2014) Transgenic mouse lines subdivide external segment of the globus pallidus (GPe) neurons and reveal distinct gpe output pathways. J Neurosci 34(6):2087–2099
Mayford M, Bach ME, Huang YY et al (1996) Control of memory formation through regulated expression of a CaMKII transgene. Science 274(5293):1678–1683

Mingote S, Chuhma N, Kusnoor SV et al (2012) The VTA dopamine neuron excitatory functional connectome. In: Champalimaud Neuroscience Symposium 2012, Champalimaud Centre for the Unknown, Lisbon, 30 Sep–3 Oct 2012

Oldenburg IA, Ding JB (2011) Cholinergic modulation of synaptic integration and dendritic excitability in the striatum. Curr Opin Neurobiol 21(3):425–432

Pennartz CMA, Berke JD, Graybiel AM et al (2009) Corticostriatal interactions during learning, memory processing, and decision making. J Neurosci 29(41):12831–12838

Seung HS (2009) Reading the book of memory: Sparse sampling versus dense mapping of connectomes. Neuron 62(1):17–29

Shepherd GM (1994) Neurobiology, 3rd edn. Oxford University Press, Oxford

Sporns O, Tononi G, Kötter R (2005) The human connectome: a structural description of the human brain. Plos Comput Biol 1(4):e42. doi:10.1371/journal.pcbi.0010042

Stuber GD, Hnasko TS, Britt JP et al (2010) Dopaminergic terminals in the nucleus accumbens but not the dorsal striatum corelease glutamate. J Neurosci 30(24):8229–8233

Stuber GD, Sparta DR, Stamatakis AM et al (2011) Excitatory transmission from the amygdala to nucleus accumbens facilitates reward seeking. Nature 475(7356):377–380

Sulzer D, Joyce MP, Lin L et al (1998) Dopamine neurons make glutamatergic synapses in vitro. J Neurosci 18(12):4588–4602

Szydlowski SN, Pollak Dorocic I, Planert H et al (2013) Target selectivity of feedforward inhibition by striatal fast-spiking interneurons. J Neurosci 33(4):1678–1683

Taverna S, Ilijic E, Surmeier DJ (2008) Recurrent collateral connections of striatal medium spiny neurons are disrupted in models of Parkinson's disease. J Neurosci 28(21):5504–5512

Tecuapetla F, Patel JC, Xenias H et al (2010) Glutamatergic signaling by mesolimbic dopamine neurons in the nucleus accumbens. J Neurosci 30(20):7105–7110

Tepper JM, Koós T, Wilson CJ (2004) Gabaergic microcircuits in the neostriatum. Trend Neurosci 27(11):662–669

Tepper JM, Tecuapetla F, Koos T et al (2010) Heterogeneity and diversity of striatal gabaergic interneurons. Front Neuroanat 4:150. doi:10.3389/fnana.2010.00150

Tritsch NX, Ding JB, Sabatini BL (2012) Dopaminergic neurons inhibit striatal output through non-canonical release of GABA. Nature 490(7419):262–266

Tritsch NX, Oh WJ, Gu C et al (2014) Midbrain dopamine neurons sustain inhibitory transmission using plasma membrane uptake of GABA, not synthesis. eLife 3:e01936. doi:10.7554/eLife.01936

Van Bockstaele EJ, Pickel VM (1995) GABA-containing neurons in the ventral tegmental area project to the nucleus accumbens in rat brain. Brain Res 682(1–2):215–221

Watabe-Uchida M, Zhu L, Ogawa SK et al (2012) Whole-brain mapping of direct inputs to midbrain dopamine neurons. Neuron 74(5):858–873

Wilson CJ (2004) The basal ganglia. In: Shepherd GM (ed) The synaptic organization of the brain, 5th edn. Oxford University Press, Oxford, pp 361–413

Xia Y, Driscoll JR, Wilbrecht L et al (2011) Nucleus accumbens medium spiny neurons target non-dopaminergic neurons in the ventral tegmental area. J Neurosci 31(21):7811–7816

Yamaguchi T, Sheen W, Morales M (2007) Glutamatergic neurons are present in the rat ventral tegmental area. Eur J Neurosci 25(1):106–118

Yin HH, Knowlton BJ (2006) The role of the basal ganglia in habit formation. Nat Rev Neurosci 7(6):464–476

Yizhar O, Fenno LE, Davidson TJ et al (2011) Optogenetics in neural systems. Neuron 71(1):9–34

Chapter 19
Potential of Optogenetics for the Behavior Manipulation of Non-human Primates

Masaharu Kinoshita and Tadashi Isa

Abstract Optogenetics, which enables the manipulation of neural activities in millisecond order by light stimulation, is new and useful technology for neuroscience research. With this technology, many important issues of neural functions have been elucidated with causal arguments in the small model animals, such as mice. For the understanding of the higher brain function, the applications of optogenetics for non-human primates, such as macaque monkeys, are expected. However, only a few studies have been reported about the optogenetic manipulation of non-human primates' behavior. In this chapter, we highlight the difficulties in applying optogenetics on non-human primates, and then review the current situations in this study field, and discuss the future perspectives, including circuit-level manipulations by optogenetics. To advance optogenetic studies in non-human primates, several technical improvements—efficient gene expression using viral vectors, cell-type-specific or pathway-specific gene expression and efficient light delivery in a large brain—are required.

Keywords Non-human primates • Macaque monkey • Behavior manipulation • Neural circuit manipulation • Pathway selective manipulation • Retrograde viral vector • ChR2 • eNpHR • Optogenetics

19.1 Introduction

Optogenetics enables us to manipulate neural activities (activation and suppression) in millisecond order. This technique has opened up a new way for analyzing neural circuit functions. At the first stage of optogenetic studies, a gene of light-sensitive

M. Kinoshita (✉)
Department of Physiology, Hirosaki University Graduate School of Medicine, Hirosaki, Japan
e-mail: kinoshita-m@umin.ac.jp

T. Isa
Department of Developmental Physiology,
National Institute for Physiological Sciences, Okazaki, Japan
e-mail: tisa@nips.ac.jp

H. Yawo et al. (eds.), *Optogenetics*, DOI 10.1007/978-4-431-55516-2_19

membrane protein (opsin) was introduced into small animals, such as mice, zebrafish, *Caenorhabditis elegans*, and *Drosophila*, by creating transgenic animals. In these studies, targeted neural activities were manipulated by light stimulation and then many important issues of neural circuit functions have been elucidated with causal arguments (Gradinaru et al. 2009; Cardin et al. 2009; Liu et al. 2012).

In the next stage of optogenetic study, applications for non-human primates (NHPs) are expected for understanding of the higher brain functions and for the translational research of human disease therapies. However, many researchers have faced difficulties in applying optogenetics in NHPs. This is due to the following reasons.

1. Efficiency of gene expression in NHPs. In small model animals, such as mice, the creation of transgenic animals is established. In transgenic animals, a higher level of expression of opsin in each cell can be expected. Conversely, for NHPs, a realistic way to introduce a gene is using a viral vector, by which gene expression is relatively weak. In order to manipulate the behavior of large brain animals such as NHPs, highly efficient gene expression in a large number of neurons in a wide brain area is required. But achievement of this is still not easy.
2. Cell-type-specific gene expression in NHPs. Even without achieving behavior manipulation, if cell-type-specific expression of an introduced gene is achieved, this technique could be useful for cell identification. However, there is a limitation in the length of gene able to be carried on the lentivirus and adeno-associated virus (AAV) vectors. Because of this length limitation, cell-type-specific expression using cell-specific promoters has not been achieved.
3. Light delivery in the large NHP brain. To achieve behavior manipulation, it is necessary to illuminate wide area of the brain with enough strength of light, even if it is at a deep area of brain. This is still not an easy task.

Even though there are several difficulties, many researchers, including ourselves, have made an effort to improve optogenetics technology. Furthermore, several studies of behavior manipulation in NHPs have been reported since 2012. With continuous effort on this innovative optogenetics method, breakthrough studies on higher brain functions in NHPs are anticipated in the near future. In this chapter, we outline the current situation in this study field.

19.2 Current Studies

There are at least two sides to the optogenetic manipulation of neural activity. One is the manipulation of an individual neuron's activity. The other is the manipulation of animal behavior.

For the manipulation of individual neurons, sufficient expression of opsin is required, even if it is limited to the local area around the vector injection site. In 2009, Boyden's team at MIT used lentivirus injected into the frontal lobe to express channelrhodopsin-2 (ChR2) in macaque monkeys. They used laser light and successfully elicit firing in individual neurons within the order of milliseconds and also found suppression in a portion of the neurons. Furthermore, Boyden's team showed long-term, stable expression without inflammation or tissue damage (Han et al. 2009). In 2011, using Arch-T opsin, this team reported the suppression of neural activity at the injection site in the cerebral cortex (Han et al. 2011). Meanwhile, in 2010, Deisseroth and Shenoy's group at Stanford University reported excitation and suppression of neural firing by introducing ChR2, enhanced halorhodopsin (eNpHR) and step-function opsin with lentiviral and AAV vectors injected into the cerebral cortex of macaque monkeys (Diester et al. 2011). However, despite these cell-level successes, as yet there was still no report about the manipulation of the behavior of NHPs.

Eventually, behavioral manipulation in NHPs began to be reported in 2012. First, Geris et al. (2012) reported the modulation of latency of visually guided saccade in macaque monkeys by light stimulation to ChR2-expressing neurons in the arcuate sulcus and also reported activity change in a specific brain region revealed via simultaneous functional magnetic resonance imaging (fMRI). Jazayeri et al. (2012) reported that light stimulation of the primary visual cortex (V1), the neurons in which expressed ChR2, can elicit the saccade to the visual field corresponding to the light stimulation site in V1 in macaque monkeys. Furthermore, Cavanaugh et al. (2012) reported that light stimulation to the mid to deep layers of the superior calculus, expressing Arch-T, caused a shift in saccade endpoints, a delay in response latency, and a decrease of saccade velocity, which are similar to the effects elicited by the γ-aminobutyric acid $(GABA)_{-A}$ agonist muscimol or lidocaine injection. At the Meeting of the Society for Neuroscience held in the same year, several studies related to the optogenetic manipulation of NHPs' behavior were reported as follows.

1. The behavior in visual detection tasks was modulated by light stimulation in V1 (Andrei et al. 2012).
2. The perceptual processes in attentional tasks was biased by light stimulation in the frontal eye field (Ohayon et al. 2012).
3. The disruption of motor preparation processes by light stimulation in the primary motor cortex (M1) (Goo et al. 2012).
4. Selection in the attentional task was biased by light stimulation in the parietal association area (Dai et al. 2012).
5. Modulation of behavioral selection strategy was achieved by light stimulation in the locus coeruleus (Cope et al. 2012).
6. The reaction time of reaching behavior was modulated by light stimulation in M1 (Wang et al. 2012).

However, all these studies were observations of subtle changes in behavior that had already been predicted, thus none of these reports provided new or innovative findings for neuroscience research. However, many improvements and new knowledge and technologies in optogenetic methods are accumulated. In the near future, the important problems in the neuroscience research, which was impossible to solve with traditional methods, may be uncovered using optogenetics in NHPs.

Between the two levels of manipulation by optogenetics, mentioned above—individual cell level and animal behavior level—there exists the other level of manipulation: 'neural circuit' level.

One way to achieve circuit-level manipulation is to test the modulation at the projection area of the neurons that express an opsin, by light stimulation to the cell bodies. Even if no behavior was modulated, such manipulation is helpful to explore the function of this projection. In these aspects, we intended to manipulate the reaching and grasping behavior of macaque monkeys by light stimulation in the M1, which expressesd opsins. We introduced eNpHR into the macaque M1 hand–arm region using AAV2 vector injection. We penetrated the metal microelectrode glued to optical fiber (optorode) connected to the light source into the M1 injection area and also recorded electromyograms (EMGs) from hand and arm muscles. Via light stimulation around the eNpHR-expressing M1 area, we found not only suppression of spontaneous M1 neuron activity (Fig. 19.1) but also suppression of the M1 activity related to reaching and grasping behavior (Fig. 19.2). However, we found no obvious modulation of reaching and grasping behavior in itself. Even though no clear behavior modulation was observed, we found a reversed effect on the averaged EMGs elicited by light stimulation against that elicited by electrical stimulation in the macaque M1 (Fig. 19.3). This result implies that optogenetic stimulation is useful to reconfirm the cortical map without the suspicion associated with traditional electrical stimulation methods, which can stimulate not only the cell body but also passing fibers near the electrode.

Another method for circuit-level manipulation is the selective gene introduction method, which allows gene expression limited to a specific pathway of neurons using retrograde transport virus vector. Lima et al. injected herpes simplex virus 1 (HSV1) carrying the ChR2 as a retrograde vector at the projecting area of target neurons. They then recorded neural activities in the area at which the cell body of the target neurons exist, and applied light stimulation. If the recorded neuron responded to light stimulation, it could be identified as the target neuron that projects to the injection area. This novel method, called 'Photostimulation-assisted Identification of Neuronal Populations' (PINP) (Lima et al. 2009), has potential as a useful method to identify cell types in vivo; however, a highly efficient retrograde gene transport vector is required (see below, HiRet).

19.3 Future Perspectives

19.3.1 Innovations for Gene Introduction and Light Stimulation

In order to achieve behavior manipulation by optogenetics in NHPs, the following improvements are necessary:

1. Higher efficiency of gene introduction and expression for a large number of neurons.

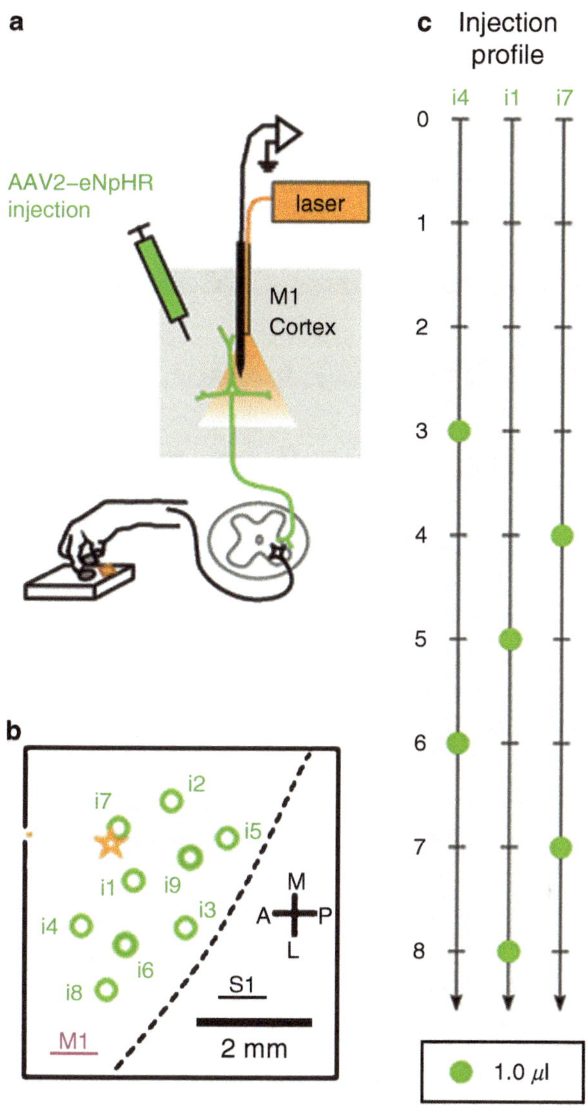

Fig. 19.1 Optogenetic suppression of neural activities in a macaque primary motor cortex (M1). (**a**) Schematic drawing of the experiment. The adeno-associated virus (AAV)-2 vector, which carried an enhanced halorhodopsin (eNpHR) gene (AAV2-eNpHR), was injected into macaque M1. By using the optorode (thin optical fiber glued with micro electrode), neural activities were recorded and light stimulation was applied around the tip of the recording electrode. (**b**) Injection map (surface view of the left hemisphere around the central sulcus). AAV2-eNpHR (titer 1.1×10^{10} particles/μl) was injected at 19 sites in nine penetrations (*green circles* i1–i9) in M1. Total injection volume was 20 μl. The *orange star* represents the example penetration shown in **d**. (**c**) The depth profile of injection sites of representative penetrations. Vector solution (1 μl/site) was injected at the site depicted by *green circles*. (**d**) Suppression of neural activities recorded at the track shown in **b** (*orange star*) by light stimulation (wavelength 561 nm laser, less than 50 mW). *Black histograms* represent averaged multi-neuron spike activities recorded at the site indicated at left depth profile without light stimulation. The *black number* at the right represents the number of averaged trials for the black histogram. *Red histograms* represent averaged activities recorded around the light stimulation. The *orange rectangle* represents the duration of light stimulation. At many recording sites, suppression of spike activities were seen during light stimulation via the optorode

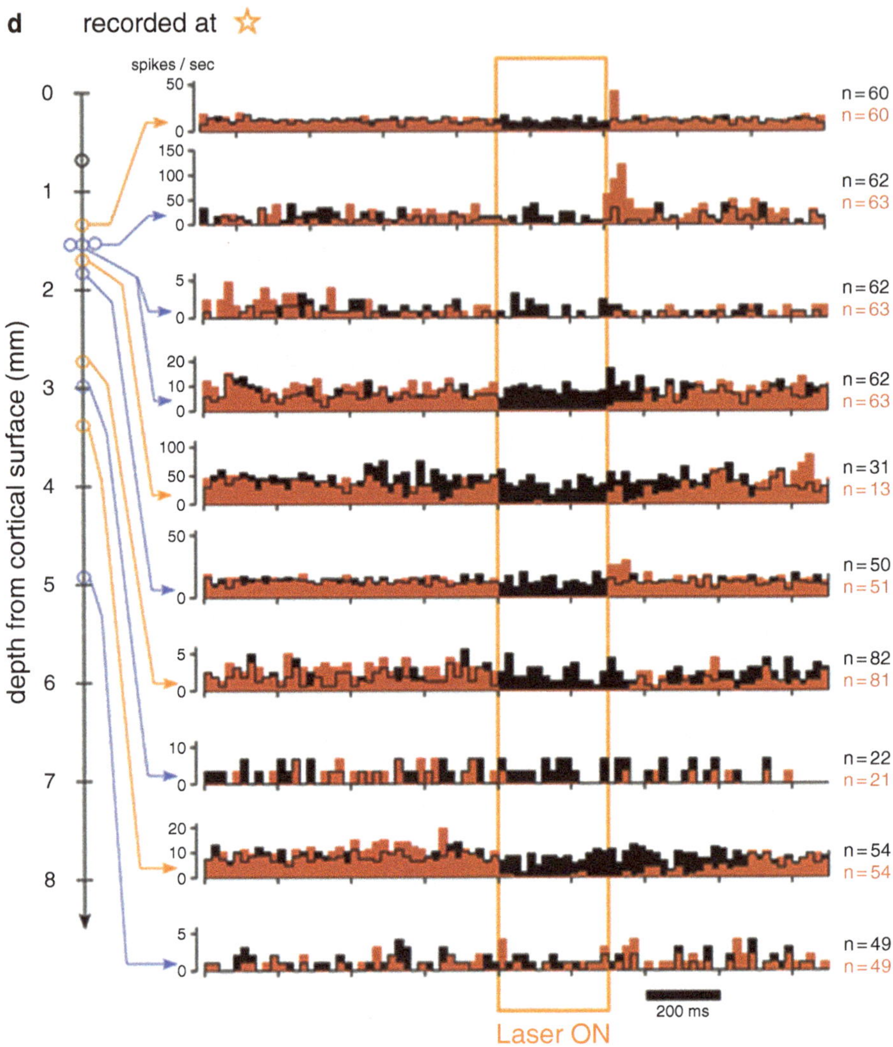

Fig. 19.1 (continued)

2. Delivery of light with enough strength for wide and deep areas of a brain.

First, the optimization of virus vectors, particularly optimization of the promoter and serotype of the AAV vector, is required. Improvement of the amplification system for gene expression, such as Tet system, may also be important. Furthermore, opsins are improving rapidly for better photosensitivity, stronger photocurrents, and useful time course of photoreaction. The light delivery techniques are also improving day by day. For example, Miyashita's group at the University of Tokyo has developed the glass-coated optorode, which has multiple optical fibers and is suitable for penetration into the deep structures of the large brain (Tamura et al. 2012).

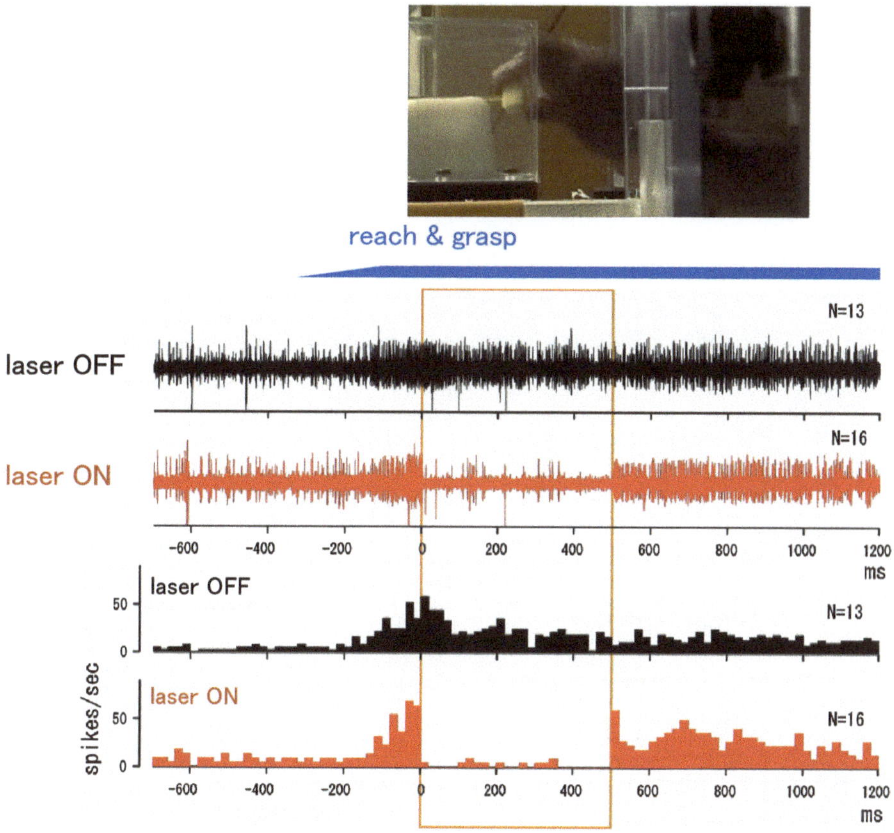

Fig. 19.2 Suppression of movement-related M1 neuron activities recorded in the awake behaving macaque monkey. (*Inset*) An example scene: the monkey performed the reach and grasp task with the right hand. (*Top*) Superimposed M1 activities around the onset of the hand movement. Traces are aligned by the timing of the infra-red beam, placed at the center panel shown in above inset photo, was cut by a monkey hand (time = 0 ms). *Black traces* were recorded without light stimulation (N = 13 trials). *Red traces* were recorded under light stimulation for 500 ms duration indicated by an *orange rectangle* (N = 16 trials). Movement-related activities were suppressed by light stimulation via the optorode. (*Bottom*) Histogram of averaged multi-neuron spike activities shown in the above traces

Lentivirus and AAV are frequently used, because of their higher efficiency for gene introduction to neurons. The lentivirus vector has been frequently utilized, because of its relative ease of creation. However, lentivirus is required to be handled under the P2 biosafety level regulation and is of relatively low efficiency of infection compared with AAV. Therefore, some AAV variants are being utilized as a first choice. On the other hand, the retrograde transport lentivirus has been developed via modification of its glycoproteins (Kato et al. 2007). Furthermore, Kobayashi, et al. improved the efficiency of the retrograde transfer rate by making the chimera glycoprotein from vesicular stomatitis virus and rabies virus. It was named ‘highly

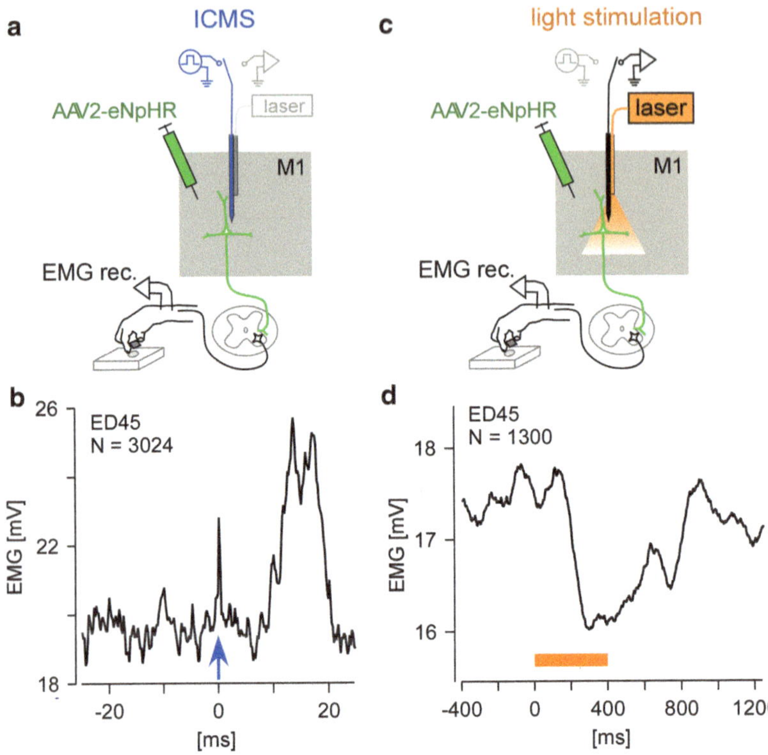

Fig. 19.3 The light stimulation in M1 affected the averaged hand muscle activities of the awake behaving macaque. (**a**) The optorode was inserted into the enhanced halorhodopsin (eNpHR) introduced M1 area, and intra-cortical micro stimulations (ICMS) were made. Simultaneously, electromyograms (EMGs) of hand muscles were recorded while the monkey performed the reaching and grasping task. Repetitive ICMS were applied asynchronously to the timing of hand movement. (**b**) The averaged EMG (ED45, extensor digitorum 4 and 5. Full wave rectified then averaged, N = 3,024 trials) aligned by the timing of ICMS (0 ms, *blue arrow*). At this optorode position, enhanced EMG activity was seen with 10 ms delay from ICMS. (**c**) At the same optorode position, light stimulation was applied repeatedly without ICMS while the monkey performed the task. (**d**) The averaged EMG (ED45, N = 1,300 trials) aligned by the light stimulation (0–400 ms, *orange bar*). The EMG activity was suppressed after the light stimulation

efficient retrograde gene transfer vector' (HiRet) (Kato et al. 2011). For pathway-selective gene introduction, such a pseudotyped lentivirus is utilized as a first choice.

In some countries, AAV can be handled under the P1 biosafety level regulation. The AAV family includes more than ten subtypes. For transfection to neurons, serotypes 1, 2, 5, 8, and 9-AAV are often utilized. Each serotype has a different affinity with the different areas of the brain, with such differences in affinities and toxicities also varying between species. Therefore, a serotype that worked efficiently within a specific brain area within a mouse would not necessarily be directly used to target the same brain area in an NHP. Furthermore, as a large amount of vector solution needs to be injected, the purification process is important to avoid an inflammatory response. Choice of a promoter is also important. Although highly versatile promoters such as cytomegalovirus (CMV) yield a high level of gene expression, issues including tissue damage and silencing have been observed with long-term use. For neurons, in general, a synapsin promoter is often chosen, while a CaMKIIα promoter is used for excitatory neurons. Currently, the authors are undertaking an investigation into the selection of the various serotypes (AAV-1, 2, 5, 8, and 9) (Watakabe et al. 2014). In previous studies, AAV-1 or 5 is most frequently used for the cerebral cortex (Masamizu et al. 2011). A high level of expression is obtained in the brain stem, spinal cord, and the retina using AAV2. While, although the infection efficiency is high at the central area of the injection site when used within the cerebral cortex, this expression does not spread much further.

A selective expression of opsins for a specific neural pathway or for a specific cell type is important for the analysis of neural circuit functions. One breakthrough in relation to such a requirement is the HiRet by Kobayashi et al. mentioned earlier. HiRet is anticipated as being useful in the identification method of neurons (PINP) as outlined earlier.

Furthermore, using HiRet, the authors developed a novel synaptic transmission blocking method that is pathway selective and reversible, and have achieved behavior manipulation in macaque monkeys (Kinoshita et al. 2012). As seen in Fig. 19.4a, we injected the HiRet, which carries tetracycline-responsive element (TRE), enhanced tetanus neurotoxin (eTeNT), and enhanced green florescent protein (EGFP) gene (HiRet-TRE-eTeNT.EGFP), into the motor nucleus innervating hand–arm muscles in the spinal cord (C6-Th1), which is the projection area of the propriospinal neurons (PNs). Afterward, we injected AAV2, which carries CMV promoter and the highly efficient reverse tetracycline transactivator (rtTAV16) developed by Watanabe et al. (AAV2-CMV-rtTAV16), into the C2–C5 spinal cord, where PN cell bodies exist. Then, only the PNs were double infected by two vectors. Under the administration of doxycycline (Dox), the synaptic transmission of PNs was blocked because of eTeNT expression in the PNs. A few days after oral administration of Dox, we found deficits in the precision grip and/or reaching behavior in the monkeys. Thus, we concluded that PNs play a role in the control of the precision grip and reaching behaviors. Such control using the Tet system is relatively slow compared with optogenetics, being within the order of days. However, one of the above-mentioned difficulties of optogenetics, the issue of light delivery into deep tissue, has been overcome through the systemic application of Dox. Therefore, this method should become beneficial for the study of the function of specific neural pathways, if the slow time course is not important. A future task is to replace eTeNT, used in the double infection method, with an opsin, as seen in Fig. 19.4b, thereby realizing pathway-selective optogenetic manipulation.

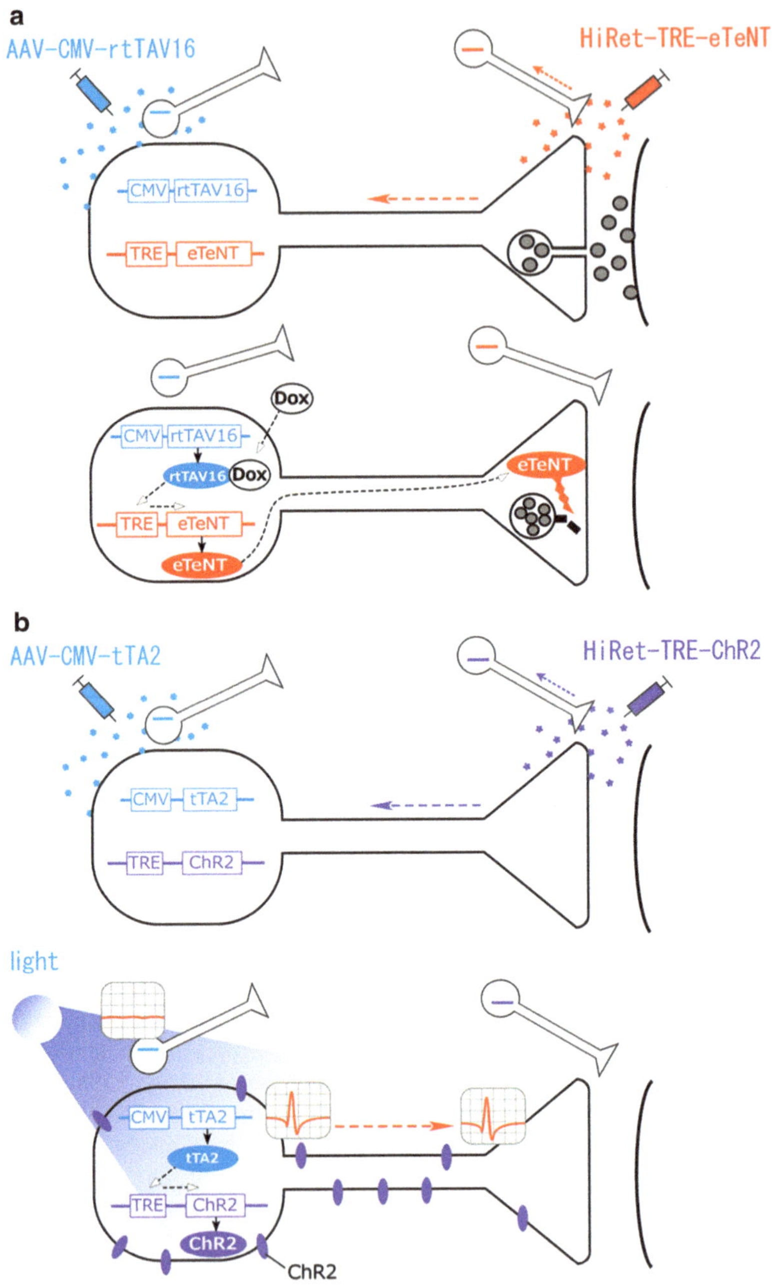

Fig. 19.4 Methods for pathway-specific and reversible manipulation of neural activity. (**a**) The method of dissecting the synaptic transmission by pathway-specific expression of enhanced tetanus neurotoxin (eTeNT). The retrograde lentiviral vector (HiRet-TRE-eTeNT) is injected at the terminal region and then the anterograde adeno-associated virus (AAV) vector (AAV-CMV-rtTAV16) is injected at the soma region. As a result, the double infection occurs only in the neurons connecting two specific regions. The expression of eTeNT can be induced only in these double-infected neurons under the application of doxycycline (Dox). (**b**) The potential method for the selective activation of neurons in a specific pathway with millisecond order. By using an anterograde vector carrying channelrhodopsin-2 (ChR2) gene instead of eTeNT, ChR2 will be expressed only in the neurons connecting two specific regions

19.3.2 Transgenic Marmoset for Optogenetics

As mentioned above, we discussed the future perspective of optogenetics in NHPs using viral vectors. Alternatively, there is another method for the use of optogenetics in NHPs with the creation of transgenic NHPs, for more robust expression of opsin and robust manipulation of neural activity. In 2009, Sasaki, at the Central Institute for Experimental Animals, and Okano et al., at Keio University, have established a lineage of transgenic marmosets that systemically express GFP (Sasaki et al. 2009). Currently, transgenic marmoset lineages, useful for neuroscience studies, are being created one after another. Using this technique, there is no reason why a marmoset lineage expressing opsins in the brain cannot be established. In comparison with a macaque, a marmoset has a small, thin, and smooth (few sulcus) cerebral cortex, which is suitable for optogenetic studies. For this reason, the establishment of a research paradigm for the analysis of higher brain function using marmosets is expected.

References

Andrei AR, Pologa S, Janz R et al (2012) Optogenetic stimulation of V1 modulates behavior in a visual detection task. In: Abstract for 42nd annual meeting of the society for neuroscience, New Orleans, 570.15

Cardin JA, Carlen M, Meletis K et al (2009) Driving fast-spiking cells induces gamma rhythm and controls sensory responses. Nature 459:663–667

Cavanaugh J, Monosov IE, McAlonan K et al (2012) Optogenetic inactivation modifies monkey visuomotor behavior. Neuron 76(5):901–907

Cope ZA, Vazey E, Moorman D et al (2012) Optogenetically mediated tonic stimulation of locus coeruleus improves maintenance of novel strategies on extradimensional shifts. In: Abstract for 42nd annual meeting of the society for neuroscience, New Orleans, 208.08

Dai J, Brooks DI, Sheinbeg DL (2012) Modulating attention by optogenetic stimulation in the non-human primate. In: Abstract for 42nd annual meeting of the society for neuroscience, New Orleans, 229.08

Diester I, Kaufman MT, Mogri M et al (2011) An optogenetic toolbox designed for primates. Nat Neurosci 14:387–397

Gerits A, Farivar BR, Rosen LL et al (2012) Optogenetically induced behavioral and functional network changes in primates. Curr Biol 22(18):1722–1726

Goo W, O'shea D, Kalanithi P et al (2012) Disruption of motor preparation by optogenetic stimulation of primate premotor cortex. In: Abstract for 42nd annual meeting of the society for neuroscience, New Orleans, 229.06

Gradinaru V, Mogri M, Thompson KR et al (2009) Optical deconstruction of Parkinsonian neural circuitry. Science 324:354–359

Han X, Qian X, Bernstein JG et al (2009) Millisecond-timescale optical control of neural dynamics in the nonhuman primate brain. Neuron 62(2):191–198

Han X, Chow BY, Zhou H et al (2011) A high-light sensitivity optical neural silencer: development and application to optogenetic control of non-human primate cortex. Front Syst Neurosci 5:18. doi:10.3389/fnsys.2011.00018

Jazayeri M, Lindbloom Z, Horwitz GD (2012) Saccadic eye movements evoked by optogenetic activation of primate V1. Nat Neurosci 15:1368–1370

Kato S, Inoue K, Kobayashi K et al (2007) Efficient gene transfer via retrograde transport in rodent and primate brains using a human immunodeficiency virus type 1-based vector pseudotyped with rabies virus glycoprotein. Hum Gene Ther 18:1141–1152

Kato S, Kobayashi K, Inoue K et al (2011) A lentiviral strategy for highly efficient retrograde gene transfer by pseudotyping with fusion envelope glycoprotein. Hum Gene Ther 22:197–206

Kinoshita M, Matsui R, Kato S et al (2012) Genetic dissection of the circuit for hand dexterity in primates. Nature 487:235–238

Lima SQ, Hromádka T, Znamenskiy P et al (2009) PINP: a new method of tagging neuronal populations for identification during *in vivo* electrophysiological recording. PLoS One 4(7):e6099. doi:10.1371/journal.pone.0006099

Liu Y, Ramirez S, Pang PT et al (2012) Optogenetic stimulation of a hippocampal engram activates fear memory recall. Nature 484:381–385

Masamizu Y, Okada T, Kawasaki T et al (2011) Local and retrograde gene transfer into primate neuronal pathways via adeno-associated virus serotype 8 and 9. Neuroscience 193:249–258

Ohayon S, Grimaldi P, Schweers N et al (2012) Perceptual bias induced by optical stimulation in the Macaque frontal eye field. In: Abstract for 42nd annual meeting of the society for neuroscience, New Orleans, 229.13

Sasaki E, Suemizu H, Shimada A et al (2009) Generation of transgenic non-human primates with germline transmission. Nature 459:523–527

Tamura K, Ohashi Y, Tsubota T et al (2012) A glass-coated tungsten microelectrode enclosing optical fibers for optogenetic exploration in primate deep brain structures. J Neurosci Method 211:29–57

Wang J, Ozden I, Lu Y et al (2012) Single and multi-element integrated optoelectronic devices for combined optical stimulation and electrophysiology in non-human primates. In: Abstract for 42nd annual meeting of the society for neuroscience, New Orleans, 610.07

Watakabe A, Ohtsuka M, Kinoshita M et al (2014) Comparative analyses of adeno-associated viral vector serotypes 1, 2, 5, 8 and 9 in marmoset, mouse and macaque cerebral cortex. Neurosci Res doi:10.1016/j.neures.2014.09.002 (in press)

Chapter 20
Activity Regulation in the Study of Neural Plasticity

Shoji Komai

Abstract Needless to say, optogenetics has been popular in the manipulation of neural activities in a specific neuron expressing a photosensitive molecule, called rhodopsin. In the neural plasticity field, researchers have been trying to use this tool to precisely manipulate neural activities spatially and chronically. At the same time, optogenetics has helped us regulate the neural activity of a specific neuron expressing a particular gene of interest, such as the inhibitory interneuron for glutamic acid decarboxylases GADs, because a transfected rhodopsin is regulated under a particular promoter of interest. The nature of this tool allows us to control the expression of rhodopsin in the specific neurons in living brain. This means optogenetics helps us to observe animals behavior, such as memory and learning, after stimulating or inhibiting the activities of a neural network in the brain. Optogenetics will be more powerful tool to elucidate the relationship between neural activity and behavior, including neural plasticity. In this chapter, I introduce you to some of examples of optogenetics in neural plasticity research.

Keywords Neural plasticity • Learning and memory • Animal behavior • In vitro • In vivo • Clinical application

20.1 Neural Plasticity

Brains, unlike computers, are self-assembled during each stage of development. The 'hardware' and 'software' in our brains, formed by neural connections, change through the course of our lives. The architecture of the network itself defines the information, resulting in a very complex network. The network is dynamic, and changes on a daily basis, while our identity is maintained. This dichotomy means that understanding how the brain works is quite complicated; it is one of most important issues that neuroscience research needs to address. In order to understand the network, it is necessary to capture the network encoding specific information

S. Komai (✉)
Graduate School of Biological Sciences, Nara Institute of Science and Technology, Nara, Japan
e-mail: skomai@bs.naist.jp

H. Yawo et al. (eds.), *Optogenetics*, DOI 10.1007/978-4-431-55516-2_20

and observe, in detail, the activity of the neural circuit that relates to a specific animal behavior. The relationship between a particular neural activity and a related behavior is likely to change over time. 'Stimulation' and 'destruction' such as electrical stimulation and burnout of the brain, in which physiologists work to identify a particular function of the neural circuit, is one method that can be used to observe this dynamic relationship. Nobel Prize-winning work by David Hubel, Torsten Wiesel, and Roger Sperry captured a snapshot of brain function in the visual cortex and described the laterality of cerebral hemispheres. This 'stimulus' and 'destruction' can be used to improve accuracy. Optogenetics could be a significant tool to observe specific neural activity in networks of interest. In recent years, optogenetic methods have improved, and today it is a standard methodology. In this chapter, we describe and discuss some important work where optogenetics has been used as an investigative tool. Optogenetics represents new hope and an opportunity to shed light on brain activity and function.

20.1.1 Hebb Theory

"The two converging cells need not have any simple anatomical or physiological relation to one another, and physiological integration would not be supposed to consist of independent closed chains". With these words, Canadian psychologist Donald Hebb proposed an idea that forms the basis of our understanding of neural plasticity (Hebb 2002, Fig. 20.1). Basically, he suggested that neural networks alter neural connections in response to brain activity. These ideas were summarized in Hebb's 1949 book, *The Organization of Behavior*. This book is still revised and published today. It is still cited, not only in papers relating to neuronal plasticity but also in papers regarding other neural functions, indicating Hebb's foresight and the continued relevance of his ideas.

20.1.2 Long-Term Depression and Long-Term Potentiation

When the phenomenon of long-term potentiation (LTP) was originally described by Terje Lømo and Timothy Bliss in the rabbit hippocampus in 1966, it was called long-lasting potentiation (Bliss and Lomo 1973). Lømo initially used anesthetized rabbits, but quickly moved on to describe LTP in the hippocampus of conscious animals (Bliss and Gardner-Medwin 1973). Since then, much work has been done to elucidate the fundamental mechanisms of LTP as a "biological model of memory" (Malinow and Malenka 2002, Fig. 20.2).

Long-term depression (LTD) is a phenomenon induced in synapses between cerebellar Purkinje neurons and parallel fibers, first described by Masao Ito (Ito et al. 1982). LTD is thought to be a physiological process that underlies particular behaviors, including the eye blink reflex (Perrett et al. 1993; Ito 2002, Fig. 20.3). The

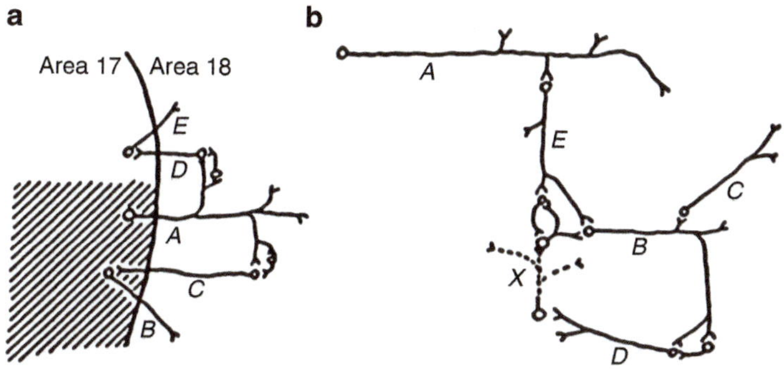

Fig. 20.1 Schematic figures depicted by Hebb. Figure (**a**) indicates that cell C in area 18 excites cell B in area 17, when cell A and B was strongly stimulated through afferents. Innervation of cell E is not strengthened, even though it is located in area 17, as it is located outside the excitation area. Figure (**b**) shows cells A, B, and C, which fire very well in response to a particular visual stimulus, and D, E, and X, which indicate plausible connections. Excitation of A fires E, and excitation of B induces firing of C and D; these firings trigger large changes, and increase the probability of coordination between each pair. (**a**) Cells *A* and *B* lie in a region of area 17 (shown by hatching) which is massively excited by an afferent stimulation. *C* is a cell in area 18 which leads back to 17, *E* is in area 17 but lies outside the region of activity. (**b**) *A*, *B*, and *C* are cells in area 18 which are excited by converging fibers (not shown) leading from a specific pattern of activity in area 17. *D*, *E*, and *X* are, *among the many cells with which* A, B, *and* C *have connections*, ones which would contribute to an integration of their activity. (From Hebb (2002). With permission)

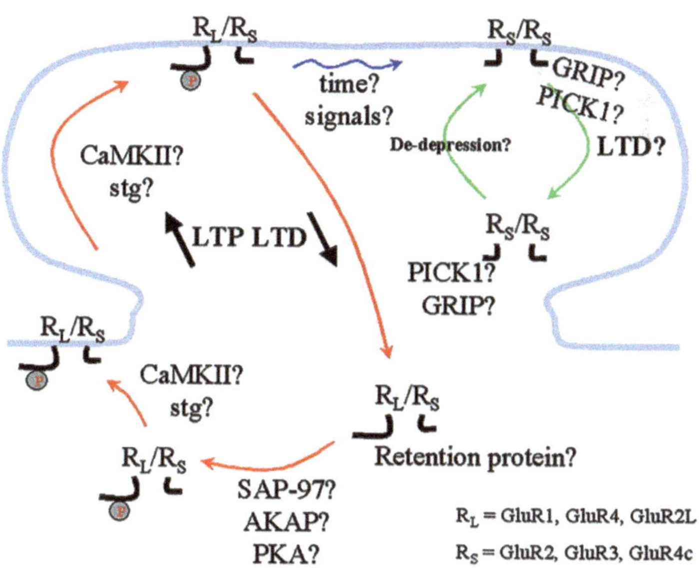

Fig. 20.2 Schematic figure depicting long-term potentiation (LTP) and long-term depression (LTD) and the molecular cascade thought to be involved in each. Loops in *red* and *green* illustrate how synaptic receptors are inserted into, or excised out of, the active zone of the synapse (From Malinow and Malenka (2002). With permission)

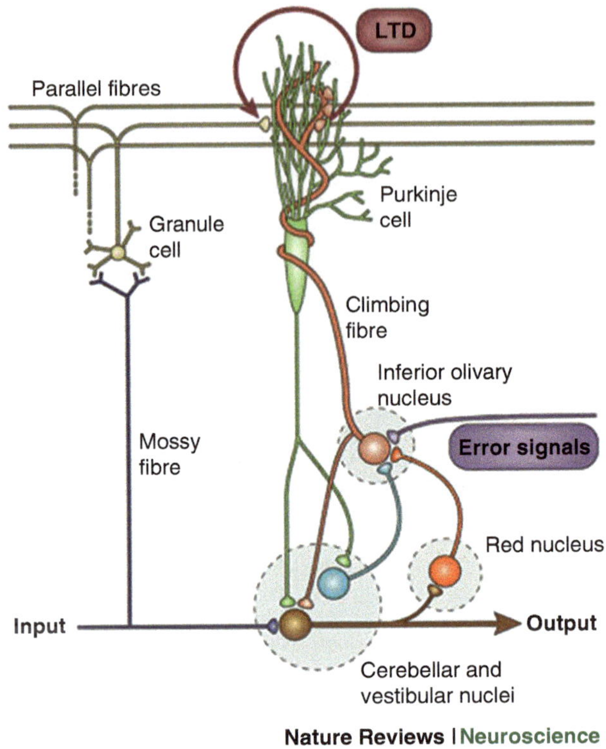

Fig. 20.3 Cerebellar long-term depression (LTD) and motor learning. Motor learning occurs when LTD develops in the synapse between parallel fibers and Purkinje cells; error signals are conveyed to Purkinje cells via climbing fibers (From Ito (2002). With permission)

discovery of LTD strongly suggested that neural plasticity was induced not only in an excitatory direction but also in an inhibitory direction; this interpretation is very important when trying to understand information processing in the brain.

20.1.3 Spike-Timing-Dependent Plasticity

Long-lasting stimulation at high or low frequencies is required for the induction of LTP or LTD. However, a different phenomenon was required to explain, physiologically, Hebb's theory of "fire together, wire together." Henry Markram solved the problem by stimulating both pre- and post-synaptic neurons at different time points (Markram et al. 1997; Markram et al. 2011, Fig. 20.4), and Mu-ming Poo and Guo-qiang Bi (Bi and Poo 1998) crystallized this idea, describing timing-dependent neuronal plasticity, or spike-timing-dependent plasticity. Both groups found that the time difference between firing of pre- and post-synaptic neurons determined

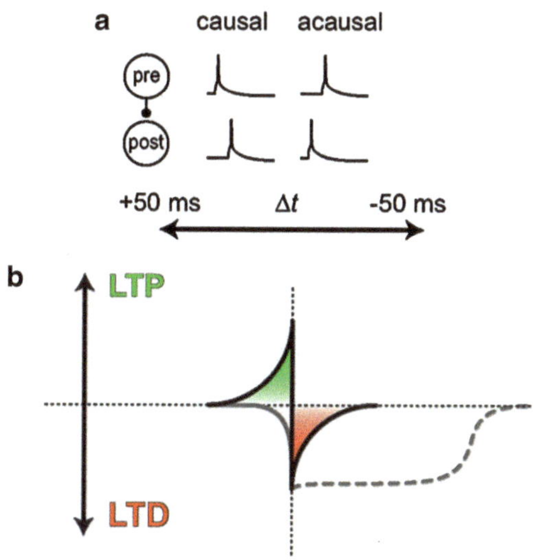

Fig. 20.4 Neural plasticity caused by firing synchronicity (spike-timing dependent plasticity). (**a**) Schematic illustration of membrane potentials pre- and post-synaptically. Pre-synaptic firing followed by post-synaptic firing is 'causal,' post-synaptic firing followed by pre-synaptic firing is 'acausal.' (**b**): Causal firing potentiates the synaptic connection (similar to long-term potentiation [LTP] indicated in *green*), whereas acausal firing inhibits it (similar to long-term depression [LTD] in *red*). *Dashed line* indicates time axis of LTD in some cortical neurons (From Markram et al. (2011). With permission)

the efficiency, or weight, of the synapse. This means that the direction of the strengthening of the synapse is determined by the difference in timing between activity of the post-synaptic cell and input from the pre-synaptic cell, known as coincidence detection. A different profile of coincidence detection has been described in each region of the brain, indicating that each brain area has its own method to determine synapse strength direction. This may explain some of the characteristics of individual brain areas.

20.1.4 Homeostatic Plasticity

The concept of homeostatic plasticity was first explained by Gina Turrigiano, after she described constancy in whole network plasticity in cultured cortical neurons. Homeostatic plasticity helped keep the network stable, avoiding conditions of over-excitement or over-inhibition (Turrigiano et al. 1998; Turrigiano and Nelson 2004, Fig. 20.5). The amplitude of synaptic potentials decreased when γ-aminobutyric acid (GABA) receptor antagonists were applied to cultured cortical neurons, suggesting the presence of a compensatory mechanism that protects the network against over-excitement. Application of tetrodotoxin, which inhibits network activity, has

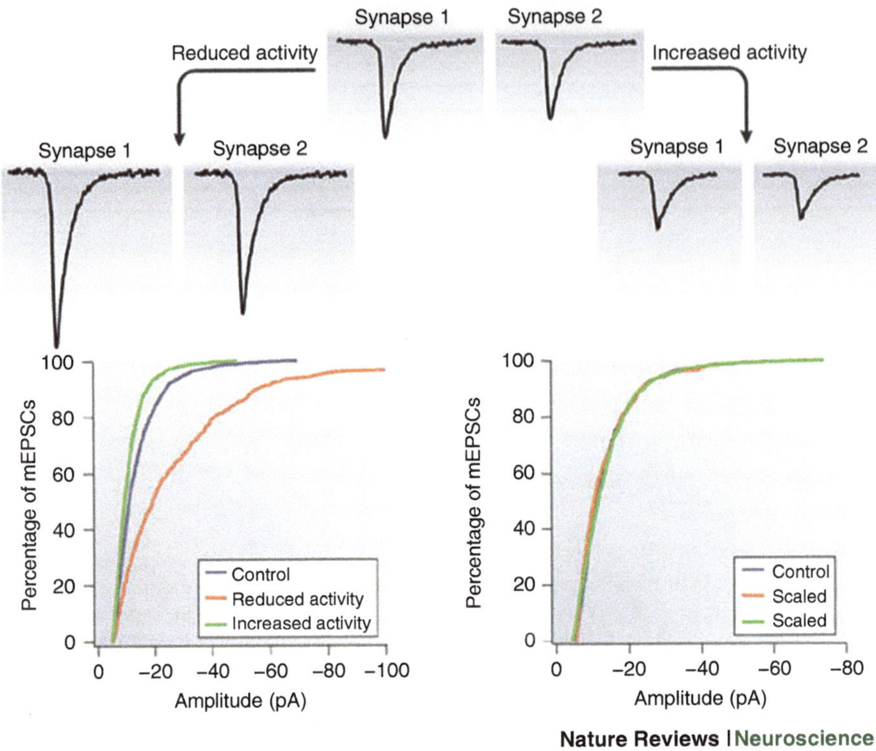

Fig. 20.5 A schematic figure representing homeostatic plasticity. Synaptic potentials are potentiated when whole network activity is weakened, regardless of the amplitude of synaptic potential (*left*). The amplitude of synaptic potentials is inhibited when whole network activity is strengthened (*right*). The graph on the *bottom left* indicates the percentage of mini excitatory post-synaptic potentials (mEPSPs) according to whole network activity. The graph on the *bottom right* shows that this percentage is not skewed if each graph is scaled (From Turrigiano and Nelson (2004). With permission)

been reported to increase the amplitude of synaptic potentials, compensating for an over-inhibition of whole network activity (Wierenga et al. 2006). These mechanisms seem to be important for keeping the whole network stable, and working in a 'normal' range. Further work is required to elucidate the actual mechanisms responsible for homeostatic plasticity.

20.1.5 Imaging Methods for Neural Plasticity Research

Two-photon laser scanning microscopes (2PLSMs), established by Winfried Denk, use two-photon absorption to analyze living tissue (Denk et al. 1990). Biological samples are exposed to pulses of near-infrared wavelength lasers that excite fluorophores in the samples. Long-wavelength lasers are likely to be weaker than short-wavelength lasers, allowing us to observe living tissue with low bleaching and phototoxicity, and at greater depths. The development of 2PLSMs, and applicable fluorescent probes, has allowed us to observe histological and physiological changes in the brain, even in deep regions.

Using 2PLSM, it was determined that α-amino-3-hydroxy-5-methyl-4-isoxazolepropionic acid (AMPA)-type glutamate receptors could be inserted into the synaptic site in order to induce LTP in hippocampal neurons (Malinow and Malenka 2002; Shi et al. 1999, Fig. 20.2). This result helped neuroscientists explain why LTP seemed to occur, whether pre- or post-synaptically.

20.2 Optogenetics

20.2.1 Conventional Stimulation and Optical Regulation of Neural Activity

Various methods have been developed to control neuronal and brain activity. Stimulation via microelectrode is still a popular technique in modern neuroscience; microelectrodes, commonly glass or metal, inserted in the tissue allow us to stimulate neurons located nearby, through application of a small current. Many other physical methods, such as temperature, light, and magnetic fields, have been developed to control particular neural activities. However, stimuli locations are restricted, as every technique is limited by probe distance and physical quantities.

At the same time, pharmacological approaches that target different molecules present in neurons have been widely used to control neural activity. It is possible to control synaptic transmission and intrinsic neuronal activity using agonists and antagonists. However, pharmacological regulation is usually specific to a certain molecule, and it can take some time for the drug to diffuse to target, as well as wash out from the site of administration.

Channelrhodopsin (ChR)-2 has been used in various areas of neuroscience, from basic retinal physiology to emotional behavior (Fenno et al. 2011). In combination with molecular tools, ChR2 helps researchers to target cells of interest and regulate both excitatory and inhibitory neuronal activity. ChR2 has an advantage over the physical and pharmacological stimulations described above as, once ChR2 is successfully expressed in a cell of interest, it is then possible to noninvasively stimulate or inhibit the activity of that neuron. Some neuroscience mysteries have been solved using this selective optical stimulation as part of a process called optogenetics. Some examples of this success are the findings that particular astrocytes regulate respiration and the regulation of behavior by glutamatergic interneurons (Gourine et al. 2010; Hagglund et al. 2010). At the single neuron level, optogenetics has enabled the analysis of complex physiological innervations through the entire cortex, and the analysis of spine activity of an individual neuron, without any changes in intracellular signal transduction (Zhang et al. 2008; Petreanu et al. 2009).

20.2.2 In Vitro Studies

ChR2-assisted circuit mapping (CRACM), proposed by Leopoldo Petreanu and Karel Svoboda, is an outstanding method to detect physiological connections and map the extensive innervation of neural circuits (Petreanu et al. 2007, Fig. 20.6). ChR2 was inserted into layer II/III neurons of the somatosensory cortex, neural activity was recorded using patch clamp, and light stimulation was applied to the contralateral side. Neurons in layer II/III of the somatosensory cortex were found to project to layers V, II/III, and VI, but not IV, of the ipsilateral side. In both sides, the connection between layers II/III and V was relatively stronger than that within layer II/III. These findings in the long-distance connections are consistent with the literature, showing local network connections layer to layer. This is one of the most significant examples of optogenetics helping to elucidate histological and physiological connections in a particular brain region.

Homeostatic plasticity is one of the phenomena that are considered to 'bridge the gap' between the static characteristics of brain function and dynamic synaptic plasticity. Neural plasticity, such as LTP, is induced in a cell- and synapse-specific manner. However, excitability can result in complementary changes in whole network activity. This is homeostatic plasticity but, to date, it has not been clear how neurons homeostatically regulate plasticity in accordance with whole network activity. AMPA and N-methyl-D-aspartate (NMDA) glutamate receptors may contribute to homeostatic

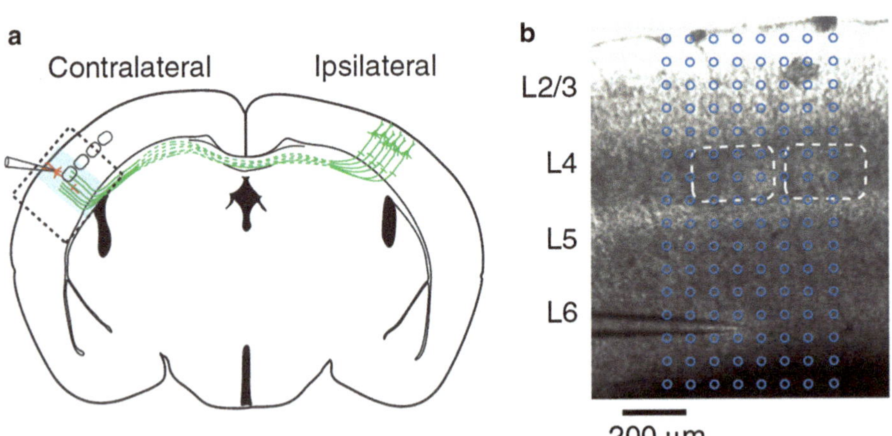

Fig. 20.6 Channelrhodopsin (ChR)-2-assisted circuit map of long-range callosal projections. ChR2 is electroporated (ipsilateral side, **a**) and patch clamp recordings taken from neurons in the contralateral side (*red* neurons, **a**). Exposure to blue light, indicated in **b**, stimulates terminals from the contralateral side. Membrane potentials are measured using a patch electrode to map inputs from the contralateral side. *L* layer (From Petreanu et al. (2007). With permission)

plasticity by decreasing the number of receptors expressed in the synapse. Optogenetics may help to resolve this uncertainty in work adapted to the CA1 region of the hippocampus (Goold and Nicoll 2010). Pharmacological studies inform us that calcium influx through L-type calcium channels activates calmodulin (CaM)-kinase kinase and CaM kinase 4, resulting in decreased sensitivity of AMPA and NMDA glutamate receptors. Only synaptic inhibition via AMPA receptors would require protein synthesis, indicating that compensatory synaptic inhibition via AMPA or NMDA receptors could be responsible for the downstream CaM kinase cascade. Optogenetics would be a powerful tool allowing us to regulate neural activity in a spatio-temporal manner, and investigate neural plasticity across the whole network.

Recently, Joel Kralj developed a fluorescent membrane potential indicator protein (VIP) (Kralj et al. 2011, Fig. 20.7). Co-expression of VIP and ChR2 allowed us to record virtual 'optical voltage clamps,' due to the speed at which current could be injected and voltage measured using different wavelengths (blue for activation of ChR2, red for indicating membrane potential, Program No. 229.02. 2012 Neuroscience Meeting Planner). Micro-electrical-mechanical system technologies, such as Digital Mirror Devices, help to improve the spatial accuracy of light stimulation to microsites, such as neurites or processes. To date, these technologies have mainly been applied in vitro; however, they are expected to be used extensively in future applications.

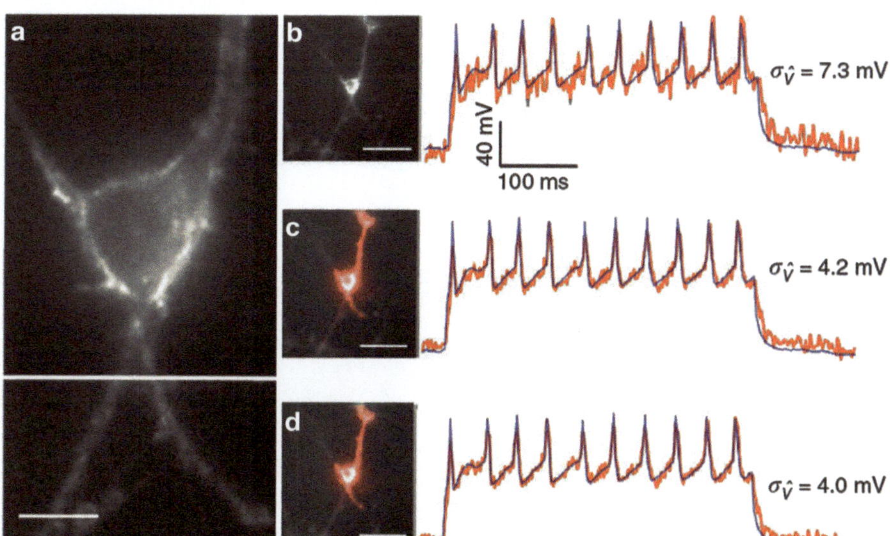

Fig. 20.7 Super-sensitive fluorescent membrane potential indicator protein (VIP). (**a**) A fluorescent image of VIP expressed in cultured hippocampal neurons. (**b**) Wide-field view of A. (**c**) Changes in membrane potential are indicated in pseudo-color (*red*). *Blue* traces indicate actual membrane potentials (From Kralj et al. (2011). With permission)

20.2.3 *In Vivo Studies*

20.2.3.1 Perceptual Discrimination

Optogenetics allows us to elucidate the impact that a neuron of interest may have on a particular behavior in vivo. ChR2 was applied to neurons in layers II/III of the primary somatosensory cortex, and mice were trained to discriminate tactile whisker stimulation. Using light stimulation, rather than 'real' tactile whisker stimulation, and training to discriminate between reinforced and non-reinforced stimulation, the size of the network involved in associative learning was estimated. Sixty neurons were found to contribute to the whiskers' network of sensory perception (Huber et al. 2008, Fig. 20.8).

The importance of inhibitory neurons has been described in the functional expression of various neural networks; however, which types of GABAergic neurons are strongly involved in processing particular sensory information is unclear. Seung-Hee Lee and Yang Dan were able to reduce the level of excitation in interneurons after application of ChR2 to various GABAergic inhibitory neurons in the primary visual cortex (Lee et al. 2012). Parvalbumin-positive GABAergic interneurons tuned the orientation selectivity of adjacent neurons and, following light stimulation, amplified their directional selectivity.

Recently, some studies have tried to discover and describe the entire brain network, the 'connectome.' In general, these research movements have been called 'omics' studies. One recent study tried to examine behaviors in *Drosophila* larvae driven by 1,054 neurons (Vogelstein et al. 2014). ChR2 was applied to certain neurons via genetic manipulation of the larvae; this allowed researchers to control neural activity and record resulting behaviors. The behaviors that resulted from activity of a particular neuron were recorded and analyzed for feature detection and catego-

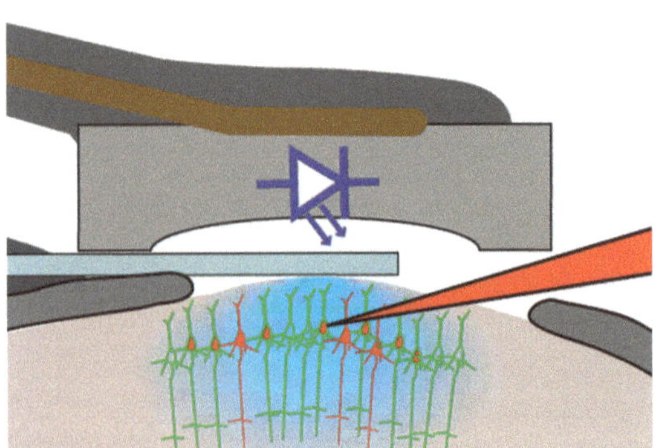

Fig. 20.8 Using a light-emitting diode, light stimulation mimics sensory perception. Schematic of the recording geometry is illustrated to show layer II/III neurons expressing ChR2–green fluorescent protein (GFP) and cytosolic red fluorescent protein (RFP). Changing the intensity of light stimulation alters the number of neurons. This method estimates the minimum number of neurons required to induce sensory perception (From Huber et al. (2008). With permission)

rization, resulting in identification of a 'cause and effect' relationship between specific neural activity and particular behaviors. Researchers identified 29 discrete, statistically distinguishable, observer-unbiased behavioral phenotypes. In this study, optogenetics worked as a powerful tool to manipulate neural activity in a specific spatio-temporal manner.

20.2.3.2 Learning and Memory

Optogenetics allows us to investigate the mechanisms of memory in precise time windows, a feat that is almost impossible using traditional pharmacological or molecular biology tools. Arch-enhanced green fluorescent protein molecules were inserted into newly synthesized neurons in the hippocampal dentate gyrus using retroviral vectors, and spatial learning was assessed using the water maze and fear-conditioning tasks (Gu et al. 2012, Fig. 20.9). Yellow light (589 nm) was applied to neurons during the spatial learning test session to inhibit their activity. Animals that received yellow light during the probe test could not retrieve the position of the platform, though there were no significant differences in swimming distance or speed in the probe or visible platform tasks. These results indicate that the dentate gyrus is closely related to spatial memory recall. In the fear-conditioning task, freezing behavior observed during contextual conditioning was reduced when yellow light was applied up to 2 min after the start of the test trial. This reduction was not observed 2 min after the test trial. The reduction of freezing behavior because of yellow light application was not evident during tone conditioning. This suggests that the dentate gyrus is involved in the short-term recall of context.

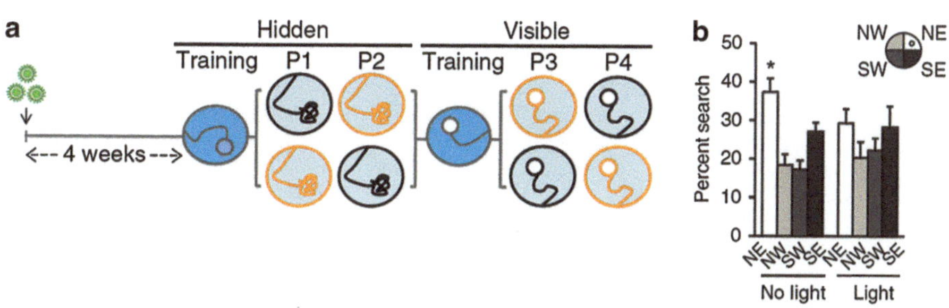

Fig. 20.9 Hippocampal function was assessed during spatial learning using the Morris water maze. Probe tests were measured over 2 days (**a**: P1–2, P3–4). Animals were split into two groups. One group was exposed to light on the first day, the other group on the second day. Light stimulation reduced time spent in the quadrant where the platform was located, indicating an inhibition of hippocampal function. There was no change in the time spent swimming, or speed of swimming, in the visible platform test. These results suggest that the hippocampus is responsible for memory recall, and that newly synthesized neurons play an important role in this function (From Gu et al. (2012). With permission)

20.3 Towards Clinical Applications

The use of optogenetics to stimulate or inhibit neural activity may be useful in clinical applications, as many neurological or psychiatric diseases are caused by abnormal neural activity. Pharmacological or electrical alterations in neural activity can cause some of the broad pathologies evident in schizophrenia, anxiety disorders, and Parkinson's disease, although whether these abnormalities are the major causes of those diseases is unclear (Kane and Correll 2010; Stevens and Pollack 2005; Kringelbach et al. 2007). Some in vivo optogenetic trials have already been conducted using animal models of these diseases. One of the best early optogenetic studies showed recovery of visual responsiveness of the retina when ChR2 was applied to the membrane of particular bipolar cell (Lagali et al. 2008). Even more recently, it has been reported that halorhodopsin (*Np*HR)-induced hyperpolarization resulted in remarkable functional recovery of neural circuits of the retina and return of function of the outer segments of cone cells (Busskamp et al. 2010).

Optogenetics may also be considered for the treatment of Parkinson's disease. Deep-brain stimulation is one of the most effective Parkinson's treatments (Gradinaru et al. 2009), but it was found to be even more effective when optogenetic tools were used to stimulate the direct pathway of the dorsal striatum in an animal model of Parkinson's disease (Kravitz et al. 2010). These examples illustrate the potential for these optical tools to help understand the mechanisms of brain disease and develop novel therapies for them.

Impairment of sensory perception is often caused by malfunction of peripheral nerves. It has been suggested that this may be due to bilateral abnormalities in the somatosensory cortex. It had been presumed that impaired inputs led to abnormal neural activity in the somatosensory cortex, which suppressed inhibitory interneurons through the corpus callosum. A recent study using optogenetics clarified this presumption (Li et al. 2011, Fig. 20.10). When *Np*HR is expressed in pyramidal cells of the

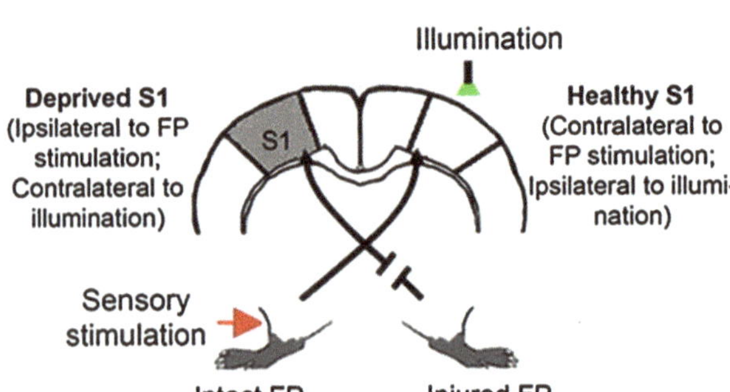

Fig. 20.10 Halorhodopsin (*Np*HR) inhibits hypersensitivity of sensory perception on the healthy side following peripheral nerve damage. When light was applied to *Np*HR, cortical activity was inhibited and neural activity in the contralateral side increased. The hypersensitivity disappeared when the damaged side was inhibited with *Np*HR and light stimulation. These results indicate that hypersensitivity following peripheral nerve damage occurs because of innervation of the contralateral side of brain, via the corpus callosum. *FP* forepaw (From Li et al. (2011). With permission)

somatosensory cortex, healthy neural activity was suppressed in cortical neurons of layer II/III, which represents the start of the corpus callosum. This treatment led to an increased response to electrical stimulation in healthy forelimbs. These results indicate that neural plasticity, via the corpus callosum, is one of the causes of paresthesia.

20.4 Application of Optogenetics in Neuroplasticity Research

To date, research on neural plasticity has been conducted relatively independently by molecular biologists, physiologists, and behaviorists. However, recently, it has become realistic to analyze function in the living brain with optical techniques. Calcium imaging and voltage measurements with 2PLSMs allow us to simultaneously record the activity of multiple neurons. At the same time, optogenetics helps us noninvasively manipulate neural activity in a spatio-temporal manner. Optogenetics has significant advantages in two rapidly developing fields, optical physics, and molecular biology. Researchers are still developing methods for the 'online' handling of neurons of interest; at present, neuronal activity can only be manipulated in neurons, which indicates their history of activity by expression of immediate early genes. This issue needs to be resolved in order to fully elucidate brain activity during cognitive behavior using these technologies. We hope a breakthrough occurs soon so that there may be further progress; overall, it is exciting to consider the possibilities for optogenetics in neuroscience research.

References

Bi GQ, Poo MM (1998) Synaptic modifications in cultured hippocampal neurons: dependence on spike timing, synaptic strength, and postsynaptic cell type. J Neurosci 18:10464–10472

Bliss TV, Gardner-Medwin AR (1973) Long-lasting potentiation of synaptic transmission in the dentate area of the unanaestetized rabbit following stimulation of the perforant path. J Physiol 232:357–374

Bliss TV, Lomo T (1973) Long-lasting potentiation of synaptic transmission in the dentate area of the anaesthetized rabbit following stimulation of the perforant path. J Physiol 232:331–356

Busskamp V, Duebel J, Balya D et al (2010) Genetic reactivation of cone photoreceptors restores visual responses in retinitis pigmentosa. Science 329:413–417

Denk W, Strickler JH, Webb WW (1990) Two-photon laser scanning fluorescence microscopy. Science 248:73–76

Fenno L, Yizhar O, Deisseroth K (2011) The development and application of optogenetics. Annu Rev Neurosci 34:389–412

Goold CP, Nicoll RA (2010) Single-cell optogenetic excitation drives homeostatic synaptic depression. Neuron 68:512–528

Gourine AV, Kasymov V, Marina N et al (2010) Astrocytes control breathing through pH-dependent release of ATP. Science 329:571–575

Gradinaru V, Mogri M, Thompson KR et al (2009) Optical deconstruction of Parkinsonian neural circuitry. Science 324:354–359

Gu Y, Arruda-Carvalho M, Wang J et al (2012) Optical controlling reveals time-dependent roles for adult-born dentate granule cells. Nat Neurosci 15:1700–1706

Hagglund M, Borgius L, Dougherty KJ et al (2010) Activation of groups of excitatory neurons in the mammalian spinal cord or hindbrain evokes locomotion. Nat Neurosci 13:246–252
Hebb DO (2002, 1949) The organization of behavior: a neuropsychological theory. L. Erlbaum Associates, Mahwah
Huber D, Petreanu L, Ghitani N et al (2008) Sparse optical microstimulation in barrel cortex drives learned behaviour in freely moving mice. Nature 451:61–64
Ito M (2002) The molecular organization of cerebellar long-term depression. Nat Rev Neurosci 3:896–902
Ito M, Sakurai M, Tongroach P (1982) Climbing fibre induced depression of both mossy fibre responsiveness and glutamate sensitivity of cerebellar Purkinje cells. J Physiol 324:113–134
Kane JM, Correll CU (2010) Past and present progress in the pharmacologic treatment of schizophrenia. J Clin Psychiatry 71:1115–1124
Kralj JM, Douglass AD, Hochbaum DR et al (2011) Optical recording of action potentials in mammalian neurons using a microbial rhodopsin. Nat Methods 9:90–95
Kravitz AV, Freeze BS, Parker PR et al (2010) Regulation of Parkinsonian motor behaviours by optogenetic control of basal ganglia circuitry. Nature 466:622–626
Kringelbach ML, Jenkinson N, Owen SL et al (2007) Translational principles of deep brain stimulation. Nat Rev Neurosci 8:623–635
Lagali PS, Balya D, Awatramani GB et al (2008) Light-activated channels targeted to ON bipolar cells restore visual function in retinal degeneration. Nat Neurosci 11:667–675
Lee SH, Kwan AC, Zhang S et al (2012) Activation of specific interneurons improves V1 feature selectivity and visual perception. Nature 488:379–383
Li N, Downey JE, Bar-Shir A et al (2011) Optogenetic-guided cortical plasticity after nerve injury. Proc Natl Acad Sci U S A 108:8838–8843
Malinow R, Malenka RC (2002) AMPA receptor trafficking and synaptic plasticity. Annu Rev Neurosci 25:103–126
Markram H, Lubke J, Frotscher M et al (1997) Regulation of synaptic efficacy by coincidence of postsynaptic APs and EPSPs. Science 275:213–215
Markram H, Gerstner W, Sjostrom PJ (2011) A history of spike-timing-dependent plasticity. Front Synaptic Neurosci 3:4
Perrett SP, Ruiz BP, Mauk MD (1993) Cerebellar cortex lesions disrupt learning-dependent timing of conditioned eyelid responses. J Neurosci 13:1708–1718
Petreanu L, Huber D, Sobczyk A et al (2007) Channelrhodopsin-2-assisted circuit mapping of long-range callosal projections. Nat Neurosci 10:663–668
Petreanu L, Mao T, Sternson SM et al (2009) The subcellular organization of neocortical excitatory connections. Nature 457:1142–1145
Shi SH, Hayashi Y, Petralia RS et al (1999) Rapid spine delivery and redistribution of AMPA receptors after synaptic NMDA receptor activation. Science 284:1811–1816
Stevens JC, Pollack MH (2005) Benzodiazepines in clinical practice: consideration of their long-term use and alternative agents. J Clin Psychiatry 66(Suppl 2):21–27
Turrigiano GG, Nelson SB (2004) Homeostatic plasticity in the developing nervous system. Nat Rev Neurosci 5:97–107
Turrigiano GG, Leslie KR, Desai NS et al (1998) Activity-dependent scaling of quantal amplitude in neocortical neurons. Nature 391:892–896
Vogelstein JT, Park Y, Ohyama T et al (2014) Discovery of brainwide neural-behavioral maps via multiscale unsupervised structure learning. Science 344:386–392
Wierenga CJ, Walsh MF, Turrigiano GG (2006) Temporal regulation of the expression locus of homeostatic plasticity. J Neurophysiol 96:2127–2133
Zhang YP, Holbro N, Oertner TG (2008) Optical induction of plasticity at single synapses reveals input-specific accumulation of alphaCaMKII. Proc Natl Acad Sci U S A 105:12039–12044

Chapter 21
Strategies to Probe Mechanoreception: From Mechanical to Optogenetic Approaches

Zhi-Gang Ji, Toru Ishizuka, and Hiromu Yawo

Abstract There remain many questions about the rules governing the neural representation of touch; how the spatio-temporal patterns of points are integrated in the brain to generate the complex senses such as shape, size, weight, movement, and texture. Conventionally, the touch sense has been artificially generated mechanically or electrically with insufficient spatio-temporal resolution. Recently, the somatosensory system has been optogenetically investigated with the expression of optogenetic molecular reagents (OMRs) such as channelrhodopsin (ChR)-2 and halorhodopsin (*Np*HR). In a transgenic rat model that expresses ChR2 in peripheral mechanoreceptive neurons in the dorsal root ganglion (DRG) and the trigeminal ganglion (TG), blue light irradiation on skin or whiskers evoked a touch-dependent response in the cortex. Thus, various and reproducible patterned touch stimulations could be made by the patterned irradiations on the skin or whiskers. Optogenetic approaches could open a new avenue to investigate the neural representation of complex touch patterns using a combination of electrophysiological and imaging techniques.

Keywords Somatosensory system • Mechanoreception • Touch sense • Neural representation • Optogenetics • Channelrhodopsin • Transgenic • fMRI

21.1 Introduction

All knowledge about the world is perceived through our sensory systems, which consist of peripheral sensory organs, sensory nerves, and the central nervous system (CNS). In principle, a sensation is classified according to its modality, a kind of

Z.-G. Ji (✉)
Department of Neurology, Baylor College of Medicine, Houston, TX 77030, USA
e-mail: zhigang.ji@bcm.edu

T. Ishizuka • H. Yawo
Graduate School of Life Sciences, Tohoku University, Sendai, Japan
e-mail: ishizuka@m.tohoku.ac.jp; yawo-hiromu@m.tohoku.ac.jp

H. Yawo et al. (eds.), *Optogenetics*, DOI 10.1007/978-4-431-55516-2_21

energy inducing physiological transduction in a specific group of sensory organs. For example, photoreceptors in the eye absorb light and translate it into electrical signals that are sent to the brain, where they are translated into an image of the world. In the skin, the natural stimulation is induced either physically or chemically. The membrane deformation caused by mechanical forces on a peripheral sensory organ complex generates a sense of mechanoreception or proprioception, whereas local chemical signals around free nerve endings evoke nociception (Fig. 21.1). These stimuli (inputs) are related to the responses (outputs) under given cause-effect rules in the neural network (O'Connor et al. 2009). One of the aims of neuroscience is to reveal the rules. However, there remain many key questions, even about the rules limited to mechanoreception, how the spatio-temporal patterns of touch points are integrated in the brain to generate complex senses such as shape, size,

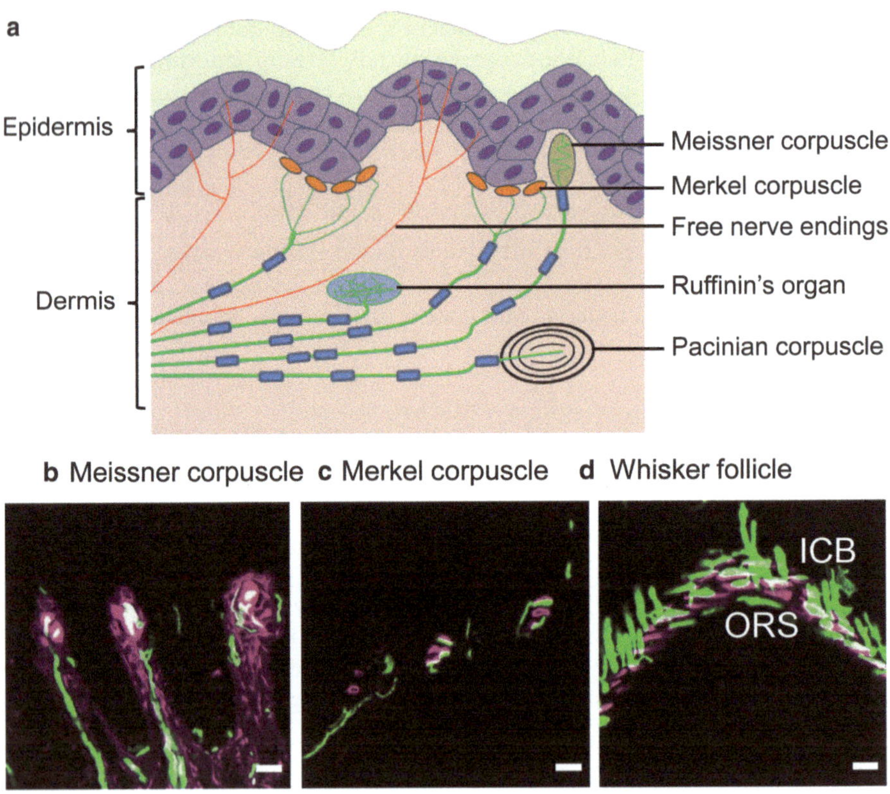

Fig. 21.1 Expression of channelrhodopsin (ChR)-2 in the peripheral mechanoreceptors. (**a**) Peripheral sensory organs in the hairless skin. (**b**). Meissner corpuscles (*magenta/gray*) innervated by ChR2-positive nerve endings (*green/ash*). (**c**) Merkel cells (*magenta/gray*) associated with ChR2-positive nerve endings (*green/ash*). (**d**) Whisker follicles are surrounded by the ChR2-positive nerve plexus (*green/ash*) the endings of which are associated with Merkel cells (*magenta/gray*) (Scales, 20 μm. (**b–c**), From Ji et al. (2012))

weight, movement, and texture. Therefore, an important experimental challenge is how to design artificial input stimuli that are optimal for analyzing the rules. In this chapter, we focus on the methods for designing artificial mechanoreception and future perspectives for optogenetic approaches.

21.2 Current Methods and the Challenges

21.2.1 Mechanical Stimulation

Classically, the touch sense is evoked in the skin by mechanical stimulation using such objects as plastic probes (Chen et al. 2007), glass rods (Warren et al. 1986), or hair brushes (Greenspan 1992; Saito et al. 2008; Björnsdotter et al. 2009; Boada et al. 2010). An air-puff is another approach that is widely used for tactile studies, especially those concerned with the whisker-barrel system. Conventionally, the device is home-made and consists of a pipette tube to deliver air to the skin or whiskers/vibrissae of rodents, a pressure device to supply air, and a device with electromagnetic valves that control the timing of pressure pulses. Typically, the pipette tube ranges from 1 to 2.5 mm in diameter (Tremblay et al. 1993; Rowland and Jaeqer 2005; Holtzman et al. 2006; O'Connor et al. 2010; Bezdudnaya and Castro-Alamancos 2011; Pidoux et al. 2011) and is used at a distance of 2–5 mm from the surface of the skin or whiskers (Tremblay et al. 1993; Rowland and Jaeqer 2005) or 10–25 mm in the case of whisker deflections (Ahissar et al. 2000; Bezdudnaya and Castro-Alamancos 2011; Pidoux et al. 2011). Artificially, the whiskers are stimulated with mechanical deflectors in various patterns such as single whisker deflection, simultaneous arc deflection (stimulation of two, three, or four whiskers of the same arc), or simultaneous row deflection (stimulation of two, three, or four whiskers of the same row) (Mirabella et al. 2001). More recently, a multiple-whisker stimulator has been reported (Krupa et al. 2001; Jacob et al. 2012; Ramirez et al. 2014). Generally, the whisker deflector consists of a plastic tube that holds each whisker and a piezoelectric bender with a lever arm that controls its deflection. Although the current multiple-whisker deflectors stimulate multiple whiskers at once, it should be improved to stimulate individual whiskers with various spatiotemporal patterns.

21.2.2 Electrical Stimulation

Electrical stimulation is frequently used for mechanoreception studies, especially when using functional magnetic resonance imaging (fMRI). The basic device is simple and consists of an electrode (to deliver current to the skin), a pulse generator, and an isolator (to control the current pulse parameters). Typically, the pulse duration ranges from 0.05 to 0.5 ms with a current intensity of 0.1–10 mA (Jennings and

Fitzgerald 1998; Li et al. 2002; Hentschke et al. 2006). Although electrical stimulation is convenient, it inevitably induces slight muscle and joint movements, even if the stimulation is instantaneous. Therefore, it is difficult to distinguish the sensorimotor effects from the somatosensory effects (Van Camp et al. 2006). Although the pulse width of electrical stimulation has been very short in previous studies, the tactile somatosensory stimuli are more prolonged with complex temporal patterns under natural conditions (Morissette and Bower 1996). Furthermore, electrical stimulation has limited targeting specificity.

21.3 Optogenetic Approaches to Studying Mechanoreception

21.3.1 Overview of Optogenetics

Upon discovery of channelrodopsin (ChR)-2 as a light-gated cation channel (Nagel et al. 2003), it was genetically expressed in vertebrate neurons to manipulate their activities and studied independently by at least three groups (Boyden et al. 2005; Li et al. 2005; Ishizuka et al. 2006). Recently, this technique has opened a new era in neuroscience, under the term 'optogenetics', which integrates optical (opto-) and genetic (genetics) methods with resulting improvements in spatio-temporal resolutions. To apply optogenetics, three requirements are necessary: selection of optogenetic molecular reagents (OMRs), targeted expression of OMRs in the neurons or regions of interest, and a light-delivery system (Yawo et al. 2013). In this chapter, we focus on the expression of OMRs in the somatosensory system.

21.3.2 Targeting OMRs in the Somatosensory System

Several ways exist to target OMRs to the neurons of interest in an in vivo somatosensory system so that mechanoreception can be evoked using an optogenetic approach. One is by utilizing transgenic animal lines that express optogenetic actuators in somatosensory neurons. So far, several transgenic animal lines of *Caenorhabditis elegans* (Li et al. 2011; Husson et al. 2012), zebrafish (Douglass et al. 2008; Arrenberg et al. 2009; Low et al. 2010; Umeda et al. 2013), and rodents (Wang and Zylka 2009; Hägglund et al. 2010; Ji et al. 2012; Daou et al. 2013), have been reported for studying the somatosensory system. For example, Mrgprd-ChR2-Venus knock-in mice mainly expressed ChR2 in a subpopulation of nonpeptidergic nociceptive neurons (Wang and Zylka 2009). In the case of Vglut2-ChR2-YFP mice, ChR2 was widely expressed in Vglut2-positive neurons in some regions such as the spinal cord and spinal trigeminal nucleus (Hägglund et al. 2010). Recently, the nociceptive pathway was optogenetically activated in a transgenic mouse in which ChR2 was conditionally targeted to mainly small- and medium-diameter $Na_v1.8$-positive dorsal root ganglion (DRG) neurons, although a subpopulation of large-diameter $Na_v1.8$-positive DRG neurons also expressed ChR2 under the Cre-loxP system (Daou et al. 2013). The significance of Merkel cells in the intact skin was optogenetically investigated using the Cre-loxP strategy; they were both found to be necessary and sufficient for sustained action-potential firing in slowly-adapting type I (SAI) afferents (Maksimovic et al. 2014).

With the rapidity of experimental implementation, viral vectors, such as lentiviruses and adeno-associated viruses (AAVs), are currently employed to target OMRs to the somatosensory systems. For example, pyramidal neurons in layer 2/3 of the barrel cortex could be made to express ChR2 using lentivirus vectors under control of the αCaMKII promoter (Mateo et al. 2011; Avermann et al. 2012). Thus, the synaptic interaction was dependent on the state of the brain in layer 2/3 of the barrel cortex. Under the control of the pan-neuronal human synapsin-1 promoter, ChR2 was retrogradely expressed throughout small DRG neurons by delivering AAV6 vectors into the sciatic nerve (Iyer et al. 2014).

In utero electroporation is another way to deliver optogenetic actuators to the somatosensory cortex (Petreanu et al. 2007; Huber et al. 2008; Hull et al. 2009). For example, with the expression of ChR2 in layer 2/3 pyramidal neurons of the mouse barrel cortex, Petreanu et al. (2007) mapped the long-range projection of ChR2-expressing neurons and demonstrated that layer 2/3 axons are in contact with neurons in layers 2/3, 5, and 6, but not with layer 4, in both the ipsilateral and the contralateral cortex.

21.3.3 Light-Evoked Somatosensory Perception

In the somatosensory systems, each mode of touch-pressure, temperature, or pain is sensed by independent sensory endings of primary afferent neurons and conducted to the specific cortical locus as nerve impulses, but integrated thereafter as a whole. However, the mammal does not sense light at its skin as its peripheral sensory organs lack photoreceptor molecules such as rhodopsins. This is classically known as 'the doctrine of specific nerve energies' formulated by Johannes Müller in the nineteenth century (Müller 1834–1840); the modality of sensation depends upon the end organ or nerve fiber stimulated and not upon the energy that stimulated them. Therefore, it has been anticipated that peripheral sensory nerve endings could become photosensitive with the ectopic expression of photoreceptive molecules, but had never been proved using mammalian systems.

Tomita et al. (2009) have previously generated several lines of transgenic rats that express ChR2 under regulation of the Thy1.2 promoter. Thy1.2 is one of the cell-surface glycoproteins of the immunoglobulin superfamily and its regulatory elements have been used to drive the expression of heterologous genes in the CNS. Interestingly, the phenotypic expression pattern of heterologous genes varies among transgenic lines, probably due to differences in the integration sites in the chromosomes and/or the number of inserted copies. Ji et al. (2012) investigated one of these transgenic rat lines, W-Tg(Thy1-COP4/YFP*)4Jfhy (W-TChR2V4, The

National BioResource Project for the Rat in Japan) and provided clear evidence that ChR2 is specifically expressed in a subpopulation of mechanoreceptive neurons in the DRG, but not in the small-sized neurons involved in nociception. Furthermore, ChR2 is also expressed in their peripheral nerve endings such as those innervating Merkel corpuscles and Meissner corpuscles, which are involved in the sense of touch (Fig. 21.1). Indeed, this transgenic rat showed a sensory-evoked behavior in response to blue flash light on their plantar skin as if it were touched by something (Fig. 21.2). However, it ignores red light, which is not sensed by ChR2. The W-TChR2V4 rat senses blue light by the mechanoreceptive nerve endings in the skin, not by the nociceptive receptors. The light-evoked signal is then transmitted to the somatosensory cortex via the spinal cord and thalamus and evokes the tactile sense in the rat (Fig. 21.3). This rat is thus the first mammal that has acquired an unusual sensory modality in so far as it senses light by its skin.

The rodent whisker-barrel system has been an ideal model for studying somatosensory representations in the cortex (Diamond et al. 2008). Rodents use their whiskers to collect information around them: locating objects (Celikel and Sakmann 2007; Knutsen and Ahissar 2009; O'Connor et al. 2010), discriminating texture (Arabzadeh et al. 2005; Petersen 2007; Aggestam and Cahusac 2007; Ritt et al. 2008; Morita et al. 2011), and sensing the form/shape of objects (Brecht et al. 1997; Diamond et al. 2008). Indeed, the whisker-barrel pathway is topographically ordered from the peripheral whisker follicle innervation to the barrel cortex (Petersen 2007; Erzurumlu et al. 2010). Widespread spatial and temporal integration in the primary

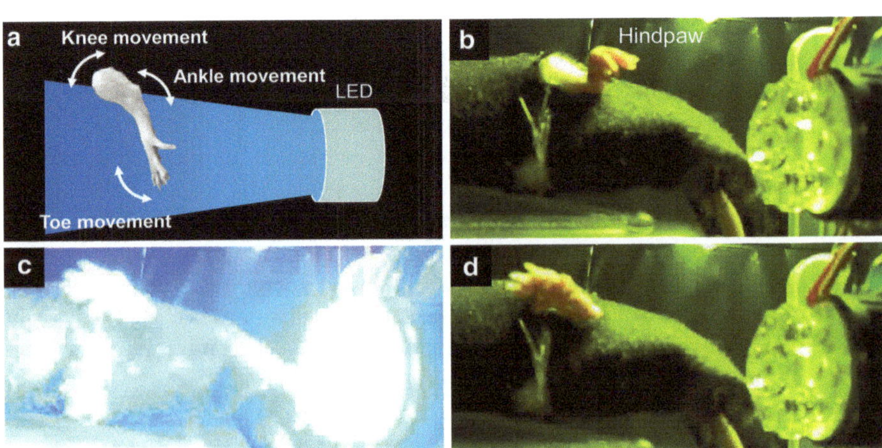

Fig. 21.2 Sensing light by the skin. (**a**) One of four paws was placed in the light path of a blue or red light-emitting diode (LED) so that the plantar skin faced the light while the rest of the body was covered with a black cloth. (**b–d**) The hindpaw before (**b**), during (**c**), and immediately after (**d**) blue light irradiation. W-TChR2V4 transgenic rat, which expresses ChR2 (From Ji et al. (2012))

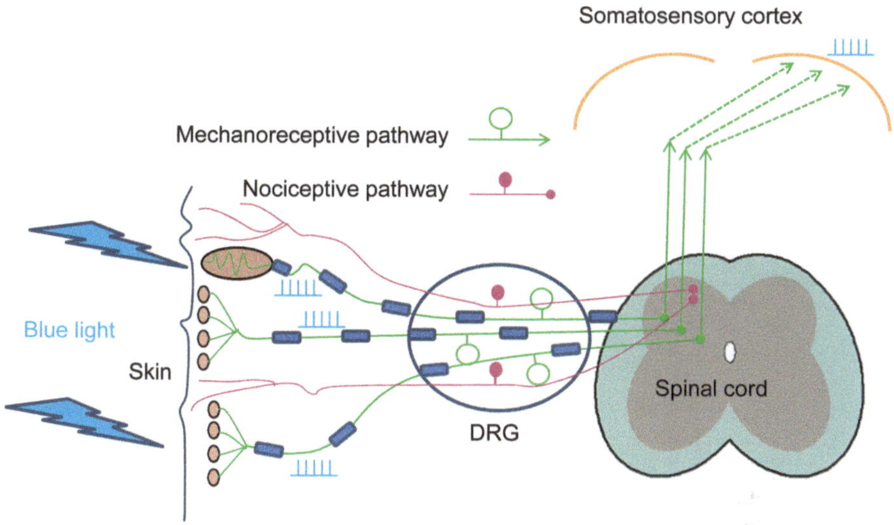

Fig. 21.3 Optogenetic generation of touch sense (From Ji et al. Global Medical Discovery, 2012)

somatosensory cortex (SI or area 3b) has been demonstrated (Simons 1985; Ego-Stengel et al. 2005; Reed et al. 2008). However, it remains a challenge to experimentally stimulate whiskers with a given pattern under spatio-temporal precision. In the W-TChR2V4 transgenic rat, ChR2 was also selectively expressed in the large mechanoreceptive neurons in the trigeminal ganglion (TG) as well as the peripheral nerve endings innervating the whisker follicles (Honjoh et al. 2014) (Fig. 21.1). Each whisker was trimmed and connected to a blue light-emitting diode (LED) by an optical fiber (Fig. 21.4). The spatio-temporal pattern of whisker irradiation thus produced a barrel-cortical response with a specific spatio-temporal pattern as evidenced by electrophysiological and fMRI studies.

Over 50 years ago, Hubel, Wiesel, and their colleagues investigated how the spatio-temporally patterned assembly of light-induced signals in the retina could be interpreted in the cerebral cortex as information about the shape, size, movement, color, and depth of objects (Nicholls et al. 2001). In their series of experiments, they measured the cortical responses while producing many sizes and shapes of spots of light and focused them on the animal's retina. Similarly, the spatio-temporal pattern of touching points is interpreted in the somatosensory cortex as information about the shape, size, weight, movement, and texture of the touching objects. Future optogenetic approaches, such as generating various patterns of touch by light, would open a new avenue for investigating unresolved questions, such as the representation of complex senses in the brain, the network mechanisms organizing complex senses from simple information of points and the way to organize complex senses by experience.

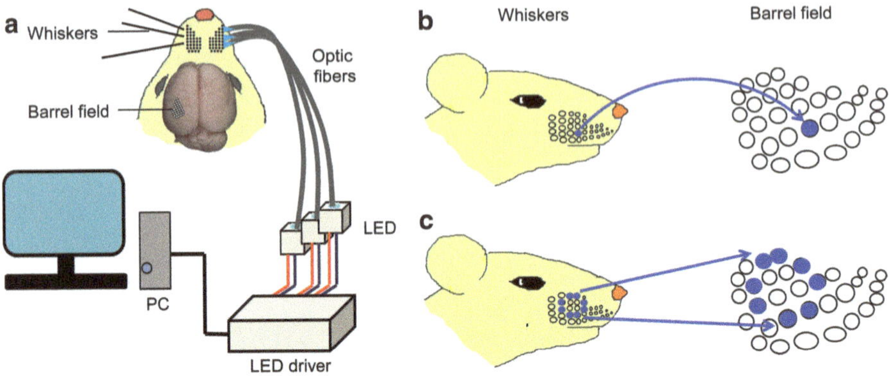

Fig. 21.4 Optogenetic tactile pattern. (**a**) Design of optogenetic approach to stimulate whiskers of a W-TChR2V4 transgenic rat with a given spatio-temporal pattern. (**b**) A simple point stimulation gives the information of stimulus point in the barrel field. (**c**) The brain representation of the spatio-temporal pattern of touch could be investigated by stimulating whiskers with a certain spatio-temporal pattern under optogenetic approaches. In this case, the information related to a form (*circle*) is made

References

Aggestam F, Cahusac PM (2007) Behavioural lateralization of tactile performance in the rat. Physiol Behav 91:335–339

Ahissar E, Sosnik R, Haidarliu S (2000) Transformation from temporal to rate coding in a somatosensory thalamocortical pathway. Nature 406:302–306

Arabzadeh E, Zorzin E, Diamond ME (2005) Neuronal encoding of texture in the whisker sensory pathway. PLoS Biol 3:e17

Arrenberg AB, Del Bene F, Baier H (2009) Optical control of zebrafish behavior with halorhodopsin. Proc Natl Acad Sci U S A 106:17968–17973

Avermann M, Tomm C, Mateo C et al (2012) Microcircuits of excitatory and inhibitory neurons in layer 2/3 of mouse barrel cortex. J Neurophysiol 107:3116–3134

Bezdudnaya T, Castro-Alamancos MA (2011) Superior colliculus cells sensitive to active touch and texture during whisking. J Neurosci 106:332–346

Björnsdotter M, Löken L, Olausson H et al (2009) Somatotopic organization of gentle touch processing in the posterior insular cortex. J Neurosci 29:9314–9320

Boada MD, Houle TT, Eisenach JC et al (2010) Differing neurophysiologic mechanosensory input from glabrous and hairy skin in juvenile rats. J Neurophysiol 104:3568–3575

Boyden ES, Zhang F, Bamberg E et al (2005) Millisecond-timescale, genetically targeted optical control of neural activity. Nat Neurosci 8:1263–1268

Brecht M, Preilowski B, Merzenich MM (1997) Functional architecture of the mystacial vibrissae. Behav Brain Res 84:81–97

Celikel T, Sakmann B (2007) Sensory integration across space and in time for decision making in the somatosensory system of rodents. Proc Natl Acad Sci U S A 104:1395–1400

Chen LM, Turner GH, Friedman RM et al (2007) High-resolution maps of real and illusory tactile activation in primary somatosensory cortex in individual monkeys with functional magnetic resonance imaging and optical imaging. J Neurosci 27:9181–9191

Daou I, Tuttle AH, Longo G et al (2013) Remote optogenetic activation and sensitization of pain pathways in freely moving mice. J Neurosci 33:18631–18640

Diamond ME, von Heimendahl M, Knutsen PM et al (2008) 'Where' and 'what' in the whisker sensorimotor system. Nat Rev Neurosci 9:601–612
Douglass AD, Kraves S, Deisseroth K et al (2008) Escape behavior elicited by single, channelrodopsin-2-evoked spikes in zebrafish somatosensory neurons. Curr Biol 18:1133–1137
Ego-Stengel V, Mello e Souza T, Jacob V et al (2005) Spatiotemporal characteristics of neuronal sensory integration in the barrel cortex of the rat. J Neurophysiol 93:1450–1467
Erzurumlu RS, Murakami Y, Rijli FM (2010) Mapping the face in the somatosensory brainstem. Nat Rev Neurosci 11:252–263
Greenspan JD (1992) Influence of velocity and direction of surface-parallel cutaneous stimuli on responses of mechanoreceptors in feline hairy skin. J Neurophysiol 68:876–889
Hägglund M, Borgius L, Dougherty KJ et al (2010) Activation of groups of excitatory neurons in the mammalian spinal cord or hindbrain evokes locomotion. Nat Neurosci 13:246–252
Hentschke H, Haiss F, Schwarz C (2006) Central signal rapidly switch tactile processing in rat barrel cortex during whisker movements. Cereb Cortex 16:1142–1156
Holtzman T, Rajapaksa T, Mostofi A (2006) Different responses of rat cerebellar purkinje cells and Golgi cells evoked by widespread convergent sensory inputs. J Physiol 574:491–507
Honjoh T, Ji ZG, Yokoyama Y et al (2014) Optogenetic patterning of whisker-barrel cortical system in transgenic rat expressing channelrhodopsin-2. PLoS One 9:e93706
Huber D, Petreanu L, Ghitani N (2008) Sparse optical microstimulation in barrel cortex drives learned behavior in freely moving mice. Nature 451:61–64
Hull C, Adesnik H, Scanziani M (2009) Neocortical disynaptic inhibition requires somatodendritic integration in interneurons. J Neurosci 29:8991–8995
Husson SJ, Costa WS, Wabnig S et al (2012) Optogenetic analysis of a nociceptor neuron and network reveals ion channels acting downstream of primary sensors. Curr Biol 22:743–752
Ishizuka T, Kakuda M, Araki R et al (2006) Kinetic evaluation of photosensitivity in genetically engineered neurons expressing green algae light-gated channels. Neurosci Res 54:85–94
Iyer SM, Montgomery KL, Towne C et al (2014) Virally mediated optogenetic excitation and inhibition of pain in freely moving nontransgenic mice. Nat Biotechnol 32:274–278
Jacob V, Petreanu L, Wright N et al (2012) Regular spiking and intrinsic bursting pyramidal cells show orthogonal forms of experience-dependent plasticity in layer V of barrel cortex. Neuron 73:391–404
Jennings E, Fitzgerald M (1998) Postnatal changes in responses of rat dorsal horn cells to afferent stimulation: a fibre-induced sensation. J Physiol 509:859–868
Ji ZG, Ito S, Honjoh T et al (2012) Light-evoked somatosensory perception of transgenic rats that express channelrhodopsin-2 in dorsal root ganglion cells. PLoS One 7:e32699
Knutsen PM, Ahissar E (2009) Orthogonal coding of object location. Trends Neurosci 32:101–109
Krupa DJ, Matell MS, Brisben AJ et al (2001) Behavioral properties of the trigeminal somatosensory system in rats performing whisker-dependent tactile discriminations. J Neurosci 21:5752–5763
Li CX, Callaway JC, Waters RS (2002) Removal of GABAergic inhibition alters subthreshold input in neurons in forepaw barrel subfield (FBS) in rat first somatosensory cortex (SI) after digit stimulation. Exp Brain Res 45:411–428
Li X, Gutierrez DV, Hanson MG et al (2005) Fast noninvasive activation and inhibition of neural and network activity by vertebrate rhodopsin and green algae channelrhodopsin. Proc Natl Acad Sci U S A 102:17816–17821
Li W, Kang L, Piggott BJ et al (2011) The neural circuits and sensory channels mediating harsh touch sensation in *Caenorhabditis elegans*. Nat Commun 2:315
Low SE, Ryan J, Sprague SM (2010) Touché is required for touch-evoked generator potentials within vertebrate sensory neurons. J Neurosci 30:9359–9367
Maksimovic S, Nakatani M, Baba Y et al (2014) Epidermal Merkel cells are mechanosensory cells that tune mammalian touch receptors. Nature 29:617–621. doi:10.1038/nature13250

Mateo C, Avermann M, Gentet LJ et al (2011) In vivo optogenetic stimulation of neocortical excitatory neurons drives brain-state-dependent inhibition. Curr Biol 21:1593–1602
Mirabella G, Battiston S, Diamond ME (2001) Integration of multiple-whisker inputs in rat somatosensory cortex. Cereb Cortex 11:164–170
Morissette J, Bower JM (1996) Contribution of somatosensory cortex to responses in the rat cerebellar granule cell layer following peripheral tactile stimulation. Exp Brain Res 109:240–250
Morita T, Kang H, Wolfe J et al (2011) Psychometric curve and behavioral strategies for whisker-based texture discrimination in rats. PLoS One 6:e20437
Müller J (1834–1840) Handbuch der Physiologie des Menschen für Vortesungen. J. Holscher, Coblenz, Translated section (Elements of Physiology, Book V. Of The Senses) in Rand B (1912) The classical psychologists. Houghton Mifflin, Boston
Nagel G, Szellas T, Huhn W et al (2003) Channelrhodopsin-2, a directly light-gated cation-selective membrane channel. Proc Natl Acad Sci U S A 100:13940–13945
Nicholls JG, Martin AR, Wallace BG et al (2001) From neuron to brain, 4th edn. Sinauer associates Inc, Sunderland
O'Connor DH, Huber D, Svoboda K (2009) Reverse engineering the mouse brain. Nature 461:923–929
O'Connor DH, Clack NG, Huber D et al (2010) Vibrissa-based object localization in head-fixed mice. J Neurosci 30:1947–1967
Petersen CC (2007) The functional organization of the barrel cortex. Neuron 56:339–355
Petreanu L, Huber D, Sobczyk A (2007) Channelrhodopsin-2-assisted circuit mapping of long-range callosal projections. Nat Neurosci 10:663–668
Pidoux M, Mahon S, Deniau JM et al (2011) Integration and propagation of somatosensory responses in the corticostriatal pathway: an intracellular study in vivo. J Physiol 589:263–281
Ramirez A, Pnevmatikakis EA, Merel J et al (2014) Spatiotemporal receptive fields of barrel cortex revealed by reverse correlation of synaptic input. Nat Neurosci 17:866–875
Reed JL, Pouget P, Qi HX et al (2008) Widespread spatial integration in primary somatosensory cortex. Proc Natl Acad Sci U S A 105:10233–10237
Ritt JT, Andermann ML, Moore CI (2008) Embodied information processing: vibrissa mechanics and texture features shape micromotions in actively sensing rats. Neuron 57:599–613
Rowland NC, Jaeqer D (2005) Coding of tactile responses properties in the rat deep cerebellar nuclei. J Neurophysiol 94:1236–1251
Saito K, Hitomi S, Suzuki I et al (2008) Modulation of trigeminal spinal subnucleus caudalis neuronal activity following regeneration of transected inferior alveolar nerve in rats. J Neurophysiol 99:2251–2263
Simons DJ (1985) Temporal and spatial integration in the rat SI vibrissa cortex. J Neurophysiol 54:615–635
Tomita H, Sugano E, Fukazawa Y et al (2009) Visual properties of transgenic rats harboring the channelrhodopsin-2 gene regulated by the thy-1.2 promoter. PLoS One 4:e7679
Tremblay N, Bushnell MC, Duncan GH (1993) Thalamic VPM nucleus in the behaving monkey. II. Response to air-puff stimulation during discrimination and attention tasks. J Neurophysiol 69:753–763
Umeda K, Shoji W, Sakai S et al (2013) Targeted expression of a chimeric channelrhodopsin in zebrafish under regulation of Gal4-UAS system. Neurosci Res 75:69–75
Van Camp N, Verhoye M, Van der Linden A (2006) Stimulation of the rat somatosensory cortex at different frequencies and pulse widths. NMR Biome 19:10–17
Wang H, Zylka MJ (2009) Mrgprd-expressing polymodal nociceptive neurons innervate most known classes of substantia gelatinosa neurons. J Neurosci 29:13202–13209
Warren S, Kelahan AM, Pubols BH Jr (1986) The somatosensory thalamus of the raccoon: properties of single neurons responsive to light mechanical stimulation of the forepaw. J Neurosci 6:308–317
Yawo H, Asano T, Sakai S et al (2013) Optogenetic manipulation of neural and non-neural functions. Dev Growth Differ 55:474–490

Chapter 22
Casting Light on the Role of Glial Cells in Brain Function

Ko Matsui

Abstract To understand the role of glia in brain function, specific manipulation of glial cell activity is required. Optogenetics was originally introduced as a tool that can manipulate cell membrane potential with light illumination, and its use was mostly limited to neuronal activity manipulation. Depolarization or hyperpolarization by itself did not seem likely to have much effect on glial cell function as these cells largely lack voltage-gated ion channels. A mostly unrecognized fact is that the main cation that crosses the plasma membrane upon light activation of channelrhodopsin-2 or archaerhodopsin is proton. Thus, these optogenetic tools can be regarded as tools that can manipulate intracellular pH. Not only Ca^{2+} but also H^{+}, Na^{+} and other intracellular ionic concentrations can dynamically change upon cell activity and can have a profound effect on cell function. Presented here is a study that shows that glutamate release from glia is triggered by intracellular acidification. The released glutamate from glia can have a profound effect on higher-ordered brain functions such as learning and behavior. It is also shown that rampant glial cell activity occurs upon pathological conditions, such as ischemia, and extreme intracellular glial acidification leads to excess glial glutamate release and excitotoxicity. Optogenetics will likely become an essential tool to study the function of cells previously categorized as 'non-excitable' cells such as the glia.

Keywords Glia • Astrocyte • Glutamate • Cerebellum • Bergmann glia • Purkinje cell • Optokinetic response • Ischemia • Acidosis • Excitotoxicity

Numerous cells constitute our body, and interactions amongst these cells produce coordination that enables us to function as integrated individuals. Our mind is considered to be the result of information processing in the brain that is realized by signals handed between neurons. However, recent studies suggest that glial cells, which constitute more or less half of the cells in the brain, also form a network of cells and they too seem to convey some sort of information. The functional significance of such glial signals is largely unknown. There is a report showing that,

---

K. Matsui (✉)
Division of Interdisciplinary Medical Science, Center for Neuroscience, Graduate School of Medicine, Tohoku University, Sendai, Japan
e-mail: matsui@med.tohoku.ac.jp

H. Yawo et al. (eds.), *Optogenetics*, DOI 10.1007/978-4-431-55516-2_22

upon application of anesthesia, neuronal activity continues but glial activity is profoundly suppressed. This suggests that our consciousness or state of mind could be controlled by glia. However, since our muscles are contacted by neurons, the final observable outcome of the information processing in the brain, which is the behavior, is undoubtedly controlled by neurons. Even if glial cells present some sort of activity, if communication between neuronal and glial circuit does not exist, then it would be impossible to prove that glial cells have any influence on our mind. In this passage, I introduce how we were able to find the process of interaction between the two circuits. We are currently trying to solve what sort of influence glial cells have on brain functions such as cognition, learning, and behavior. These studies have been promoted by the use of optogenetic tools, which enabled us to control the activity of glial cells at will, which was not previously possible with other conventional tools. We aim to ultimately address the reason for such interacting parallel circuits in the brain.

22.1 Half of the Brain Is Glia

In modern Western history, it was probably Franz Joseph Gall in early nineteenth century Vienna who first clearly recognized the location of the mind to be within the substance contained inside the skull. Gall proposed that superiority of the functions of the mind manifests on the bumps on the skull surface and thus founded phrenology. His theory is now generally despised, especially because it was later exploited in eugenics and in racial discrimination; however, his insight in identifying that the mind, which was at that time defined to be created by God, was to be found within the matter that we call the brain was definitely novel. The approach to searching for the correlation between what we can observe and the function of the mind is in fact similar to what we do in current brain science, although we use more sophisticated tools such as magnetic resonance imaging (MRI) and positron-emission tomography (PET) as against calipers. In the early twentieth century, a Spanish histologist, Santiago Ramón y Cajal, dug deeper into the brain and found that the cells that constitute the brain could be divided into two categories. Using the Golgi staining method to visualize individual cells, he found cells that have polarity; on one end it seemed to be receiving information (dendrites) on the other it seemed to be sending information (axons). He correctly drew an arrow in this direction and named such cells neurons and viewed neurons as the entity that conveys the mind. The other cells were glial cells and such cells with no apparent polarity were considered to be nothing more than ‘glue’ that holds the brain structure together. In the 1960s, Stephen Kuffler stuck an electrode into these glial cells and found that these cells have much more negative resting membrane potential than neurons; however, he was unable to find much large and rapid fluctuation in the membrane potential, which was in contrast with neurons, which display rapid spikes of depolarization followed by hyperpolarization, a sequence called action potential. As glial cells were found to be electrically silent in general, they have long been considered to be inert cells in the brain without any contribution to information processing (Kuffler et al. 1966).

22.2 Glial Cells Were Not Silent

We are now facing a paradigm shift. Current studies found active signals in glial cells (Thrane et al. 2012). Although the electrical membrane response has been considered an indication of active cells, the change of intracellular ionic concentration can also be regarded as 'activity'. In particular, calcium concentration inside the cell is normally kept at a nanomolar level, which is less than one ten-thousandth (10^{-4}) of that in the extracellular environment. Intracellular calcium concentration can reach micromolar level transiently. Since these changes in calcium concentration trigger various biological processes, they are thought to be one of the most important intracellular signals. Other ions such as proton and sodium can also change their concentrations during activity, suggesting that they could also be involved in some biological processes.

Ionic concentrations can now be optically detected using fluorescent indicator dyes. A recent study shows that calcium concentration in glial cells in the visual cortex increases upon visual stimuli presented to the animal, and sets of glial cells were found to respond to specific direction of movement in the visual field (Schummers et al. 2008). Of course, these glial cells in the cortex did not detect visual stimuli directly, as they do not send their processes to the retina. Signals initiated in the photoreceptors in the retina are sent to the cortex via relaying neurons; thus, it is most likely that glial cells in the cortex are carefully monitoring neuronal activity in their vicinity. When compared with the millisecond time scale of electrical signals in neurons, these glial calcium signals are much slower, often taking seconds to rise and decay. However, some glial cells such as Bergmann glial cells in the cerebellum are known to express fast-responding AMPA-type glutamate receptors, the same type that is also found in neurons. When highly sensitive electrophysiological recordings were made from these cells, it was found that electrical signals in Bergmann glial cells generated in response to glutamate release from neurons have a time scale comparable to that in neurons (Matsui and Jahr 2003, 2006).

It should be noted that glial cells were not completely neglected in history. As previously mentioned, Cajal also observed glial morphology with interest and hypothesized that the fine processes of glial cells may thrust themselves in between sites of neuronal signal transfer; i.e., the synapse (García-Marín et al. 2007). He then went further to propose that the resulting shutdown of synaptic signal transmission may be the underlying mechanism of sleep. Cajal's imagination was phenomenal, considering that he was only able to observe a static view of the brain structure of chemically fixed tissues. We now know that the mechanism of sleep is not via synapse shutdown by glial processes. However, current technology also shows that glial processes can dynamically and rapidly change their form within minutes in live brain tissue (Wake et al. 2009; Nishida and Okabe 2007). A report showed that the fine process of glial cells in the hypothalamic supraoptic nucleus (SON) can change its shape upon lactation in female rats (Oliet et al. 2001). Glial processes do not go in between pre- and post-synaptic contacts but they do closely surround many synaptic contacts. Upon heightened synaptic activity, synaptically released neurotransmitter molecules can spill over from individual synaptic borders (Trussell et al. 1993; Barbour and Häusser 1997; Matsui et al. 1998; Budisantoso et al. 2012). In most cases, these excess transmitters are largely taken up to glial cells by membrane transporters. Therefore, the independence of signal transfer among neighboring synapses is often maintained. However, upon lactation, the glial process in SON retracts from the vicinity of synapses. This allows the spillover of transmitter and extrasynaptic receptors to be activated (Oliet et al. 2001). Upon witnessing these glial activities, researchers now doubt the initial intuition of regarding glial cells as just a form of 'glue'.

22.3 Difficulties of Glial Study

Cell-to-cell signaling is often accomplished by transmitters released into the extracellular space. Electrophysiological methods were mostly used to study the mechanisms of signal transmission from neuron to neuron (Matsui et al. 1998; Budisantoso et al. 2012) and from neuron to glial cell (Matsui and Jahr, 2003, 2006). We have been particularly focusing on the diffusion process of neurotransmitter glutamate in the extracellular space and analyzing the spatiotemporal spread of signals between cells (Budisantoso et al. 2013). Studying the signals emitted from glial cells was a challenge. Previous studies have shown that glial cells can receive information from neurons, but evidence for glia being able to release rapidly acting transmitters was scarce. Such evidence was necessary to show that glial cells can indeed participate in the information processing and in higher brain functions.

Conventionally, the neuron-to-neuron signal transmission has been studied using extracellular-stimulating electrodes. However, recent studies have showed that such electrodes can stimulate both neurons and glial cells at the same time. For example, the same frequency and strength of stimulation that is known to activate parallel fibers in the cerebellum and induce synaptic plasticity also generated a vigorous rise of intracellular calcium concentration in Bergmann glial cells (Beierlein and Regehr 2006; Piet and Jahr 2007). The brain tissue is tightly packed with various cell components and it is nearly impossible to selectively stimulate neurons or glial cells using extracellular electrodes.

Electrodes can be patch clamped to single glial cells, and current can be injected into individual cells; however, one of the significant features of glial cells is its low input resistance. Electrical signals initiated into a portion of the cell cannot propagate much further from the initiated site. By contrast, in the case of the high-input-resistant neurons, electrical signals can often spread throughout the whole cell, allowing information to be shared or integrated in single cells. This feature alone suggests that a single glial cell may not operate as a single computational unit but that each fine process may work individually (Kanemaru et al. 2014). In any case, controlling the electrical activity beyond the cell body, where the electrode is often experimentally located, is difficult. In addition, glial cells often lack voltage-gated calcium and other ion channels; thus, even if the voltage is changed, it may not initiate much important signal that can trigger biological processes, such as transmitter release. Experimentally controlling glial activity was necessary to understand the causal relationship between glial activity and neuronal response, but until recently, this technology was lacking.

22.4 Glial Optogenetics

As soon as optogenetics was founded (Boyden et al. 2005), we started trying to use this new technology to study the function of glia. Since optogenetical tools such as channelrhodopsin (ChR)-2 or archaerhodopsin-T (ArchT) would be expressed throughout the whole cell, this would allow control of activity even in the fine processes far from the cell soma. How the glial cells respond to changes in membrane potential or changes in the intracellular ionic environment induced by ChR2 or ArchT activation was uncertain. However, glial optogenetics seemed to be an ideal method to finally address the question of how glial cells are involved in brain function.

A repertoire of transgenic mice lines with various optogenetic tools expressed on different populations of cells was developed (Tanaka et al. 2012). Among these combinations, mice expressing a modified version of the ChR2 (C128S) (Berndt et al. 2009) in various types of glia; the astrocytes, the oligodendrocytes, and the microglia were produced. We are focusing on the astrocytes, although other glial cell types would be equally of interest.

An optical fiber was attached onto the surface of the skull, just above the cerebellum of mice expressing ChR2 in astrocytes, including Bergmann glial cells in the cerebellum (Sasaki et al. 2012). Since glial cells could be responsive to surgical damage, no hole was drilled into the skull. The following experiments showed that light penetrating the skull was sufficient to activate the ChR2 expressed in astrocytes. This was probably because we used the C128S modified version of the ChR2, which has been reported to have more than 100-fold higher sensitivity to light than the wild type. A single shot of light stimulation was given, and the mice were perfused and fixed only 10 min later. Using in situ hybridization, we stained for *c-fos*, which is known to be increased in neurons upon heightened activity. Surprisingly,

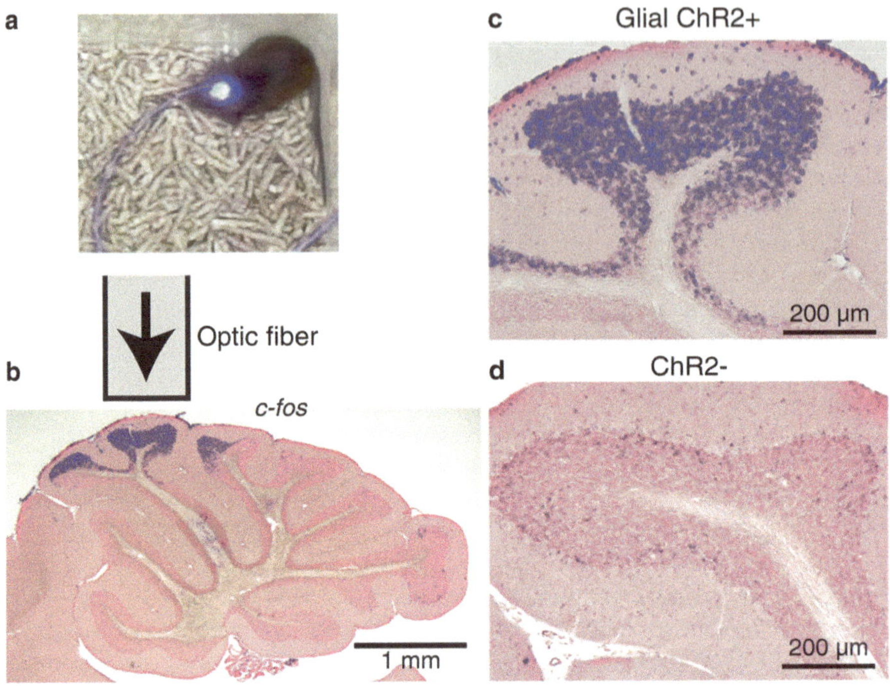

Fig. 22.1 Glial channelrhodopsin (ChR)-2 photostimulation induces neuronal activity. (**a**) In vivo photostimulation through the skull in freely moving transgenic mice with ChR2 expression in astrocytes. (**b**) In situ hybridization of *c-fos* shows increased expression directly beneath the optical fiber. (**c**) Close-up of the affected region. *c-fos* was increased in neurons. (**d**) Similar photostimulation did not evoke any response in mice not expressing ChR2 (Reproduced from Sasaki et al. 2012. With permission)

c-fos was actually increased in 'neurons' although the astrocytes were selectively photostimulated (Fig. 22.1). This shows that, upon photostimulation, some sort of signal gets emitted from astrocytes, which activates the neurons. This demonstrates the presence of glia-to-neuron signaling.

To understand the mechanism of signal transmission, acute brain slice recordings were made. Acutely prepared cerebellar slices with 250 μm thickness were superfused with artificial cerebellar spinal fluid (ACSF) containing TTX and Cd^{2+} to inhibit neuronal transmitter release. Even in such conditions, photostimulation of ChR2-expressing astrocytes resulted in excitatory current input to Purkinje neurons. This current was completely blocked by an AMPA-type glutamate receptor antagonist, showing that glutamate is likely to be released upon glial ChR2 photostimulation (Fig. 22.2). This was also interesting as neurons use glutamate as a transmitter to send excitatory signals to other neurons but here we showed that astrocytes can also use the same transmitter.

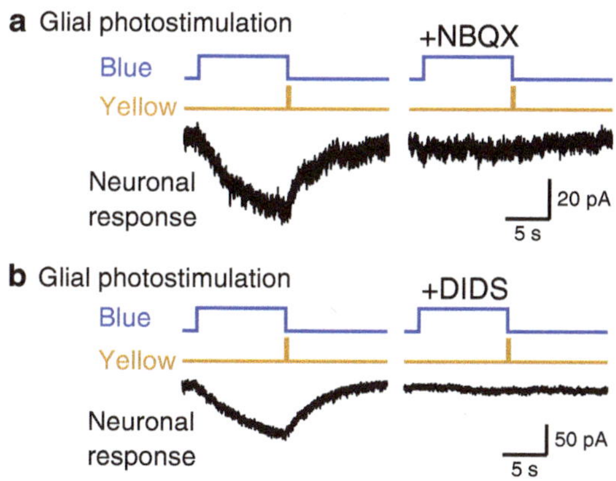

Fig. 22.2 Glial photostimulation induces glutamate release via DIDS-sensitive anion channels. (**a**) Acute cerebellar slice recordings. Astrocytes expressing C128S modified versions of channel-rhodopsin (ChR)-2 were photostimulated and excitatory currents were recorded from Purkinje cells. Note that it is known that ChR2(C128S) can be activated by blue light, but yellow light also needs to be applied to stop the activation. The excitatory current was completely blocked by an AMPA receptor-specific blocker (NBQX), suggesting that the released gliotransmitter is glutamate. (**b**) Glutamate release was suppressed by an anion channel blocker, DIDS (Reproduced from Sasaki et al. 2012. With permission)

22.5 Acceleration of Learning Via Glial Activation

We next assessed whether glial glutamate might affect neuronal synaptic transmission. Paired electrical stimuli were given to the cerebellar parallel fibers and characteristic paired-pulse facilitation of the excitatory post-synaptic current (EPSC) was recorded from a Purkinje cell. A single incident of glial photostimulation caused depression of the parallel fiber to Purkinje cell EPSCs (Fig. 22.3). The paired-pulse ratio remained the same, suggesting a post-synaptic site of action. This synaptic depression was long lasting and persisted for 30 min. This ‘long-term depression’ (LTD), was blocked by a metabotropic glutamate receptor type 1 (mGluR1) antagonist. LTD at this synapse is reported to be crucial for cerebellar motor learning and, as glial-photostimulation could also induce LTD, it became probable that glia may participate in the learning.

Synaptic plasticity, learning, and memory were all thought to be mediated solely by interactions between neurons. In the cerebellum, it is usually assumed that high-frequency activity of parallel fibers leads to glutamate spillover, which activates mGluR1 receptors. Activation of mGluR1 triggers subsequent events that lead to the reduction of AMPA receptors on the post-synaptic specializations. However, when the mGluR1 localization was carefully examined with immunogold morphological measurements, mGluR1 receptors on Purkinje cells were found to be located at the peripheral of synapses and they were actually excluded from the post-synaptic

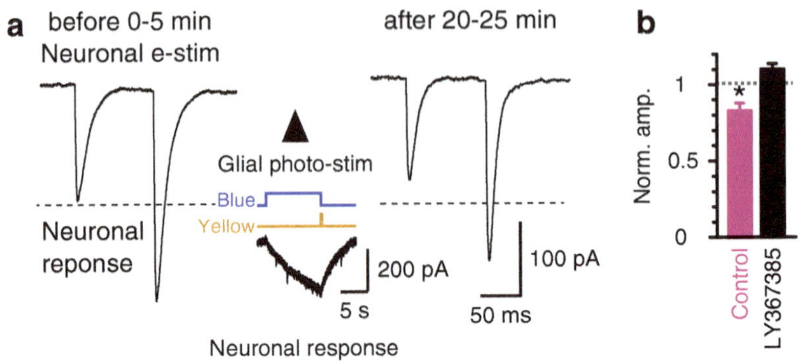

Fig. 22.3 Glial photostimulation induces synaptic plasticity. (**a**) Parallel fibers were electrically stimulated and excitatory post-synaptic currents (EPSCs) were recorded from a Purkinje cell. A single incident of glial channelrhodopsin (ChR)-2 photostimulation induced long-term depression of the EPSCs. (**b**) Long-term depression (LTD) was not induced in slices perfused with an mGluR1 antagonist (LY367385) (Reproduced from Sasaki et al. 2012. With permission)

specialization where synaptic vesicle release primarily occurs (Baude et al. 1993). This shows that mGluR1s are actually located directly across glial cell membrane. If glial cells were able to release glutamate, this would seem to be much more efficient than relying on spilled-over glutamate in activating these receptors.

Since glial photostimulation led to glutamate release and such glial glutamate were shown to affect synaptic transmission, in vivo experiments were performed to see whether glial activity could also affect animal behavior and learning. Horizontal optokinetic response, the HOKR paradigm, was studied. An un-anesthetized mouse was head fixed, and its eye was tracked using an infra-red camera. By horizontally moving a screen presented in front of the mouse back and forth, the eye involuntarily follows the screen. The location of the eye can be tracked in real time. The amplitude of the eye movement increases with repeated trials; this is called HOKR learning. Optic fibers were placed into the cerebellum to photostimulate glial cells. The flocculus region of the cerebellum is known to be related to HOKR learning, thus the optic fiber was inserted deep to ensure flocculus photostimulation. Glial cells were photostimulated during an HOKR learning paradigm. The gain of the horizontal eye movement was compared before and after the glial photostimulation. The HOKR amplitude increased, demonstrating that the eye pursues the screen better after glial photostimulation (Fig. 22.4).

22.6 Mechanism of Glial Glutamate Release

We have shown that glial glutamate can affect the neuronal circuit and accelerate learning. We next addressed the question of how glutamate is released from the astrocytes. In neurons, glutamate is packed at a very high concentration in synaptic

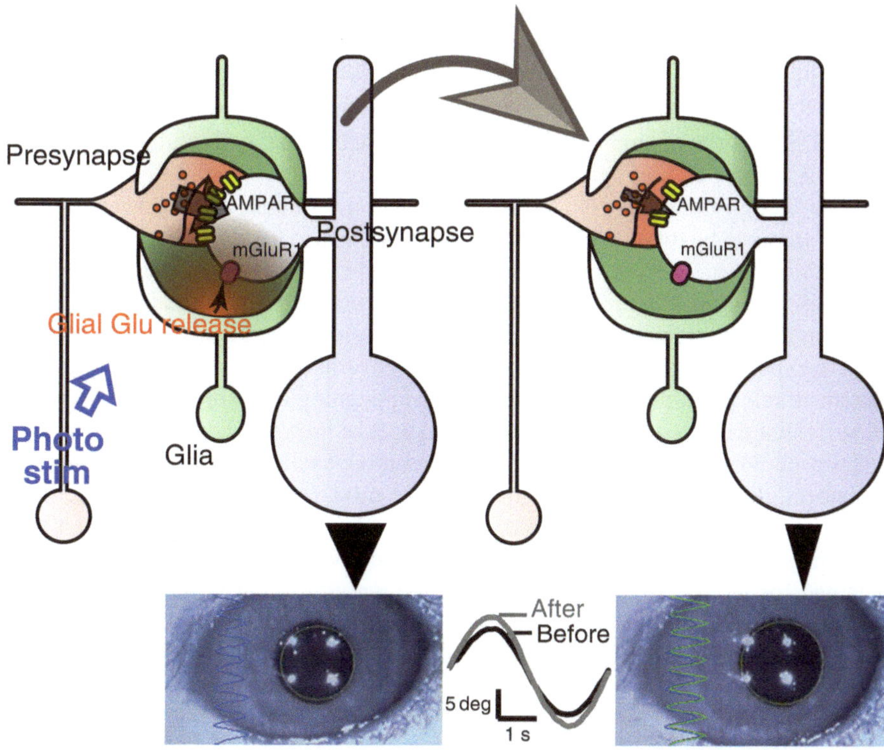

Fig. 22.4 Glial participation in learning and memory. Glial release of glutamate induced mGluR1 activation and long-term depression of parallel fiber to Purkinje cell synapse in the cerebellum. Cerebellar dependent motor (HOKR) learning was accelerated in vivo when glial cells were photostimulated using optic fibers inserted into the cerebellum (Reproduced from Sasaki et al. 2012. With permission)

vesicles and the content of the vesicles are released by exocytosis upon neuronal activation. DIDS, an anion channel/transporter blocker (Liu et al. 2009; Cavelier and Attwell 2005), was found to inhibit glutamate release from astrocytes (Fig. 22.2). Glutamate is an anion. This result suggests that cytosolic glutamate in astrocytes gets released via anion channels/transporters.

Our recent study also showed that glial ChR2-photostimulated glutamate release was not inhibited by suppressing an intracellular calcium concentration rise. Light-sensitive channels or pumps, originally found in algae or archaebacteria, can be exogenously expressed in mammalian cells and have been used in recent studies as tools for controlling membrane potential. However, the fact largely ignored by the neuroscientists applying the optogenetic techniques is that the major cation that flows in response to the widely used ChR2 and ArchT photoactivation is proton. Thus, ChR2 and ArchT could be regarded as optogenetic tools for instant intracellular acidification and alkalization, respectively. We showed that intracellular pH was indeed acidified upon ChR2 photoactivation. The magnitude of acidification can increase when intracellular pH buffering capacity is artificially lowered, even though the amount of proton influx via ChR2 remains the same. In this condition, glutamate release detected by the Purkinje cell was increased. These results suggest that acidification caused by proton influx via ChR2 photoactivation is the trigger for the DIDS-sensitive anion channel/transporter activation and glutamate release (Beppu et al. 2014).

Even though the same glutamate is used as a transmitter, the mechanism of its release seems to be totally different between neurons and astrocytes. About 50 years ago, during the era of Katz and Miledi, there was the calcium hypothesis and voltage hypothesis for transmitter release at the neuromuscular junction. At that time, the idea of transmitter release mediated by exocytosis of vesicles was also starting to emerge. Here, we show that a totally different principle governs glial glutamate release. Glial glutamate release is not via exocytosis but via opening of anion channels, and the release is not triggered by calcium but by changes in proton concentration.

Intracellular pH is known to change with neuronal activity in the brain, thus, pH-dependent glial glutamate release may occur to modulate the state of brain activity. Studying of such glial modulation of the state of brain and mind is one of our ultimate aims for the future research. However, before tackling such a problem, we searched for an extreme situation where glial activity might most affect neurons. Such extreme conditions may occur upon pathological situations.

22.7 Glial Glutamate Release upon Ischemia

Ischemia is a situation known to occur when blood flow supply is halted by stroke or other conditions. It is well known that the affected brain tissue undergoes acidosis and glutamate excitotoxicity. Excess amounts of glutamate cause over-excitation of neurons and the cells eventually die. Where the excess glutamate comes from and the mechanism underlying glutamate release is unknown. We hypothesized that acidosis in the glia is the direct cause of glial glutamate release upon ischemia.

Upon stroke, oxygen and glucose supply is halted. Glycogen stored predominantly in astrocytes of the glial cell population is converted into lactate, but lactate cannot be converted to energy as the aerobic metabolism is shut down. The produced lactate accumulates, leading to severe acidosis. We confirmed that acidosis does occur in acute cerebellar slices upon oxygen and glucose deprivation (OGD) with pH imaging of Bergmann glial cells. We next confirmed that glutamate release does occur upon OGD in acute slices. Purkinje cells were recorded in voltage clamp mode in the presence of picrotoxin to isolate the excitatory drive to Purkinje cells and in the presence of TTX and Cd^{2+} to suppress neurotransmitter release. Within 10–20 min from switching to OGD superfusate, a huge excitatory drive emerged that would be the prelude to cell death. This current was mostly blocked by an AMPA receptor blocker, GYKI, or DIDS, an anion channel blocker. These results suggest that OGD also triggers glutamate release via anion channels. These results, along with the ChR2 experiments, indicate that glial acidosis is the direct trigger of excess glutamate release upon ischemia. How can we prove this?

If intracellular pH is really triggering glial glutamate release upon ischemia, extruding the excess proton should stop glutamate release. ArchT is a proton pump that extrudes proton upon yellow light stimuli. We sought to use ArchT to counteract the acidosis induced by ischemia. Transgenic mice expressing ArchT in astrocytes were developed. Using pH imaging, it was confirmed that ArchT activation causes intracellular alkalization. We evaluated whether glial glutamate release could be stopped by optogenetic ArchT activation. The OGD-induced inward current in Purkinje cells was markedly suppressed by optogenetic activation of glial ArchT. The suppression was occluded by a glutamate receptor and transporter blocker cocktail. These data show that glial glutamate release upon ischemia is triggered by glial intracellular acidification, and such glial glutamate release can be stopped by countering acidosis with glia-specific alkalization.

To see whether glial alkalization can truly suppress ischemic brain damage, we again stuck an optic fiber onto the transgenic mouse. Cerebellar ischemia was created by injecting rose bengal and creating artificial thrombosis. Large structural damage was observed after 3 h from the creation of photothrombosis. In one group of animals, ArchT was intermittently activated after the creation of thrombosis. In this group, the structural damage was significantly reduced (Fig. 22.5). Note that

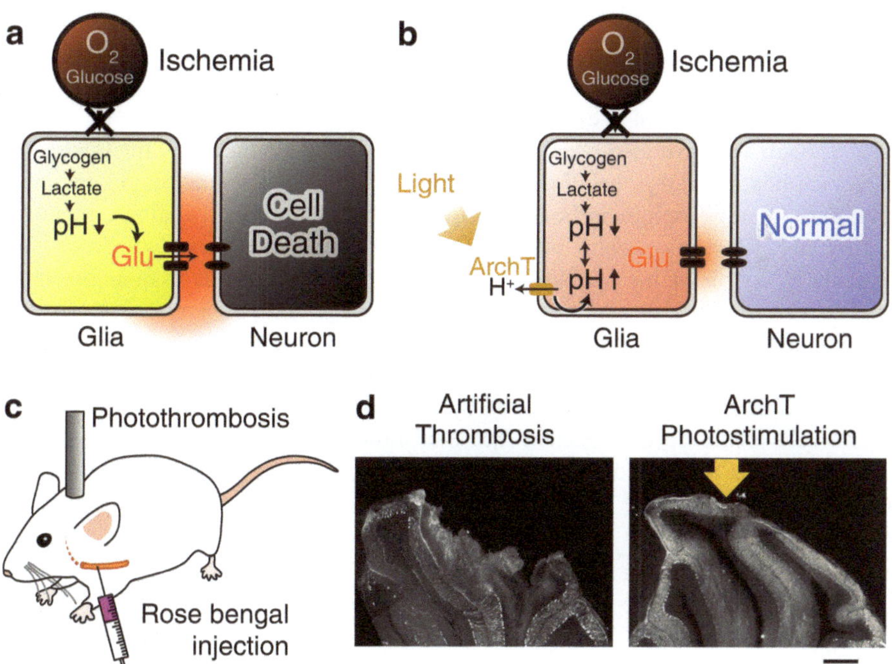

Fig. 22.5 Glial glutamate release via intracellular acidification and its role in ischemic damage. (**a**) Upon ischemia, intracellular acidification occurs, which triggers excess glial glutamate release. (**b**) Glial ArchT activation can neutralize the pH of the cytosol, leading to the cessation of glial glutamate release. (**c**) Artificial thrombosis was created in vivo. (**d**) Intermittent glial ArchT photoactivation resulted in relief of ischemic brain damage (Reproduced from Beppu et al. 2014. With permission)

blood flow is still stopped in the ArchT-activated mice. This shows that the reason the cell quickly dies upon ischemia is primarily not due to the lack of energy but because of excitotoxicity produced by an excess of glutamate.

Upon brain ischemia, application of thrombolytic agents to induce reperfusion is the only proven effective therapeutic intervention. Liberation of excess glutamate causes excitotoxicity; however, the cellular origins of glutamate and its release mechanisms upon ischemia remained unknown. Here, we show that oxygen and glucose deprivation causes rapid acidification of glia, and this is the key trigger for glial glutamate release. Photoactivation of transgenetically expressed light-activated proton pumps in glia led to glial alkalization, instant cessation of glutamate release, and relief of ischemic brain damage. Control of glial pH could be a new therapeutic target for intervention of ischemic cell death.

22.8 Discussions

We have shown that glial cells monitor neuronal activity, and here we show that glia can be activated by photostimulation and that glutamate release occurs as a consequence. Such glutamate affects synaptic plasticity and can ultimately affect learning and behavior. To understand information processing in the brain, we not only have to characterize the neuronal circuit but also the glial circuit, and the interaction between these two circuits needs to be characterized. Optogenetics has become a necessary tool to examine these interactions.

Although we have also shown the potential use of optogenetics for treatment, direct application of this technique in patients is limited as it would require preemptive gene therapy before ischemia. However, based on these findings, methods aimed at delivering a strong intracellular pH buffer, development of drugs that enhance H^+ extrusion mechanisms, or designing of glutamate-releasing anion channel blockers could be sought to alleviate brain damage upon ischemia. One can also imagine a future where induced pluripotent stem cells (iPS cells) taken from your own body is differentiated into glial cells with transgenetic inclusion of optogenetic tools, with their expression controlled by doxycycline or other inducible methods. These cells could be transplanted into the brain, and optogenetic tools could be activated by the implanted optical fibers. Such situations may be used to control the local environment of the brain tissue to treat diseases or possibly enhance brain cognitive functions.

References

Barbour B, Häusser M (1997) Intersynaptic diffusion of neurotransmitter. Trends Neurosci 20:377–384

Baude A, Nusser Z, Roberts JD et al (1993) The metabotropic glutamate receptor (mGluR1 alpha) is concentrated at perisynaptic membrane of neuronal subpopulations as detected by immunogold reaction. Neuron 11:771–787

Beierlein M, Regehr WG (2006) Brief bursts of parallel fiber activity trigger calcium signals in Bergmann glia. J Neurosci 26:6958–6967

Beppu K, Sasaki T, Tanaka KF et al (2014) Optogenetic countering of glial acidosis suppresses glial glutamate release and ischemic brain damage. Neuron 81:314–320

Berndt A, Yizhar O, Gunaydin LA et al (2009) Bi-stable neural state switches. Nat Neurosci 12:229–234

Boyden ES, Zhang F, Bamberg E et al (2005) Millisecond-timescale, genetically targeted optical control of neural activity. Nat Neurosci 8:1263–1268

Budisantoso T, Matsui K, Kamasawa N et al (2012) Mechanisms underlying signal filtering at a multisynapse contact. J Neurosci 32:2357–2376

Budisantoso T, Harada H, Kamasawa N et al (2013) Evaluation of glutamate concentration transient in the synaptic cleft of the rat calyx of held. J Physiol 591:219–239

Cavelier P, Attwell D (2005) Tonic release of glutamate by a DIDS-sensitive mechanism in rat hippocampal slices. J Physiol 564:397–410

García-Marín V, García-López P, Freire M (2007) Cajal's contributions to glia research. Trends Neurosci 30:479–487

Kanemaru K, Sekiya H, Xu M et al (2014) In vivo visualization of subtle, transient, and local activity of astrocytes using an ultrasensitive Ca(2+) indicator. Cell Rep 8:311–318

Kuffler SW, Nicholls JG, Orkand RK (1966) Physiological properties of glial cells in the central nervous system of amphibia. J Neurophysiol 29:768–787

Liu HT, Akita T, Shimizu T et al (2009) Bradykinin-induced astrocyte-neuron signalling: glutamate release is mediated by ROS-activated volume-sensitive outwardly rectifying anion channels. J Physiol 587:2197–2209

Matsui K, Jahr CE (2003) Ectopic release of synaptic vesicles. Neuron 40:1173–1183

Matsui K, Jahr CE (2006) Exocytosis unbound. Curr Opin Neurobiol 16:305–311

Matsui K, Hosoi N, Tachibana M (1998) Excitatory synaptic transmission in the inner retina: paired recordings of bipolar cells and neurons of the ganglion cell layer. J Neurosci 18:4500–4510

Nishida H, Okabe S (2007) Direct astrocytic contacts regulate local maturation of dendritic spines. J Neurosci 27:331–340

Oliet SH, Piet R, Poulain DA (2001) Control of glutamate clearance and synaptic efficacy by glial coverage of neurons. Science 292:923–926

Piet R, Jahr CE (2007) Glutamatergic and purinergic receptor-mediated calcium transients in Bergmann glial cells. J Neurosci 27:4027–4035

Sasaki T, Beppu K, Tanaka KF et al (2012) Application of an optogenetic byway for perturbing neuronal activity via glial photostimulation. Proc Natl Acad Sci U S A 109:20720–20725

Schummers J, Yu H, Sur M (2008) Tuned responses of astrocytes and their influence on hemodynamic signals in the visual cortex. Science 320:1638–1643

Tanaka KF, Matsui K, Sasaki T et al (2012) Expanding the repertoire of optogenetically targeted cells with an enhanced gene expression system. Cell Rep 2:397–406

Thrane AS, Rangroo Thrane V, Zeppenfeld D et al (2012) General anesthesia selectively disrupts astrocyte calcium signaling in the awake mouse cortex. Proc Natl Acad Sci U S A 109:18974–18979

Trussell LO, Zhang S, Raman IM (1993) Desensitization of AMPA receptors upon multiquantal neurotransmitter release. Neuron 10:1185–1196

Wake H, Moorhouse AJ, Jinno S et al (2009) Resting microglia directly monitor the functional state of synapses in vivo and determine the fate of ischemic terminals. J Neurosci 29:3974–3980

Part IV
Medical Optogenetics

Chapter 23
Towards Understanding the Neural Mechanism of Behavioral Phenotypes Seen in Psychiatric Disorders

Nobuhiro Nakai, Ofer Yizhar, and Toru Takumi

Abstract Patients with psychiatric disease are diagnosed by psychiatrists based on the information of non-quantitative objective parameters, including behavioral phenotypes. However, how any neural mechanism affects such behavioral phenotypes in patients is still unclear. Recent functional studies suggested the alteration in brain neural/network activity responds to subjected stimuli in some brain regions of psychiatric patients, indicating that excitatory/inhibitory (E/I) imbalance occurs in local neural circuits responsible for regional activities. Moreover, in human genetics, a large number of genetic variations, including single nucleotide variation (SNV) and copy number variation (CNV), have been found in psychiatric patients. Such variations must be causes of a psychiatric behavioral phenotype, while understanding of the relationship between genetic variations and neural mechanisms underlying psychiatric behavior remains poor due to the heterogeneity in genetic variations. Functional and molecular analyses with SNV and CNV suggest the mutations of synaptic genes might contribute to the abnormal neural activity due to synaptic dysfunction. To overcome the sparse knowledge of psychiatric neural phenotypes, we can choose two ways: one is to detect the abnormalities of neural function in animal models with the genetic variations found in human genetics, which means construct validity of an animal model, and another is to reproduce the behavioral phenotypes seen in psychiatric disorders by artificially controlling neural functions, referred to as face validity. Analyzing the neural activity in animal models with construct and face validities would help us understand the neural state in psychiatric patients.

N. Nakai
RIKEN Brain Science Institute, Wako, Saitama 351-0198, Japan

O. Yizhar
Department of Neurobiology, Weizmann Institute of Science, Rehovot, Israel

T. Takumi (✉)
RIKEN Brain Science Institute, Wako, Saitama 351-0198, Japan

JST, CREST, Tokyo, Japan
e-mail: toru.takumi@riken.jp

H. Yawo et al. (eds.), *Optogenetics*, DOI 10.1007/978-4-431-55516-2_23

Keywords E/I balance • Medial prefrontal cortex • Humanoid model • Social behavior • Psychiatric disorder

23.1 Construction of Animal Models for Psychiatric Disorders

One use of animal models is to understand human social behaviors. Since the genetical/biological markers in psychiatric disorders such as autism, schizophrenia, and major depression are still unknown, the behavioral diagnosis for psychiatric disorders has been via interviews with patients by clinical experts. Patients with autism and schizophrenia have deficits in social behavior. However, we have less information for the pathophysiology and functional mechanisms in the brains of these patients. Modeling behavioral abnormalities that resemble psychiatric disorders is an initial step to understanding the underlying mechanisms of psychiatric states; then we need to comprehensively study the animal models.

Recently, many genetic components have been found in genetic studies of psychiatric disorders, including autism spectrum disorder (ASD) (Liu and Takumi 2014; Malhotra and Sebat 2012; Sullivan et al 2012). In the human genome, a few dozen patients have one to several copy number variations (CNVs) and/or single nucleotide variations (SNVs) that might affect the abnormal state in psychiatric disorders (Cross-Disorder Group of the Psychiatric Genomics et al. 2013; Schaaf and Zoghbi 2011). CNVs and SNVs are able to alter the gene expression and/or physiological functions that contribute to any change in neural function. However, the relationship between CNVs/SNVs and neural abnormalities is not well understood due to heterogeneity. A chromosome-engineering technique provided a mouse model of ASD that recapitulates human CNVs and has construct validity (Nakatani et al 2009). There have been several models of ASD with construct validity. In these animal models for ASD, the behavioral phenotypes have some variations, while there is a specific trend of deficits in sociability (Blundell et al 2009; Chao et al 2010; Dani et al 2005; Etherton et al 2011; Gibson et al 2008; Gkogkas et al 2013; Han et al 2012; Hines et al 2008; Peca et al 2011; Poulopoulos et al 2009; Schmeisser et al 2012; Tabuchi et al 2007; Wallace et al 2012; Won et al 2012; Won et al 2013; Yang et al 2012). The findings in these genetic models by meta-analysis would help our understanding of pathology of psychiatric disorders.

Another validity is face validity, which indicates that a model should have phenotypes similar to those shown in human patients. By controlling a specific neural function of model animals using optogenetics, we may understand the brain mechanism of psychiatric disorders with higher spatio-temporal resolution. Recent papers using optogenetics show underlying mechanisms of the relationship between neural functions at brain circuit levels and behavioral phenotypes.

23.2 Excitatory/Inhibitory Balance in Psychiatric Disorders

Neural synchrony is regulated by excitatory and inhibitory (E/I) balance and controls responses to external stimuli from the environment. It is hypothesized that psychiatric patients have dysfunctional E/I balance. Reduction in expression of γ-aminobutyric acid (GABA)-ergic interneuron markers is reported in ASD patients. Electroencephalography (EEG) and magnetoencephalography (MEG) studies show abnormalities in a synchronized oscillatory activity in cortical regions of the patients with ASD and schizophrenia (Uhlhaas and Singer 2010). These abnormalities might arise from dysfunction in inhibitory systems. In the rodent studies, cortical E/I imbalance might be a probable cause of autistic behavior such as deficits in sociability. Yizhar et al. reported that enhanced excitability in medial prefrontal cortex (mPFC) induced social dysfunction but not exploratory behavior (Yizhar et al 2011). To regulate neural excitability, a novel optogenetic tool, 'stable step-function opsin' (SSFO) is used. SSFO can depolarize neurons for longer periods of over 30 min that allow for any behavioral tests in mice. Region-selective and optogenetic-controlled mice show increased high-frequency power of gamma oscillation and deficits in social behavior in the case of regulated mPFC but not the primary visual area, suggesting that a region-specific activity in the prefrontal cortical area can produce the outputs as specific social behaviors. The enhanced E/I balance also affects learning and memory. The mice with elevated E/I in the mPFC showed significant reductions in a freezing rate at contextual condition and tone stimuli after fear conditioning. It is considered that optogenetic enhancement in parvalbumin (PV) neurons can regulate the inhibition of regional activity, but the inhibitory output from PV neurons does not change any behavior level. Repressed E/I balance in the cortical region caused by enhanced inhibitory network might not be pathophysiology in social dysfunction. Importantly, the decreased number of PV neurons is consistently reported in the study of animal models for ASD, suggesting that dysfunction in the inhibitory system might enhance the excitatory state and directly/indirectly induce autistic behavioral phenotypes (Gogolla et al 2009). In electrophysiological analysis, E/I imbalance is found in many ASD models, while the direction about increased or decreased E/I ratio has been inconsistent, probably due to heterogeneity in genetic variation (Table 23.1). However, it is important to identify a behavioral output caused by E/I imbalance as an assessment of face validity. In a post-mortem study of patients with ASD, the expression of glutamic acid decarboxylases (GAD65/67) that synthesize GABA is decreased in parietal and cerebellar cortices (Fatemi et al 2002). GAD67 regulates synaptogenesis of PV neurons, and the expression is a differentiation marker of PV neurons. Reduced activity of PV neurons might elicit profound impairment in cellular information processing and induce specific behavioral change associated with network excitability in the cortical area. Recent optogenetics studies suggest that not excitatory but inhibitory inputs can affect the generation of gamma oscillation (Cardin et al 2009; Carlen et al 2012).

Table 23.1 Excitatory/inhibitory balance in autism spectrum disorder models

ASD model	Region	Age	mEPSC	s/eEPSC	mIPSC	s/eIPSC	Other information	Reference
Nrxn1a KO			↓(freq.)		Normal			*PNAS*. 2009 **106** 17998-8003
Nlgn3 R451C	SSC	P13–16	normal	e: normal	↑(freq.)	e: ↑(amp.)	VGAT puncta number↑	*Science*. 2007 **318** 71-6
	CA1	P28–40	↑(freq.)		normal		VGAT puncta number↑	*PNAS*. 2011 **108** 13764-9
Nlgn3 KO	SSC	P13–16			normal	e: normal		
	CA1	P28–40	↓(freq.)		↓(freq.)			
	CB	P21–28	↓(amp.)	e: ↑(CF)			PF mGluR-LTD: lost, WB: mGluR1α↑, IHC: VGlut2↑	*Science*. 2012 **338** 128–132.
Nlgn2 Tg	mPFC	Adult	Normal		↑(freq.)		VGluT puncta number↑, VGAT puncta number↑	*J Neurosci*. 2008 **28** 6055-67
Nlgn2 KO	CA1, CA3	Adult					VGAT puncta number↓	*Genes Brain Behav*. 2009 **8** 114-26
	CA1	Adult			↓(amp., freq.)	s: ↓(amp., freq.)	No decrease in VGAT puncta number	*Neuron*. 2009 **63** 628-42
	VLM	P2–4	Normal	s: normal	↓(amp., freq.)	s: ↓(amp., freq.)		
Shank3 C-term. Del	F-ctx, CA1	3–4w	Normal				NMDA/AMPA↓, LTP↓	*Cell*. 2011 **145** 758-72

Shank3B –/–	STR	6–7w	↓(amp., freq.)				NMDA/AMPA: normal, PPR: normal	*Nature*. 2011 **472** 437–442
	CA1	6–9w	Normal					
Shank3 KO	CA1	4–6w					LTP↓	*J Neurosci*. 2012 **32** 6525–41
Shank2 KO	CA1						NMDA/AMPA↓, LTP↓, LTD↓	*Nature*. 2012 **486** 261-5
ProSAP/Shank2 KO	CA1	P21–28	↓(freq.)		Normal		NMDA/AMPA↑, LTP↑	*Nature*. 2012 **486** 256-60
MeCP2 -/y	SSC	2–5w	↓(amp)		Normal		Firing rate↓	*PNAS*. 2005 **102** 12560-5
VIAAT-MeCP2 -/y	SSC	P25–35	Normal		↓(amp.)		CA1: LTP↓	*Nature*. 2010 **468** 263-9
Fmr1 KO	SSC	2–4w	Normal		Normal		EPSC to FS cell↓(amp)	*J Neurophysiol*. 2008 **100** 2615-26
Ube3a p+/m-	VC	P80	↓(freq.)		↓(freq.)	e: ↓(freq.)	Recovery of IPSC↓	*Neuron*. 2012 **74** 793-800
Scn1a +/–	PFC, CA1	P21–25	Normal	s: ↑(freq.)	Normal	s: ↓(freq.)	Rescued by benzodiazepine clonazepam	*Nature*. 2012 **489** 385-90

amp. amplitude, *CB* cerebellum, *F-ctx* frontal cortex, *freq*. frequency, *mEPSC* miniature EPSC, *mIPSC* miniature IPSC, *mPFC* medial prefrontal cortex, *s/eEPSC* spontaneous/evoked EPSC, *s/eIPSC* spontaneous/evoked IPSC, *SSC* somatosensory cortex, *STR* striatum, *VC* visual cortex, *VLM* ventrolateral medulla

Gamma oscillation on local neural circuits in the somatosensory cortex is generated by optogenetic stimulation of PV neurons at 30–50 Hz, while pyramidal neurons are unable to induce gamma-range oscillation at the same stimulation frequency. These findings indicate that inhibitory input by fast-spiking interneurons is a key regulator of high-frequency oscillation in the cortical area and support a hypothesis that E/I imbalance due to inhibitory dysfunction may cause ASD and schizophrenia. The mice receiving N-methyl-D-aspartate (NMDA) antagonists (PCP, MK801, ketamine) exhibit schizophrenia-like behavior and NMDA receptors preferentially drive the activity of cortical interneurons. It is considered that NMDA receptors are strongly associated with the emergence of gamma oscillation (Wang and Carlen 2012). PV neuron-specific NR1-deficient mice (PV-Cre/NR$1^{f/f}$) show up-regulation of gamma oscillation at a resting state. Optogenetic stimulation at gamma-range frequency induces the enhancement of gamma oscillation in the somatosensory cortex of a control group and reduction of gamma oscillation in a PV-Cre/ NR$1^{f/f}$ group. This PV-specific NR1 knockout does not affect spatial memory in a water maze test but decreases the working memory in a T-maze test. PV neuron activity via NMDA receptor is sufficient to enhance gamma oscillation, and the hypoactivity is implicated in the deficit in information processing underlying cognitive dysfunction in schizophrenia.

Chronic exposure to stress is associated with psychiatric disorders. Major depression is an acquired and debilitating syndrome. Social defeat stress (SDS) is a reliable model for depression. A population of chronic SDS-conditioned mice is susceptible to stress and exhibits depression-like behaviors, whereas another population exhibits resilient behavior to stress (Krishnan et al 2007). This difference between susceptible and resilient mice is variable by stimulating the projection pathway from the ventrotegmental area (VTA) to nucleus accumbens (NAc). VTA dopamine neurons show two firing patterns: low-frequency tonic and high-frequency phasic firing, and phasic firing is upregulated by SDS. Optogenetic phasic activation, but not tonic activation, of the VTA–NAc pathway induces the susceptibility to SDS, while optogenetic repression of this pathway using *Natronomonas pharaonis* halorhodopsin (*Np*HR) reduces the susceptibility (Chaudhury et al 2013). In the VTA–mPFC pathway, optogenetic activation does not change the susceptibility but the optogenetic repression of this pathway increases the social defeat state. This effect might be consistent with a low firing rate of the VTA neurons projecting in the mPFC in susceptible mice (Chaudhury et al 2013). The reduction in expression of immediate early genes is found in the mPFC of susceptible mice and patients with depression (Covington et al 2010). The phenotype of susceptible mice also includes low sociability in a social interaction test. Interestingly, increased mPFC activity by robust optogenetic stimulation at 100 Hz is able to recover the sociability of susceptible mice. Deep-brain stimulation has been applied to human patients and shows effective results for severe treatment-resistant depression (Lozano et al 2008). The antidepressive effect of optogenetic stimulation might be applied in human studies.

Anxiety phenotypes are common in many psychiatric disorders, including generalized anxiety disorder, panic disorder, and post-traumatic stress disorder (PTSD).

The amygdala is a responsible region for emotional and anxiety behavior. A recent optogenetic study has reported that the specific projection from the basolateral amygdala (BLA) modulates anxiety expression. Optogenetic stimulation of neural axons of the BLA projecting to the central nucleus of the amygdala (CeA) produces an anxiolytic effect, and stimulation of BLA cell bodies produces an anxiogenic effect (Tye and Deisseroth 2012). Moreover, selective repression of BLA–CeA axons with eNpHR3.0 induces an anxiogenic effect, suggesting that this cellular projection is capable of bidirectional modulation of anxiety. BLA neurons project many target areas, including not only the CeA but also NAc, the PFC, and the bed nucleus of stria terminalis. These brain regions might have intricately interacting effects on anxiety. Thus, optogenetic targeting can resolve the roles for distinct pathways from the amygdala in anxiogenic or anxiolytic production.

23.3 Connection of Construct Validity with Face Validity

Optogenetics is a useful tool for understanding neural mechanisms in psychiatric disorders since we can reproduce the behavioral phenotype by optogenetic modulation of neural function at high spatio-temporal resolutions. We can check the effect of optogenetic modulation in animal models in vivo whether the modification of neural function is really related to the behavioral phenotype in the model that shows abnormal behavior. As mentioned above, there is no consistent phenotype in ASD models at the level of in vitro electrophysiological analysis. These electrophysiological abnormalities can change E/I balance in microcircuit and as a result, dysfunction in microcircuit output occurs in ASD models. We need to check the input/output relationship between the input stimuli on the social event (direct interaction, smells of others, and/or visual information) and the output behavioral phenotype (approach or preference to others).

We know that a large number of CNVs are found in ASD and schizophrenia patients. SNVs and CNVs potentially produce alteration of gene expression and change molecular and/or cellular function in neurons. As a result, the neurons can display some abnormal actions. A recent paper has reported that there are many mosaic CNV variations in human neurons (McConnell et al 2013). The authors investigated neural CNVs in post-mortem frontal cortex and found that 13–41 % of neurons have at least one megabase-scale de novo CNV. This finding makes the genetic complexity in neurons more intricate. Application of a light-inducible transcriptional effectors (LITE) system for the modification of gene expression is a new and useful tool for achieving construct validity in neurons. The LITE system is an optogenetic two-hybrid system integrating the customizable TALE DNA-binding domain with the light-sensitive cryptochrome 2 protein and its binding partner CIB1 (Konermann et al 2013). The system establishes an optogenetic control of endogenous cellular processes and enables direct testing of the causal roles of genetic and epigenetic regulation in normal biological processes and disease states (Konermann et al 2013). It makes it possible to change gene expression with modulating gene dosages by targeting neurons selectively. The gene dosage can modulate neural function and might be a key factor for behavioral phenotype. In rodent studies, social hierarchy is controlled by synaptic efficacy in the mPFC (Wang et al 2011). The mice with upregulated expression of genes that enhance cellular excitability show an increase in social rank. On the other hand, mice with upregulated expression of genes that reduce cellular excitability show a decrease in social rank. This implicates that social behavior is originally regulated by gene expression level. The LITE system helps in understanding of the flow from gene expression to behavioral phenotypes.

By using optogenetic technology, we can modulate the regulation from gene to cell at the specific neural circuit or any brain region. But it is still difficult to control several neural circuits simultaneously. There is a limitation in selectivity of cell type because the promoter expression system using optogenetic tools is restricted. More sophisticated genetic methods for enhancing cell-type selectivity might help increase the number of controlled neural circuits. Optogenetics-based technology will be an invaluable tool to understand behavior related to mental disorders at genetic, cellular, and neural circuits and in vivo levels.

Acknowledgment Our work was supported in part by the Japan Society of Promotion of Science and Ministry of Education, Culture, Sports, Science, and Technology KAKENHI, Strategic International Coorperative Program and CREST, Japan Science and Technology Agency and the Israel Ministry of Science and Technology.

References

Blundell J, Tabuchi K, Bolliger MF et al (2009) Increased anxiety-like behavior in mice lacking the inhibitory synapse cell adhesion molecule neuroligin 2. Genes Brain Behav 8:114–126

Cardin JA, Carlen M, Meletis K et al (2009) Driving fast-spiking cells induces gamma rhythm and controls sensory responses. Nature 459:663–667

Carlen M, Meletis K, Siegle H et al (2012) A critical role for NMDA receptors in parvalbumin interneurons for gamma rhythm induction and behavior. Mol Psychiatry 17:537–548

Chao HT, Chen H, Samaco RC et al (2010) Dysfunction in GABA signalling mediates autism-like stereotypies and Rett syndrome phenotypes. Nature 468:263–269

Chaudhury D, Walsh JJ, Friedman AK et al (2013) Rapid regulation of depression-related behaviours by control of midbrain dopamine neurons. Nature 493:532–536

Covington HE 3rd, Lobo MK, Maze I et al (2010) Antidepressant effect of optogenetic stimulation of the medial prefrontal cortex. J Neurosci 30:16082–16090

Cross-Disorder Group of the Psychiatric Genomics (2013) Genetic relationship between five psychiatric disorders estimated from genome-wide SNPs. Nat Genet 45:984–994

Dani VS, Chang Q, Maffei A et al (2005) Reduced cortical activity due to a shift in the balance between excitation and inhibition in a mouse model of Rett syndrome. Proc Natl Acad Sci U S A 102:12560–12565

Etherton M, Foldy C, Sharma M et al (2011) Autism-linked neuroligin-3 R451C mutation differentially alters hippocampal and cortical synaptic function. Proc Natl Acad Sci U S A 108:13764–13769

Fatemi SH, Halt AR, Stary JM et al (2002) Glutamic acid decarboxylase 65 and 67 kDa proteins are reduced in autistic parietal and cerebellar cortices. Biol Psychiatry 52:805–810

Gibson JR, Bartley AF, Hays SA et al (2008) Imbalance of neocortical excitation and inhibition and altered UP states reflect network hyperexcitability in the mouse model of fragile X syndrome. J Neurophysiol 100:2615–2626
Gkogkas CG, Khoutorsky A, Ran I et al (2013) Autism-related deficits via dysregulated eIF4E-dependent translational control. Nature 493:371–377
Gogolla N, Leblanc JJ, Quast KB et al (2009) Common circuit defect of excitatory-inhibitory balance in mouse models of autism. J Neurodev Disord 1:172–181
Han S, Tai C, Westenbroek RE et al (2012) Autistic-like behaviour in Scn1a+/– mice and rescue by enhanced GABA-mediated neurotransmission. Nature 489:385–390
Hines RM, Wu L, Hines DJ et al (2008) Synaptic imbalance, stereotypies, and impaired social interactions in mice with altered neuroligin 2 expression. J Neurosci 28:6055–6067
Konermann S, Brigham MD, Trevino AE et al (2013) Optical control of mammalian endogenous transcription and epigenetic states. Nature 500:472–476
Krishnan V, Han MH, Graham DL et al (2007) Molecular adaptations underlying susceptibility and resistance to social defeat in brain reward regions. Cell 131:391–404
Liu X, Takumi T (2014) Genomic and genetic aspects of autism spectrum disorder. Biochem Biophys Res Commun 452:244–253
Lozano AM, Mayberg HS, Giacobbe P et al (2008) Subcallosal cingulate gyrus deep brain stimulation for treatment-resistant depression. Biol Psychiatry 64:461–467
Malhotra D, Sebat J (2012) CNVs: harbingers of a rare variant revolution in psychiatric genetics. Cell 148:1223–1241
McConnell MJ, Lindberg MR, Brennand KJ, Piper JC et al (2013) Mosaic copy number variation in human neurons. Science 342:632–637
Nakatani J, Tamada K, Hatanaka F et al (2009) Abnormal behavior in a chromosome-engineered mouse model for human 15q11-13 duplication seen in autism. Cell 137:1235–1246
Peca J, Feliciano C, Ting JT et al (2011) Shank3 mutant mice display autistic-like behaviours and striatal dysfunction. Nature 472:437–442
Poulopoulos A, Aramuni G, Meyer G et al (2009) Neuroligin 2 drives postsynaptic assembly at perisomatic inhibitory synapses through gephyrin and collybistin. Neuron 63:628–642
Schaaf CP, Zoghbi HY (2011) Solving the autism puzzle a few pieces at a time. Neuron 70:806–808
Schmeisser MJ, Ey E, Wegener S et al (2012) Autistic-like behaviours and hyperactivity in mice lacking ProSAP1/Shank2. Nature 486:256–260
Sullivan PF, Daly MJ, O'Donovan M (2012) Genetic architectures of psychiatric disorders: the emerging picture and its implications. Nat Rev Genet 13:537–551
Tabuchi K, Blundell J, Etherton MR et al (2007) A neuroligin-3 mutation implicated in autism increases inhibitory synaptic transmission in mice. Science 318:71–76
Tye KM, Deisseroth K (2012) Optogenetic investigation of neural circuits underlying brain disease in animal models. Nat Rev Neurosci 13:251–266
Uhlhaas PJ, Singer W (2010) Abnormal neural oscillations and synchrony in schizophrenia. Nat Rev Neurosci 11:100–113
Wallace ML, Burette AC, Weinberg RJ et al (2012) Maternal loss of Ube3a produces an excitatory/inhibitory imbalance through neuron type-specific synaptic defects. Neuron 74:793–800
Wang X, Carlen M (2012) Optogenetic dissection of cortical information processing-shining light on schizophrenia. Brain Res 1476:31–37
Wang F, Zhu J, Zhu H et al (2011) Bidirectional control of social hierarchy by synaptic efficacy in medial prefrontal cortex. Science 334:693–697
Won H, Lee HR, Gee HY et al (2012) Autistic-like social behaviour in Shank2-mutant mice improved by restoring NMDA receptor function. Nature 486:261–265
Won H, Mah W, Kim E (2013) Autism spectrum disorder causes, mechanisms, and treatments: focus on neuronal synapses. Front Mol Neurosci 6:19
Yang M, Bozdagi O, Scattoni ML et al (2012) Reduced excitatory neurotransmission and mild autism-relevant phenotypes in adolescent Shank3 null mutant mice. J Neurosci 32:6525–6541
Yizhar O, Fenno LE, Prigge M et al (2011) Neocortical excitation/inhibition balance in information processing and social dysfunction. Nature 477:171–178

Chapter 24
Establishment of Gene Therapy Using Channelrhodopsin-2 to Treat Blindness

Eriko Sugano and Hiroshi Tomita

Abstract The retinal photoreceptor cells are specialized neurons capable of phototransduction. The photoreceptor cells translate light stimuli to regulate glutamate and transmit the stimuli to retinal ganglion cells (RGCs). Although photoreceptor degeneration causes blindness, histological studies have shown that a subset of retinal cells survive after the degeneration.

We tested whether a treatment using channelrhodopsin (ChR)-2, which is a unique light-perceptive ion channel, could restore vision in rats with photoreceptor degeneration. Our hypothesis was that the RGCs surviving post-degeneration can function as photosensitive cells by expressing ChR2. In neurons expressing ChR2, exposure to light results in an influx of Na^+ into the cells, thereby causing depolarization, and subsequent cellular excitation.

We transfected the *ChR2* gene into an adeno-associated virus (AAV) that was injected intravitreously into the eyes of Royal College of Surgeons rats. We studied them for 16 months after the injection. To analyze the visual function, visual evoked potential was recorded and optical behavior was examined. The results revealed that ChR2 therapy restored the visual function. We also performed a local and systemic analysis for assessing the immune reaction to the virus, which showed that none occurred to cause immune rejection and inflammation following the intravitreous injection.

To conclude, we think that ChR2 therapy is an important and practical method for treating photoreceptor degeneration.

E. Sugano
Laboratory of Visual Neuroscience, Department of Chemistry and Bioengineering, Faculty of Engineering, Graduate School of Engineering, Iwate University, 4-3-5 Ueda, Morioka, Iwate 020-8551, Japan
e-mail: sseriko@iwate-u.ac.jp

H. Tomita (✉)
Laboratory of Visual Neuroscience, Department of Chemistry and Bioengineering, Faculty of Engineering, Graduate School of Engineering, Iwate University, 4-3-5 Ueda, Morioka, Iwate 020-8551, Japan

Clinical Research, Innovation and Education Center, Tohoku University Hospital, 2-1 Seiryou-machi, Sendai 980-8574, Japan
e-mail: htomita@iwate-u.ac.jp

H. Yawo et al. (eds.), *Optogenetics*, DOI 10.1007/978-4-431-55516-2_24

Keywords Channerlrhodopsin-2 • Retinitis pigmentosa • Photoreceptor degeneration • Adeno-associated virus • RCS • Immune reaction

Abbreviations

AAV	Adeno-associated virus
AMD	Age-related macular degeneration
ChR2	Channelrhodopsin-2
LGN	Lateral geniculate nucleus
PDE	Phosphodiesterase
RCS	Royal College of Surgeons
RGCs	Retinal ganglion cells
RP	Retinitis pigmentosa
RPE	Retinal pigment epithelium
VEP	Visually evoked potential

24.1 Property of Channelrhodopsin-2

Channerlrhodopsin (ChR)-2 is a light-sensitive protein in green algae *Chlamydomonas reinhardtii*. ChR2 is a seven-transmembrane receptor and is a unique light-perceptive ion channel (Fig. 24.1). In neurons expressing ChR2, light exposure results in the transit of Na^+ into cells, causing depolarization, which subsequently leads to excitation of the cells (Nagel et al. 2003).

24.2 Application of ChR2 in a Therapy for Retinal Degeneration

24.2.1 Light Absorption in the Retina

In the retina, photoreceptors are the only cells to receive light signals. Photoreceptors have a photosensitive protein, rhodopsin. Rhodopsin is a well-known G-protein-coupled receptor and contains 11-*cis* retinal as a chromosphere. After photoreceptors receive a light signal, 11-*cis* retinal coupled to rhodopsin isomerizes to all-*trans* retinal. Consequently, rhodopsin conformation is changed, causing the uncoupling of retinal and rhodopsin. The free rhodopsin works on a stable construct of G-protein transducin, αβγ and alters the formation to two separate subunits, α and βγ. These events activate phosphodiesterase (PDE) and lead to the hydrolysis of cGMP. The cGMP-gated cation channel is closed in response to cGMP degradation.

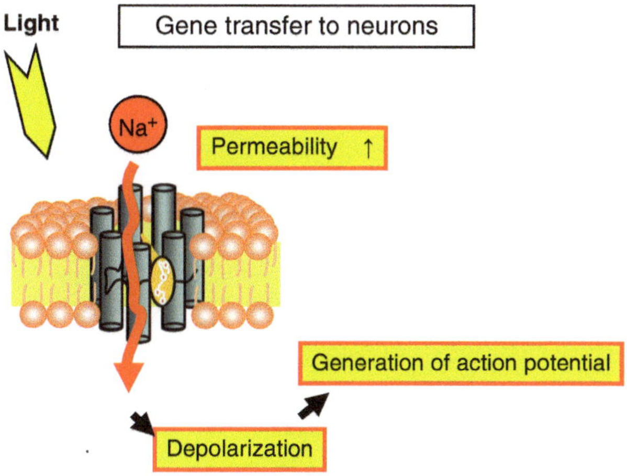

Fig. 24.1 Excitation caused by neuron cells expressing channelrhodopsin (ChR)-2. Expression of ChR2 on the cell membrane enables neurons to open cation-gated channels according to light stimuli and allow Na^+ into the cells. Thus, neurons are depolarized to generate active potentials

Photoreceptors result in hyperpolarization. Thus, photoreceptors translate a light signal into a biological signal such as a neurotransmitter release, finally transmitting the signal to inner retinal neurons.

In the inner retina, the signal is modified by the release of glutamate in bipolar, horizontal, and amacrine cells, and glutamate is finally received via the synapse of retinal ganglion cells (RGCs). RGCs open a cation-gated channel and cause the influx of Na^+. At the final step, ganglion cells are depolarized and transmit the signal to lateral geniculate nucleus (LGN) via the optic nerve (Fig. 24.2).

24.2.2 Degeneration of Photoreceptors

Photoreceptor cells translate light signals to regulate glutamate as described above and consume a high amount of oxygen. As such, photoreceptors are sensitive to oxidative stress, a major factor that leads to the degeneration of photoreceptors. To prevent this stress, retinal pigment epithelium (RPE) cells beneath photoreceptor cells produce anti-oxidants, including glutathione peroxidase, glutathione-s-transferase, catalase, and superoxide dismutase. Dysfunction of RPE cells leads to photoreceptor degeneration. RPE cells also work to maintain cellular homeostasis by phagocytosis and digestion of photoreceptors, which are shed every day. The dysfunction of phagocytosis-related genes such as *Mertk* also leads to photoreceptor degeneration. In the visual cycle, RPE cells are involved in many steps and it is essential that RPE cells and photoreceptors are both in good physical condition for photoreceptor survival. Photoreceptor cells perceive light as the first step in the

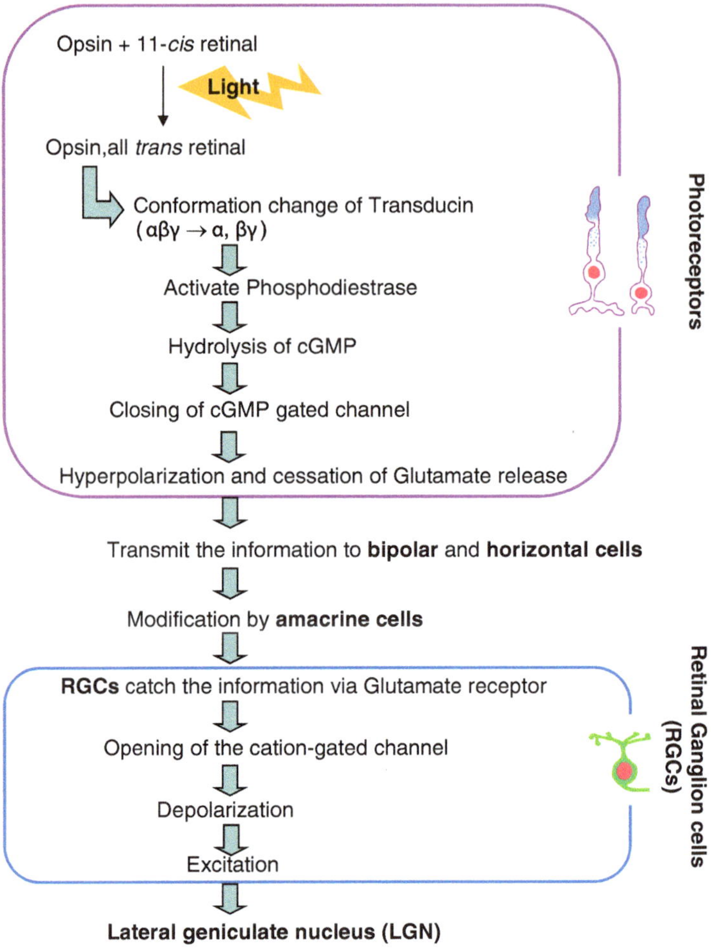

Fig. 24.2 System for absorption of light into the retina. The photoreceptors, rhodopsin coupled with chromophore, absorb light and transmit information to the inner retinal neurons through regulation of glutamate. During this transmission, the information is amplified and regulated. It is believed that the photoreceptor signal is amplified 100,000 times when it reaches horizontal cells. In the final step, the information is transmitted from retinal ganglion cells (RGCs) to the lateral geniculate nucleus (LGN)

retina, thus photoreceptor degeneration causes loss of vision, such as in retinitis pigmentosa (RP) and age-related macular degeneration (AMD).

Recently, molecular techniques have made it possible to predict retinal degeneration through a homology search of genes involved in RP, a familial disease. However, clinical strategies for the treatment of these diseases are still under investigation and patients continue to suffer. Many therapies exist for corneal disorder, cataracts, and glaucoma but not for RP. Photoreceptor degeneration is the final large problem to be resolved in vision therapy, and many scientists in the world are trying to design a viable treatment.

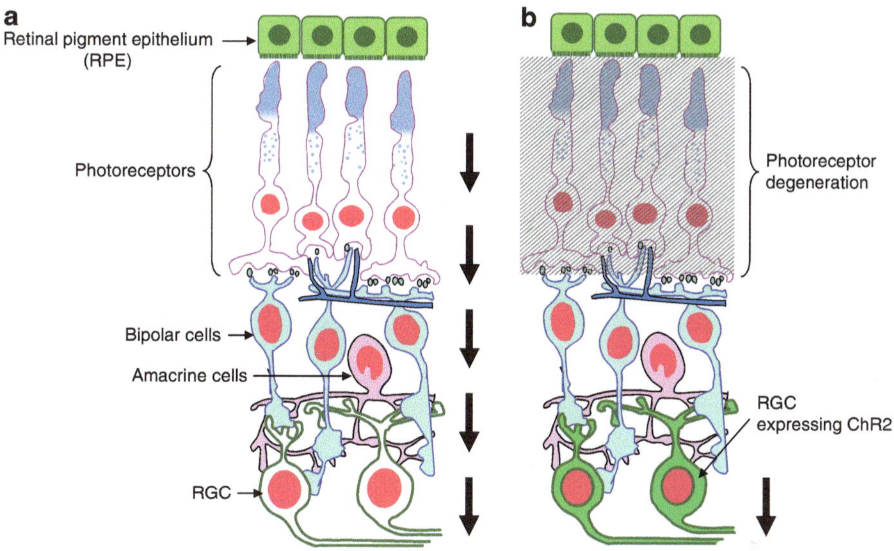

Fig. 24.3 The new pathway of transmission of light signal by channelrhodopsin (ChR)-2 therapy. The light signal is usually transmitted from photoreceptors to inner retinal neurons and retinal ganglion cells (RGCs) transmit the information to the lateral geniculate nucleus (LGN) (**a**). The visual system restored by ChR2 therapy; RGCs absorb light and transmit the information to the LGN (**b**)

24.2.3 Restoration of Vision by ChR2

Histological studies indicate a subset of retinal cells that survive after photoreceptor degeneration (Santos et al. 1997). Recently, studies on artificial vision showed that excitation of retinal neurons by electrical stimuli can translate the light signal to the visual cortex. However, the number of electrodes that stimulate the retina is limited, resulting in low resolution. The other problem is the size of the electrode, which cannot attach to merely one cell, and results in the disruption of the ideal condition, which consists of an equal electrode-to-cell ratio.

To overcome these problems, we are trying to utilize ChR2 to restore vision. Our concept is that the surviving RGCs can themselves work as photosensitive cells by expression of ChR2 (Fig. 24.3). This method carries many advantages. First, this method can give the ability of light acquisition to each cell and avoid the concurrent stimulation by one electrode in the case of retinal prosthesis. The second point is that the excitation induction occurs in the same manner to the native human ganglion cells and depolarization is by influx of cations. The third point is to use all-*trans* retinal as the chromophore, which needs no complicated synthesis steps like human 11-*cis* retinal regeneration (Fig. 24.4). All-*trans* retinal changes to 13-*cis* retinal via light absorption and 13-*cis* retinal automatically changes to the all-*trans* form because of the stabilization caused by energy absorption (Hubbard 1966). Therefore, ChR2 therapy does not need healthy photoreceptors and RPE cells, and can be utilized for the restoration of vision after photoreceptor cell death.

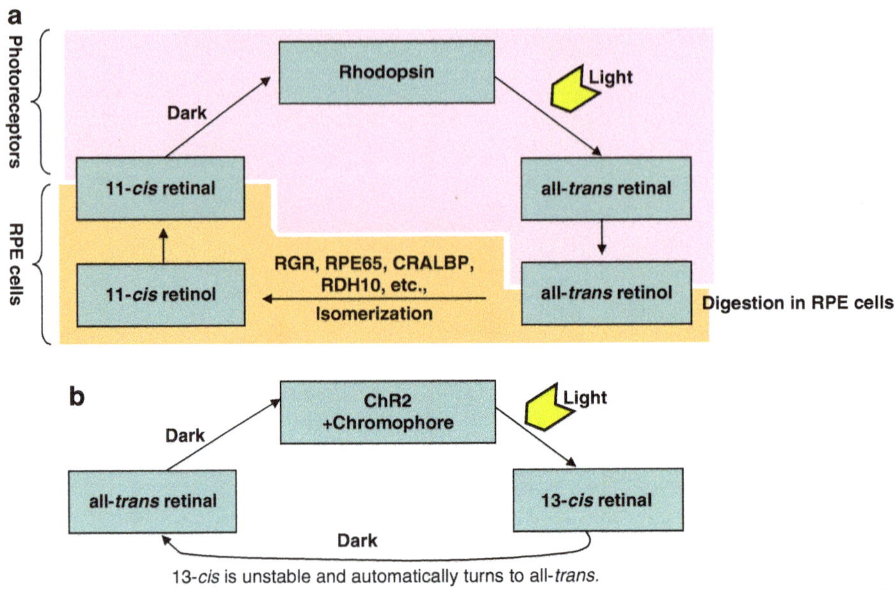

Fig. 24.4 Regeneration of retinal. In humans, 11-*cis* retinal is used as the chromophore. The regeneration of 11-*cis* retinal from all-*trans* retinal needs enzymatic reactions in photoreceptors and retinal pigment epithelium (RPE) cells (**a**). ChR2 uses all-*trans* retinal as the chromophore, which does not depend on complicated reactions for the regeneration (**b**)

24.3 Restoration of Vision in Photoreceptor-Degenerated Animals

24.3.1 Gene Transfer of ChR2 Coding Gene and Its Expression

The *ChR2* gene was inserted into an adeno-associated virus (AAV), which is derived from the non-pathogenic virus by omitting proliferation sequences. AAV has 11 different serotypes. In our experiment, we used type 2, which has been widely used for clinical studies. Recently, type 2 AAV has been used in clinical studies for Leber congenital amaurosis, targeting the *RPE65* gene, and has demonstrated efficacy (Maguire et al. 2008). Type 2 AAV also has high affinity to RGCs following intravitreal injection. In gene therapy, not only the gene of interest but also the toxicity of vectors needs to be considered for safety. Therefore, it is beneficial to choose a vector that has been used for other clinical trials.

In our experiments, *ChR2* was fused with a fluorescent protein sequence and inserted into AAV2 (AAV-ChR2V). AAV-ChR2V was injected intravitreously into the eyes of Royal College of Surgeons (RCS) rats, which have same mutation as human RP. At 2 weeks of injection, expression of ChR2V was first detected in the retina, and maximum expression was observed at 8 weeks. Following adeno virus

Fig. 24.5 Expression profile of channelrhodopsin (ChR)-2 in the retinal. Fluorogold was injected into superior colliculus and retinal ganglion cells (RGCs) were labeled via retrograde transportation (blue fluorescence). The expression of Venus (green fluorescence) indicated that the main cells targeted by the treatment were in the retinal ganglion cell layer; however, there was some expression in the inner plexiform cells layer

(AV) gene transfer, gene expression peaked at 3–7 days after injection and disappeared by 2 weeks. Compared with AV, AAV is a promising vector for long-term gene transfer. We showed that ChR2 gene expression was sustained even at 16 months after injection. Our study demonstrated that RCS rats expressed the gene for the entirety of their lives even after only one injection of AAV-ChR2V.

We observed rats for 16 months after injection and conducted no experiments over this period, because their life span is almost 2 years. In our study, RCS rats received an injection at 10 months old, the time at which they lost photoreceptors. Our histological studies demonstrated that ChR2V was mainly expressed in RGCs with inner retinal cells (Fig. 24.5). ChR2V was expressed at 28 % of the total number of RGCs (Tomita et al. 2007; Isago et al. 2012). This initially appears to be an inefficient gene transfer; however, because there are about one million RGCs in the human retina, 300,000 cells can become photosensitive following one injection of AAV. Compared with the utility of a retinal prosthesis, which can only stimulate 1,000 cells, ChR2-mediated gene therapy is more practical and demonstrates high resolution.

24.3.2 Analysis by Retinal Electrical Physiology

Visually evoked potential (VEP) was recorded after AAV injection. At 1 week after injection, light response was recorded, which peaked at 8 weeks. It was then maintained for the lifespan of the rats without any loss of function (Tomita et al. 2007). We observed that ChR2 caused shortening of the P1 latency (Fig. 24.6). The

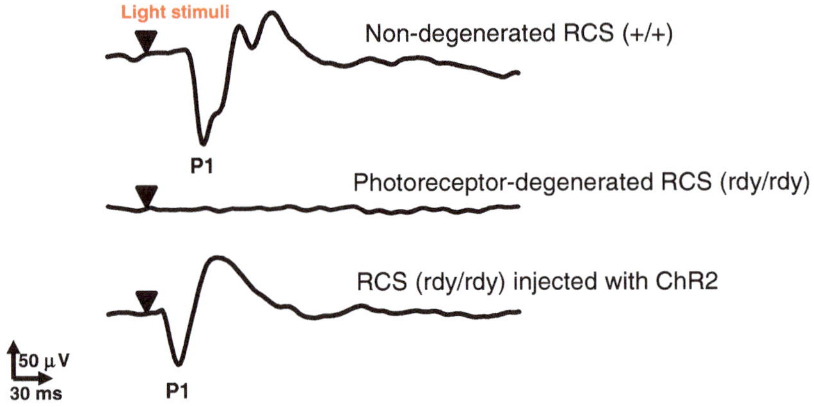

Fig. 24.6 Recording of visually evoked potential (VEP) after channelrhodopsin (ChR)-2 therapy. VEPs were recorded after ChR2 therapy, which had no recordable response before treatment. P1 latency was shortened after ChR2 therapy compared with normal rats due to the differences in the mechanism of light absorption in the retina

shortening of the P1 latency is derived from a short-cut in the transmission of inner retinal cells (Tomita et al. 2009b). These results indicate that ChR2 is a useful tool for restoring vision. However, ChR2 can only translate a signal under 460 nm. This means that ChR2 can perceive blue light signals but cannot respond to green and red light signals. To resolve the problem, glass lenses converting the information to a blue signal may be a useful tool.

24.3.3 Optical Behavioral Test

To study the visual function obtained by ChR2, we performed optomotor tests. We set up the analysis system in the same manner to the grating card method used for visual tests of infants under the age of 3 years (Fig. 24.7). As our results indicate, RCS rats injected with AAV-ChR2V can chase the rotated stripes despite the loss of photoreceptors (Tomita et al. 2010; Tomita et al. 2009a).

24.4 Aim for Gene Therapy

24.4.1 Immune Reaction in the Eye

Eyes are considered immune privileged. However, because ChR2 is not of human origin, it might be perceived as a foreign antigen and may induce an immune reaction. In our experiments, we examined adhesive leukocytes in retinal blood vessels

Fig. 24.7 Analysis of the optical behavior by rotate column test. A blue-black grating pattern (peak; 460 nm), which is the wavelength property of channelrhodopsin (ChR)-2, was displayed on computer monitors arranged in a square around a platform. In the experiments, the grating patterns with variable spatial frequency and contrast were displayed in front of rats on the platform, and optomotor responses were analyzed

(Sugano et al. 2011). As our results indicate, there was no obvious inflammation following the injection of AAV vector (control) or AAV-ChR2V (Fig. 24.8). Histochemical analysis also revealed that no leukocytes and macrophages migrated into the retina and vitreous, symptoms indicating retinal inflammation. The histology of the retina also showed no changes (Fig 24.9).

24.4.2 Systemic Analysis for Immune Reaction

To study the systemic immune response, peripheral blood was collected at several periods after injection. Analysis of lymphocyte subsets revealed that the ratio of CD4+/CD8+ T cells increased at 1 week after the injection (Sugano et al. 2011). However, the ratio immediately returned to baseline and there was no change at any future time point. Kim et al. reported that CD4+/CD8+ T cells were upregulated by bacterial infection (Kim and Lim 2002). The increment of CD4+/CD8+ T cells might be caused by recognition of the virus by the cell-mediated immunity as T-cell ratio also increased following the injection of the control virus, AAV-Venus. As our results indicate, the increment of CD4+/CD8+ T cell ratio was transient and did not result in sustained inflammation.

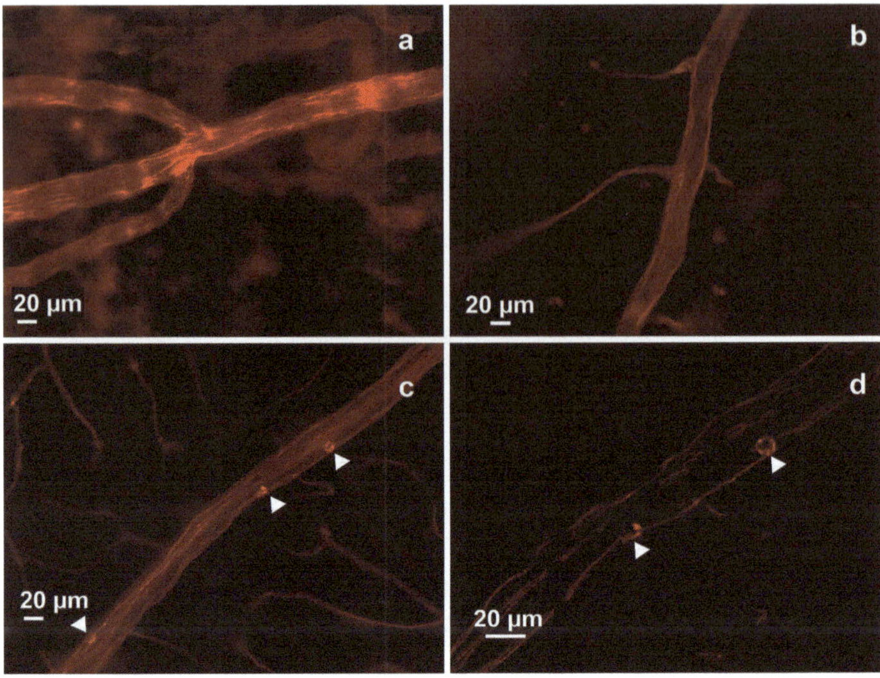

Fig. 24.8 Analysis for inflammation in the retinal blood vascular. Adhesion of leukocytes to retinal blood vessels was studied by administration of the lectin, concanavalin A, and whole-mounted retinal samples were observed by fluorescence microscopy. No adhesion was observed in untreated control (**a**) and AAV-ChR2V injected Royal College of Surgeons (RCS) (+/+) (RCS)(rdy/rdy) rats, at 1 year after injection (**b**). As a model of inflammation, non-degenerated RCS (+/+) rats received an intravitreous injection of lipopolysaccharide (LPS) and were examined. The results show adherence of leukocytes(**c**). High-magnification photography of the LPS-treated retinal vessels shows leukocyte adhesion more clearly (**d**). ▲: leukocyte

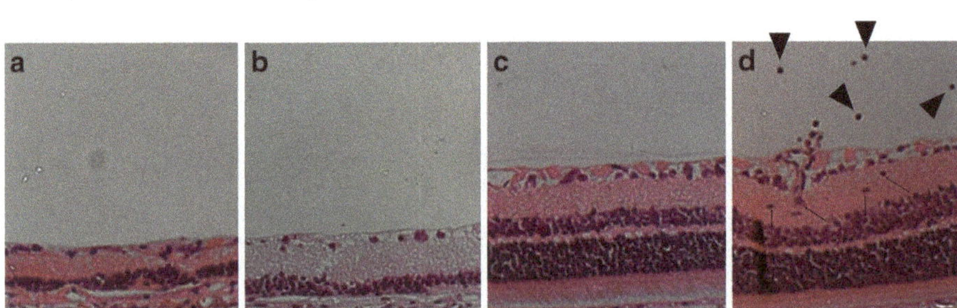

Fig. 24.9 Retinal histology. Retinal histology of the Royal College of Surgeons (RCS) rats (rdy/rdy) was examined after 64 weeks of injection of the AAV-ChR2V (**a**) and compared with the contralateral eye, which did not receive an injection (**b**). No differences were observed between these retinas. The histology of the inflammatory model: RCS (+/+) rats received injected lipopolysaccharide (LPS) (**d**), show infiltration of macrophages and leukocytes (↑) and migration of these cells into vitreous (▼). (**c**) Representation of the histology of the RCS (+/+) rats with no treatment as the control

Humoral immunity was also studied. Very few antibodies were generated against AAV and ChR2, indicating that AAV and ChR2 are safe to use (Sugano et al. 2011). To analyze the systemic scatter of the administered gene, messenger RNA (mRNA) expression in organs was analyzed by reverse transcriptase polymerase chain reaction (RT-PCR). There was no expression in the brain and there was a little expression in heart, lungs, and intestines in some cases. None of the treated rats died during the observation period and nor did they develop any form of cancer. These results suggest that there is no immune reaction to cause immune rejection and inflammation by AAV-ChR2 in the animal model of RP.

24.5 Conclusions

We have previously studied vision maintenance through the release of neurotropic factors and with the use of retinal prostheses. However, we now predict that ChR2 therapy is the most practical and applicable method for the treatment of blindness. We have studied the long-term function and safety of ChR2 treatment. We are still a long way from this technique being accepted for gene therapy, mainly because little information is available to certify the validity of the therapy. Presently, more experiments and safety data are needed to establish gene therapy procedures. In this manuscript, rats were used for analysis. However, data from animals genetically similar to humans will be necessary for consideration of the human-specific immunity.

ChR2 is a problem that is limited by the wavelength of light absorption. We have already designed a new ChR to react with visible light and have studied its function and safety for therapy.

References

Hubbard R (1966) The stereoisomerization of 11-*cis*-retinal. J Biol Chem 241(8):1814–1818

Isago H, Sugano E, Wang Z et al (2012) Age-dependent differences in recovered visual responses in Royal College of Surgeons rats transduced with the channelrhodopsin-2 gene. J Mol Neurosci 46(2):393–400. doi:10.1007/s12031-011-9599-y

Kim SA, Lim SS (2002) T lymphocyte subpopulations and interleukin-2, interferon-gamma, and interleukin-4 in rat pulpitis experimentally induced by specific bacteria. J Endod 28(3):202–205, doi: S0099-2399(05)60613-9 [pii] 10.1097/00004770-200203000-00014

Maguire AM, Simonelli F, Pierce EA et al (2008) Safety and efficacy of gene transfer for Leber's congenital amaurosis. N Engl J Med 358(21):2240–2248, doi: NEJMoa0802315 [pii] 10.1056/NEJMoa0802315

Nagel G, Szellas T, Huhn W et al (2003) Channelrhodopsin-2, a directly light-gated cation-selective membrane channel. Proc Natl Acad Sci U S A 100(24):13940–13945

Santos A, Humayun MS, de Juan E et al (1997) Preservation of the inner retina in retinitis pigmentosa. A morphometric analysis. Arch Ophthalmo 115(4):511–515

Sugano E, Isago H, Wang Z et al (2011) Immune responses to adeno-associated virus type 2 encoding channelrhodopsin-2 in a genetically blind rat model for gene therapy. Gene Ther 18(3):266–274, doi: gt2010140 [pii] 10.1038/gt.2010.140

Tomita H, Sugano E, Yawo H et al (2007) Restoration of visual response in aged dystrophic RCS rats using AAV-mediated channelopsin-2 gene transfer. Invest Ophthalmol Vis Sci 48(8):3821–3826

Tomita H, Sugano E, Fukazawa Y et al (2009a) Visual properties of transgenic rats harboring the channelrhodopsin-2 gene regulated by the thy-1.2 promoter. PLoS One 4(11):e7679. doi:10.1371/journal.pone.0007679

Tomita H, Sugano E, Isago H et al (2009b) Channelrhodopsins provide a breakthrough insight into strategies for curing blindness. J Genet 88(4):409–415

Tomita H, Sugano E, Isago H et al (2010) Channelrhodopsin-2 gene transduced into retinal ganglion cells restores functional vision in genetically blind rats. Exp Eye Res 90:429–436, doi: S0014-4835(09)00353-4 [pii] 10.1016/j.exer.2009.12.006

Chapter 25
Optogenetic Approaches to Restoring Intrinsic Visual Processing Features in Retinal Ganglion Cells

Zhuo-Hua Pan, Anding Bi, and Qi Lu

Abstract The severe loss of photoreceptor cells caused by degenerative diseases of the retina could result in partial or complete blindness. The optogenetic strategy to restoring vision involves genetically converting the surviving inner retinal neurons to photosensitive cells, thus restoring light sensitivity to the retina after photoreceptor degeneration. Proof-of-concept studies in animal models have already demonstrated that it is possible to create photosensitivity in inner retinal neurons and restore visually guided behaviors. Multiple approaches would need to be developed regarding rendering photosensitivity to particular retinal layers or cell types depending on retinal degenerative conditions. For severe retinal degenerative conditions, rendering photosensitivity to retinal ganglion cells might be the only option. This would also require restoring intrinsic visual processing features such as ON and OFF light responses, sustained and transient light responses, and center-surround antagonistic receptive fields to restore vision at the highest quality. Significant progress has been made toward achieving these goals, although challenges still remain.

Keywords Optogenetics • Vision restoration • Channelrhodopsins • Halorhodopsin • Adeno-associated virus • Retinal ganglion cells • ON response • OFF response • Center-surround receptive field

Z.-H. Pan (✉)
Department of Ophthalmology, Kresge Eye Institute, Wayne State University School of Medicine, 4717 St. Antoine, Detroit, MI, USA

Department of Anatomy/Cell Biology, Wayne State University School of Medicine, 540 E. Canfield Avenue, Detroit, MI, USA
e-mail: zhpan@med.wayne.edu

A. Bi
Department of Ophthalmology, Kresge Eye Institute, Wayne State University School of Medicine, 4717 St. Antoine, Detroit, MI, USA

Q. Lu
Department of Anatomy/Cell Biology, Wayne State University School of Medicine, 540 E. Canfield Avenue, Detroit, MI, USA

H. Yawo et al. (eds.), *Optogenetics*, DOI 10.1007/978-4-431-55516-2_25

25.1 Introduction

Vision begins in the retina when rod and cone photoreceptor cells respond to light and convert the light to electrical signals, which are then conducted through second-order (bipolar cells and horizontal cells) and third-order (amacrine cells and ganglion cells) retinal neurons to higher visual centers in the brain. The severe loss of photoreceptor cells that occurs in many inherited and acquired retinal degenerative diseases, such as retinitis pigmentosa (RP), could result in partial or complete blindness (Weleber 1994). At present, no effective treatment is available for restoring vision once the photoreceptor cells have been lost. Several therapeutic strategies have been devised in an attempt to restore vision after photoreceptor degeneration, including cell transplantation (Lamba et al. 2009; West et al. 2009), retinal device implants (Yanai et al. 2007; Stingl et al. 2013), and, more recently, optogenetic approaches.

The principle behind using optogenetics as a therapeutic approach to blindness involves restoring light-sensitivity to retinas that lack photoreceptors. This is accomplished by converting light-insensitive inner retinal neurons (i.e., the second- and third-order retinal neurons) into photosensitive cells by introducing genetically encoded light sensors. The feasibility of this strategy requires the availability of a simple light sensor. The discovery of light-gated cation channels called channelrhodopsins (ChRs) (Nagel et al. 2002, 2003), particularly ChR2, has prompted researchers to explore the possibility of using this novel strategy to restore vision (Bi et al. 2006; Tomita et al. 2007; Lagali et al. 2008; Lin et al. 2008).

25.2 Proof-of-Concept Studies

Soon after the discovery of ChR2 (Nagel et al. 2003), Bi et al. (2006) began to examine the functional expression of ChR2 in inner retinal neurons in vivo and assess its ability to restore visual responses after photoreceptor degeneration. The experiments were conducted both in normal rats and mice and in *rd1/rd1* mice. The latter is a commonly used mouse model of retinal degeneration (McLaughlin et al. 1993). Because retinal neurons do not divide, a viral-based delivery method was found to be the most effective way to express the channelopsin (Chop)-2 gene in inner retinal neurons in vivo. Specifically, an adeno-associated virus (AAV)-2 vector was used to deliver a gene construct in which the ubiquitously expressed CAG promoter drives a Chop2-green fluorescent protein (GFP) fusion construct. Approximately 1 month after this viral vector was injected into the intravitreal space of the eyes, Chop2-GFP expression was detected in inner retinal neurons, predominately in cells located in the ganglion cell layer (Fig. 25.1). ChR2 (Chop2 plus a chromophore)-mediated spike activity was recorded from Chop2-GFP-expressing retinal ganglion cells from *rd1/rd1* mice in the absence of an exogenous supply of the chromophore, all-*trans* retinal (Fig. 25.2). These ChR2-mediated light responses were primarily sustained because a component of the ChR2-mediated current was not inactivated. Furthermore, the study demonstrated that ChR2 expression could restore visually evoked potentials in the visual cortex.

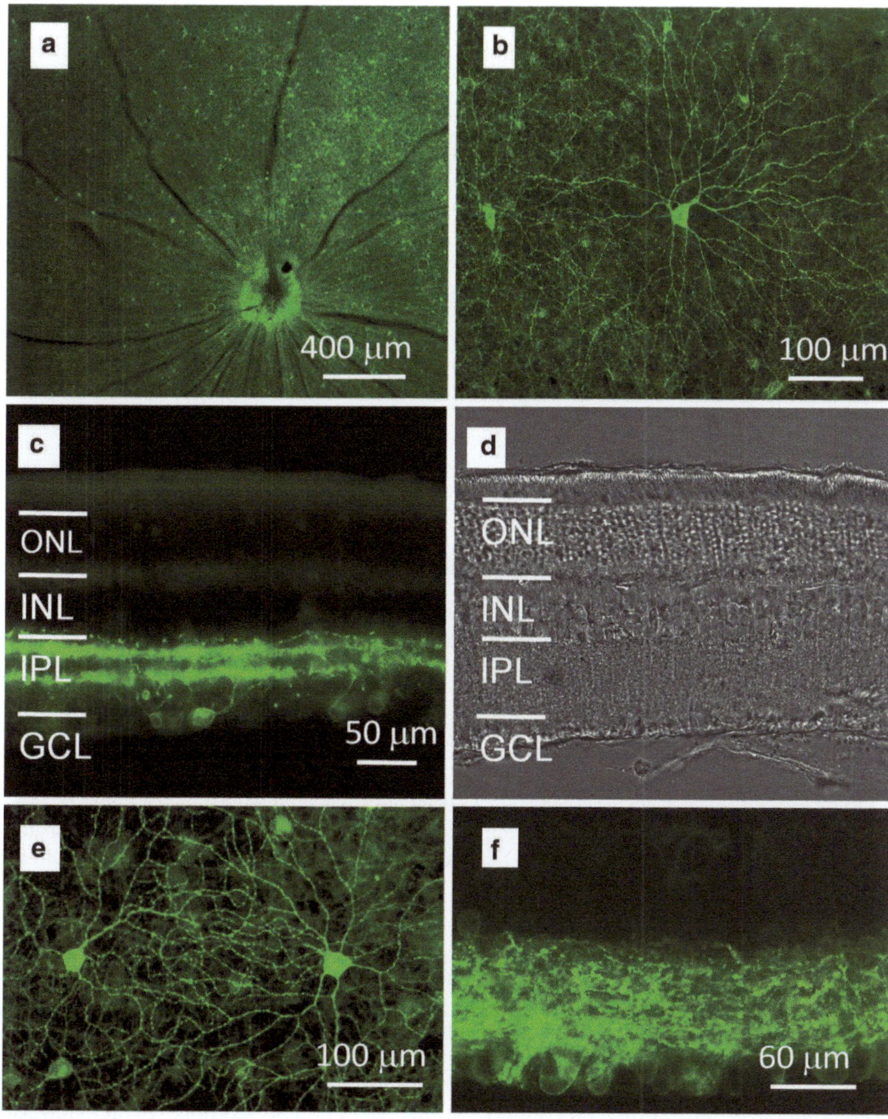

Fig. 25.1 Adeno-associated virus (AAV)-mediated in vivo expression of chop2-green fluorescent protein (GFP) in retinal neurons. (**a** and **b**) Chop2-GFP expression in whole mount retinas from wild-type rats at low (**a**) and high (**b**) magnifications. (**c** and **d**) Fluorescence image (**c**) and phase-contrast image (**d**) taken from vertical slice sections of normal rat retinas that express Chop2-GFP. (**e** and **f**) Fluorescence images taken from whole mount retinas (**e**) and vertical slice sections (**f**) of *rd1/rd1* mice that express Chop2-GFP. *ONL* outer nuclear layer, *INL* inner nuclear layer, *IPL* inner plexiform layer, *GCL* ganglion cell layer (Modified from Bi et al. 2006. With permission from Elsevier)

A similar study was also conducted using Royal College of Surgeons (RCS) rats, which is another rat model for inherited retinal degeneration. Using a similar AAV vector to deliver Chop2 to inner retinal neurons, Tomita et al. (2007, 2010) observed

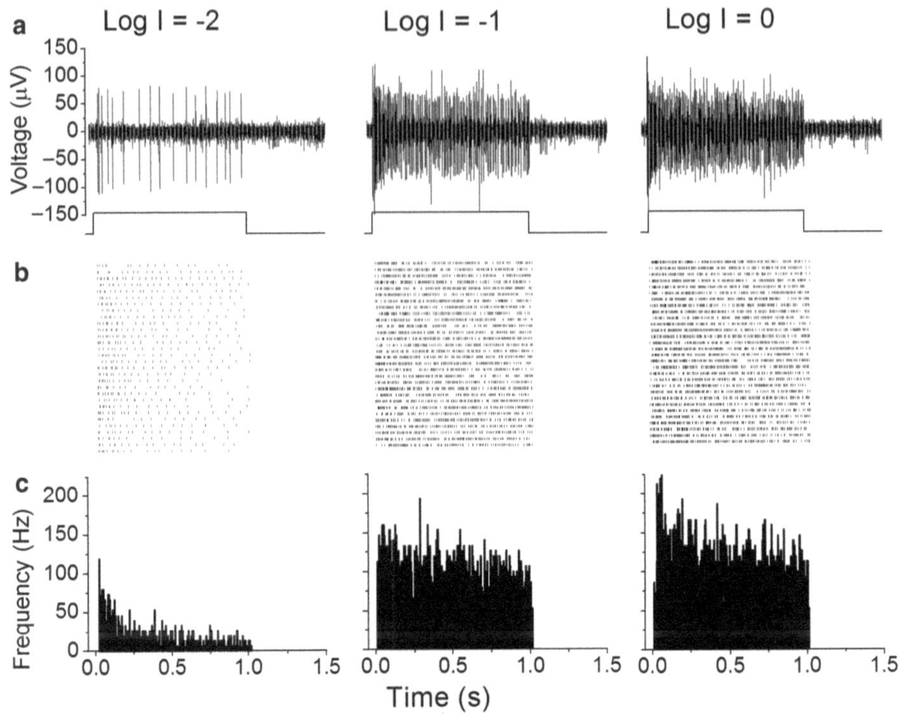

Fig. 25.2 Multi-electrode array recordings of ChR2-expressing retinas from *rd1/rd1* mice. (**a**) sample recording of light-evoked spike activity in response to three incremental light intensities. (**a**) Light-evoked spikes were recorded from a single electrode at three incremental light intensities. (**b**) Raster plots were constructed from 30 consecutive light-elicited spikes that originated from a single neuron. (**c**) Histograms that demonstrate the average spike rates (Modified from Bi et al. 2006. With permission from Elsevier)

that both visually evoked potentials and visually guided behaviors based on optomotor responses could be restored in these animals. Furthermore, the targeted expression of ChR2 in ON-type retinal bipolar cells using an mGluR6 promoter via in vivo electroporation or viral-mediated subretinal delivery restored visual functions both in *rd1/rd1* mice and in other mouse models with photoreceptor degeneration (Lagali et al. 2008; Doroudchi et al. 2011). The ectopic expression of melanopsin, an endogenous photopigment that is present in a small population of intrinsic photosensitive retinal ganglion cells, has also been reported to restore retinal light sensitivity and visual functions in *rd1/rd1* mice (Lin et al. 2008).

Follow-up studies in *rd1/rd1* mice have shown that stably expressing functional ChR2 in mouse retinal neurons by a single administration of viral vectors can last for the animals' entire life (Ivanova and Pan 2009; and unpublished results). Finally, ChR2 expression does not appear to cause neurotoxicity or elicit any significant immunological responses in vivo (Ivanova and Pan 2009; Sugano et al. 2011). Functional expression of ChR2 can also be achieved using a viral delivery system in non-human primates, although the transduction efficiency is much lower (Ivanova et al. 2010). Taken together, these studies convincingly demonstrate that using optogenetic approaches to restore vision has been successful in animal models.

25.3 Intrinsic Visual Processing in the Retina and Possible Optogenetic Approaches

One advantage of using optogenetic approaches is the possibility of restoring intrinsic visual processing features in the retina. This is important because the retina is not only the site of initiating phototransduction but also plays an important role in processing visual information before it is transmitted to higher visual centers in the brain (Wässle 2004). Visual signals are processed in the retina through multiple parallel pathways, such as the ON and OFF pathways, the scotopic (rod) and photopic (cone) pathways, and pathways that are responsible for edge detection, motion and directional selectivity, and color coding. All photoreceptor cells become hyperpolarized in response to light. Bipolar cells and ganglion cells are divided into ON and OFF cells that depolarize and hyperpolarize, respectively, in response to increased light intensity (Fig. 25.3). The segregation of visual signals into ON and OFF pathways is believed to be important in enhancing contrast sensitivity. Light responses in inner retinal neurons, especially retinal ganglion cells, can also be divided into sustained and transient based on their temporal properties. These sustained and transient light responses are thought to code for distinct aspects of visual information, such as acuity and motion, respectively. Furthermore, horizontal cells and amacrine cells provide lateral inhibition that underlies the formation of center-surround antagonistic receptive fields in both bipolar and retinal ganglion cells (Fig. 25.3). This center-surround antagonistic receptive field is an essential feature of visual information processing that enables edge detection by enhancing spatial contrast. Thus, restoring these intrinsic visual processing features in the retina would result in a better outcome of restoredvision in patients with retinal degeneration.

There are several ways in which optogenetics might be used to induce photosensitivity in particular retinal layers or cell types. One major advantage of inducing more distal retinal neurons, such as bipolar cells or even surviving cone photoreceptors, to become photosensitive (Lagali et al. 2008; Doroudchi et al. 2011; Busskamp et al. 2010) is that these cells could still utilize the remaining retinal circuitry and thereby partially retain their intrinsic retinal processing pathways. This approach might also increase the light sensitivity and spatial resolution of the retina due to the signal convergence from photoreceptor cells to bipolar cells and from bipolar cells to ganglion cells. On the other hand, a major drawback of this approach concerns the degree of retinal remodeling caused by photoreceptor death in degenerative diseases. Previous studies have suggested that retinal degeneration results in progressive and time-dependent remodeling of inner retinal neurons and circuits (Strettoi et al. 2000; Marc et al. 2003). The extent of this remodeling varies between the

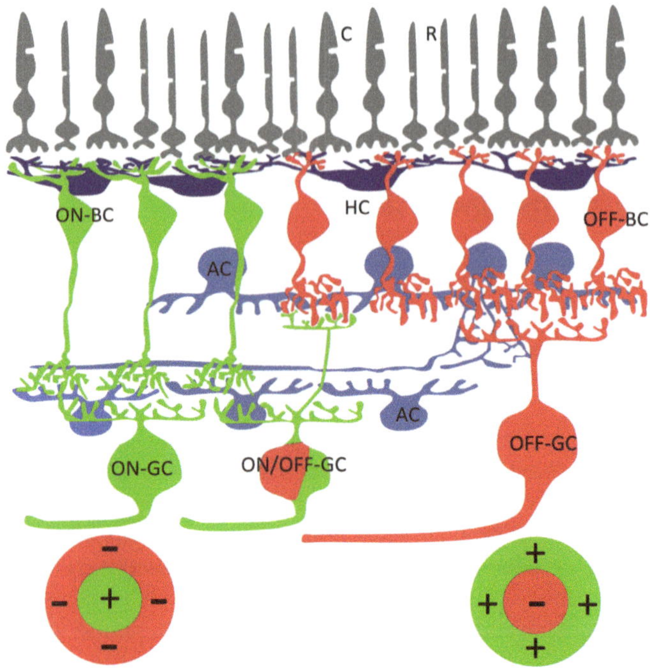

Fig. 25.3 Visual information processing in the retina. This schematic diagram outlines the visual information processing features of the retina, specifically ON and OFF pathways and center-surround antagonistic receptive fields. ON and OFF light responses originate at the level of bipolar cells and are carried downstream to retinal ganglion cells. The surround of the center-surround antagonistic receptive fields in retinal ganglion cells is formed by lateral inhibitions from horizontal and amacrine cells. The ON and OFF region of the receptive fields are indicated by + and −, respectively. *R* rod photoreceptors, *C* cone photoreceptors, *HC* horizontal cells, *AC* amacrine cells, *GC* ganglion cells

different degenerative diseases, but studies in RP animal models have reported that retinal ganglion cells are most resistant to remodeling (Mazzoni et al. 2008; Damiani et al. 2012). Therefore, retinal ganglion cells might be the only possible target for retina-based optogenetic strategy in many cases of advanced retinal degeneration.

25.4 Restoration of OFF Light Responses in the Retina

The ability to create both ON and OFF light responses is essential in restoring the intrinsic features of visual processing. As described above, the ectopic expression of ChR2 can effectively convert inner retinal neurons, including retinal ganglion cells, into sustained ON cells that respond to light with sustained membrane depolarization or spiking (Bi et al. 2006). However, creation of OFF cells in inner retinal neurons would require expressing a light sensor that can produce membrane hyperpolarization or suppress spiking activity when stimulated with light. Several microbial opsins, such as the light-driven outward proton pump bacteriorhodopsin (Oesterhelt and Stoeckenius 1973) or the light-driven inward chloride ion pump halorhodopsin (*Np*HR) (Lanyi 1986), could be used for such a purpose (Han and Boyden 2007; Zhang et al. 2007).

The study to examine whether the expression of hyperpolarizing light sensors could restore OFF light responses in the retina was carried out using *Np*HR (Zhang et al. 2009). In this study, again the AAV2 vector with a ubiquitous cytomegalovirus (CMV) promoter was used to drive a fusion-NpHR-mCherry construct. The viral vector was injected into the intravitreal space of the eyes of adult wild-type or *rd1/rd1* mice. Robust *Np*HR-mCherry expression could be observed in inner retinal neurons located predominantly in the retinal ganglion cell layer. More importantly, OFF light responses could be recorded in these *Np*HR-expressing retinal ganglion cells. As shown in Fig. 25.4, spiking activity was suppressed during light stimulation, while rebound spikes were observed after the light was turned off. Furthermore, the amplitude of these OFF responses increased with increasing light intensity. Thus, the expression of *Np*HR can effectively convert light-insensitive inner retinal neurons, including retinal ganglion cells, into OFF cells in both normal and photoreceptor-deficient retinas. Interestingly, both sustained and transient *NpHR-mediated* OFF responses were observed, and these responses were found to be correlated to the intrinsic membrane properties of retinal ganglion cells.

25.5 Subcellular Targeting to Create Center-Surround Antagonistic Receptive Fields

Another key feature of the visual processing of retinal ganglion cells is the antagonistic center-surround organization of their receptive fields. There are two types of center-surround receptive fields: ON center and OFF center. ON ganglion cells have a receptive field where the center is ON and the surround is OFF, while the receptive fields of OFF ganglion cells is the reverse (i.e., OFF center and ON surround). Because the center-surround receptive field is generated by lateral inhibition from horizontal and amacrine cells, it is lost for the approach of directly rendering photosensitivity in retinal ganglion cells,

However, it may be possible to recreate center-surround receptive fields in retinal ganglion cells by differentially expressing the depolarizing and hyperpolarizing optogenetic sensors, such as ChR2 and *Np*HR, at the center and peripheral regions of the dendritic trees of retinal ganglion cells, respectively. This specificity in expression could be achieved by employing protein-targeting motifs, which are known to be one of the mechanisms underlying the localization of membrane proteins to specific subcellular regions (Lai and Jan 2006). In support this idea, a previous study demonstrated that the differential expression of ChR2 and *Np*HR using ankyrin-G and postsynaptic density (PSD) motifs achieved by biolistic particle

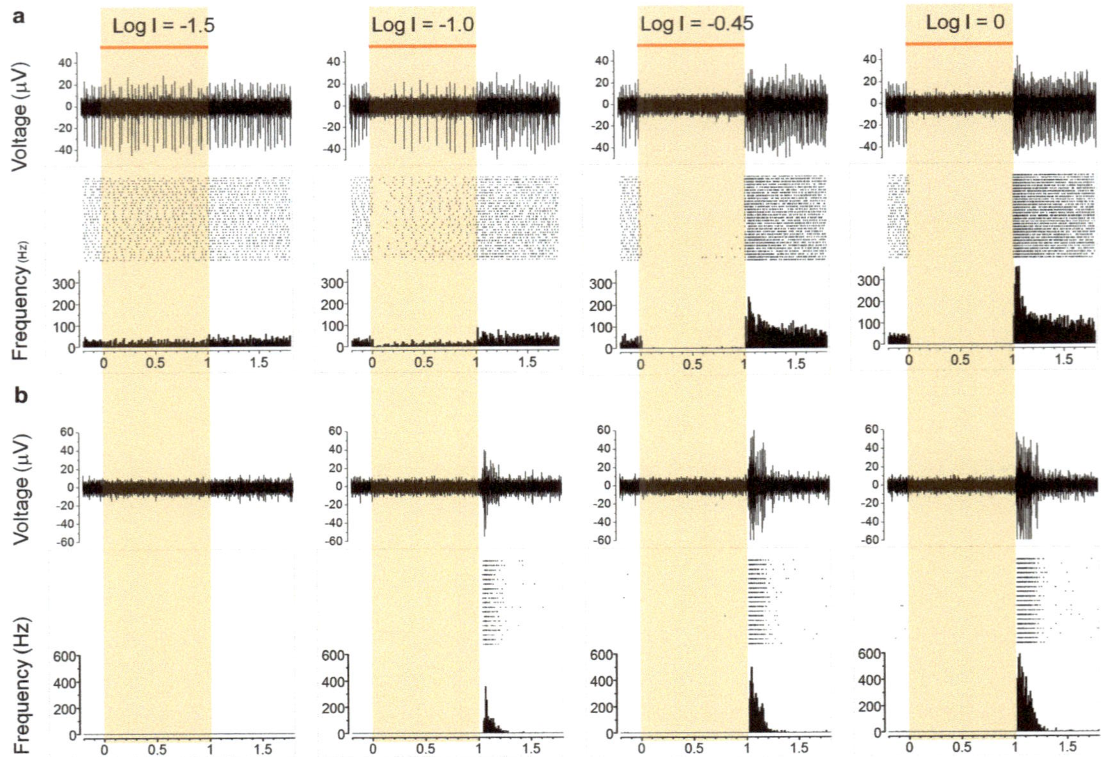

Fig. 25.4 Multi-electrode array recordings of halorhodopsin (*Np*HR)-mediated spike responses from *rd1/rd1* retinas. (**a**) Sample recordings of *Np*HR-mediated spike activity from a sustained-OFF-type cell in response to four incremental light intensities. In each panel, a single trace of light-evoked spike activity, a raster plot of 30 consecutive recordings, and an averaged spike rate histogram are shown in the top, middle, and bottom panels, respectively. (**b**) Sample recordings of *Np*HR-mediated spike activity from a transient-OFF-type cell were recorded under the same conditions as in (**a**) (Modified from Zhang et al. 2009)

delivery can consequently generate a center-surround receptive field in retinal ganglion cells (Greenberg et al. 2011). However, potential therapeutic applications would require that the motif targeting be achieved through a viral-mediated delivery system, such as AAV vectors. In addition, motifs that decrease axonal expression would be important because the responses generated by retinal ganglion cell axons could interfere with the retinotopic mapping in higher visual centers. Two motifs have been identified to be suitable for the AAV-mediated targeting in mice (Wu et al. 2013). One motif from the voltage-gated K^+ channel 2.1 (Kv2.1) was shown to be able to target ChR2 or *Np*HR to cell soma and proximal dendrites of retinal ganglion cells, which would render it useful to establish a center. A second motif from neuroligin (NLG)-1 targets the somatodendritic region of retinal ganglion cells, which would be used to establish a surround. It should be noted that, although the surround-targeting motif (NLG1) was expressed in the entire dendritic tree in addition to the cell soma, which overlaps with the expression of the center-targeted motif (Kv2.1), a center-surround antagonistic receptive field could still be generated as long as net excitatory and inhibitory zones were produced. For both motifs, the expression in distal axons was significantly reduced. Figure 25.5 shows the center-targeting ability of the Kv2.1-motif in a retinal ganglion cell that was co-transduced by two viral vectors carrying ChR2-mCherry without the motif and NpHR-YFP with the Kv2.1-motif. The dendritic field size expressing *Np*HR-YFP with the Kv2.1 motif is markedly reduced compared with that expressing ChR2-mCherry without the motif. Additionally, the reduced axonal expression of *NpHR-YFP* with the Kv2.1-motif can be noted (indicated by arrowheads in Fig. 25.5). Finally, when a light bar was used to map receptive fields, a reduction in the physiological receptive field size of ChR2- or *Np*HR-mediated light responses can be observed when the Kv2.1 motif is used in comparison with either the NLG-1 motif or a control (Fig. 25.6).

25.6 Future Directions

For the retinal ganglion cell-based optogenetic approach, although ubiquitously expressing either depolarizing or hyperpolarizing light sensors to convert retinal ganglion cells to entirely ON or entirely OFF cells might be sufficient to restore useful vision, more fine tuning may be necessary to restore a higher quality of vision. Although studies have already shown that the AAV-mediated expression of ChR2 and *Np*HR can produce ON and OFF light responses, and AAV-mediated subcellular motif targeting can be used to create center-surround receptive fields in retinal ganglion cells, a number of further developments are required to achieve that goal. First, further development of optogenetic light sensors that can better fit into the need for vision restoration will be required. Specifically, the development of more operational light-sensitive optogenetic sensors is needed because the operational light sensitivity of currently available light sensors, especially hyperpolarizing light sensors, is several log units lower than that of cone photoreceptors. Although the

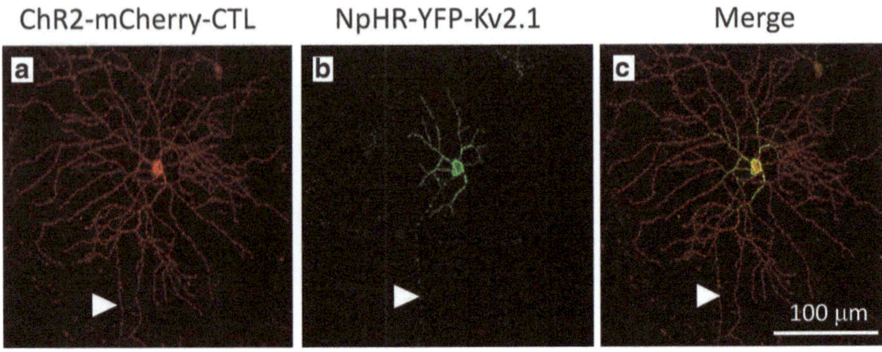

Fig. 25.5 Co-expression of channelrhodopsin (ChR)-2-mCherry and halorhodopsin (*Np*HR)-YFP-Kv2.1 in retinal ganglion cells. (**a–c**) Eyes were co-injected with a viral vector carrying either ChR2-mCherry or *Np*HR-YFP-Kv2.1. The images of ChR2-mCherry (*red*; **a**) and *NpHR-YFP-Kv2.1* (*green*; **b**) were then merged to demonstrate a center-surround type organization in the dendritic field of a single retinal ganglion cell (RGC) (**c**). The axon is indicated by *arrowheads* (Reprinted from Wu et al. 2013)

limitation of low-light sensitivity could be partially compensated for by using an extra ocular imaging device, the development of light sensors that can work under near-ambient light would eliminate the need for using a bright or unnatural stimulus light that could potentially cause tissue damage or other complications. For depolarizing light sensors, the light sensitivity gap is being narrowed because a number of more operational light-sensitive ChRs have been reported (Wang et al. 2009; Mattis et al. 2011; Berndt et al. 2011; Kleinlogel et al. 2011; Prigge et al. 2012). In contrast, a wide gap still exists for hyperpolarizing light sensors. Recently, several new light-driven outward proton pumps, such Arch and ArchT, have been reported to show larger light-evoked currents when compared with *Np*HR (Chow et al. 2010; Han et al. 2011). More recently, light-gated chloride channels converted from ChR by site-directed mutagenesis have been reported (Wietek et al. 2014; Berndt et al. 2014). However, the expression and functional light-response properties of these newly reported hyperpolarizing light sensors in retinal neurons will need to be examined.

Second, the ability to restore sustained and transient light-response properties in retinal ganglion cells should be further investigated. Because the expression of ChR can only produce sustained ON light responses, a method to produce transient ON light responses still needs to be developed. Additionally, although the expression of the hyperpolarizing light sensor *Np*HR can produce both sustained and transient OFF light responses, it is unclear whether these cells are correlated to the intrinsic sustained and transient retinal ganglion cells.

Furthermore, restoring light-response properties of retinal ganglion cells to mimic intrinsic ON and OFF pathways requires that the depolarizing and hyperpolarizing light sensors be targeted to ON and OFF retinal ganglion cells, respectively.

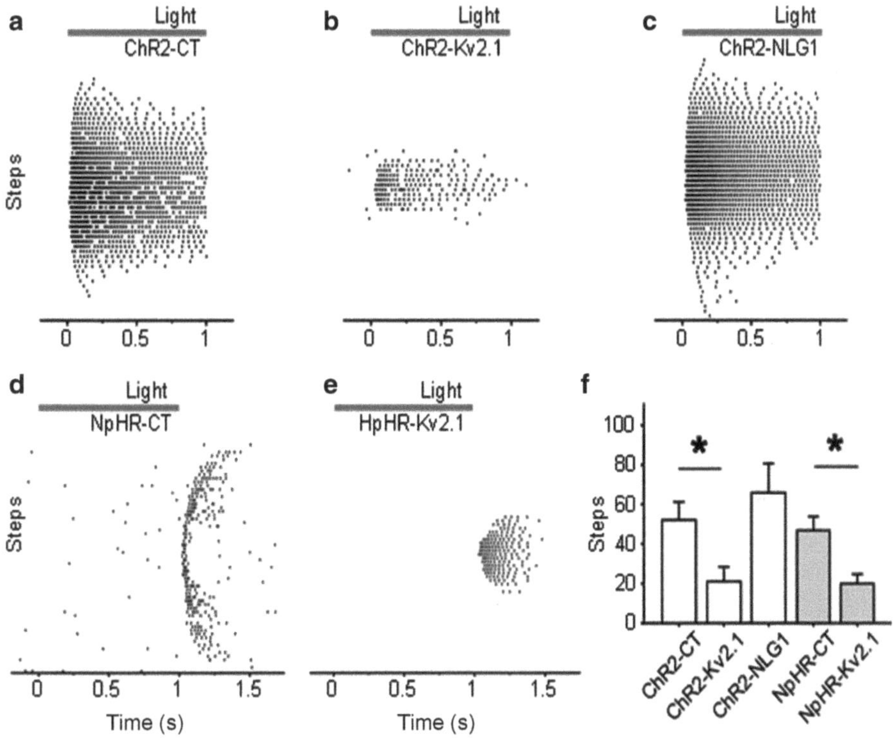

Fig. 25.6 Motif-targeted physiological response field size assessed by multi-electrode array recordings. (**a–c**) Sample multielectrode array (MEA) recording traces from ganglion cells that express channelrhodopsin (ChR)-2-YFP (**a**), ChR2-YFP with a Kv2.1 motif (**b**), ChR2-YFP with a neuroligin (NLG)-1-motif (**c**), halorhodopsin (*Np*HR)-YFP (**d**), and *Np*HR-YFP with a Kv2.1-motif (**e**) in response to a 200 μm light bar stimulus at 20 μm increments. Each row of dots represents spiking activity elicited by a 1-s light pulse, and sequential rows represent sequential steps through the cell's receptive field. (**f**) Quantitative comparison of the response field sizes of ChR2-YFP- and *Np*HR-YFP-expressing retinal ganglion cells. For each cell, the physiological receptive field size was estimated from the number of steps that elicited light-driven spiking activity. When compared with the ChR2-YFP-control, ChR2-YFP/Kv2.1 motif expression elicited a significantly smaller receptive field size. Similarly, the *Np*HR-mediated receptive field sizes between the *NpHR-YFP* control and the *Np*HR-YFP/Kv2.1 motif are significantly different (Reprinted from Wu et al. 2013)

Thus, further studies are needed to identify promoters that can selectively target ON and OFF retinal ganglion cells. Additionally, these cell-type-specific promoters will need to be small enough to fit into the maximum capacity of the viral genome, which is 4.7 kb for AAV vectors. Finally, creating center-surround receptive fields requires that the expression and properties of the depolarizing and hyperpolarizing light sensors be balanced. This balance would need to take into account the protein expression level, operational light sensitivity, and spectral properties of both light sensors.

References

Berndt A, Schoenenberger P, Mattis J et al (2011) High-efficiency channelrhodopsins for fast neuronal stimulation at low light levels. Proc Natl Acad Sci U S A 108:7595–7600

Berndt A, Lee SY, Ramakrishnan C et al (2014) Structure-guided transformation of channelrhodopsin into a light-activated chloride channel. Science 344:420–424

Bi A, Cui J, Ma Y-P et al (2006) Ectopic expression of a microbial-type rhodopsin restores visual responses in mice with photoreceptor degeneration. Neuron 50:23–33

Busskamp V, Duebel J, Balya D et al (2010) Genetic reactivation of cone photoreceptors restores visual responses in retinitis pigmentosa. Science 329:413–417

Chow BY, Han X, Dobry AS et al (2010) High-performance genetically targetable optical neural silencing by light-driven proton pumps. Nature 463:98–102

Damiani D, Novelli E, Mazzoni F et al (2012) Undersized dendritic arborizations in retinal ganglion cells of the rd1 mutant mouse: a paradigm of early onset photoreceptor degeneration. J Comp Neurol 520:1406–1423

Doroudchi MM, Greenberg KP, Liu J et al (2011) Virally delivered channelrhodopsin-2 safely and effectively restores visual function in multiple mouse models of blindness. Mol Ther 19:1220–1229

Greenberg KP, Pham A, Werblin FS (2011) Differential targeting of optical neuromodulators to ganglion cell soma and dendrites allows dynamic control of center-surround antagonism. Neuron 69:713–720

Han X, Boyden ES (2007) Multiple-color optical activation, silencing, and desynchronization of neural activity, with single-spike temporal resolution. PLoS One 2:e299

Han X, Chow BY, Zhou H et al (2011) A high-light sensitivity optical neural silencer: development and application to optogenetic control of non-human primate cortex. Front Syst Neurosci 5:18

Ivanova E, Pan Z-H (2009) Evaluation of virus mediated long-term expression of channelrhodopsin-2 in the mouse retina. Mol Vision 15:1680–1689

Ivanova E, Hwang G-S, Pan Z-H et al (2010) Evaluation of AAV-mediated expression of chop2-GFP in the marmoset retina. IOVS 51:5288–5296

Kleinlogel S, Feldbauer K, Dempski RE et al (2011) Ultra light-sensitive and fast neuronal activation with the Ca^{2+}-permeable channelrhodopsin CatCh. Nat Neurosci 14:513–518

Lagali PS, Balya D, Awatramani GB et al (2008) Light-activated channels targeted to ON bipolar cells restore visual function in retinal degeneration. Nat Neurosci 11:667–675

Lai HC, Jan LY (2006) The distribution and targeting of neuronal voltage-gated ion channels. Nat Rev Neurosci 7:548–562

Lamba DA, Gust J, Reh TA (2009) Transplantation of human embryonic stem cell-derived photoreceptors restores some visual function in Crxdeficient mice. Cell Stem Cell 4:73–79

Lanyi JK (1986) Halorhodopsin: a light-driven chloride ion pump. Annu Rev Biophys Biophys Chem 15:11–28

Lin B, Koizumi A, Tanaka N et al (2008) Restoration of visual function in retinal degeneration mice by ectopic expression of melanopsin. Proc Natl Acad Sci U S A 105:16009–16014

Marc RE, Jones BW, Watt CB et al (2003) Neural remodeling in retinal degeneration. Prog Retin Eye Res 22:607–655

Mattis J, Tye KM, Ferenczi EA et al (2011) Principles for applying optogenetic tools derived from direct comparative analysis of microbial opsins. Nat Methods 9:159–172

Mazzoni F, Novelli E, Strettoi E (2008) Retinal ganglion cells survive and maintain normal dendritic morphology in a mouse model of inherited photoreceptor degeneration. J Neurosci 28:14282–14292

McLaughlin ME, Sandberg MA, Berson EL et al (1993) Recessive mutations in the gene encoding the beta-subunit of rod phosphodiesterase in patients with retinitis pigmentosa. Nat Genet 4:130–134

Nagel G, Ollig D, Fuhrmann M et al (2002) Channelrhodopsin-1: a light-gated proton channel in green algae. Science 296:2395–2398

Nagel G, Szellas T, Huhn W et al (2003) Channelrhodopsin-2, a directly light-gated cation-selective membrane channel. Proc Natl Acad Sci U S A 100:13940–13945

Oesterhelt D, Stoeckenius W (1973) Functions of a new photoreceptor membrane. Proc Natl Acad Sci U S A 70:2853–2857

Prigge M, Schneider F, Tsunoda SP et al (2012) Color-tuned channelrhodopsins for multiwavelength optogenetics. J Biol Chem 287:31804–31812

Stingl K, Bartz-Schmidt KU, Besch D et al (2013) Artificial vision with wirelessly powered subretinal electronic implant alpha-IMS. Proc Biol Sci 280:20130077

Strettoi E, Pignatelli V (2000) Modifications of retinal neurons in a mouse model of retinitis pigmentosa. Proc Natl Acad Sci U S A 97:11020–11025

Sugano E, Isago H, Wang Z et al (2011) Immune responses to adeno-associated virus type 2 encoding channelrhodopsin-2 in a genetically blind rat model for gene therapy. Gene Ther 18:266–274

Tomita H, Sugano E, Yawo H et al (2007) Restoration of visual response in aged dystrophic RCS rats using AAV-mediated channelopsin-2 gene transfer. Invest Ophthalmol Vis Sci 48:3821–3826

Tomita H, Sugano E, Isago H et al (2010) Channelrhodopsin-2 gene transduced into retinal ganglion cells restores functional vision in genetically blind rats. Exp Eye Res 90:429–436

Wang H, Sugiyama Y, Hikima T et al (2009) Molecular determinants differentiating photocurrent properties of two channelrhodopsins from chlamydomonas. J Biol Chem 284:5685–5696

Wässle H (2004) Parallel processing in the mammalian retina. Nat Rev Neurosci 5:747–757

Weleber RG (1994) Retinitis pigmentosa and allied disorders. In: Ryan SJ (ed) Retina. Mosby, St. Louis, pp 335–466

West EL, Pearson RA, MacLaren RE et al (2009) Cell transplantation strategies for retinal repair. Prog Brain Res 175:3–21

Wietek J, Wiegert JS, Adeishvili N et al (2014) Conversion of channelrhodopsin into a light-gated chloride channel. Science 344:409–412

Wu C, Ivanova E, Zhang Y et al (2013) AAV-mediated subcellular targeting of optogenetic tools in retinal ganglion cells. PLoS One 8:e66332

Yanai D, Weiland JD, Mahadevappa M et al (2007) Visual performance using a retinal prosthesis in three subjects with retinitis pigmentosa. Am J Ophthalmol 143:820–827

Zhang F, Wang LP, Brauner M et al (2007) Multimodal fast optical interrogation of neural circuitry. Nature 446:633–639

Zhang Y, Ivanova E, Bi A et al (2009) Ectopic expression of multiple microbial rhodopsins restores ON and OFF light responses in the retina after photoreceptor degeneration. J Neurosci 29:9186–9196

Chapter 26
On Optogenetic Tissue Engineering on Visual Cells: A Review on Its Development, Practices and Application

Misato Ichise and Seigo Yamada

Abstract In the field of optogenetic tissue engineering, the most recent and widely accepted clinical research method is for a channelrhodopsin to be expressed in nerve cells through exposing these cells to light since the subjected nerve cells become stimulated, or inhibited, by light (Kehoe, Science 273:1409–1412, 1996; Bhaya, Proc Natl Acad Sci USA 98:7540–7545, 2001; Terauchi, Mol Microbiol 51:567–577, 2004). Other clinical methodologies are also in development whose purposes are to be used in controlling interactions between various proteins through exposing the said subjected nerve cells to light. When one considers applying these methods of optogenetic tissue engineering to any part of the human organs, the most relevant and effective subject organ is the eye. Visual cells of the eyes possess visual pigments made of photoreceptor proteins, and those proteins have high photo sensitivity and responsiveness. It is also a known fact that collection of data from clinical application for analysis is relatively easy since the tissues are marked in the process of tissue engineering (Corredor, J Neural Eng 10.1088/1741-2560/6/5/055001, 2009).

Keywords Visual cells • Retina • Tissue engineering

26.1 Structure of the Retina

Light arrives at the retina after passing through the cornea, the pupil, and the crystalline lens. The thickness of the retina is about 120 μm. Visual cells that can sense light have rod cells and cone cells. The rod cells function as a sensor for brightness and darkness. In contrast, the cone cells function as a sensor for color. The retina has about 120 million rod cells and 5 million cone cells. The rod cells, which are sensitive

M. Ichise (✉)
Faculty of Applied Information Technology, Nagasaki Institute of Applied Science, Aba-machi 536, Nagasaki, Nagasaki Prefecture 851-0193, Japan
e-mail: ICHISE_Misato@NiAS.ac.jp

S. Yamada
Center for General Education, Nagasaki Institute of Applied Science, Nagasaki, Japan

H. Yawo et al. (eds.), *Optogenetics*, DOI 10.1007/978-4-431-55516-2_26

to brightness and darkness, can sense extremely dim light, but brighter light is required for the cone cells to sense a color. The cone cells need light in large quantities. The required quantity must be about 100 times more than the rod cells because of their low light sensitivity. The number of ganglion cells and optic nerve fibers are about 1 and 1.2 million, respectively, whereas the visual cells have large quantities of ganglion cells. This suggests that signals are integrated by binding many visual cells.

The size of a rod cell is about 50×3 μm and it has slim shape. The upper part of the rod cell that is exposed to light is separated by the plasma membrane. Discrete flat saccular disks, because hundreds or thousands of pieces of them become accumulated, they construct a stratified formation. The rhodopsins are included in these disks, and they also react to light. A rhodopsin is a spherical membrane protein whose molecular weight is 30,000–60,000 Da, and this protein has an area called the 'opsin'. A rhodopsin is a chemical complex with the opsin protein and vitamin A and is called 'retinal'. In rod cells, there are rhodopsins that absorb 500 nm of light at a maximum.

A rhodopsin is a G-protein-coupled receptor that is a typical seven-transmembrane protein type with seven alpha helixes. In the rhodopsins, the opsins get binded to the retinals, and the retinals do not exist in solitude. A retinal that is bound to the opsin has a shape called '*cis* retinal' in the beginning of the phase, but this shape changes its structure to the '*trans* retinal' when triggered by light, and therefore, the form of the molecules changes. Furthermore, because of the structural change of the protein, the retinals cannot be bound to the opsins; as a result, the retinals that become free get released in cytoplasm. In cytoplasm, the '*trans* retinal' returns to *cis* retinal without light. As a result, the *cis*-retinal can bind to the opsins again.

As with a rod cell, a cone cell structure has a slim shape, and the size of a cone cell is about 60×1.5 μm. The form of the tip of the cell is conical and contains iodopsins. There are three kinds of iodopsins in a cone. These iodopsins are also a complex made of the retinals and opsins, but amino acid sequences of bound opsins have a different complex. Because of the structure of protein here in that the retinals are connected is different, three kinds of cone cells exist: S, M, and L. The absorption wavelength of the S cone is 420–480 nm, of the M cone is 500–550 nm, and of the L cone is 620–660 nm. Each cone can sense the colors blue, green, and red. These three kinds of cone cells make it possible for the eyes to possess sensitivity for blue, green, red, and various mixes of these three colors.

26.2 Characteristics of the Visual Cell

Not only in visual cells, but also in an extremely small space whose width is 2–4 nm. This space exists in the cell membrane between cells that are adjacent to surrounding cells. In this space, the edges of identical protein complexes that are included in magnitude in cell membranes form a bridge structure. This protein complex is called 'connexon'. A passage that connects cell membranes is made from binding the ends of connexons, and inorganic ions and water-soluble small molecules can move between the cytoplasm of cells directly through this small passage. This phenomenon produces electrical and metabolic conjugation among cells.

In many human organizations, the gap junction opens and closes depending on a signal from outside of cells. For example, dopamine is a neurotransmitter that reduces signal transduction in the gap junction in certain retina neurons when the intensity of light increases. When their permeability into gap junction decreases, electric signals change in intensity accordingly during its transmission process, and photoreceptors that the retina uses are replaced by cone cells that are able to distinguish various colors and delicate details in dim or bright light from rod cells.

A connexon is formed by six subunit proteins. Therefore, the mutation of the subunit protein gene is thought to cause abnormalities in the visual cells. It is also thought that retinal movements between rod cells and cone cells in visual cells are affected and changed by the structure of connexons. This change is precipitated by mutation of the subunit protein gene, and some disorders occur from signal transduction in the gap junction.

Moreover, many genetic mutations trigger " retinitis pigmentosa" that is one of the opthalmic diseases. This disease is mainly caused by genetic mutations that related to specific functions of the retina. But some mutation in gene of spliceosome may trigger this disease (McKie et al. 2001; Vithana et al. 2001; Chakarova et al. 2002). In this disease, proteins performing splicing are produced, but at a lower quantity than normal, because genetically mutated and normal copies are produced individually. The influence of the component size in the splicing mechanism is smaller than usual. This specifically influences the retina. And, on the rod cells, high metabolism level is caused by rhodopsins which are visual pigments (McNally et al. 1999; Concepcion et al. 2002). In other words, the demand for opsin production is very high, and the splicing mechanism must supply opsins that are lost by resolution. Therefore, it is presumed that sensitivity for partial defects in splicing is higher for the retina than other organizations to produce specific proteins at a high level without burden. According to this presumption, it can be said that we can activate to overproduce each opsin for rhodopsins and iodopsins by various wavelengths of light in visual cells.

26.3 Expectations for Medical Application

26.3.1 Ophthalmic Diseases

In ophthalmic disease such as retinitis pigmentosa, illnesses are caused by gene deletions, and other underlying diseases such as diabetic retinopathy may become complicated.

Retina stem cells produce the retina cells necessary for sight on the ring-shaped black part around the iris; this stem cell is one that we can multiply in human cells in vitro with relative ease. These adult human retina stem cells are a source of differentiation for rod cells and cone cells, which are visual cells, and retina pigment epithelium cells. When the retina stem cells differentiate themselves to various cells at the time of fetus outbreak, it is thought that the cells change to a differentiated dormancy state. Therefore, unlike blood and some organs of humans, the retina stem cells participating in the formation of eyesight do not regenerate retina cells like the visual cells that are differentiated in adult human cells. Therefore, until recently, it has been concluded that restoration of the retina was impossible was after injury.

However, stem cells were discovered in the retina and subsequently retinas were produced from embryonic stem (ES) cells (Wallace 2011). It is hoped that blindness and deteriorating eyesight caused by denatured eyes can improve with the practical use of stem cell therapy (Bull and Martin 2011). In addition, restoration is aimed at abnormal neovascularization caused by a response to cicatrization and phlogosis. With this method, adult human stem cells obtained from human marrow are used. When these stem cells are injected into the eyeball, blood vessel cells are differentiated in such a way that can stabilize existing blood vessels that have been denatured without treatment (Otani et al. 2002). Furthermore, it is thought that these stem cells play a role in uniformity to support nerve cells that surround the tissue.

Such effects offer the possibility that stem cells are applicable for treatment for other ophthalmic diseases such as retinitis pigmentosa (Rama et al. 2010). Retinitis pigmentosa is a progressive ophthalmic disease in that rod cells of the retina have some disorders, and it is also a genetic disease in that visual cells and retina pigment epitheliums are affected in protopathic and progressive ways. The most common subjective symptom is nyctalopia, but further deterioration results in a gradual decrease in visual acuity and complicated color blindness, eventually resulting in blindness. However, there are individual differences in each symptom, and decreased visual acuity and color blindness may be detected at an early stage, and awareness of nyctalopia may be poor. In addition, there is an extreme individual difference in the speed of progress, and the prevalence rate of the disease is thought to be one in 4,000. This disease is thought to be caused by gene abnormalities, but the number of patients who have clear tendency of heredity is less than 50 % of all patients with retinitis pigmentosa (Fahim et al. 2000–2013).

In recent years, stem cells such as ES cells or induced pluripotent stem (iPS) cells have drawn much attention in the medical field and have already provided so much remarkable evidence (El Yakoubi et al. 2012; West et al. 2012; Ballios et al. 2012), but clinical application of this evidence has been very limited. Researchers who applied these cells for tissue engineering have faced various problems such as ethical issues around cloning with a therapeutic purpose. However, iPS cells acquired from patients with the genome for specific genetic diseases will help us discover effective medicine for curing these diseases using stem cells. It is also thought that stem cells will help clarify the genetic mechanism of the diseases.

26.3.2 Developmental Disorders

Some developmental disorders that are mental diseases are thought to have visual abnormality as a cause of the disorder (Kaga et al. 1998). In particular, children with developmental disorders with a high degree of frequency experience a functional decline of sight, and there have been quite a few cases with sight-related clinical symptoms. For example, a reading disorder called 'developmental dyslexia' is known as a learning disorder. The visual scope of children with this developmental dyslexia is often smaller than that of control children. When dyslexic children read letters, only the center of the field of vision may be activated. In other words, they can see the center of a letter, but their consciousness does not act on the domain of the upper or lower part of the letter. Therefore, they cannot recognize detailed parts of the letter. As a result, it is thought that it becomes difficult to keep perspective of the letter under cognitive control. One of these causes includes a possibility that it is an error in visual information processing based on cerebral dysfunction (Murakami and Cavanagh 1998; Stein 2003). With cerebral morphological studies on developmental dyslexia, neuronal localized allopatry and minute developmental abnormality for neuron shapes are actually confirmed under microscopy (Galaburda and Kemper 1979; Livingstone et al. 1991).

In recent years, candidate genes related to developmental dyslexia have been identified: DYX1C1, ROBO1, DCDC2, and KIAA0319 (Galaburda et al. 2006; Caylak 2007; Gabel et al. 2010). These various candidate genes have been reported and the incidence differs between males and females (Paracchini et al. 2007). In developmental dyslexia, this disorder is thought to be caused not by a single gene disease, but by multiple genes compositely. If these genes influence the formation of optic nerve system neurons in the cerebrum, we can obtain considerable information from mouse models not only for dyslexia but also other human diseases. It is thought that optogenetical application for mouse models with learning disorders greatly contributes to molecular biological studies for developmental disorders.

References

Ballios BG, Clarke L, Coles BL et al (2012) The adult retinal stem cell is a rare cell in the ciliary epithelium whose progeny can differentiate into photoreceptors. Biol Open 1:237–246

Bhaya D, Takahashi A, Grossman AR (2001) Light regulation of type IV pilus-dependent motility by chemosensor-like elements in Synechocystis PCC6803. Proc Natl Acad Sci U S A 98:7540–7545

Bull ND, Martin KR (2011) Concise review: toward stem cell-based therapy for retinal neurodegenerative diseases. Stem Cells 29:1170–1175

Caylak E (2007) A review of association and linkage studies for genetical analyses of learning disorders. Am J Med Genet B Neuropsychiatr Genet 144:923–943

Chakarova CF, Hims MM, Bolz H et al (2002) Mutations in HPRP3, a third member of pre-mRNA splicing factor genes, implicated in autosomal dominant retinitis pigmentosa. Hum Mol Genet 11:87–92

Concepcion F, Mendez A, Chen J (2002) The carboxyl-terminal domain is essential for rhodopsin transport in rod photoreceptors. Vision Res 42:417–426
Corredor RG, Goldberg JL (2009) Electrical activity enhances neuronal survival and regeneration. J Neural Eng. doi:10.1088/1741-2560/6/5/055001
El Yakoubi W, Borday C, Hamdache J et al (2012) Hes4 controls proliferative properties of neural stem cells during retinal ontogenesis. Stem Cells 30:2784–2795
Fahim AT, Daiger SP, Weleber RG (2000–2013) Retinitis pigmentosa overview gene reviews® [Internet]
Gabel LA, Gibson CJ, Gruen JR et al (2010) Progress towards a cellular neurobiology of reading disability. Neurobiol Dis 38:173–180. doi:10.1016/j.nbd.2009.06.019
Galaburda AM, Kemper TL (1979) Cytoarchitectonic abnormalities in developmental dyslexia: a case study. Ann Neurol 6:94–100
Galaburda AM, LoTurco J, Ramus F et al (2006) From genes to behavior in developmental dyslexia. Nat Neurosci 9:1213–1217
Kaga M, Inagaki M, Uno A (1998) Event related potentials in children with leaming disabilities visual event related potentials in specific Kanji writing disabilities, Perat MV (ed.) New Developments in Child Nerology, Monduzzi Editore, Bologna, 627–633
Kehoe DM, Grossman AR (1996) Similarity of a chromatic adaptation sensor to phytochrome and ethylene receptors. Science 273:1409–1412
Livingstone MS, Rosen GD et al (1991) Physiological and anatomical evidence for a magnocellular defect in developmental dyslexia. Proceedings of the National Academy of Sciences USA, 88:7943–7947
McKie AB, McHale JC, Keen TJ et al (2001) Mutations in the pre-mRNA splicing factor gene PRPC8 in autosomal dominant retinitis pigmentosa (RP13). Hum Mol Genet 10:1555–1562
McNally N, Kenna P, Humphries MM et al (1999) Structural and functional rescue of murine rod photoreceptors by human rhodopsin transgene. Hum Mol Genet 8:1309–1312
Murakami I, Cavanagh P (1998) A jitter after-effect reveals motion-based stabilization of vision. Nature 395:798–801
Otani A, Kinder K, Ewalt K et al (2002) Bone marrow-derived stem cells target retinal astrocytes and can promote or inhibit retinal angiogenesis. Nat Med 8:1004–1010
Paracchini S, Scerri T, Monaco AP (2007) The genetic lexicon of dyslexia. Annu Rev Genomics Hum Genet 8:57–79
Rama P, Matuska S, Paganoni G et al (2010) Limbal stem cell therapy and long-term corneal regeneration. N Engl J Med 363:147–155
Stein J (2003) Visual motion sensitivity and reading. Neuropsychologia 41:1785–1793
Terauchi K, Montgomery BL, Grossman AR et al (2004) RcaE is a complementary chromatic adaptation photoreceptor required for green and red light responsiveness. Mol Microbiol 51:567–577
Vithana EN, Abu-Safieh L, Allen MJ et al (2001) A human homolog of yeast pre-mRNA splicing gene, PRP31, underlies autosomal dominant retinitis pigmentosa on chromosome 19q13.4 (RP11). Mol Cell 8:375–381
Wallace V (2011) Concise review: making a retina – from the building blocks to clinical applications. Stem Cells 29:412–417
West EL, Gonzalez-Cordero A, Hippert C et al (2012) Defining the integration capacity of embryonic stem cell-derived photoreceptor precursors. Stem Cells 30:1425–1435

Part V
Optogenetics and Opto-electronics

Chapter 27
CMOS-Based Neural Interface Device for Optogenetics

Takashi Tokuda, Toshihiko Noda, Kiyotaka Sasagawa, and Jun Ohta

Abstract Optical and electronic neural interface devices based on complementary metal-oxide-semiconductor (CMOS) technology are presented. Concept, design strategy, and fabrication of the CMOS-based optoelectronic neural interface devices are described. Functional demonstrations of the devices are also presented.

Keywords CMOS image sensor • GaInN LED array • Local light stimulation • On-chip imaging • Implantable electronics • Stimulator for optogenetics • CMOS-based intelligent stimulator • Opto-electronic multional stimulator

27.1 Introduction

The evolution of optogenetics has led to the realization of various methods to understand neural systems. A great number of new approaches have continuously been demonstrated around the world.

In the same way as electronics technology, such as electrodes and amplifiers, has played an important role in the history of electrophysiology, engineering is expected to provide devices and systems to stimulate and measure neural activity in the field of optogenetics. The particular device function required in optogenetics is local and addressable light stimulation. Spatio-temporally-limited stimulation realized with a combination of the genetic introduction of rhodopsin and optical stimulation device technology is one of the significant advantages of the optogenetic approach.

Currently, a number of research groups are developing optical stimulation devices applicable for in vitro and in vivo optogenetics. We are developing a complementary metal-oxide-semiconductor (CMOS)-based optoelectronic neural interface device based on implantable CMOS imaging device technology. In this article, we present the concept, design, and functional evaluations of a CMOS-based implantable neural interface device for optogenetics.

T. Tokuda (✉) • T. Noda • K. Sasagawa • J. Ohta
Graduate School of Materials Science, Nara Institute of Science and Technology, 8916-5 Takayama-cho, Ikoma, Nara 630-0192, Japan
e-mail: tokuda@ms.naist.jp

H. Yawo et al. (eds.), *Optogenetics*, DOI 10.1007/978-4-431-55516-2_27

27.2 Concept of CMOS-Based Neural Interface Device for Optogenetics

27.2.1 Status of Optical Stimulation Devices for Optogenetics

Optical stimulation device technology, along with the modification of the rhodopsin family of proteins and a genetic strategy to introduce the proteins into the target animal, plays an important role in optogenetics. Because most of the in vitro experiments are performed using a microscope-based platform, light projection onto a target using the optics of a microscope is widely used. Using a laser or light-emitting diode (LED) array, systems in which one can perform local illumination have been realized (Poher et al. 2008; Grossman et al. 2010). For in vivo applications, broad illumination was realized with optical fiber (Aravanis et al. 2007) or an LED fixed on a skull (Huber et al. 2008).To perform more specially localized light stimulation, an approach to fix an anesthetized animal under microscopy optics has been widely adopted. Currently, device technologies for local and addressable optical stimulation for freely moving animals are under development. We proposed and have been developing an optoelectronic neural stimulation device for both in vitro and in vivo optogenetics. We integrate a GaInN LED array on an implantable CMOS image sensor to realize a device that can handle both light and electricity. Figure 27.1 shows the classifications of the optical stimulation device technologies and significance of our CMOS-based optoelectronic neural interface device.

27.2.2 Implantable CMOS Imaging Device Technology

We have developed a CMOS-based implantable neural imaging device (Ng et al. 2006; Tagawa et al. 2009; Kobayashi et al. 2012). Figure 27.2 shows the typical appearance of a CMOS-based implantable imaging device. We designed a small dedicated CMOS image sensor that can be operated with a minimum number of interconnections and mounted it on a polyimide flexible substrate. For fluorescence imaging, we integrated a color filter layer over the CMOS image sensor. LEDs are integrated surrounding the CMOS image sensor for illumination (regular imaging) or excitation (fluorescence imaging). In most cases, we use GaInN LEDs with peak emission wavelengths ranging from near ultraviolet (UV) to blue light.

Figure 27.3 shows an example result obtained with the implantable imaging device (Kobayashi et al. 2012). Fluorescence imaging using voltage-sensitive dye (VSD) was performed on the visual cortex of a mouse. Figure 27.3a shows a normal image obtained with external illumination and (b), (c) show VSD images. KCl injection (K^+ stimulation) was performed at an area indicated with a circle. We can clearly see blood vessels in Fig. 27.3a, and the response of the brain tissue in Fig. 27.3c.

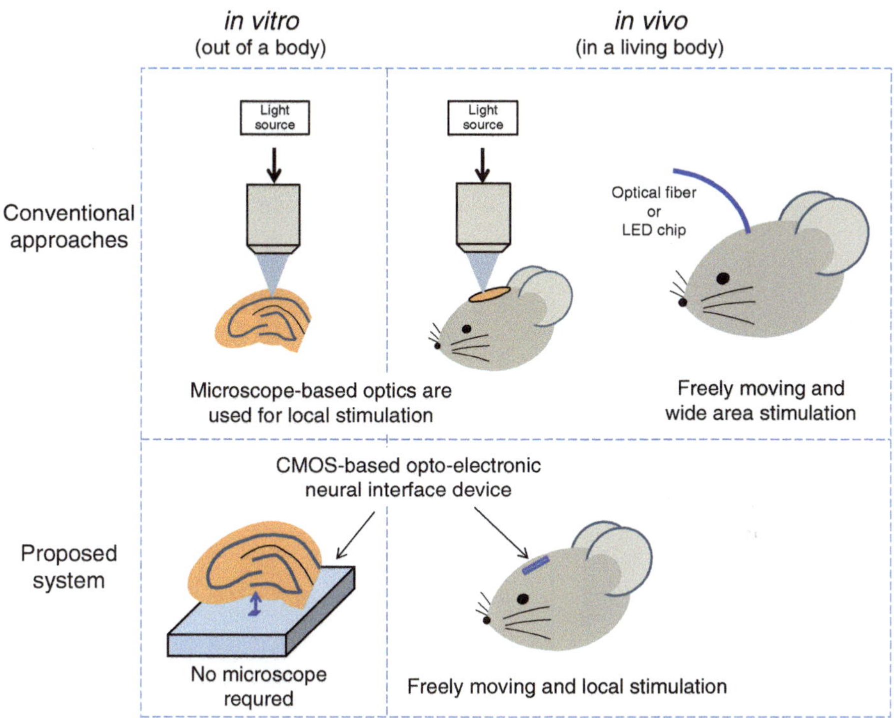

Fig. 27.1 Classifications of optical stimulation devices for optogenetics

27.2.3 Contact-Type Implantable Optical Stimulator Using GaInN LED as Light Source

GaInN LEDs are small, power-efficient light sources that can emit at wavelengths near UV to green light. The typical stimulation intensity used in optogenetics experiments is within 10 mW/mm^2. On the other hand, a GaInN blue LED chip with a size smaller than 500×500 μm and typical emission flux larger than 10 mW is commercially available. This comparison suggests that if we place a GaInN blue LED near the target neural cells, we can perform optical stimulation of rhodopsin proteins.

The implantable neural imaging device (see Fig. 27.2) includes GaInN LEDs, which we can use for optical stimulation in optogenetic experiments. However, all the LED are serially connected and simultaneously operated. To realize more specially localized and addressable optical stimulation, we proposed a novel CMOS-based implantable optoelectronic neural interface device (Tokuda et al. 2006).

Figure 27.4 shows the concept and schematic structure of the proposed optoelectronic neural stimulator. The core of the device is a multifunctional CMOS image sensor, which we have developed in previous works (Tokuda et al. 2007a, b, 2012;

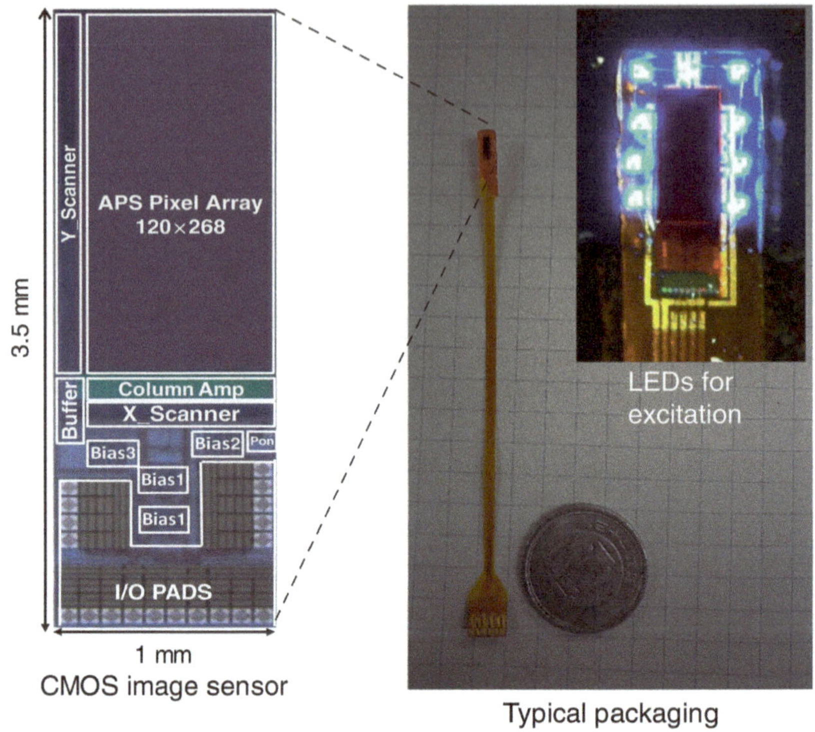

Fig. 27.2 Appearance of complementary metal-oxide-semiconductor (CMOS)-based implantable neural imaging device

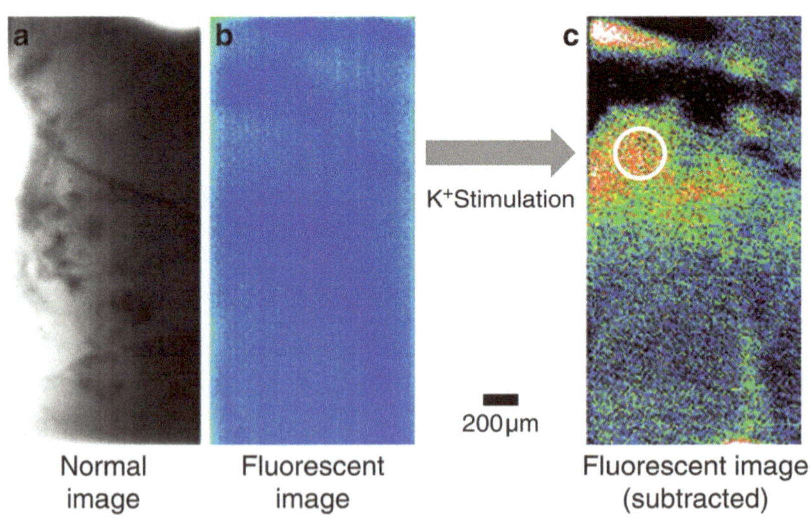

Fig. 27.3 An example result of brain imaging using the implantable imaging device (Kobayashi et al. 2012)

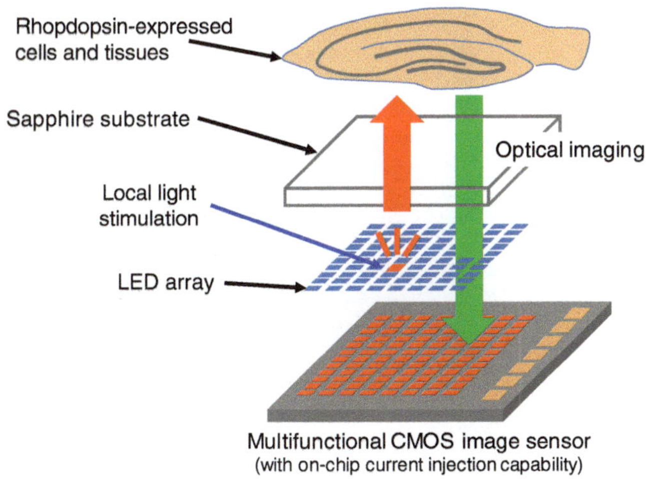

Fig. 27.4 Concept of implantable complementary metal-oxide-semiconductor (CMOS)-based optoelectronic neural stimulator for optogenetics

Nakajima et al. 2012; Sawadsaringkarn et al. 2013). The multifunctional CMOS image sensor technology was originally proposed as a solution for various on-chip imaging applications. A pixel on the multifunctional CMOS image sensor has an on-chip electrode, and we can use the pixel not only as a light sensor but also as an electric potential sensor or current injection electrode. In the proposed neural stimulator, we integrate the multifunctional CMOS image sensor and GaInN LED array wafer with a flip-chip boding technique. Taking advantage of the addressable electric connection capability, we expect to operate arbitrary LEDs in the array and perform localized, patterned optical stimulation. Because GaInN is almost transparent to visible light, an on-chip imaging capability is also expected. In the following section, the design, packaging, and functionality of the proposed CMOS-based implantable neural stimulator are described.

27.3 Design and Fabrication of Implantable CMOS-Based Optoelectronic Neural Interface Device

27.3.1 Design of Multifunctional CMOS Image Sensor

A multifunctional CMOS image sensor that acts as the core of the proposed neural interface device was designed and fabricated using a 0.35 μm, 2-poly, 4-metal standard CMOS process. Figures 27.5 and 27.6 show the layout and block diagram of the multifunctional CMOS image sensor, respectively. Table 27.1 shows the specifications of the sensor. The sensor has a 260×244 pixel array. The size of the pixels

Fig. 27.5 Layout of the multifunctional complementary metal-oxide-semiconductor (CMOS) image sensor

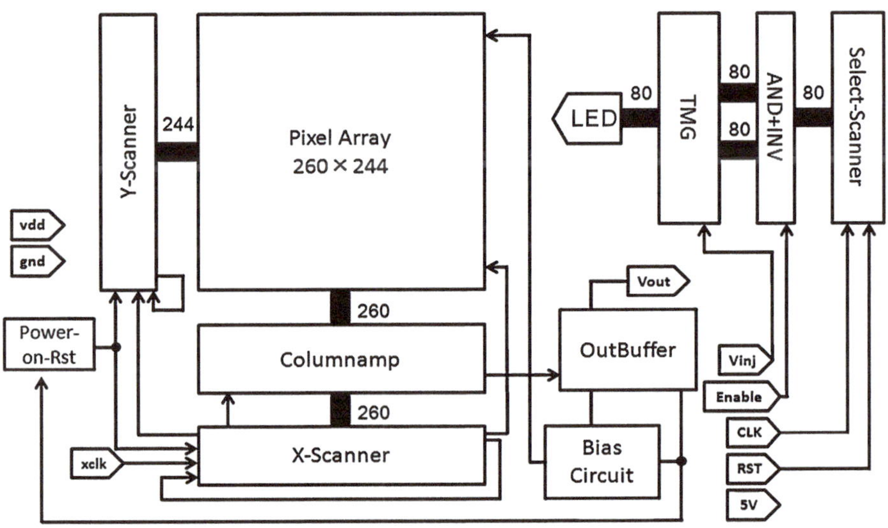

Fig. 27.6 Block diagram of the multifunctional complementary metal-oxide-semiconductor (CMOS) image sensor

Table 27.1 Specifications of the multifunctional complementary metal-oxide-semiconductor (CMOS) image sensor

Process	0.35 μm, 2-poly, 4-metal standard CMOS
Chip size	2,200 × 2,500 μm
Array size	260 × 244
Pixel size	7.5 × 7.5 μm
Pixel circuitry	3-Tr active pixel sensor
LED channel	1
Electrode for LEDs	8 × 10
Operation voltage	3.3 V (optical imaging) 5 V (LED operation)

in the image sensor is 7.5×7.5 μm. The pixel circuitry includes a three-transistor active pixel sensor, which is a commonly used circuit for conventional CMOS image sensors. The operational voltage of the sensor is 3.3 V.

As a unique feature of the sensor, addressable electrodes for current injection and electronic sensing were implemented over the imaging pixel array. The electrodes were designed with a top metal layer and exposed to the air to establish an electrical connection to the on-chip measurement target or bonded devices such as an LED array. The electrode has openings to introduce light to pixel photodiodes beneath the electrode. Thus, we introduced an electrical function without omitting the light-sensing function of the CMOS image sensor. However, it should be mentioned that the image quality was degraded when we bonded the LED array wafer on the CMOS sensor using flip-chip bonding with Au bumps. The electrodes on the GaInN LEDs and Au bumps cause a shadow on the captured image.

The sensor can drive the GaInN LED array bonded on the pixel array. We use the switch array (shown as a combination of CMOS switches and a scanner in Fig. 27.6) to establish an electrical connection between the cathode electrode of the GaInN LED to an input line of the CMOS sensor prepared for LED operation. The current injection electrode was internally configured as a linear switchable electrode array. We can select one of the electrodes by applying reset and clock pulses to the electrode scanner unit. Transmission gate circuitry was used as the switch between the on-pixel electrodes to the common current injection line. In addition, we implemented a master switch that can enable/disable the electrode selection. Because the typical operational voltage of a GaInN LED is approximately 3.0 V, we used 5.0 V-compatible MOS transistors for the current injection circuitry. The I/O pads of the multifunctional CMOS image sensor were gathered on one side of the CMOS sensor.

27.3.2 GaInN LED Array Chip

Because Si has an indirect gap semiconductor, no LED is available with current CMOS fabrication technologies. It is a realistic strategy to integrate LEDs fabricated from other materials to realize an on-chip light-emitting function for an optical neural stimulator. We adopted GaInN LEDs because they cover a wavelength region from near UV to green, and the device is almost transparent to visible light. We prepared an 8×10 GaInN LED array with a peak emission wavelength of 470 nm. The size of each LED and LED array is approximately 200×230 μm and 1.8×1.9 mm, respectively. Each LED has its own anode (P-type) and cathode (N-type) electrodes. We can operate the LED by injecting current between these two electrodes.

27.3.3 Wafer Bonding and Device Packaging

All the electrodes of the GaInN LED array are on the surface LED array chip. The current injection electrodes of the multifunctional CMOS image sensor are also on the top surface of the CMOS chip. Therefore, to realize the addressable optical stimulator, we bonded the GaInN array chip and CMOS sensor chip using a flip-chip bonding technique. This structure is advantageous because the bottom side of the GaInN array wafer is used as the contact surface to target tissue or cells. The substrate of the GaInN array is sapphire, which is chemically stable and shows minuscule toxicity to biological targets. Prior to the flip-chip bonding process, we formed Au bumps on the electrodes of GaInN LED array. Two chips are bonded with the help of anisotropic conducting paste (ACP), which is commonly used for the flip-chip bonding process. Figure 27.7a shows the appearance of the bonded chips. As shown on the rightmost

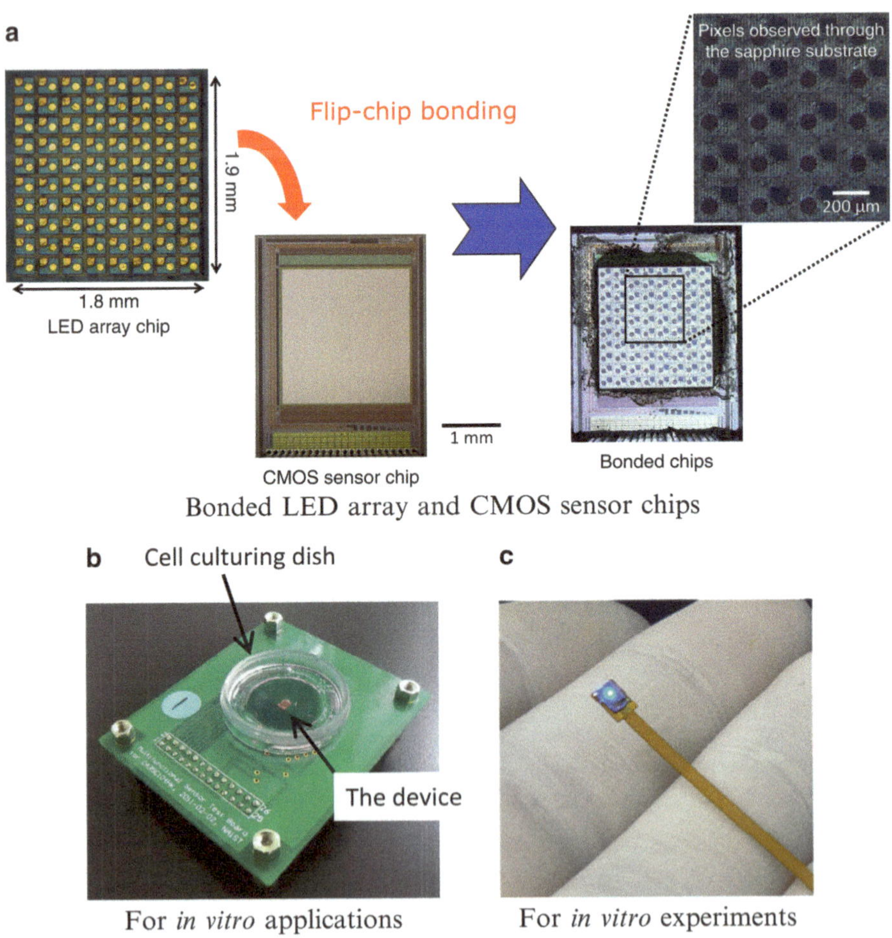

Fig. 27.7 Device packaging

image of Fig. 27.7a, the pixels of the CMOS sensor are visible through the GaInN array wafer. This figure suggests that optical imaging through the GaInN array chip is expected as a future functional extension. The device shown in Fig. 27.7a can be used for various optogenetic experiments in both in vitro and in vivo situations. Figure 27.7b and c shows typical device packages for in vitro and in vivo experiments, respectively.

27.3.4 *Control and Data Acquisition System*

Power and control signals to operate the proposed optoelectronic neural interface device were supplied on the multifunctional CMOS image sensor. The imaging function integrated on the CMOS sensor is compatible with our previous works (Ng et al. 2006; Tagawa et al. 2009; Kobayashi et al. 2012). We can operate the imaging function using only four I/O lines: vdd and gnd for power supply, a clock signal for image sensor operation (xclk), and a signal output line (Vout). For LED operation, we need to apply a 5 V power line (see Fig. 27.6) as well as a scanner reset (RST) and scanner clock (CLK) to reset and increment the electrode selection, respectively. We also need to apply a master signal (Enable) to enable/disable the current injection into an LED from the selected electrode and a current injection line (Vinj) to operate the selected LED.

We prepared an interconnection board equipped with conventional digital and analog buffers and analog-to-digital converter (ADC) chips between the neural stimulator and Windows PC. We also developed GUI-based control software on a Windows PC to control the device.

27.4 Functional Evaluation of the CMOS-Based Optoelectronic Neural Interface Device

27.4.1 *Verification of LED Operation*

The function of the local and addressable optical stimulation was characterized. Figure 27.8 shows both external images during LED operation and images captured with the imaging function of the multifunctional CMOS image sensor. Figure 27.8a and b were taken during single LED operation. We successfully selected and operated an LED. It was also confirmed that we can monitor the situation of LED operation from the image taken by the CMOS sensor during the LED operation (Fig. 27.8b). This result suggests that the imaging function can be used not only for observation of the on-chip biological target but also for monitoring the LED operation. This capability will be quite advantageous to monitor the device function, particularly in the cases in which the device is completely implanted into a freely moving animal.

Because the present multifunctional CMOS image sensor is equipped with a single current injection channel for LED operation, we can only operate one LED. However, by scanning the LED, we can perform quasi two-dimensional patterned stimulation. Figure 27.8c and d shows a case when we perform quasi two-dimensional LED operation (showing 'NAIST' characters with rapid scan).

Figure 27.9 shows the emission flux as a function of the injected current in single LED operation. A corresponding intensity of 47 mW/mm^2, which is significantly larger than the typical stimulation intensity range to drive a typical rhodopsin protein (~10 mW/mm^2), was obtained. This result suggests that the proposed device is capable of performing the optical stimulation required in most optogenetic experiments.

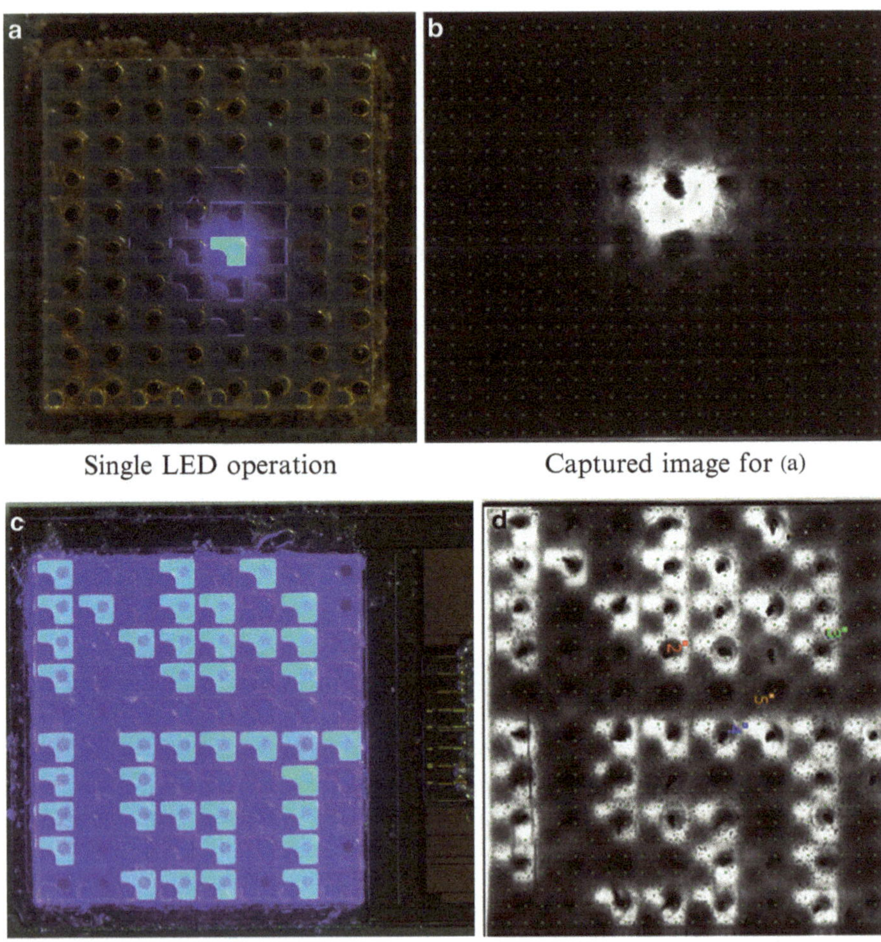

Fig. 27.8 Images taken during (**a**), (**b**) single and (**c**), (**d**) patterned light-emitting diode (LED) operation

Fig. 27.9 Emission flux in single light-emitting diode (LED) operation

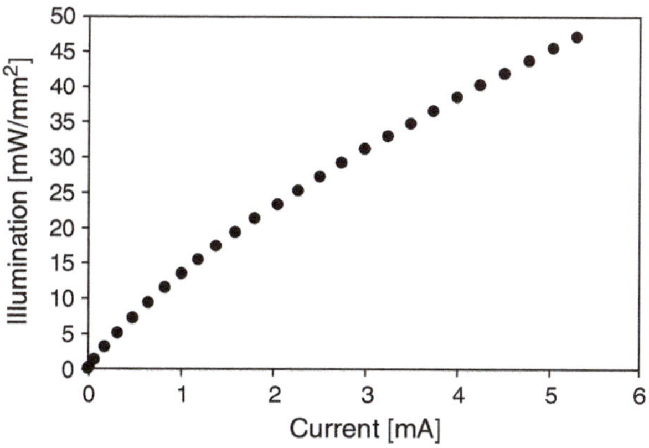

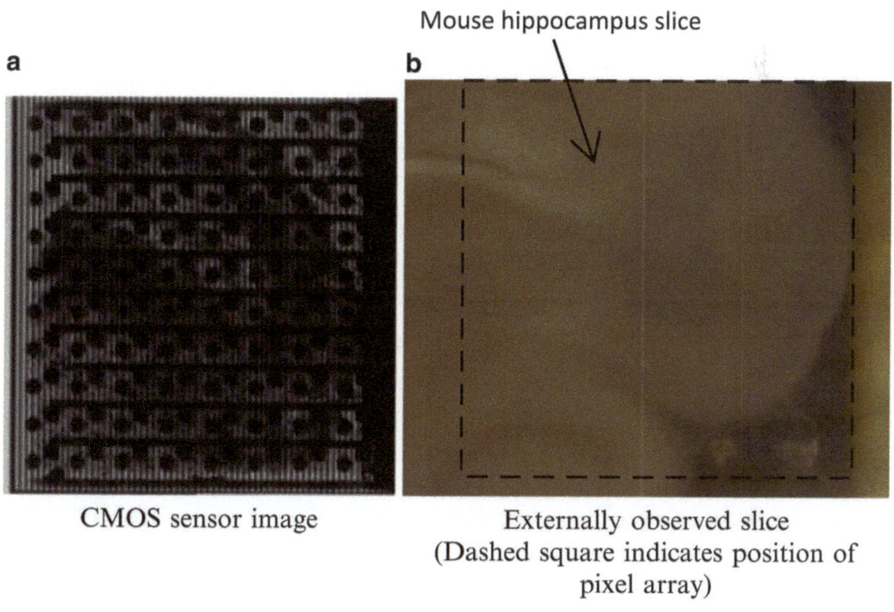

Fig. 27.10 Example of result obtained in an on-chip imaging trial

27.4.2 Evaluation of Optical Imaging Function

As mentioned in the previous subsection, we can monitor the LED operation by taking advantage of the optical imaging function of the multifunctional CMOS image sensor. However, from the viewpoint of the on-chip imaging of the biological target, i.e., observing tissues or cells contacting the device surface, there is an issue regarding image quality. Figure 27.10 shows an example of an imaging trial to observe a brain slice directly placed on the device surface. Compared with the image obtained with an external microscope, we can roughly observe the position and silhouette of the brain slice from the image captured with the CMOS sensor. However, because of shadows caused by the anode and cathode electrodes and Au bumps, the image quality is significantly degraded. We will improve the image quality by adopting transparent electrodes and a bumpless bonding process as a future improvement in device functionality.

27.4.3 Cell Culturing on the Device Surface

To use the proposed device for various on-chip optogenetic experimental applications, the device surface must be biocompatible. Moreover, it should be confirmed that cell adhesion and culturing is available on the device surface. Because the device has a polished sapphire surface, the toxicity to the biological targets is minuscule and thus negligible for acute and semichronic applications. To improve the adhesion of the cells on the sapphire surface, we confirmed that poly-L-lysine treatment is effective and succeeded in culturing Neuro2a cells on the device, as shown in the next section.

27.5 Functional Verification of Optical Stimulation on ChR2-Expressed Cultured Cell

27.5.1 Experimental Setup and Procedure of Optical Stimulation Experiment

The functionality of the optical stimulation was verified in an in vitro experiment using channelrhodopsin (ChR)-2-expressed Neuro2a cells. ChR2 is the most commonly used rhodopsin protein in optogenetics. Neuro2a is a neuron-like cell originating in mouse neuroblastoma. We introduced ChR2 into neuro2a cells cultured on the device by means of lipofection. We performed a patch clamp (voltage clamp) measurement to measure the response of the neuro2a cells caused by optical stimulation.

27.5.2 Observation of Neural Response Caused by Optical Stimulation

Figure 27.11 shows (a) the experimental setup and (b) a microscopic image obtained during the in vitro experiment. We performed optical stimulation on a cell on which the voltage clamp was established. Figure 27.11c shows an image obtained by the CMOS image sensor during the experiment. Although the cell position cannot be

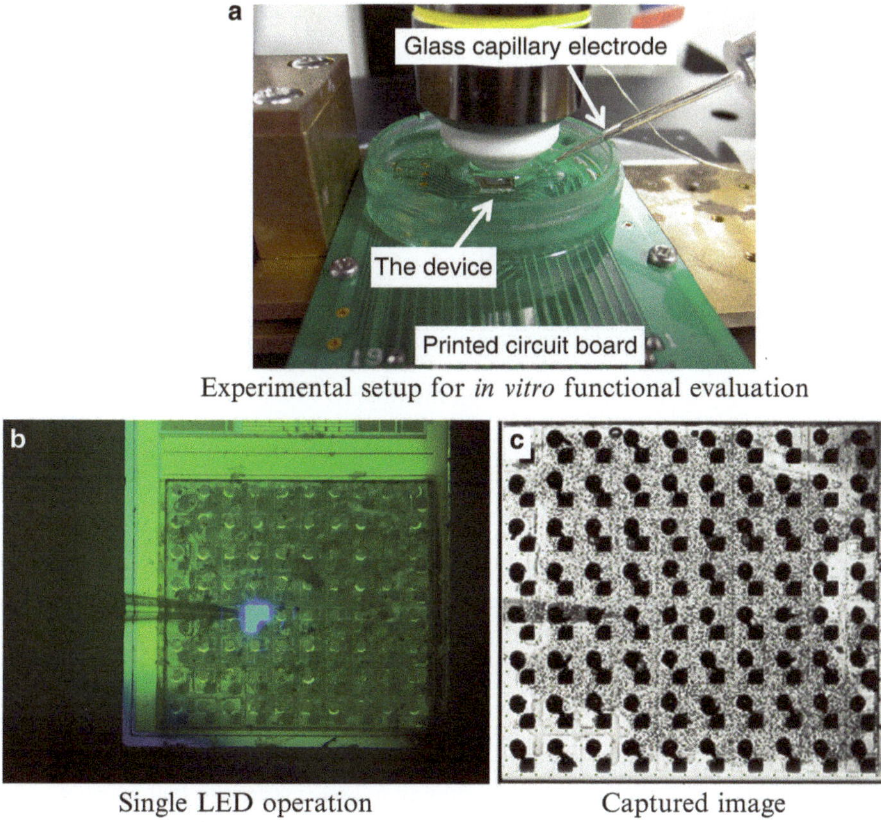

Fig. 27.11 Experimental setup and images observed during in vitro experimental verification

observed clearly in the image, the position of the glass electrode can be monitored using the imaging function.

Figure 27.12 shows the response observed in the membrane current caused by optical stimulation trials from the proposed CMOS-based optoelectronic neural interface device. An increase in membrane current was clearly observed. These results suggest that the proposed device can drive ChR2 and is applicable for optical stimulation in optogenetic experiments.

27.6 Conclusions and Future Work

A novel CMOS-based optoelectronic neural interface device was proposed and fabricated. The device was designed for optical stimulation and neural observation in optogenetic experiments. We integrated a multifunctional CMOS image sensor

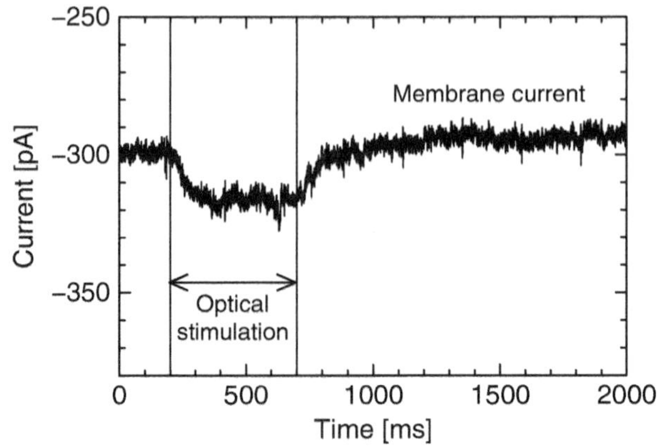

Fig. 27.12 Response in membrane current caused by optical stimulation by the device

with on-chip addressable current injection capability and a GaInN LED array chip formed on a sapphire substrate. The device function to select and operate an LED in the array was evaluated, and it was confirmed that the device has the capability to perform optical stimulation with sufficient intensity. Optical stimulation on ChR2-expressed neuron-like cells was successfully performed.

As future studies, we are working on various improvements of the device functions. An improvement in image quality by reducing the shadow caused by electrodes on an LED array chip and integration of an on-chip optical filter for fluorescence imaging are expected. We are also developing a device structure that is capable of electrical stimulation and measurement.

Acknowledgments All the work presented in this article was achieved under collaboration with members of the Photonic Device Science Laboratory in the Graduate School of Materials Science, Nara Institute of Science and Technology. The optical stimulation experiment using ChR2-expressed cells was performed with great help from Professor Sadao Shiosaka and Assistant Professor Yasuyuki Ishikawa (currently with the Maebashi Institute of Technology) at the Graduate School of Bioscience, Nara Institute of Science and Technology.

This work was partly supported by the Japan Science and Technology Agency, Precursory Research for Embryonic Science and Technology (JST-PRESTO), Grant-in-Aid for Challenging Exploratory Research #50314539 from the Japan Society for the Promotion of Science, and the VLSI Design and Education Center (VDEC) at the University of Tokyo, in collaboration with Cadence Design Systems, Inc.

References

Aravanis AM, Wang L-P, Zhang F et al (2007) An optical neural interface: in vivo control of rodent motor cortex with integrated fiberoptic and optogenetic technology. J Neural Eng 4:S143–S156. doi:10.1088/1741-2560/4/3/S02

Grossman N, Poher V, Grubb MSMS et al (2010) Multi-site optical excitation using ChR2 and micro-LED array. J Neural Eng 7:16004. doi:10.1088/1741-2560/7/1/016004

Huber D, Petreanu L, Ghitani N et al (2008) Sparse optical microstimulation in barrel cortex drives learned behaviour in freely moving mice. Nature 451:61–64. doi:10.1038/nature06445

Kobayashi T, Motoyama M, Masuda H et al (2012) Novel implantable imaging system for enabling simultaneous multiplanar and multipoint analysis for fluorescence potentiometry in the visual cortex. Biosens Bioelectron 38:321–330. doi:10.1016/j.bios.2012.06.035

Nakajima A, Kimura H, Sawadsaringkarn Y et al (2012) CMOS image sensor integrated with micro-LED and multielectrode arrays for the patterned photostimulation and multichannel recording of neuronal tissue. Opt Express 20:6097–6108

Ng DC, Tamura H, Tokuda T et al (2006) Real time in vivo imaging and measurement of serine protease activity in the mouse hippocampus using a dedicated complementary metal-oxide semiconductor imaging device. J Neurosci Methods 156:23–30. doi:10.1016/j.jneumeth.2006.02.005

Poher V, Grossman N, Kennedy GT et al (2008) Micro-LED arrays: a tool for two-dimensional neuron stimulation. J Phys D Appl Phys 41:094014. doi:10.1088/0022-3727/41/9/094014

Sawadsaringkarn Y, Miyatani T, Noda T et al (2013) A CMOS optoelectronic neural interface device based on an image sensor with on-chip light stimulation and extracellular neural signal recording for optogenetics. ITE Trans Media Technol Appl 1:184–189. doi:10.3169/mta.1.184

Tagawa A, Higuchi A, Sugiyama T, et al (2009) Development of complementary metal oxide semiconductor imaging devices for detecting green fluorescent protein in the deep brain of a freely moving mouse. Jpn J Appl Phys 48:04C195. doi: 10.1143/JJAP.48.04C195

Tokuda T, Yamamoto A, Kagawa K et al (2006) A CMOS image sensor with optical and potential dual imaging function for on-chip bioscientific applications. Sensors Actuators A Phys 125:273–280. doi:10.1016/j.sna.2005.08.023

Tokuda T, Tanaka K, Matsuo M et al (2007a) Optical and electrochemical dual-image CMOS sensor for on-chip biomolecular sensing applications. Sensors Actuators A Phys 135:315–322. doi:10.1016/j.sna.2006.08.027

Tokuda T, Kadowaki I, Kagawa K et al (2007b) A new scheme for imaging on-chip dry DNA spots using optical/potential dual-image complementary metal oxide semiconductor sensor. Jpn J Appl Phys 46:2806–2810. doi:10.1143/JJAP.46.2806

Tokuda T, Kimura H, Miyatani T et al (2012) CMOS on-chip bio-imaging sensor with integrated micro light source array for optogenetics. Electron Lett 48:312. doi:10.1049/el.2011.4087

Chapter 28
Use of Channelrhodopsin in the Development of Neural-Network High-Throughput Screening Devices

Tsuneo Urisu

Abstract Channelrhodopsin (ChR)-2 and halorhodopsin are powerful tools for the function analysis of in vivo and in vitro neural networks. However, in this chapter, another useful application of these optogenetic tools is introduced. Ion-channel biosensors have important, unique applications due to their high sensitivity and high selectivity, and the planar patch clamp, which is a planar ion-channel device, is already in commercial use as a high-throughput screening device for new drug developments. However, currently available planar patch clamp devices cannot be used for neural network analysis in which the incubation function is necessary. In this work, we have developed an incubation-type planar patch clamp by considering the application of a planar patch clamp to high-throughput screening devices using a neural network, which is considered to be extremely useful for cause analysis and new drug development for intractable brain neural network diseases. The weak point of the incubation-type planar patch clamp is extremely high baseline noise due to low seal resistance, which makes it almost impossible to use in practical devices. In this work, we have succeeded in decreasing the noise to a sufficiently low level by using a salt-bridge stable electrode. We have successfully measured the ion-channel current from a neural network incubated in a planar patch clamp biosensor for the first time. In this development of the incubation-type planar patch clamp biosensor, the excellent performance of the ChR-wide receiver with high time resolution was quite useful.

Keywords Channelrhodopsin • ChRWR • TRPV1 • Biosensor • Planar patch clamp. high throughput screening • Neural network

T. Urisu (✉)
Green Mobility Collaborative Research Center, Nagoya University,
Furo-cho, Chikusaku, Nagoya 464-8603, Japan
e-mail: t.urisu@gvm.nagoya-u.ac.jp

H. Yawo et al. (eds.), *Optogenetics*, DOI 10.1007/978-4-431-55516-2_28

28.1 Introduction

The incubation-type planar patch clamp, the first operation of which was reported in 2008 (Urisu et al. 2008), is a unique way of realizing a neural-network high-throughput screening device with an ion-channel current measurement function. Such a device that can measure ion-channel currents from neural networks would be useful for detecting preclinical abnormalities of many neurodegenerative diseases (van Zundert et al. 2012).

The recently developed light-gated ion channel is especially useful for investigating the functions of both in vivo and in vitro neural networks (Chow et al. 2010; Zhang et al. 2006; Nagel et al. 2003; Wang et al. 2009; Wen et al. 2010) by using a pipette patch clamp. Here though, we report another application of the light-gated ion channel, the excellent time and spatial resolutions and ease of handling of which make the light-gated ion channel useful as a simple performance test of ion-channel biosensor devices. However, for this application, several technical challenges exist: due to the long incubation time, it is difficult to achieve a sufficiently high seal resistance (gigaohm seal), which is important to have with a patch clamp, and the sensor cell must be positioned at the micropore for a long incubation time while maintaining its viability. All of these problems have been solved in the present work. In particular, low seal resistance causes large baseline fluctuations through fluctuations of the solid–liquid and liquid–liquid interface potentials. As a remedy, we have developed a stable electrode with a salt bridge structure. Concerning the cell-positioning technology, we have developed a plastic (polymethylmethacrylate [PMMA]) substrate and formed a cell-trapping pattern with a microfluidic structure on the surface of the sensor chip. By combining the stable electrode and cell-trapping pattern technologies, we succeeded in dramatically increasing the success probability (number of devices showing good performance/total number of devices constructed) from 1–2 % to 60–80 %. These successful results are owed to the utilization of the light-gated ion channel. This section describes the current status and future prospects of these applications in neural-network high-throughput screening devices.

28.2 Incubation Type of Planar Patch Clamp Biosensor

The pioneering work on ion-channel biosensors was conducted using the pipette patch clamp, and the recently developed planar patch clamp has opened the way to high-throughput screening applications. The pipette patch clamp technique has been used to investigate many electro-physiological subjects, including cell biological function, neural cell signal transduction characteristics, and the effects of chemical compounds on ion channels (Sakmann and Neher 1995; Schmidt et al. 2009; Neher 1992; Denyer et al. 1998; Gonzalez et al. 1999; Xu et al. 2001; Abe et al. 1994) . The theory and technique of this extremely powerful method are well established.

However, its application to high-throughput screening is hampered by its large size. The pipette patch clamp is also easily disturbed by mechanical vibrations, and a high level of skill is required to make measurements with it. On the other, hand, the planar patch clamp has been developed for high-throughput screening (Stett et al. 2003; Matthews and Judy 2006; Fertig et al. 2002; Pantoja et al. 2004; Sordel et al. 2006; Li et al. 2006; Zhang et al. 2008), and it has certain advantages over the pipette patch clamp. It is relatively easy to operate and is insensitive to mechanical vibrations; it is thus suitable for long-term measurements. In fact, high-throughput screening devices based on the planar patch clamp are already commercially available (Dunlop et al. 2008). However, these devices cannot be used in a system that requires long incubation times.

We have successfully measured the ion channel current with a neural network by using the incubation type of planar patch clamp after long incubation. This is a breakthrough that will enable a high-throughput screening device consisting of a neural network with ion-channel current measurements.

The incubation type of planar patch clamp has a weak point in that it is not easy to obtain a high seal resistance. The seal resistances in this sort of clamp are usually several to dozens of MΩ, which is very low compared with the pipette patch clamp (>1 GΩ), in which the tip of the pipette contacting the cell membrane surface is extremely smooth. On the other hand, in the case of the incubation type of planar patch clamp, since the substrate surface must be coated with extracellular matrix (ECM) molecules, which cause the surface to become rough, and furthermore, cell-membrane debris formed during the long incubation time also creates roughness, the seal resistance decreases significantly. Thus, a small change in the interface potential of the electrode induces a large fluctuation in the baseline of the channel current measurements. We successfully reduced the baseline fluctuation by developing a salt bridge type of stable electrode.

28.3 Sensor Chip and Ion Channel Biosensor

Sensor chip materials of the planar patch clamp include glass, Si, polydimethylsiloxane (PDMS), and plastic materials. In this study, a Si-on-insulator (SOI) substrate was used for a single-channel device. A thermal oxidation layer about 1 μm thick was formed on the substrate surface via wet thermal oxidation at 900 °C using a water-saturated O_2 flow. A large well was then made via diamond drilling on the reverse surface. A pyramid-shaped hole that reached the buried SiO_2 layer was formed by etching in 8 % (v/v) tetramethylammonium hydroxide at 90 °C for about 40 min. A micropore with a diameter of 1–2 μm was made in the silicon membrane using a focused ion beam (FIB) machine (Seiko Instruments Inc.) from the reverse side, as shown in Fig. 28.1. The SOI structure was extremely effective at reducing the parasitic capacitance that causes noise in the channel current measurements. Furthermore, we have developed a sensor chip using a PMMA substrate. Although the SOI substrate has several advantages, a PMMA substrate seems to be more

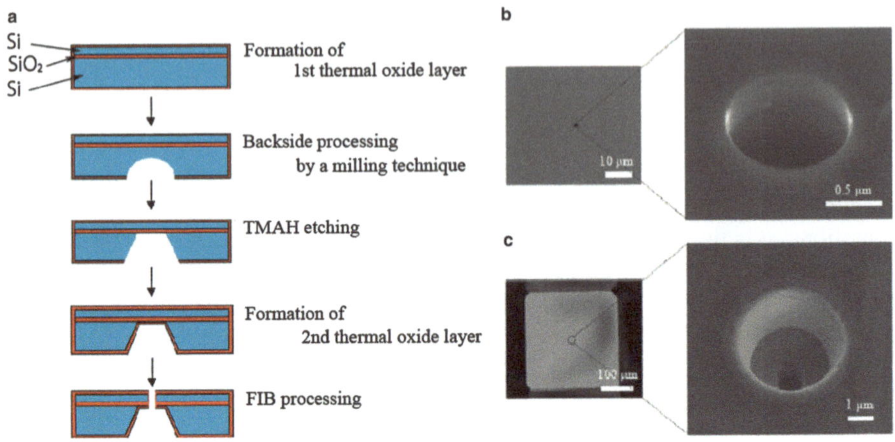

Fig. 28.1 (**a**) Development of silicon-on-insulator (SOI) sensor chip. Microscopic images of a chip from (**b**) the front side and (**c**) the back side taken using an optical microscope (*left*) and scanning electron microscope (*right*)

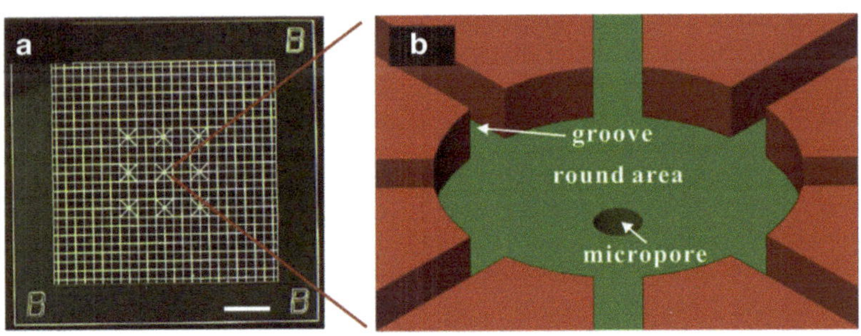

Fig. 28.2 Structure of the cell trapping patterns (**a**) and (**b**)

attractive for developing a high-throughput screening device using the incubation type of planar patch clamp. SOI substrate makes precise microfabrication and integration with electrical circuits easy, and it has a nanometer-level flat surface suitable for a higher seal resistance. On the other hand, the PMMA substrate has a lower cost and enables microfabrication using hot embossing. Moreover, its transparency, which enables observation by inverse optical microscopy, is especially useful in cell imaging.

In the case of the incubation type of planar patch clamp, we must keep the sensor cell on the micropore at the sensing area on the chip surface during the long incubation time. For this purpose, we formed a lattice pattern with round cell trapping areas at the crossing points by using negative photoresist SU8, as shown in Fig. 28.2. We designed this pattern not only for the experiments reported here but also for future applications such as multisensory point measurements and neural network formation. A schematic drawing of the biosensor is shown in Fig. 28.3a. The sensor chip is sandwiched between PDMS plates.

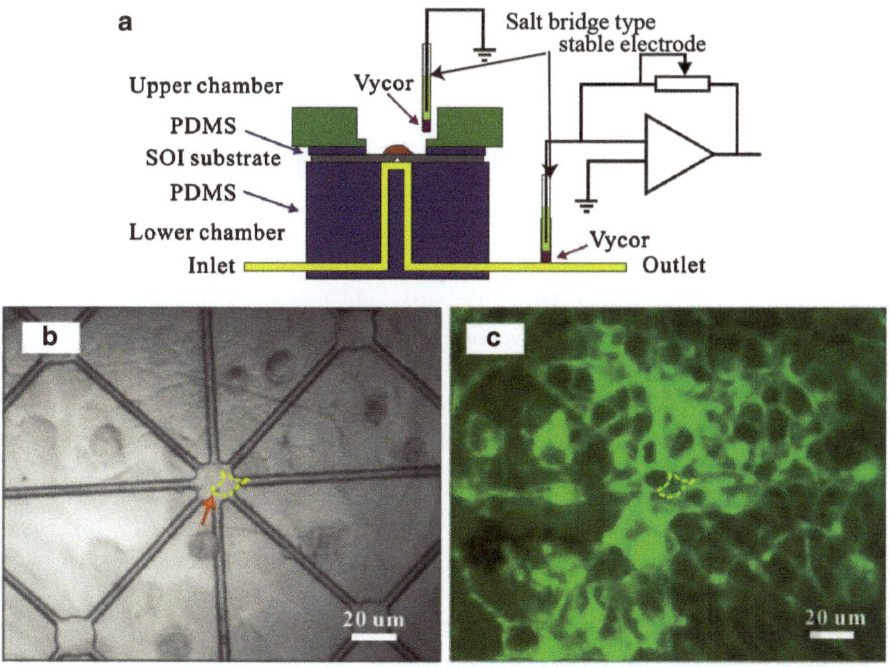

Fig. 28.3 (**a**) Schematic view of the ion-channel biosensor. (**b**) and (**c**) Bright and fluorescence microscopy images of patterned surface. A single monolayer colony is apparent. The cell on the micropore is outlined with the dotted line. The micropore is indicated by the *red arrow* in (**b**)

28.4 Stable Electrode

AgCl/Ag electrodes are usually used in patch clamp experiments. The reaction at the junction between the AgCl/Ag wire and the buffer solution is $Ag+Cl^- \leftrightarrow AgCl+e^-$. Since the pipette solution has a low Cl concentration, the interface becomes nonreversible, and the electrode potential is vulnerable to drift with time. Similarly, the electrode potential is easily changed by the addition of ligand molecules to the bath solution. A change of this kind can easily be several millivolts, even during measurements lasting 1–2 h. This level of potential change is not so serious if the seal resistance is sufficiently high (gigaohm seal). However, in the case of the incubation type, if the seal resistance is assumed to be 10 MΩ, a potential change of several millivolts results in a baseline change of several hundreds of picoamps, which is equal to or larger than the signal current. To overcome this problem, we chose to use the salt bridge structure for the electrode (Hassel et al. 1999; Shao and Feldman 2007). Here, an AgCl/Ag wire, which is formed by painting AgCl ink (BAS Inc.) onto the surface of the Ag wire (0.3 mm diameter), is inserted into a Pyrex glass tube filled with saturated KCl and AgCl solution, and the tip is sealed with Vycor glass (Gikenkagaku Co., Ltd.). The AgCl/Ag wire is connected to a gold-plated electrode pin, as shown in Fig. 28.3a. We use the salt bridge type stable electrode in the following channel current measurements.

28.5 Expression of ChRWR and TRPV1 on HEK293

ChRWR is the name given to the chimeric molecule ChR(ABCDEfg) in Wang et al. (2009), and its plasmid construction is reported in detail in that document. For the experiment described below, HEK293 cells, which were a generous gift from Minoru Wakamori of Tohoku University, were cultured at 37 °C and 5 % CO_2 in Dulbecco's modified Eagle's medium (DMEM) (Sigma-Aldrich Co.) supplemented with 10 % fetal bovine serum (FBS, Biological Industries Ltd) and transfected using Effectene transfection reagent (Qiagen, Tokyo, Japan) according to the manufacturer's instructions. After cloning twice with the addition of G418 (Gibco) in a 10 cm dish, single colonies with bright Venus (Nagai et al. 2002) fluorescence were selected by using a cloning cylinder (IWAKITE-32, Asahi Glass Co., Ltd.) and cultured in a medium containing G418 until they were confluent in the dish. TRPV1-expressing HEK293 was a gift from Professor Makoto Tominaga of the Okazaki Institute for Integrative Bioscience. Details of the transfection procedure are reported in Caterina et al. (1997).

28.6 Culture in Biosensor

The surfaces of both the SOI and PMMA sensor chips were coated with ECMs of poly-L-lysine (PLL; Sigma-Aldrich Co.), which performed better in the preliminary experiments on the culture in the biosensor than the fibronectin used in our previous experiments (Asano et al. 2008; Urisu et al. 2008). A 50 μl solution was dropped onto the substrate surface, followed by incubation for 1–2 days at room temperature. At this stage, the surface densities of the ECM were about 3–5 $\mu g/cm^2$. After removal of excess solution, the substrate was rinsed with sterilized water, dried under a gentle nitrogen stream, and kept sterile until use. Cells were cultured in dishes filled with the medium under conventional incubating conditions, i.e., 37 °C and 5 % CO_2. The culture medium was DMEM supplemented with 10 % (v/v) FBS, 1 % (v/v) GlutamaxTM (Gibco), and 0.5 % (v/v) penicillin/streptomycin (Gibco). After detaching cells from the culture dishes, the cell suspension was seeded at a density of 100–300 $cells/mm^2$ on the ECM-coated chip. The channel current was measured after 5 days of culturing, at which point about 70 % confluence was reached in the case of the SOI chip with no cell-trapping pattern. In the PMMA sensor chip with the cell-trapping pattern, the channel current was measured after about 3 days, when the monolayer colony around the micropore had formed (Fig. 28.3b and c).

28.7 Ion Channel Recording

The channel current was measured using a patch-clamp amplifier (Axonpatch 200B, Axon Instruments) and a data acquisition system (Digidata 1440A, Axon Instruments) with a whole cell mode at room temperature. All data were recorded using a 2-kHz low pass filter at an output gain of 1 mV/pA and analyzed using pClamp 10.2 software (Axon Instrument). Just before the channel current measurement, the media in the upper and lower chambers of the biosensor were replaced with a bath solution (BS) and pipette solution (PS), respectively.

The electrolyte solution, intracellular solution (PS), and extracellular solution (BS) that were used for the measurements are listed in Table 28.1. After removing impurities such as unresolved nystatin in the PS through a syringe filter of 0.2 μm, the BS and PS were injected into the upper and lower chambers, respectively.

For the whole-cell current recordings, sub-nanometer conductive pores through the cell membrane, which electrically connected the inside of the cell to the lower chamber, were formed by applying nystatin (Sigma) solutions to the lower chamber (Akaike and Harata 1994). The nystatin stock solution was prepared by dissolving nystatin in 1 ml of methanol and successively adding 45 μl of HCl (1 M) and 45 μl of NaOH (1 M), after which the solution was diluted with the lower chamber solution to final concentrations of 100–200 μg/ml before use. The formation of the whole-cell arrangement was confirmed by the observation of a capacitance increase of about 10 pF from 5 to 10 min after the addition of the nystatin solution to the lower chamber. In the case of the laser-evoked channel current measurements, the beam from a semiconductor laser with a 473 nm peak wavelength and 3.2 mW maximum output power (Sumitomo Osaka Cement Co., Ltd) was guided by optical fiber and focused with a microlens with a 26.5 mm focal length under the fluorescence microscope's objective lens (Olympus). The beam diameter at the focal point was 30–100 μm.

Table 28.1 Compositions of extracellular and intracellular solutions

<For TRPV1>		<For ChR2,ChR-WR>	
BS: pH 7.4		BS: pH 7.4	
NaCl	140 mM	NaCl	140 mM
KCl	5 mM	KCl	3 mM
HEPES	10 mM	HEPES	10 mM
$MgCl_2$	2 mM	$MgCl_2$	1.25 mM
$CaCl_2$	2 mM	$CaCl_2$	2.5 mM
Glucose	10 mM	Glucose	10 mM
PS: pH 7.4		PS: pH 7.4	
KCl	140 mM	CsCl	40 mM
EGTA	5 mM	$CsCH_3SO_4$	80 mM
HEPES	10 mM	$MgCl_2$	1 mM
		MgATP	2.5 mM
		HEPES	10 mM
		Na_2EGTA	0.2 mM

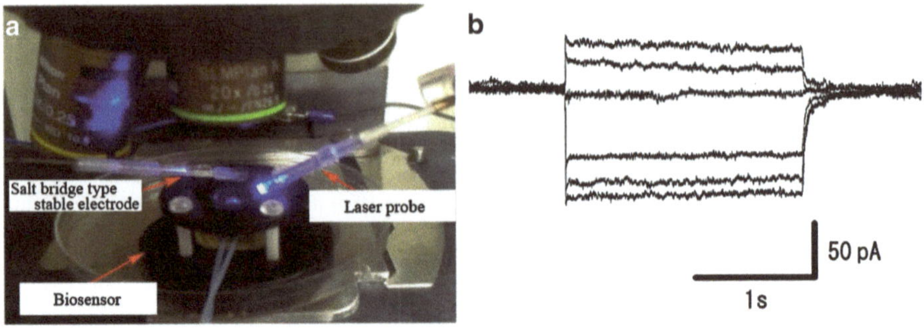

Fig. 28.4 Biosensor and channelrhodopsin (ChR)-WR channel current recording result. (**a**) Set-up of laser stimulation under a microscope. (**b**) Channel current record obtained by laser ($\lambda = 473$ nm output power $= 1.5$ mW) stimulation of ChR-WR expressed on HEK293. Sequentially from the bottom, the membrane potential is –60 mV, –40 mV, –20 mV, 0 mV, 20 mV, and 40 mV

The HEK293 cells expressed by ChRWR were used as sensor cells. The dependence of the channel current on the membrane potential shown in Fig. 28.4 was observed under laser stimulation. The sharp rise and slow fall in channel current were in good agreement with the measured results for a pipette patch clamp (Wang et al. 2009) and indicated that the channel current measurements could be trusted even though the seal resistance was low (10~30 MΩ).

As a reference, C2C12 myoblast cells expressed by channelrhodopsin (ChR)-2 were also used as sensor cells. The channel current measurement was conducted under laser stimulation, as shown in Fig. 28.4a.

Figure 28.5 shows the pulse interval dependence of the channel current and sharp desensitization of the rise time. These characteristics can be explained by the presence of an intermediate state with a long lifetime, although the mechanisms are not yet determined. They are not preferable from the viewpoint of network functional analysis of nerve cells. A significant improvement was made by developing the chimeric molecule ChRWR (Wang et al. 2009) (Fig. 28.4b).

28.8 Detection of Capsaicin by TRPV1-Expressing HEK293 Biosensor

The ion-channel biosensor has diverse applications ranging from the detection of biological warfare agents to high-throughput screening devices for pharmaceutical development, thanks to its extremely high selectivity and sensitivity. In this study, we constructed the single-channel biosensor shown in Fig. 28.6a by using TRPV1-expressing HEK293 cells to detect capsaicin, which is the main pungent ingredient in hot chili peppers and elicits a sensation of burning pain by selectively

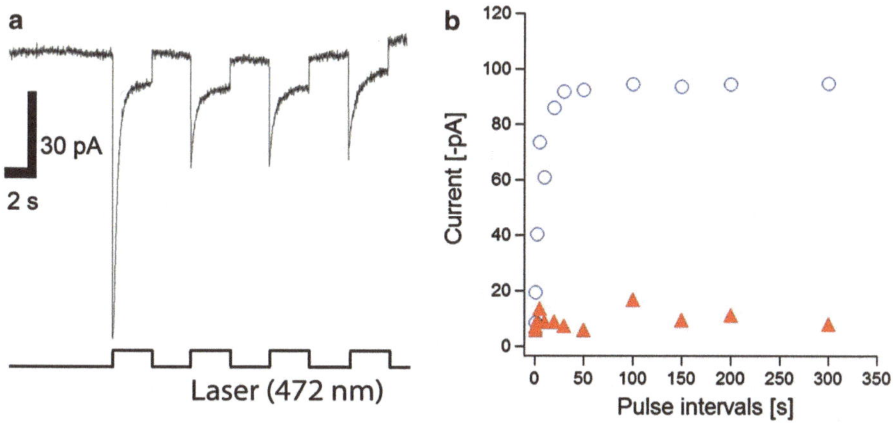

Fig. 28.5 Channelrhodopsin (ChR)-2 channel-current recording. Pulse interval dependence of the laser stimulation ionic channel current waveform at the time of using ChR2-expressing C2C12 cell as a sensor cell. (**a**) Pulse dependence of channel current waveform at a pulse interval of 2.5 s. (**b**) Stimulation–response relationship of the peak (○) and the plateau (▲) current as a function of the pulse interval

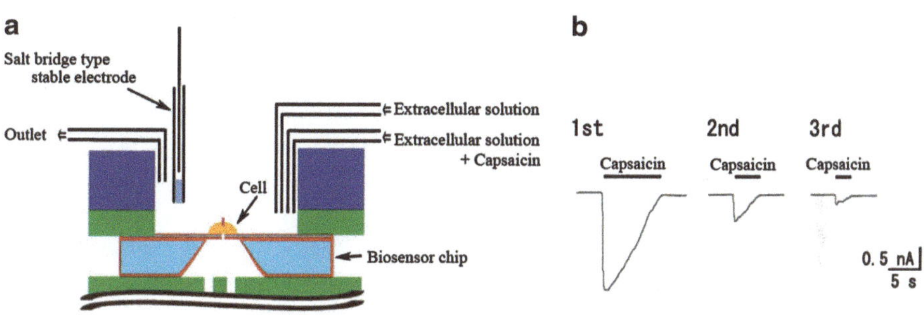

Fig. 28.6 (**a**) Schematic view of the ion-channel biosensor. The upper chamber has a system to supply and discharge the solution. (**b**) Capsaicin-stimulated ion-channel current recordings when Ca^{2+} ions were in the bath solution. The fixed membrane potential was –40 mV, and the seal resistance was 14 MΩ

activating sensory neurons. To investigate sensitivity for capsaicin, we replaced the bath solution with no capsaicin with solutions containing capsaicin at certain concentrations by using a syringe pump (KD Scientific Inc.) at a pumping speed of 12 ml/h. The measurements were carried out at room temperature. Unique desensitization was observed when Ca^{2+} ions were present in the bath solution, and it was not observed with a Ca^{2+}-free bath solution (Fig. 28.6b). These results match data measured with a pipette patch clamp (Caterina et al. 1997). The lower detection limit was about 0.01 μM. Highly selective and sensitive detection of neurotransmitter molecules and biologically active molecules have been reported using an ion-channel biosensor with a pipette patch clamp (Hazama et al. 1998; Schwartz 1987; Kataoka and Ohmori 1994). In those studies, the noise level including the baseline fluctuations was 5–20 pA (p-p). The noise level of 10–20 pA (p-p) in the present case indicted that an ion-channel biosensor having almost equivalent performance to that of a pipette patch clamp can be constructed using an incubation-type planar patch clamp equipped with a salt-bridge stable electrode, so long as it is in the whole-cell mode. In this experiment, the use of a stable electrode is essential. The success probability increased from 1 % to 2 % with a simple AgCl/Ag electrode to about 60 % with the stable electrode.

28.9 Development of Multi-channel Biosensor and Application to Neural Network

In vitro neural networks are useful for elucidating the causes and drug discovery of the nerve intractable diseases such as neurodegenerative diseases. In particular, it should be possible to build a patient-specific disease model chip by forming a neural network through differentiation of the patient's induced pluripotent stem (iPS) cells. By using the incubation type planar patch clamp, it is possible to analyze the neural network function for high-throughput screening applications using information on the ion channel current. We are developing a multi-channel incubation-type planar patch clamp biosensor that takes into account the design of the disease model chip. The chip's materials are PMMA and polycarbonate (PC). The pattern formed by hot embossing is shown in Fig. 28.7. A fine through-hole (micropore) with a diameter of 1.5~2 μm for the ion-channel current measurements is formed by focused ion beam (FIB) or deep X-ray lithography (DXL) processing. In the case of FIB processing, Cu thin films are deposited on both surfaces of the PMMA substrate to reduce charge-up effects during FIB processing, The Cu films are easily removed using a diluted HNO_3 solution after the micropore formation.

We used an incubation type of planar patch clamp to form a neural network from rat hippocampal nerve cells on the sensor chip and succeeded for the first time in measuring the ion channel current due to the spontaneous neurotransmitter release at the synapses (Fig. 28.8). At present, it is difficult to explain the origin of the observed channel current in detail, since the neural network structure is too complex due to the aggregation of the neural cells. We are developing a new technology to form a two-dimensional neural network that is as homogenous as possible. At any rate, the observation of the ion channel current in the neural network indicates that the incubation type of planar patch clamp, which has the potential for multi-point observations, is a powerful technology to realize a neural-network high-throughput screening device.

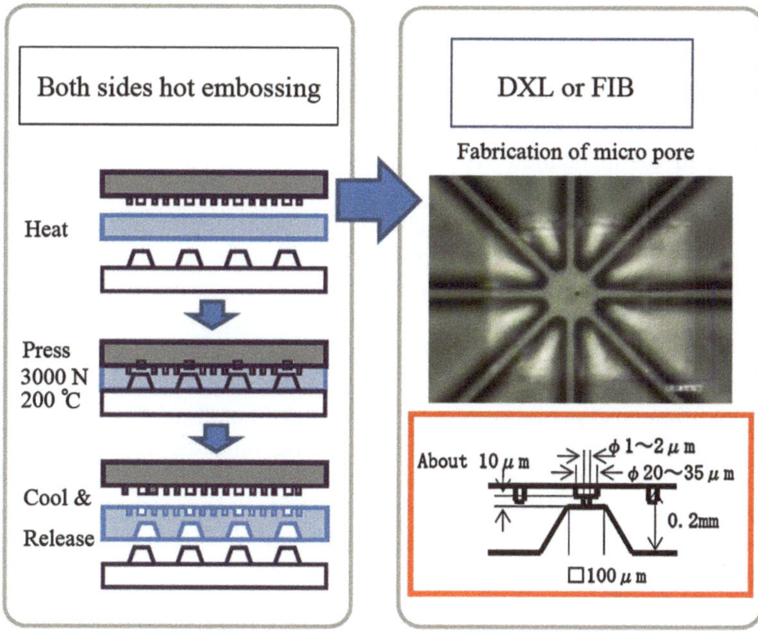

Fig. 28.7 Fabrication process of a polymethylmethacrylate (PMMA) sensor chip. The upper side micro fluidic structures and the back side well structures (pipette solution well) are formed by hot embossing from both sides. A micropore with diameter of 1~2 μm is formed via deep X-ray lithography (DXL) or focused ion beam (FIB) processing after hot embossing

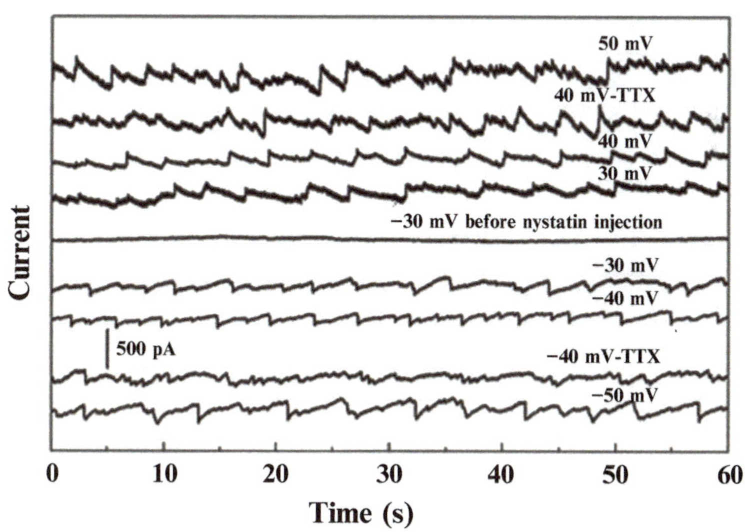

Fig. 28.8 Spontaneous channel current measurements in a network of in vitro rat hippocampal neurons. The neuron network was formed on a flat surface on the silicon-on-insulator (SOI) sensor chip after 10 days of culture. A spontaneous channel current was measured for several membrane potentials. Effect of adding tetrodotoxin (TTX) was investigated for membrane potentials of −40 mV and +40 mV. The smooth line around the center is the recording at a membrane potential −30 mV before the nystatin injection into the pipette solution. All other recordings were measured after the nystatin injection

28.10 Conclusion

We developed an incubation type of planar patch-clamp biosensor for neural network function analysis. The challenge was that the seal resistance was low compared with that of the pipette patch clamp. By developing a stable salt bridge AgCl/Ag electrode, we achieved a response and S/N at the same levels as the pipette patch clamp in laser-stimulated ion channel current measurements using an incubation type of planar patch clamp biosensor in which a ChRWR-expressed HEK293 cell was used as the sensor cell. In this development, a light-gated ion channel was a powerful tool for conducting a simple performance test of the device due to its excellent time and spatial resolution and ease of handling. The observation of the ion channel current from the neural network indicates that this clamp is a powerful tool for neural-network functional analysis and the development of innovative drugs for treating neurologically intractable diseases. For high-throughput screening applications, we developed a sensor chip that works on a plastic PMMA substrate.

References

Abe Y, Furukawa K, Itoyama Y et al (1994) J Neurophysiol 72:1530–1537
Akaike N, Harata N (1994) Jpn J Physiol 44:433–473
Asano T, Uno H, Shibasaki K et al (2008) Trans Mater Res Soc Jpn 33:767–770
Caterina MJ, Schumacher MA, Tominaga M et al (1997) Nature 389:816–824
Chow BY, Han X, Dobry AS et al (2010) Nature 463:98–102
Denyer J, Worley J, Cox B et al (1998) Drug Discov Today 3:323–332
Dunlop J, Bowlby M, Peri R et al (2008) Nat Rev Drug Discov 7:358–368
Fertig N, Blick RH, Behrends JC (2002) Biophys J 82:3056–3062
Gonzalez JE, Oades K, Leychkis Y et al (1999) Drug Discov Today 4:431–439
Hassel AW, Fushimi K, Seo M (1999) Electrochem Commun 1:180–183
Hazama A, Hayashi S, Okada Y (1998) Pflugers Archiv Eur J Physiol 437:31–35
Kataoka Y, Ohmori H (1994) J Physiol 477:403–414
Li XH, Klemic KG, Reed MA et al (2006) Nano Lett 6:815–819
Matthews B, Judy JW (2006) J Microelectromech Syst 15:214–222
Nagai T, Ibata K, Park ES et al (2002) Nat Biotechnol 20:87–90
Nagel G, Szellas T, Huhn W et al (2003) Proc Natl Acad Sci U S A 100:13940–13945
Neher E (1992) Science 256:498–502
Pantoja R, Nagarah JM, Starace DM et al (2004) Biosens Bioelectron 20:509–517
Sakmann B, Neher E (1995) Single channel recording. Plenum Press, New York
Schmidt C, Mayer M, Vogel H (2009) Angew Chem Int Ed 39:3137–3140
Schwartz EA (1987) Science 238:350–355
Shao XM, Feldman JL (2007) J Neurosci Methods 159:108–115
Sordel T, Garnier-Raveaud S, Sauter F et al (2006) J Biotechnol 125:142–154
Stett A, Burkhardt C, Weber U et al (2003) Receptors Channels 9:59–66
Urisu T, Asano T, Zhang ZL et al (2008) Anal Bioanal Chem 391:2703–2709

van Zundert B, Izaurieta P, Frits E et al (2012) J Cell Biochem 113:3301–3312
Wang H, Sugiyama Y, Hikima T et al (2009) J Bio Chem 284:5685–5696
Wen L, Wang H, Tanimoto S et al (2010) PLoS One 5:e12893
Xu J, Wang XB, Ensign B et al (2001) Drug Discov Today 6:1278–1287
Zhang F, Wang LP, Boyden ES et al (2006) Nat Methods 3:785–792
Zhang ZL, Asano T, Uno H et al (2008) Thin Solid Films 516:2813–2815

Index

H. Yawo et al. (eds.), *Optogenetics*, DOI 10.1007/978-4-431-55516-2

Kolibri 360 GmbH
Ferdinand-Jühlke-Straße 7
99095 Erfurt, Deutschland
produktsicherheit@kolibri360.de